T0296862

Ecology and Conservation of the Sirenia

Dugongs and Manatees

Dugongs and manatees, the only fully aquatic herbivorous mammals, live in the coastal waters, rivers and lakes of more than 80 subtropical and tropical countries and territories. Symbols of fierce conservation battles, sirenian populations are threatened by multiple global problems.

Providing comparative information on all four surviving species, this book synthesises the ecological and related knowledge pertinent to understanding the biology and conservation of the Sirenia. It presents detailed scientific summaries, covering sirenian feeding biology; reproduction and population dynamics; behavioural ecology; habitat requirements and threats to their continued existence.

Outlining the current conservation status of the sirenian taxa, this unique study will equip researchers and professionals with the scientific knowledge required to develop proactive, precautionary and achievable strategies to conserve dugongs and manatees.

Supplementary material is available online at: www.cambridge.org/9780521888288.

HELENE MARSH is Distinguished Professor of Environmental Science and Dean of Graduate Research Studies at James Cook University, Townsville, Australia. She is an international authority on the conservation biology of dugongs, sea turtles and coastal cetaceans, co-chairs the IUCN Sirenia Specialist Group and led the team that developed the United Nations Environment Programme's Global Dugong Action Plan.

THOMAS J. O'SHEA is Scientist Emeritus at the US Geological Survey, Fort Collins, Colorado. A mammalogist, he conducted research on manatees for many years and has served on the federal Florida Manatee Recovery Team, the Committee of Scientific Advisors to the US Marine Mammal Commission and the IUCN Sirenia Specialist Group.

JOHN E. REYNOLDS III is a Senior Scientist at Mote Marine Laboratory, Florida. He served as Chairman of the US Marine Mammal Commission between 1991 and 2010, is a former chair of the IUCN Sirenia Specialist Group and has recently worked closely with the United Nations Environment Programme to develop and implement a Caribbean-wide Marine Mammal Action Plan.

Frontispiece. One of the traditional ways of hunting dugongs in Torres Strait was by building a platform over the mud flats at low tide. A skilled hunter could tell, by looking for feeding trails, where a dugong would return over the next few nights and constructed a platform (nath) over this site with much magic and ritual. The rope was coiled up on the platform and then let out, running across to the shoreline where other villagers secured it, awaiting the kill. When a dugong came close, the hunter saw the phosphorescent glow in the water and aimed his harpoon slightly in front of it. The hunter is depicted in the form of a spirit with a moon-shaped head and feet for flying through the air. Nath (State II). Linocut – Kaidaral.

Conservation Biology

This series aims to present internationally significant contributions from leading researchers in particularly active areas of conservation biology. It focuses on topics where basic theory is strong and where there are pressing problems for practical conservation. The series includes both authored and edited volumes and adopts a direct and accessible style targeted at interested undergraduates, postgraduates, researchers and university teachers.

Ecology and Conservation of the Sirenia

Dugongs and Manatees

HELENE MARSH
James Cook University, Queensland, Australia

THOMAS J. O'SHEA
Glen Haven, Colorado, USA (US Geological Survey, retired)

JOHN E. REYNOLDS III
Mote Marine Laboratory, Florida, USA

CAMBRIDGE
UNIVERSITY PRESS

CAMBRIDGE
UNIVERSITY PRESS

University Printing House, Cambridge CB2 8BS, United Kingdom

One Liberty Plaza, 20th Floor, New York, NY 10006, USA

477 Williamstown Road, Port Melbourne, VIC 3207, Australia

314-321, 3rd Floor, Plot 3, Splendor Forum, Jasola District Centre, New Delhi - 110025, India

79 Anson Road, #06-04/06, Singapore 079906

Cambridge University Press is part of the University of Cambridge.

It furthers the University's mission by disseminating knowledge in the pursuit of
education, learning and research at the highest international levels of excellence.

www.cambridge.org
Information on this title: www.cambridge.org/9780521716437

© H. Marsh, T. J. O'Shea and J. E. Reynolds III 2011

First published 2012

A catalogue record for this publication is available from the British Library

Library of Congress Cataloging in Publication data
Marsh, H. (Helene), 1945–.
Ecology and conservation of the Sirenia : dugongs and manatees / Helene Marsh, Thomas
J. O'Shea, John E. Reynolds III.
   p.   cm. – (Conservation biology)
Includes bibliographical references and index.
ISBN 978-0-521-88828-8
1. Sirenia.   I. O'Shea, Thomas J. 1948–   II. Reynolds, John Elliott,
1952–   III. Title.   IV. Series.
QL737.S6M37   2011
599.55–dc22
                                                                    2011008024

ISBN 978-0-521-88828-8 Hardback
ISBN 978-0-521-71643-7 Paperback

Additional resources for this publication at www.cambridge.org/9780521888288

Contents

Foreword

On the surface of it, it seems preposterous to confuse mermaids and sirens with a 500 kg marine mammal with a face that only another sirenian might love. But knowledge and information have their own imperatives, and the more we know, the more fascinating these creatures become, and the more we are driven to protect them.

Knowledge itself is one of the tools that we can use to do that. Despite the difficulties associated with studying manatees and dugongs – they are inhabitants of turbid water, surface unpredictably and lie quietly on the bottom for long periods – researchers have pieced together much of their basic biology. Puzzles remain, however: Why do male dugongs have tusks? Is reproductive behaviour in sirenians characterised by scramble competition among males, or might they actually form mating leks? However, a rich understanding of these creatures has emerged from careful science.

Sirenians are biologically more different than most. They are the only herbivorous mammals that are fully aquatic. Their evolutionary affinity with elephants and hyraxes is intriguing. Even with only four extant species, there is wonderful variety: the Amazonian manatee lives in freshwater rivers and lakes, while the dugong is marine. Dugongs specialise on seagrasses, while manatees are much more eclectic. Long-term social bonds seem to be restricted to mother–offspring relations, but sirenians can gather comfortably in aggregations of hundreds.

Manatees and dugongs live in rivers, lakes and coastal areas, which brings them together with our species. Throughout history, sirenians have inspired cultural responses, provided people with meat, hides and oils, and stimulated myths and stories. But today, artisanal fisheries, direct hunting, declines in seagrasses, recreational boating and pollution are unrelenting pressures on sirenian populations.

Knowledge is necessary for effective conservation action, and information informs action. It is not coincidence that the populations for which we have the most scientific data – Florida manatees and Australian dugongs – have the strongest management programmes and are the most likely to persist. This justifies further efforts, especially research to understand the interaction with people and the effectiveness of different management regimes on wild populations. But knowledge by itself is not sufficient; governmental agencies ultimately must use that knowledge to forge policies, and people must implement the practices that will allow us to co-exist with these wonderful animals.

John G. Robinson
Chief Conservation Officer,
The Wildlife Conservation Society, New York

Preface

This book was born out of passion, frustration, excitement and hope. Readers will perceive all these emotions throughout the following pages. If the hard-won lessons recounted here lead to improved conservation of some remarkable species and their ecosystems, we will have fulfilled our aim.

We have cumulatively spent well over a century studying the various members of the Order Sirenia. Our passions include learning about and trying to conserve manatees and dugongs. The word 'unique' has lost its force through inaccuracy and over-use, but we want to reclaim its original meaning here. The sirenians possess suites of morphological, ecological and physiological adaptations that allow them, truly, to hold a unique place in the animal kingdom. Although once a more diverse group, the sirenians are limited now to only four species, albeit with remarkably wide geographic ranges (especially the dugong). Reduced in numbers through much of that range, and with myriad threats to their long-term survival, the sirenians have nonetheless demonstrated tenacity and resilience, hopeful signs that they will persist, if given a chance.

We hope that this book will stir the passion of those best placed to achieve the survival of the four sirenian species, particularly the rising generation of scholars and conservationists. We have all had the privilege of spending some of our professional lives working with research students. We are proud of the fact that these young scholars have the tools and insights to conserve sirenians and other wildlife species and their habitats.

If passion has driven us, frustration has tempered us. The world has many gifted and enthusiastic people trying to achieve a better balance between human activities and conservation of wild, living resources. All too often, human values and actions have led to the loss of wildlife species and natural ecosystems. Some species of marine mammals have become extinct at the hands of people in the past 60 years. Sadly, all of these extinctions were preventable.

Human population and consumption of natural resources expand every year. The profit motive generally trumps conservation. This situation frustrates those of us who champion the natural environment. Nonetheless, we sense a sea change in people's attitudes, prompting excitement and hope that a new generation of scientists and conservationists may reverse the trends of our own generation and achieve better preservation of wildlife and ecosystems.

This book will fail if it does not instil passion, frustration, excitement and hope for the Sirenia. However, our purpose was also to document and encourage the meticulous acquisition and use of scientific data to ensure that decisions are evidence based. Passion and hope are great motivators; but information, effective partnerships and respect for cultural differences are essential for conservation to succeed. This book is a scholarly synthesis that provides detailed, complete information and perspectives on the ecology and conservation of sirenians, and acknowledges gaps in our understanding. In reviewing the literature for the book, we were struck by the amount of new information about the Sirenia that has been published in a wide variety of outlets over the last decade. By synthesising and distilling this information we hope to help readers more effectively access the diverse primary literature.

As described in Chapter 1, the dugong is at the heart of an international political, economic, military, cultural and environmental conflict. These animals have added fuel to the controversy over the presence of a US military base on Japanese soil, in Okinawa. Indeed, as we note later, dugongs and manatees are increasingly being used as 'flagship species' to represent larger environmental causes.

Chapter 2 introduces frustration, as we describe the demise of the largest sirenian species ever; one of the largest creatures ever to inhabit our planet. The tragic loss of Steller's sea cow at the hands of Russian explorers and hunters only 27 years after its discovery by Europeans exemplifies the triumph of human greed over Nature. Sadly, Steller's sea cow is not an isolated example, as people persist in 'doing business' in a manner that jeopardises wildlife and the natural world.

Chapter 3 provides the evolutionary history of the Order Sirenia. That history is peppered with unusual beasts and considerable diversity. Many sirenian species co-existed throughout tropical and subtropical waters, but today that diversity is limited. We are concerned that factors such as human population growth, habitat deterioration, changes in fishing technology and even recreational pursuits make the accelerated extinction of the remaining species of sea cows more likely. Societies must develop better conservation values and people must adjust their behaviours to ensure that does not occur.

Chapters 4–6 provide a thorough review and synthesis of the scientific information about sirenian feeding and foraging; habitat use; behaviour; life history; and population dynamics. Optimal efforts to conserve species and their habitats take advantage of all that is known about the species' needs and capabilities. Our detailed review of the science of these three broad areas sets the scene for our closing chapters on threats, status and conservation opportunities.

Chapters 7–9 get to the heart of our concerns for the future of the sirenians. Chapter 7 provides an assessment of the threats to sirenian populations from environmental factors such as climate change and harmful algal blooms, and human-related factors such as habitat destruction, directed hunting and incidental fishing take. Understanding and mitigating threats are vital to conservation. Chapter 8 examines how we understand the 'status' of a species or population and addresses hard-to-study factors that would ideally be included in a status assessment, but which are often difficult or impossible to integrate because of lack of funds or logistic difficulties. We then provide details of the conservation status of sirenian populations around the world. Some populations are actually doing quite well, whereas others have been lost or greatly reduced and are likely to disappear in a matter of years.

Although Chapters 7 and 8 contain information that may discourage those who wish to promote conservation of sirenians and their environment, hope returns in Chapter 9. This chapter is titled 'Conservation Opportunities' because we believe that new approaches, new tools and new perceptions about science, values and partnerships will equip motivated people with what they need to make a difference. In part, the change has occurred because people and governments are increasingly acknowledging that traditional ways of doing science and conservation do not always work; this admission can be liberating, as it encourages people to seek novel ways of working together.

We do *not* consider that conservation has become formulaic, and that 'plugging in' particular components or steps will automatically lead to success. Nonetheless, we believe that the time is right for informed, passionate, dedicated people to prevail. We hope that the conservation opportunities we describe will be useful not only to people concerned with sirenians, but for all conservation efforts, especially in developing countries.

There is much to be done to ensure that sirenians and their ecosystems are conserved. We trust that this book conveys both humility and urgency and provides a foundation for future successes.

Helene Marsh
Tom O'Shea
John Reynolds

Acknowledgements

This book is the outcome of many years of research, during which our families have tolerated both lengthy absences for fieldwork and many hours spent in front of computers, especially recently. Our loving thanks to our spouses (Lachlan, Sherry and Kristen) and our children (Roderick, Duncan, Mary, Jackie and Jack) for their unfailing support.

The generosity of an insistently anonymous donor with a passion for dugongs enabled us to employ Shane Blowes as an invaluable research assistant, purchase the rights to use key photographs, pay for the production of the illustrations and maps and for assistance in editing and checking, all of which were central to the success of the project. John Robinson generously and graciously found the time to write the Foreword on his study leave in Cambridge. The US Marine Mammal Commission, the Friends of the Mote Marine Laboratory Library, Marine Mammal Science and the Convention of Migratory Species' Dugong Secretariat generously assisted in getting copies of this volume into the libraries of developing counties in the ranges of dugongs and manatees.

For introducing us to dugongs and manatees early in our careers we thank: George Heinsohn and Alister Spain; Robert Brownell and Galen Rathbun; Blair Irvine and Daniel Odell. We are grateful to our numerous other colleagues, including our graduate students, for teaching us much about sirenians over many years.

James Cook University underwrote periods of study leave for Helene Marsh in 2006 (hosted by Mote Marine Laboratory in Sarasota) and 2008 (hosted by University of Melbourne, Australia). These periods provided valuable time to focus on this book.

We gratefully acknowledge the assistance of the many people who helped with the production of the book as listed below:

Cartography: Adella Edwards, Alana Grech, Julian Lawn, Antoine Riou.
Colleagues: Nicole Adimey, Lem Aragones, Cathy Beck, Bob Bonde, Peter Corkeron, Daryl Domning, Alana Grech, Ellen Hines, Amanda Hodgson, Carol Knox, Donna Kwan, Janet Lanyon, Ivan Lawler,

Lynn Lefebvre, Miriam Marmontel, James Sheppard, Tony Preen, Katie Tripp, Jim Valade, Leslie Ward, Graham Worthy.

Editorial assistance: Dominic Lewis, Sharon Read, Gary Smith, Charlotte Thomas, Liz Tynan, Megan Waddington, Kristen Weiss.

General assistance: Sheri Barton, Alvaro Berg-Soto, Willie Buchholz, Erin Easter, Rie Hagihara, Jennifer Helseth, Allison Honaker, Molly Jessup, Marisa Kaminski, Milena Kim, Rachel Krause, Cathy Marine, Lachlan Marsh, Molly McGuire, Kim Miller, Lucas Nell, Kerri Scolardi, Emma Witherington

Illustrations: Brandon Bassett, Shane Blowes, Catherine Collier, Robert Donaldson, Chip Deutsch, Brett Jarrett, Rhondda Jones, Butch Rommel, Anneke Silver, State Library of Victoria, Gareth Wild.

Expert review: Eduardo Mores Arraut, Cathy Beck, David Blair, Bob Bonde, Greg Bossart, Vic Cockcroft, Aurélie Delisle, Daryl Domning, Josh Donlan, Jodie Gless, Alana Grech, Ellen Hines, Amanda Hodgson, Hans de Iongh, Lucy Keith, Cathy Langtimm, Ivan Lawler, Frank Loban, Jackie Lorne, Winifred Perkins, Bill Perrin, Nick Pilcher, Tony Preen, Liz Tynan, Natalie Stoeckl, Dipani Sutaria, Tim Ragen, Chris Wilcox.

Personal communications: Nicole Adimey, Ibelice Anino, Eduardo Mores Arraut, Sheri Barton, Brian Beatty, Cathy Beck, David Blair, Vic Cockcroft, Daryl Domning, Glenn Dunshea, Holly Edwards, Rie Hagihara, William Haller, Wayne Hartley, Lucy Keith, Jeremy Kiszka, Donna Kwan, Cathy Langtimm, Janet Lanyon, Frank Loban, Benjamin Morales, Tony Preen, James Reid, Kherson Ruiz, Caryn Self Sullivan, Ketan Tatu, Lindsey West, Graham Worthy.

Photographs: Kanjana Adulyanukosol, Anselmo d'Affonseca, Australian Navy, Cathy Beck, Bob Bonde, Olivier Born, Alexander Burdin, the late Chip Clark, Rob Coles, Tomas Diagne, Martine deWit, Mandy Etpison, Luc Faucompré, Florida Fish and Wildlife Conservation Commission, Florida Power & Light Company, Alana Grech, Rachel Groom, Jay Herman (NASA), Amanda Hodgson, Hendrik Hoeck, Indigenous Art Network, Pierre Larue, Col Limpus, Ian McNiven, Chris Marshall, Benjamin Morales, Miriam Marmontel, Len McKenzie, Cynthia Moss, Mote Marine Laboratory's Manatee Research Program, Dennis Nona, Grant Pearson, Charley Potter, Bob Prince, Galen Rathbun, James Reid, Peter J. Stephenson, Andrew Taylor, US Geological Survey, Karen Willshaw.

Project management: Shane Blowes.

Traditional owner permissions: Terrence Whap.

Introduction

In 2008 a US federal court judge ruled that the Defense Department's plans to construct an offshore marine airbase on the island of Okinawa, Japan contravened the US National Historic Preservation Act of 1966 (Tanji 2008). The rationale for the court's decision, known as *Dugong* v. *Gates*, was that construction plans for the base failed to protect the dugong, one of the animals that are the subjects of this book. The dugong is listed as *critically endangered* by the Japanese Ministry of the Environment and as a National Monument on the Japanese Register of Cultural Properties because of its high cultural value to the people of Okinawa. In 2005 a companion court case (*Dugong* v. *Rumsfeld*) had established the legitimacy of declaring an animal to be a historically significant 'property' under US legislation, ensuring that the US National Historic Preservation Act of 1966 applied.

The outcome of these court cases does not guarantee the future of the dugong in Japan, where it is subject to multiple threats in addition to the airbase (Marsh *et al.* 2002; Ikeda and Mukai 2012; Chapter 8). Indeed, the court decision seems unlikely to prevent the construction of the airbase.

However, this controversy and the associated landmark court cases highlight the cultural importance of animals such as dugongs and manatees (species known collectively as sirenians) and the value of addressing their cultural significance at national and international levels (King 2006). Although these court cases were ostensibly about the conservation of the dugong and its associated cultural values, the underlying conflict results from the polarised attitudes of Japanese people to US military bases in Okinawa. Fourteen bases occupy 18% of the area of the main island, and two large bases are close to residential areas. People who support the bases give preservation of dugongs low priority (Tanji 2008).

The conflict over the Okinawan dugong is extraordinary, because it involves the environmental organisations, courts and governments of two powerful nations in a localised conservation issue. Nonetheless, high-profile national conflicts over the conservation of sirenians are manifestations of similar clashes of culture. The most obvious parallels occur in other developed countries in the ranges of manatees and dugongs: the United States and Australia. The conflict between the conservation lobby and marine industry and coastal development interests over the management of human impacts on the Florida manatee has festered for decades (Reynolds 1999; Tripp 2006) and is an ongoing source of controversy. In Australia, there have been analogous conflicts over dugong conservation and fisheries management, particularly in the Great Barrier Reef region (Marsh 2000), where the dugong is explicitly listed as one of the World Heritage values (GBRMPA 1981). In addition, cultural conflict simmers between Australia's indigenous peoples and the wider community over traditional hunting of dugongs (Marsh *et al.* 2004).

Conflicts in the developing countries[1] that constitute most of the ranges of sirenians receive far less publicity but are much more intractable than the examples from the developed world. The ranges of the dugong and the three species of manatee collectively span more than 80 subtropical and tropical countries on five continents. In many such countries, conservation is seen as clashing with food security and the development associated with rapid human population increase. Here the challenge of sirenian conservation is a consequence of some of the world's major environmental problems: human population increase, the movement of people to coastal areas and the destruction of tropical and subtropical habitats, especially aquatic systems.

Globally, all four extant species of sirenians are considered to be in danger of extinction (IUCN 2009; Table 1.1). Nonetheless, in contrast to many other species, enough knowledge has accumulated about sirenian biology and the threats to their populations for governments to take

Table 1.1 *Scientific and common names and the IUCN global conservation status of the Recent members of the Order Sirenia as of October 2010: dugongs and manatees*

Species	Subspecies	Common name	IUCN global conservation status
Family Dugongidae			
Dugong dugon		Dugong	Vulnerable
Hydrodamalis gigas		Steller's sea cow	Extinct
Family Trichechidae			
Trichechus inunguis		Amazonian manatee	Vulnerable
Trichechus manatus		West Indian manatee	Vulnerable
	latirostris	Florida manatee	Endangered
	manatus	Antillean manatee	Endangered
Trichechus senegalensis		West African manatee	Vulnerable

effective steps towards their conservation, if the political will exists to do so. However, much of the required information on sirenian ecology and conservation is difficult to access and has not been synthesised. This book attempts to address this deficiency.

THE SIRENIA

The dugong and the Amazonian, West African and West Indian manatees are the only extant members of the mammalian Order Sirenia (or sea cows). Two subspecies of the West Indian manatee are recognised: the Florida manatee and the Antillean manatee.[2] The fifth Recent sirenian, Steller's sea cow, the largest mammal to exist in historic times other than the great whales, was exterminated by humans in the eighteenth century (Chapter 2).

The marine mammals include approximately 87 species in the Order Cetacea: the whales and dolphins; approximately 35 species in the Order Carnivora, Suborder Pinnipedia: the sea lions, walrus and seals; three species of fissipeds (members of the Order Carnivora that have separate digits): the sea otter, the marine otter and the polar bear; and the six Recent members of the Order Sirenia, if both subspecies of West Indian manatee are considered separately (Committee on Taxonomy 2009). The sirenians are exceptional as the only large herbivorous mammals that are strictly aquatic, justifying their common name: sea cows. Like whales and dolphins but unlike seals, sea lions, otters and polar bears, manatees and dugongs spend their entire lives in the water and do not return to land to give birth and suckle young.

The major groupings of marine mammals have evolved separately from the different groups of terrestrial mammals and are not closely related to each other (Reynolds *et al.* 1999). Nonetheless, they share many superficially similar morphological adaptations to their aquatic environments and the cetaceans, pinnipeds and sirenians were often grouped together in past centuries, even when it was recognised that the grouping was anomalous (Box 1.1).

The sirenians have a long evolutionary history extending back more than 50 million years (Chapter 3). Their closest contemporary relatives are the elephants and hyraxes, and the three groups have been linked together as the clade Paenungulata (see Chapter 3). At least 35 named species of sirenians have existed through time, ranging in mass from small sea cows weighing an estimated 150 kg to Steller's sea cow at a body mass estimated to be more than 10 000 kg. Some early sirenians walked on land with sturdy hind limbs, but fed on aquatic vegetation. Later forms evolved a variety of

> **Box 1.1. Herman Melville on the taxonomic position of sirenians**
>
> 'I am aware that down to the present time, the fish styled Lamantins and
> Dugongs (Pig-fish and Sow-fish of the Coffins of Nantucket) are included by
> many naturalists among the whales. But these pig-fish are a nosy, contemp-
> tible set, mostly lurking in the mouths of rivers, and feeding on wet hay, and
> especially as they do not spout, I deny their credentials as whales; and have
> presented them with their passports to quit the Kingdom of Cetology.'
>
> Melville (1851, p. 138)

foraging strategies, adapting to changing climatic, geologic, oceanographic
and biological conditions within the bounds of aquatic megaherbivory
(Chapter 3).

The Recent sirenians are classified into two families. The three species
of manatees are grouped in the Family Trichechidae; the dugong and
Steller's sea cow in the Family Dugongidae (Table 1.1). The ancestors of
these families diverged some 25–40 million years ago (see Chapter 3).
Despite this long separation, trichechids and dugongids look remarkably
alike. Pictures labelled as dugongs often depict manatees, sometimes with
dugong tails attached by photo editing. For example, in Torres Strait
between Australia and Papua New Guinea, the dugong is one of the most
significant cultural symbols (see Frontispiece). Nonetheless, some souvenir
T-shirts from Torres Strait feature manatees mislabelled as dugongs.

The external form of all the extant sirenians (Figure 1.1) reflects their
adaptations to a life of swimming, diving and eating aquatic flowering
plants. Thus the sirenian body plan is radically different from that of large
terrestrial herbivorous mammals, including the other paenungulates.
However, because they live in relatively shallow waters, do not dive deeply
for food, and apparently lack the complex social systems of some cetaceans,
sirenians are considered to be less specialised for an aquatic life than whales
or dolphins (Reynolds *et al.* 1999).

Nonetheless, sirenians are highly derived mammals. Their digestive
tract is more similar to that of elephants and horses than to most other
mammals (and has features in common with koalas, wombats and beavers;
see Chapter 4). There are two mammary glands, each opening via a single
teat situated in the axilla or 'armpit', a position that has some resemblance
to that of the breasts of human females, which may explain links between
(mythical) mermaids and sirenians – at least in the minds of sailors in days
of old. (Indeed, the etymology for the name Sirenia dates from the sirens of

Figure 1.1. Illustrations of the four extant species of sirenians (from top: dugong, Amazonian manatee, West African manatee, West Indian manatee) drawn against the scale of a human diver. The sirenians were drawn by Brett Jarrett and are reproduced with permission. The human diver was drawn by Gareth Wild.

Greek mythology, who purportedly drew ships to wreck with their alluring songs.) Like most other marine mammals, the manatees and dugongs have pectoral flippers. Hind limbs are absent, with vestigial pelvic bones a reminder of their terrestrial forebears. Even though all extant sirenians have the same characteristic body plan, there are notable external differences (Table 1.2 and Figure 1.1).

The most obvious difference between dugongs and manatees is the shape of the fluke. Manatees have a spatulate fluke (Figure 1.1). Dugongs have whale-like flukes with a median notch and are more streamlined and cetacean-like than manatees, with a smooth fusiform body. Dugongs look like a cross between a walrus and a dolphin – or like a manatee that goes to

Table 1.2 *Differences in the external appearance of the three species of manatee and the dugong. Jefferson et al. (2008) provide a comprehensive guide to their identification*

Species	Adult body length	Body mass	Skin colour	Shape of tail fluke	Flipper nails	Other distinguishing features
Trichechus inunguis	2.8–3.0 m	Up to 450 kg[1]	Black; may have white belly patch	Rounded	No	Black or dark grey, often with white or pink belly or chest patches; more slender than other manatees
Trichechus manatus	Up to 3.5 m	Up to 1620 kg	Dark greyish-brown	Rounded	Yes, 3–4	
Trichechus senegalensis	Up to 3.5 m	Up to 460 kg[1]	Dark greyish-brown	Rounded	Yes[2], 3–4	Very similar to the West Indian manatee, but more slender and perhaps with slightly more bulging eyes
Dugong dugon	Up to 3.3 m	Up to 570 kg[3]	Grey-brown; old animals may have white 'scar backs'	Whale-like	No	Tusks erupt in mature males and some old females, but do not extend beyond end of premaxilla

[1] Few data on body mass are available.
[2] There are reports of animals without nails (Jefferson *et al.* 2008).
[3] There is an implausible report of one animal 4.1 m long and 1016 kg (Mani 1960).

> **Box 1.2.   David Attenborough on the external appearance of a West Indian manatee**
>
> 'On land she was not a pretty sight. Her head was little more than a blunt stump, garnished with an extensive but spare moustache on her huge blubbery upper lip. Her minute eyes were buried deep in the flesh of her cheek and would have been almost undetectable if they had not been suppurating slightly. Apart from her prominent nostrils, therefore, she possessed no feature which could give her any facial expression whatsoever. From her nose to the end of her great spatulate tail she was just over seven feet long. She had two paddle-shaped front flippers, but no rear hind limbs and where she kept her bones was a mystery for, robbed of the support of water, her great body slumped like wet sand.'
>
> Attenborough (1956, p. 180)

the gym. Manatees are more rotund, and their bodies, especially that of the Florida manatee, often exhibit numerous folds and wrinkles. In addition, the flippers of manatees and dugongs differ. Manatees have long flexible flippers that are used to manipulate their food plants (Chapter 4). Like those of Steller's sea cow (Chapter 2), the flippers of dugongs are short and, unlike those of the West Indian and West African manatees, lack nails. Amazonian manatees also do not have nails.

Externally, the heads of the three species of manatees and the dugong are very similar: small with no discernible neck, reminiscent of the head of a walrus without the protruding tusks. The eyes are small and the external ears are tiny holes in the sides of the head. The two nostrils, located dorsally and at the cranial end of the snout, enable a sirenian to surface discreetly with only its nostrils out of the water, making the animals hard to see and census (see below and Chapter 8, especially Figure 8.1). The sirenian face is endearing but not beautiful from a human perspective, as attested by David Attenborough's description of a West Indian manatee (Box 1.2).

It is difficult to imagine how sailors mistook sirenians for seductive mermaids, especially when the sounds they make are bird-like chirps that are inaudible above water (Chapter 5). The link is likely to have been more about lust than likeness – stories abound about dugongs being used by sailors as surrogate human females. Kingdon (1971) claimed: 'To this day fishermen in Zanzibar who have caught a female dugong have to swear they have not interfered with it' (p. 198).

The most striking feature of the faces of dugongs and manatees is the fleshy oral disk, the greatly expanded region between the mouth and nose,

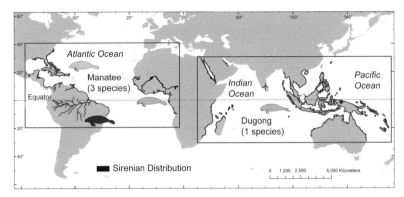

Figure 1.2. An outline of the distributions of the three species of manatees and the dugong, contrasting the Atlantic distribution of the manatees with the Indo-West Pacific distribution of the dugong. Drawn by Adella Edwards, with icons by Gareth Wild; reproduced with permission.

which is covered with vibrissae. The sirenian oral disk is an elaborate sensory–muscular complex that enables manatees and dugongs to find and manipulate food even in dark or murky environments (Chapters 4 and 5).

All sirenians have very sparse, short, fine, sensory body hairs. Roger Reep and his co-workers consider that sirenian body hairs constitute a tactile array equivalent to the lateral line systems of fish (Reep *et al.* 2002; Chapter 5), and have speculated that this may be an important aid to navigation at night and in shallow, turbid environments where visual acuity is of little value and acoustic communication limited to short distances.

Despite similarities in appearance, the three species of manatees and the dugong are unlikely to be confused in the wild, simply because their ranges have little or no overlap (Figure 1.2). All occur within the tropics, and the ranges of the Florida manatee and the dugong extend to the subtropics. The dugong occurs in the Indo-West Pacific Ocean, where its huge range spans the coastal and island waters from East Africa to the Solomon Islands and Vanuatu. Manatees occur on both sides of the Atlantic Ocean. The Amazonian manatee occurs in the freshwater systems of the vast Amazon River basin. The Florida manatee occupies coastal and riverine waters and inland lakes of the south-eastern United States. The Antillean subspecies has an extensive range from the wider Caribbean region, extending to south of the mouth of the Amazon River in Brazil, and may interbreed with the Amazonian manatee in areas where ranges abut or overlap (Chapter 3). West African manatees occur in most of the coastal marine waters, brackish

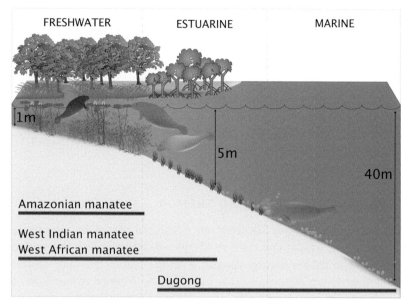

Figure 1.3. Schematic diagram illustrating the habitats and food plants of the four extant species of sirenians. Nowhere in the world do any of the four species occur together, except for the overlap in the ranges of the Amazonian manatee and the Antillean manatee near the mouth of the Amazon River. Drawn by Catherine Collier with assistance from Gareth Wild; reproduced with permission.

estuaries and adjacent rivers along the coast of West Africa from 16° N to 18° S, sometimes penetrating far inland. Chapter 8 gives further details of the ranges of each species.

The habitats of Amazonian manatees and dugongs are more specialised than those of the other two extant sirenians (Figure 1.3; Chapters 4 and 5). The dugong is marine, while Amazonian manatees occur only in fresh water, whereas West Indian and West African manatees are habitat generalists, occurring in lakes, rives, estuaries and shallow coastal waters. All four species show morphological adaptations that allow them to take advantage of the flowering plants available in their respective habitats (Chapter 4). The degree of snout deflection reflects where in the water column each species of sirenian feeds most efficiently (Domning 1982a; Domning and Hayek 1986; Chapters 3 and 4). Dugongs are benthic foragers that specialise in feeding on seagrass communities and have the most deflected snouts, whereas Amazonian and West African manatees have the least deflected snouts, a likely adaptation for feeding on natant and emergent vegetation.

The snout deflection of West Indian manatees is intermediate, reflecting their generalist foraging niche (Chapter 4). The horny plates in the mouths of manatees and dugongs help them masticate their food. Unlike almost all other mammals, manatees also have a system of constant tooth replacement, which enables them to eat plants with abrasive silica particles. In contrast, dugongs – which are considered to have the least abrasive diet of any extant sirenian – have simple peg-like molars that wear quickly. Fortunately, the dugong's last two molars are open-rooted and grow throughout life (Chapters 3 and 4).

POPULATION SIZES AND TRENDS

The first question people ask about any threatened species is: 'How many are there?' As will be explained in Chapter 8, this question has proved impossible to answer for any sirenian, even the Florida manatee, despite millions of dollars being spent over more than 30 years on research on its population biology. Scientifically defensible estimates of abundance greater than 1000 only exist for two species across five regions of the world; namely, dugongs in: (1) the Red Sea, (2) the Arabian Gulf, (3) New Caledonia, and (4) Australia; and (5) the Florida manatee. Australia is the only country where such estimates are in the tens of thousands (Chapter 8). In most developing countries, estimates of West Indian manatee and dugong populations either do not exist or are typically crude, but at best are in the low hundreds. There are no rigorously derived estimates of population sizes for Amazonian and West African manatees; only unverifiable expert opinions are available.

The second question people ask about threatened species is: 'How are they doing?' The International Union for Conservation of Nature (IUCN 2009) concluded that populations of the Amazonian and West Indian manatees are decreasing and that the population trends of the West African manatee and the dugong are unknown. Despite these uncertainties, the primary reason for categorising sirenians as at risk is not disputed. Their intrinsic rates of increase are low: less than 8% per annum (Chapter 6). Sirenians are several years old when they have their first calf; have a single offspring at intervals of several years; and experience low natural mortality rates (Chapter 6). Such species require very high and stable levels of adult survival to maintain their numbers and can sustain only very limited levels of mortality from human causes. Ways in which mortality rates might be reduced are discussed in Chapter 9.

HUMAN–SIRENIAN INTERACTIONS

The reason the US courts were able to intervene on behalf of the dugong in Okinawa was because of its status as a 'National Monument', a reflection of its high cultural value. This situation is not peculiar to Okinawa. Sirenians have been important to human cultures for thousands of years, although their cultural values have been remoulded by the changing nature of the interactions between people and sirenians.

Dugongs and manatees are very good to eat. Their muscle tastes like veal or pork; local people claim they can distinguish seven different flavours of Amazonian manatee meat (Marmontel *et al.* 2012). A manatee or a dugong represents a windfall of meat to an indigenous hunter or impoverished fisher. Sirenians have also been a source of a range of other products, including hides, oil, bones and teeth, often used as medicines and love potions (see Chapter 7). In the eighteenth century, Steller described the fat of the giant sea cow that bears his name as 'far preferable to that of any other quadruped' (see Chapter 2). These products have been put to a wide range of uses. For example, the coverings of the biblical Tabernacle and Ark of the Covenant are believed to have been made from dugong skin (Cansdale 1970).

Because sirenians occur in coastal and freshwater habitats, they are accessible to hunters with relatively simple equipment: traps, canoes and harpoons. The dugong hunting culture in the Middle East is at least 6000 years old (Méry *et al.* 2009). Hunters did not rely only on their physical prowess and technical skills to catch sirenians. As illustrated by the frontispiece, magic and rituals were also used to boost performance (e.g. McNiven and Feldman 2003; McNiven 2010). These practices have undoubtedly enhanced the cultural values of sirenians.

In places where sirenians were abundant, European colonisation tended to be followed by commercial exploitation of sirenians; this occurred most notably for manatees in Brazil (Domning 1982b) and dugongs in Australia (Daley *et al.* 2008). Western-style commercial exploitation has ceased, except as a by-product of other enterprises such as the shark-fin trade (Chapter 7). Nonetheless, although hunting sirenians is banned in most countries, poaching is still a major source of mortality, and the meat of dugongs and manatees is the aquatic equivalent of bush meat. In many developing countries, hunting with harpoons has been wholly or partially replaced by catching animals in fishing nets, often monofilament nylon gill nets provided by Western aid to address issues of food security (Chapter 7).

Hunting of sirenians is still legal in a few countries, most notably Australia, where it is a Native Title right (see Frontispiece and Chapter 7).

Pressure on the coastal and riverine habitats of manatees and dugongs is escalating as human populations increase, with concomitant habitat loss, fragmentation and change, increased pollution and demand for resources. The situation has been exacerbated by the displacement and urbanisation of rural human populations. All sirenians are adversely affected by such impacts (Chapter 7).

OBJECTIVES OF THIS BOOK

The conservation of the Sirenia presents a complex challenge for policy makers, scientists and the general public, and to be effective conservation must be informed by research. Over the last four decades we have observed simultaneous increases in the complexities of this challenge and the scientific knowledge about the Sirenia. One of the most stunning discoveries and cautionary tales exists in Steller's eighteenth-century account of the giant sea cow (Chapter 2). At the time, Steller served not only as a zoologist, but also as a botanist, physician and theologian. The days when one person could cover the breadth of knowledge in such an array of disciplines are long gone. Indeed, most sirenian specialists now further divide their efforts into just a few subdisciplines. Future generations of the experts required to inform the conservation of Sirenia could easily be overwhelmed upon entry to the field. In this book, we attempt to synthesise the current research base pertinent to the ecology and conservation of this unique group of mammals. We hope to provide a platform for future efforts to develop scientifically robust and socially, economically and politically achievable strategies to conserve dugongs and manatees throughout the world.

NOTES

1. There is no established convention for the designation of 'developed' and 'developing' countries or areas in the United Nations system. The United Nations Development Programme produced the Human Development Index (HDI), a composite index that measures the average achievements of a country for which data are available in three dimensions of human development: life expectancy at birth, years of schooling, and gross national income per capita (UNDP 2010; see Chapter 8). We have defined sirenian range states with a 'very high' HDI as 'developed'; the remaining range states as 'developing'.
2. Throughout this book, we use the term West Indian manatee to refer collectively to both subspecies, and we use names of the subspecies when our reference is more precise.

Steller's sea cow

Discovery, biology and exploitation of a relict giant sirenian

INTRODUCTION

The culmination of the evolution of the hydrodamaline lineage in the North Pacific Ocean that began in the Miocene (Chapter 3) ended abruptly in historic times. *Hydrodamalis gigas*, once distributed widely around the region, was restricted to coastal fringes of two islands near the Kamchatka Peninsula of Siberia, where it was discovered by Europeans in the middle of the eighteenth century. This magnificent, though bizarre, beast was quickly exploited for its meat to supply workers in the expanding fur trade, and could not hold out for long after it was found. Perhaps typical of its time, this exploitation was wasteful and opportunistic. Given the restricted distribution, limited population and likely low rates of potential population growth (see Chapter 6 for discussion of life histories of extant sirenians), the pressures of exploitation could only be withstood by *H. gigas* for some 27 years before it was rendered extinct. This extinction was a great loss of a biological treasure, and is sometimes cited as a warning of the vulnerability of sirenian populations (among others) to over-exploitation. In appreciation of the uniqueness of this animal, and of the importance of the lesson its loss teaches us, we highlight the historical background to its discovery during a truly epic voyage; describe the demise of the last population; and provide a summary of both historical and modern findings regarding its biology. Over the years, previous authors have also given accounts of the sea cow's discovery and extirpation (e.g. Goodwin 1946; Scheffer 1973; Haley 1980; Reynolds and Odell 1991; Dietz 1992; Anderson and Domning 2002). However, only one scientist ever observed and dissected *Hydrodamalis gigas*, and so throughout this chapter we provide the reader with excerpts of Steller's descriptions through direct translations of his own words rather than attempt to interpret them anew.

GEORG WILHELM STELLER AND THE VOYAGE OF THE *ST PETER*

Steller's sea cow (*Hydrodamalis gigas*) was named for Georg Wilhelm Steller, the only scientist ever to observe this extraordinary animal alive. He discovered a relict population of the giant sirenian when he was stranded with crew mates on an uninhabited island in the eastern Bering Sea in 1741. Steller was an excellent scientist of his time, particularly in botany, and was physically strong and well-suited to the spartan life of a frontier naturalist (Stejneger 1936). He was born in 1709 in Germany, where his university education emphasised theology, anatomy, medicine and botany (he eventually lectured in the last of these). His early career took him to Russia, where he changed the spelling of his name (originally Stöhler or Stöller) to better suit the Russian language (Stejneger 1936). Steller received an appointment from the Academy of Sciences at St Petersburg, which sent him to the Kamchatka Peninsula of far eastern Siberia as a botanist. Soon thereafter he joined the second expedition of Vitus Bering (for whom the Bering Sea is named) on which he served as naturalist, physician and pastor. Captain-Commander Bering was a Dane, commissioned by Peter the Great to explore passages at the eastern frontiers of Russia and beyond to North America. The expedition sailed from Kamchatka in two ships, the *St Peter* and the *St Paul* (Figure 2.1).

They embarked in late May 1741, with Bering and Steller on board the *St Peter*. The crew of the *St Peter* lost sight of the *St Paul* in bad weather on 20 June, and the two vessels never regained contact. (The *St Paul* reached the coast of Alaska and sent men ashore on 17 July; this shore party was lost, and the *St Paul* set sail back to Kamchatka on 27 July, returning in October.) Bering, Steller and the crew onboard the *St Peter* saw what is now known as Mount St Elias in Alaska on 16 July and made landfall on the nearby shores of North America. Steller thus became the first scientist to set foot in Alaska. Other familiar animals that now bear his name include Steller's jay (*Cyanocitta stelleri*), Steller's sea lion (*Eumetopias jubatus*) and Steller's sea eagle (*Haliaeetus pelagicus*). A comprehensive biography of Steller's life and discoveries has been provided by Stejneger (1936), and a recent synopsis of the voyage and of Steller's career is given by Egerton (2008).

The Bering expedition to North America was chronicled firsthand by Steller in his journal, published after his death (see Steller 1925, 1988 for translations) and by Sven Waxell (1962), Bering's second-in-command; the historian Golder (1968) also provides a detailed analysis of the voyage. After exploring the coast and making contact with native inhabitants in the

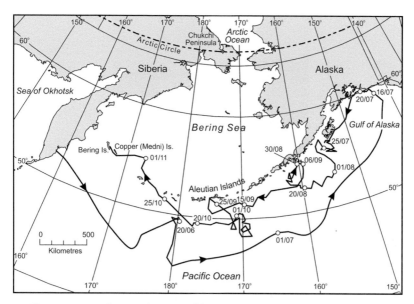

Figure 2.1. Map showing the route of the *St Peter*. Redrawn by Adella Edwards from Steller (1988), with permission from Stanford University Press.

Aleutians, the *St Peter* set sail in August 1741 for the westward return to Kamchatka across the Bering Sea (Figure 2.1). The planned return trip was slower than expected, was late in the short season of favourable North Pacific weather, experienced strong winds and gales, and encountered much hardship. More than 30 of the crew of 78 eventually died, most from scurvy (Stejneger 1936). With 49 sick men, unable to repair or control the ship, the *St Peter* anchored off an uninhabited treeless island (Figure 2.2) on 5 November 1741, soon to run aground and wreck in the ensuing bad weather. The site is now known as Bering Island, which together with the smaller adjacent Copper Island make up the two principal Commander (Komandorskiye) Islands. The crew over-wintered in huts constructed from driftwood and salvaged sails, but Bering died on the island on 8 December 1741.

The survivors from the *St Peter* eventually rebuilt a smaller ship from the wreckage and launched the return voyage to Kamchatka on 13 August 1742. It was during the interim period that the 32-year-old Steller made his observations of the great northern sea cow (see boxed text throughout this chapter). Steller's expert knowledge, in what today might be referred to as ethnobotany, was disregarded by many of his shipmates. However, the various teas and preparations from native plants he and a few of his friends

Figure 2.2. Rib of Steller's sea cow on the beach near Podutesnaya Bay, Bering Island in July 2006. Photograph by Alexander Burdin; reproduced with permission.

consumed allowed them to escape the scurvy that decimated others, prior to the general recognition of the nutritional basis of the disease by Western science. His relatively good health probably also facilitated his observations of these sirenians. Steller never returned to view his sea cows, and died of fever in Tyumen, Russia, in November 1746 at the age of 37. His shipmates on the voyage found him to be a highly critical intellectual, at times difficult to live with (Michael 1962), a trait that is clearly seen in his published journal. Although reportedly 'the greatest single factor in causing his death was strong drink' (Golder 1968, p. 4), Steller was a skilled observer and tireless worker who accomplished much in his short life. Many of his findings were never published, or were published after his death.

In addition to Steller's firsthand accounts, science is indebted to the work of Daryl Domning (Figure 2.3) and Leonhard Stejneger for our knowledge of Steller's sea cow. Domning's masterful body of modern work (see Chapter 3) has put Steller's sea cow in a larger evolutionary context, and increased our understanding of its functional morphology. Stejneger had a long career in the late 1800s and early 1900s as a vertebrate biologist at the Smithsonian Institution's US National Museum of Natural History, and for

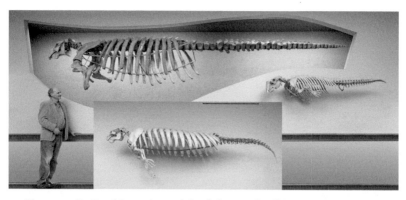

Figure 2.3. Dr Daryl Domning and the skeletons of Steller's sea cow (top), an adult male dugong (right) and a Florida manatee (bottom) at the Smithsonian Museum in Washington in 2010. The scale bar represents 1 metre. The Steller's sea cow skeleton is a composite from several individuals and appears somewhat smaller than an adult Steller's sea cow, which is estimated to have reached lengths of up to 7.9 m, as explained in the text. Photographed by the late Chip Clark with the assistance of Charley Potter and Daryl Domning, reproduced with permission.

many years was Head Curator of Biology. Stejneger made multiple trips to Bering Island, living there in 1882–1883 (Wetmore 1946). He translated Steller's journal and was Steller's chief biographer, spending some 50 years preparing the definitive volume reconstructing Steller's life and discoveries (Stejneger 1936). While on Bering Island, he finagled possession of a number of unearthed skulls and partial skeletons of Steller's sea cow left behind from the 1700s (Stejneger 1893; Scheffer 1973), which remain available for study at the US National Museum today. This collection constitutes the largest number of osteological specimens of this remarkable animal in any museum in the world (Mattioli and Domning 2006). Stejneger (1884, 1886, 1887) also carefully reconstructed the period of exploitation and extirpation of Steller's sea cow in response to claims by others that the sea cow may have survived later than described.

DISCOVERY AND DESCRIPTIONS OF SEA COW BIOLOGY AND NATURAL HISTORY

Despite his many skills, Steller was not an artist and Bering could not honour Steller's request to bring an illustrator to document his natural history discoveries because of limitations of space on board the *St Peter*. Thus posterity has been provided almost nothing in the way of anatomically detailed firsthand sketches or paintings of the sea cow, although a few

Figure 2.4. Drawing of Steller's sea cow published by Pallas in 1811, believed by Stejneger (1936) to be one of Plenisner's lost sketches, made during the stranding of the crew of the *St Peter* on Bering Island. No other illustration is known to exist that was made by anyone who actually saw the creature. Redrawn by Gareth Wild.

sketches (Figure 2.4) were made by Steller's friend Friedrich Plenisner, a surveyor listed as a 'conductor' on the ship's roster (Stejneger 1936; Golder 1968). Plenisner's sketches were lost, although one may have been published by Pallas (1811). Nonetheless, Steller provided enticing verbal descriptions in his notes and journals (Steller 1925, 1988). He also subsequently prepared accounts of his observations on North Pacific marine mammals in a monograph titled *De Bestiis Marinis* (*The Beasts of the Sea*), most of which was written on Bering Island (Stejneger 1936). Because plates intended for this work (presumably by Plenisner) were lost, no drawings were published in the monograph (Stejneger 1936). Nonetheless, Steller's *De Bestiis Marinis* (see Boxes 2.1–2.12) provides his most detailed description of the species. It was published posthumously in Latin in St Petersburg in 1751 (Steller 1751), later translated into English by Miller and Miller and published by David Starr Jordan in 1899 (Jordan *et al.* 1899). From these writings we are provided with the only firsthand knowledge of the species attained by any scientist.

On 8 November 1741, shortly after first arriving at Bering Island, Steller was walking along the coast when 'here and there he observed a huge blackish back like an overturned boat moving slowly about, and every few minutes a snout in front of it emerged for a moment and drew breath with a noise like a horse's snort' (Stejneger 1936, p. 318; see Box 2.1). At the time Steller was unsure of what he was seeing, but later realised they were sirenians, which he and the others then referred to as 'manatis' (Steller was familiar with early accounts of manatees by explorers in the New World tropics, particularly the observations by the privateer William Dampier around the Caribbean). As a result, a prominent point of land on the south-eastern end of Bering Island is now referred to as Cape Manati.

Box 2.1.   On feeding and effects of winter

'Half of the body – the back and sides – projects above the water. While they feed, the gulls are wont to perch upon their backs and to feast upon the vermin that infest their skin, in the same way as crows do upon the lice of hogs and sheep. The manatees do not eat all seaweeds without distinction, but especially (1) *Crispum Brassicae Sabaudicae* [sic], with cancellate leaf; (2) that which has the shape of a club; (3) that which has the shape of an ancient Roman whip [Miller and Miller mistranslate *scutica* as 'shield'; see Domning 1978a, p. 112]; (4) a very long seaweed with a wavy ruffle along the stalk. Where they have stopped, even for a day, great heaps of roots and stems are to be seen cast upon the shore by the waves. When their stomachs are full some of them go to sleep flat on their backs, and go out a distance from the shore that they may not be left on the dry sand when the tide goes out. In winter they are often suffocated by the ice that floats about the shore and are cast upon the beach dead. This also happens when they get caught among the rocks and are dashed by the waves violently upon them. In the winter the animals become so thin that, besides the bones of the spine, all the ribs show.'

Excerpt from Steller's (1751) observations on form and function of the sea cow
(from the Miller and Miller translation of *De Bestiis Marinis*)

The stranded crew at Bering Island first subsisted on the meat of foxes, sea otters and pinnipeds: it was not until late June 1742 that they first killed and ate a sea cow (Steller 1925, 1988). They highly preferred this meat and developed a method of hunting with a small boat (a yawl; see Box 2.2).

The most detailed notes on the sea cow's anatomy (Box 2.3) and gross measurements came when Steller knew they would soon set sail in the rebuilt ship for the return to Kamchatka, and so he set out to make the careful dissection described in *De Bestiis Marinis* (Figure 2.5; Box 2.5). The task required additional labour because of the large size of the beast (four men with rope were required to remove the food-packed stomach alone), but helpers were unwilling to assist in this task and demanded payment in tobacco. Steller could afford to pay only for a few hours' work, and he explains in his records that this hampered a complete anatomical description. Steller also preserved a sea cow calf's skin stuffed with dried grass as a museum specimen, but was forced to leave this and other valuable material behind because of limited space on the return voyage.

How large was Steller's sea cow? Other than the great whales, it was likely the largest mammal to exist in historic times. Steller (1751) reported the female he dissected as 296 inches (7.5 m) in 'length from the extremity of the upper lip to the extreme right cornu of the caudal fork'. Elsewhere, Steller gives different body lengths, and there has been some difficulty in

Box 2.2. On hunting

'A harpooner and five other persons for rowing and steering took their places in it. They had a very long rope lying in it in proper order, in exactly the manner as in Greenland whaling, one end of which was fastened to the harpoon, with the other held on shore in the hands of the other forty men. They rowed very quietly towards the animals, herds of which were foraging for food along the coast in the greatest security. As soon as the harpooner had struck one of them, the men on shore started to pull it to the beach while those in the yawl rowed toward it and by their agitation exhausted it even more. As soon as it had been somewhat enfeebled, the men in the yawl thrust large knives and bayonets in all parts of its body until, quite weak through the large quantities of blood gushing high like a fountain from its wounds, it was pulled ashore at high tide and made fast. As soon as the tide had receded and it was stranded on the dry beach, we cut off meat and fat everywhere in large pieces, which with great pleasure we carried to our dwellings. A large part we stored in barrels. Another part, especially the fat, we hung up on racks. And at long last, we found ourselves suddenly spared all trouble about food and capable of continuing the construction of the new ship . . . .'

Excerpt from Steller's (1751) observations on form and function of the sea cow (from the Miller and Miller translation of *De Bestiis Marinis*)

Box 2.3. On the shape of the body

'From the shoulders toward the umbilical region it grows rapidly wider, and from there on to the anus it again grows rapidly slender; the sides are roundish and paunched like a belly which is swollen with a great mass of intestines . . . . When the animals are fat, as they are in spring and summer, the back is slightly convex; but in winter, when they are thin, the back is flat and excavated at the spine with a hollow on either side, and at such times all the vertebrae with their spinous processes can be seen . . . . The tail grows perceptibly thinner toward the fin. It is not so much flat as rather somewhat quadrangular . . . the tail itself gets the form of a square oblong with obtuse angles . . . . For the rest the tail is thick, very powerful, and ends in a very hard, stiff, black fin . . . . With a gentle sidewise motion of its tail it swims gently forward; with an up-and-down motion of the tail it drives itself violently forward and struggles to escape from the hands of enemies who are trying to draw it in.'

Excerpt from Steller's (1751) observations on form and function of the sea cow (from the Miller and Miller translation of *De Bestiis Marinis*)

Figure 2.5. Depiction of Steller with two paid assistants making the first measurement of a sea cow. Bering Island, 12 July 1742. The date and measurements are given in excerpts from Steller's (1751) observations on form and function of the sea cow, *De Bestiis Marinis* (from the Miller and Miller translation pp. 294–296). Redrawn by Gareth Wild from a reconstruction by L. Stejneger published by Golder (1925).

deriving the units of measurement used in various historical writings. Domning (1978a) reviewed this literature, which provides a range of estimates of body length from 7.3 m to 10.6 m, and based on the size of the skull calculated that the largest Steller's sea cow at Bering Island was probably about 7.9 m long, twice the length of the largest modern manatee and more than 2.5 times the length of the largest dugong. Steller's estimates for the body mass of the sea cow were also disparate and likely inaccurate. Calculations by Scheffer (1972) suggest the mass of an 8 m sea cow to be about 10 000 kg. This estimate was revised downward to about 6300 kg by Domning (1978a), who suggests that the sea cows at Bering Island were in a suboptimal habitat and may have been somewhat small for the species compared with past populations in other areas of its former range. Assuming that some Steller's sea cows indeed reached a plausible maximum length of 10 m, Domning (1978a) estimated that such large specimens would have a body mass of 11 196 kg, about seven times the mass of the largest manatee. The gigantism of Steller's sea cow was likely a result of selection for the heat retention value of a small surface-to-volume ratio in cold northern waters (Domning 1978a; Chapter 3).

Figure 2.6. Rostral horny mouth plate of an adult Steller's sea cow. From a specimen (probably saved by Steller) in the Museum of the Leningrad Academy of Science. The dimensions of the horny plate from which the original drawing was made were approximately: 196 mm in length and 81 mm maximum width. Redrawn by Gareth Wild from Brandt (1846).

In addition to its exceedingly large size, Steller described several morphological specialisations of his sea cow that stand apart from those of the dugong and the manatees. These were primarily adaptations for grazing on algal kelps (which are softer than vascular plants) in rocky nearshore habitats. The sea cow had no teeth. Instead, it cropped and crushed its food using the prominent rostral masticating pads found in other sirenians (described as two strong white 'molar bones' by Steller), but modified with backward extensions of the lower pad and with diagonal ridges, which likely improved the mashing of kelp (Domning 1978a). Steller described and collected specimens of these, which were characterised in detail a century later by Brandt (1846; Figure 2.6). Domning (1978a) provided further analysis of the myological and osteological specialisations of the head, neck and mouth.

Unlike living sirenians in which the front flippers are used for balance, sculling and 'bottom-walking' during aquatic locomotion (Chapter 5), the forelimbs of Steller's sea cow were highly modified to serve as holdfasts and for propulsion in the rocky, nearshore areas subject to strong waves. Steller also reported that they were used to scrape or pull seaweed from the rocks (Box 2.4). The forelimbs had thickened coverings and stiff bristles, with no phalanges, and the metacarpals formed a somewhat claw-like manus.

Box 2.4. On the forelimbs

'The strangest feature of all . . . is its arms, or, if you please, its front feet; for
two arms . . . consisting of two articulations, are joined immediately to the
shoulders at the neck . . . the ulna and radius terminate bluntly with tarsus
and metatarsus [sic]. There are no traces of fingers, nor are there any of nails
or hoofs; but the tarsus and metatarsus are covered with solid fat, many
tendons and ligaments, cutis and cuticle, as an amputated human limb is
covered with skin . . . underneath they are flat and hollowed out in a way, and
rough with countless very closely set bristles . . . with these he walks on the
shallows of the shore, as with feet; with these he braces and supports himself
on the slippery rocks; with these he digs out and tears off the algae and
seagrasses from the rocks . . . with these he fights, and when taken with a
hook and dragged from the water upon dry land he resists so vehemently that
the cuticle surrounding these arms is often torn and pulled off in pieces; and
finally with these the female when smitten with the sting of passion, swimming prone upon her back, embraces her covering lover and holds him and
permits herself in turn to be embraced.'

Excerpt from Steller's (1751) observations on form and function of the sea cow
(from the Miller and Miller translation of *De Bestiis Marinis*)

Modifications of the humerus and shoulder joint and supporting structures
favoured pulling of the body in a parasagittal (longitudinal) plane of motion
(described in detail by Domning 1978a).

Other internal anatomical descriptions given by Steller generally agree
with those of the living sirenians, including the large horizontal lungs and
diaphragm, placement of the kidneys, the large presumed cardiac gland off
the huge stomach, a capacious caecum and a lengthy hindgut (Box 2.5).

The unusual buoyancy of Steller's sea cow and its likely inability to
submerge may be attributable to its specific gravity being less than that of
the typically neutrally buoyant sirenian (Domning 1978a), at least in part
due to natural selection for its large size. (Volume increases as a cubic
function of length, whereas mass of the heavy bones increases as little more
than the square of linear dimensions in large marine animals; Domning
1978a). Additionally, there was natural selection for a large, 'paunchy' belly
to accommodate the lengthy intestines reported by Steller ('roundish and
paunched like a belly which is swollen with a great mass of intestines';
Box 2.5), a greater lung volume and the large layering of blubber (Box 2.6;
Steller 1925). The latter two factors probably accounted for the typical
observations of Steller's sea cow with its back seemingly always above
water (Domning 1978a).

> **Box 2.5. On the digestive tract**
>
> 'The stomach is of stupendous size, 6 feet long, 5 feet wide, and so stuffed with food and seaweed that four strong men with a rope attached to it could with great effort scarcely move it from its place and drag it out ... what was most peculiar, and perhaps incredible to many, is that I found contained in the stomach, and not far from the entrance of the oesophagus into the stomach, an oval gland as large as a man's head ... this gland opened through the villous coat with many pores and openings and exuded into the cavity of the stomach a great quantity of whitish liquid .... The inner coat of the stomach was perforated by white worms half a foot long, with which the whole stomach, pylorus and duodenum, swarmed; .... The pylorus was so large and tumid that at first sight I took it for a second stomach .... There are more intestines in this animal than in any other .... The abdominal cavity was so full that the abdomen was tumid and swollen like an inflated skin. Hence, when the common coverings and muscles of the abdomen were removed and the peritoneum received ever so slight a wound, the wind came out with such a whistle and hum as it is wont to come from an aeolipile .... When the peritoneum is cut the intestines gush out violently, and without any outside assistance they move from their original place, because they are found always so tightly stuffed that from oesophagus to anus they make a solid pack without any open space. The thin intestines are smooth, rolled up in a great amount of fat; they are round and 6 inches broad in diameter. If only a very slight aperture should be made with the point of a knife, the liquid excrement (a ridiculous thing to behold) would squirt out violently like blood from a ruptured vein; and not infrequently the face of the spectator would be drenched by this springing fountain whenever some one opened a canal upon his neighbor opposite, for a joke. The coecum was very large, as was also the colon ... the intestines were different from a horse's in size and capacity alone, but not in structure. And so the final product of this workshop is so like the excrement of horses, in shape, size, smell, and color, and all other attributes, that it would deceive the most expert stable boy .... The whole intestinal tract, from gullet to anus, when this Augean stable was thoroughly cleansed, measured fully 5968 inches, and so the intestines are twenty and a half times as long as the whole living animal.'
>
> Excerpt from Steller's (1751) observations on form and function of the sea cow (from the Miller and Miller translation of *De Bestiis Marinis*)

The thickened epidermis described by Steller (Box 2.7) probably was protective against abrasion on the ice and rocks as he indeed observed, and may also have helped prevent drying of the exposed back (Domning 1978a). This bark-like hide was also home to unique ectoparasites of unknown affinities (Brandt 1846; Anderson and Domning 2002).

How many of these unique sea cows were present in the Commander Islands at the time the *St Peter* stranded? Unfortunately, there is no direct

Box 2.6.   On the blubber

'The fat underlies the cuticle and the skin and covers the whole body to the depth of a span, and in some parts is almost 9 inches thick. It is glandulous, stiff, and white, but when exposed to the sun it becomes yellow like May butter (*butyri maialis*). Its odor and flavor are so agreeable that it can not be compared with the fat of any other sea beast. Indeed, it is by far preferable to that of any other quadruped. Moreover, it can be kept a very long time, even in the hottest weather, without becoming rancid or strong. When tried out it is so sweet and fine flavored that we lost all desire for butter. In flavor it approximates nearly the oil of sweet almonds and can be used for the same purposes as butter. In a lamp it burns clear, without smoke or smell. And, indeed, its use in medicine is not to be despised, for it moves the bowels gently, producing no loss of appetite or nausea, even when drunk from a cup.'
Excerpt from Steller's (1751) observations on form and function of the sea cow
(from the Miller and Miller translation of *De Bestiis Marinis*)

Box 2.7.   On the appearance of the skin

'It is covered with a thick hide, more like unto the bark of an ancient oak than unto the skin of an animal ... black, mangy, wrinkled, rough, hard, and tough; it is void of hairs, and almost impervious to an ax .... From the nape to the caudal fin the surface is uneven with nothing but circular wrinkles, but the sides are exceedingly rough, especially about the head .... This cuticle which surrounds the whole body like a crust ... seems given to the animal for two purposes: (1) that, inasmuch as the animal is compelled, for the sake of getting a living, to live continuously in rough and rugged places, and in the winter among the ice, it may not rub off the skin, or that it may not be beaten by the heavy waves and bruised with the stones, and when pursued it is protected by this coat of mail; (2) that the natural heat may not be dissipated in the summer ... or completely counteracted by the cold of winter .... I have observed in the case of many that were cast up dead upon the shore of the sea, that the cuticle had been broken off on one side or the other, and that that had been the cause of their death; and this happens principally in the winter time, from the ice .... While this cuticle is wet it is tawny black, like the skin of a smoked ham .... In certain animals it is marked with rather large white spots and zones, and this color penetrates clear to the cutis.'
Excerpt from Steller's (1751) observations on form and function of the sea cow
(from the Miller and Miller translation of *De Bestiis Marinis*)

Box 2.8.   On habitat

'These animals are fond of shallow sandy places along the seashore, but they like especially to live around the mouths of rivers and creeks, for they love fresh running water, and they always live in herds.'
Excerpt from Steller's (1751) observations on form and function of the sea cow
(from the Miller and Miller translation of *De Bestiis Marinis*)

information on this topic in the firsthand accounts. If Steller's sea cow had survived, it would likely have been by far the easiest sirenian for which an estimate of abundance (even perhaps a total count) could be obtained, due to its seeming inability to submerge and its occupancy of a narrow coastal zone (Box 2.8).

A reasonable guess on population size was made by Stejneger (1887). Given that Stejneger was an excellent scientist, lived on Bering Island for over a year in the 1880s, salvaged a good proportion of the known specimen material and was Steller's greatest biographer and a translator of his journal, Stejneger's statement on the likely number of sea cows alive at the time of discovery may be the best expert opinion we will ever have. He wrote:

> Unfortunately, Steller, in describing this animal and its habits, only says that he found it numerous and in herds, without stating exactly how numerous or in how large herds. We are thus left to guess at their probable number when first found . . .. I should regard fifteen hundred as rather above than below the probable number . . .. There are hardly more than fifteen places on the island which could afford them suitable grazing-grounds, and if each of these were regularly visited by an average of one hundred animals, one would easily be impressed by their number, especially if divided up into five to ten herds of from ten to twenty individuals.
> (Stejneger 1887, p. 1049)

Domning (1978a) speculated that there were perhaps as many as 2000 at the time of discovery.

Whatever the number, this was a last remnant of a much more widespread population of Steller's sea cow, with a former distribution that extended from central California (only 20 000 years ago, and from Baja California earlier, a range co-extensive with the sea otter) in a northward arc to the archipelago of Japan (Domning 1978a; Chapter 3). It is highly likely that hunting of sea cows throughout this range was practised by early humans, whose coastal distributions overlapped that of Steller's sea cow for most of the last 20 000 years. A tenable hypothesis has been advanced that past hunting by earlier native people elsewhere was the major factor in the ultimate reduction of the species to the uninhabited Bering and Copper Islands; indeed, the hunting of sea cows may have been a first step towards aboriginal whaling by people of the North Pacific (Domning 1972, 1978a; Domning *et al.* 2007). Following this period of early exploitation, it took only a short time from the point of discovery of the final relict population for people to complete the job of relegating Steller's sea cow to extinction through unmanaged hunting.

EXTIRPATION OF STELLER'S SEA COW

The fur trade was a major driver of the Russian economy for centuries prior to the voyage of the *St Peter* (Crownhart-Vaughan 1972). The discovery of a passage to North America by the Bering expedition in 1741 opened up a new frontier for exploitation of furs. Results were far-reaching. They included the near elimination of the California sea otter, and major population reductions in fur-bearing marine mammals over extensive areas of the west coast of North America, which continued until the international fur seal treaty of 1911 (Ogden 1941; Kenyon 1969). Stejneger (1887) reviewed historical accounts and ships' records on the taking of Steller's sea cows for provisions in the Commander Islands and concluded that the last one was killed in 1768, just 27 years after Steller first observed the species. This approximate date was also substantiated by Brandt's (1846) analysis and later affirmed by Domning (1978a) and others.

Steller's sea cows were simple to acquire because of their seeming inability to submerge, their tolerance of humans and their restriction to the limited nearshore area (Box 2.9).

Steller's sea cow meat was prized for its taste, and a single animal could supply large quantities. Initially, hunting crews that set out from Kamchatka specifically targeted the Commander Islands as a place to obtain furs, beginning with the winter of 1743–1744, and fur hunters stayed there for about 8–10 months nearly every year thereafter until 1763. During these stays the crews 'lived almost exclusively on the meat of the sea cow' (Stejneger 1887, p. 1049). In addition, Stejneger (1887) pointed out that many other expeditions first stopped at the Commander Islands to stock their larders with enough preserved sea cow meat to supply them for further outbound journeys of 2–3 years across the North Pacific. How many sea cows were taken for these purposes? Stejneger (1887) reported on the ships' records of those crews that wintered there for at least eight months and calculated a total of about 670 men. At least ten other vessels with a total

---

Box 2.9.   On reactions to humans

'When the tide came in they came up so close to the shore that I often hunted them with my stick or lance, and sometimes even stroked their backs with my hand. If they were badly hurt, they did nothing but withdraw to a distance from the shore, and after a short time they would forget their injury and come back.'

'These animals are very voracious and eat incessantly, and because they are so greedy they keep their heads always under water, without regard to life and safety. Hence a man in a boat, or swimming naked, can move among them without danger and select at ease the one of the herd he desires to strike – and accomplish it all while they are feeding.'

Excerpts from Steller's (1751) observations on form and function of the sea cow (from the Miller and Miller translation of *De Bestiis Marinis*)

---

of about 400 men prepared provisions of sea cow meat to last at least 12 months on forays farther abroad. Records (particularly the diaries of the mining engineer Jakovleff; see Domning 1978a) examined by Stejneger further showed that by 1754 all sea cows had been extirpated on nearby Copper Island. The methods of hunting were wasteful: unlike the method of using many men to pull the struggling sea cows to shore employed by the stranded crew of the *St Peter*, some of these later hunters simply harpooned the animals and waited for them to die and wash ashore. Thus many carcasses were lost at sea or were not cast ashore until they had putrefied and were unsuitable as food.

Jakovleff recognised the folly of these practices and petitioned the authorities in Kamchatka to stop the waste, but nothing was done. Stejneger (1887) reckoned that five times as many sea cows as were butchered for meat were killed but lost. Based on Jakovleff's report that each sea cow supplied enough meat for 33 men for one month, Stejneger calculated that 495 sea cows were needed to sustain both the crews that spent winters on the island as well as those that stopped off to provision long trips farther away. Given the large number of sea cows that were wasted, well over 2000 sea cows were probably slaughtered for provisions. After 1763 fewer ships stopped at Bering Island (fur-bearing resources on the island had also been exhausted). None of the records of these ships, nor the written observations in 1772 specifically enumerating the marine mammals of the island, mention the presence of sea cows (Brandt 1846; Stejneger 1884). This amazing animal was extinct.

MODERN INVESTIGATIONS AND SPECULATIONS ON SEA COW BIOLOGY

A few contemporary authors have speculated about the biology, ecology and life history of Steller's sea cow based on the eighteenth-century accounts. Some highly intriguing possibilities have emerged from these analyses. However, such interpretations should be made with care: although Steller was likely very accurate in many of his descriptions, some of his statements are based on supposition. For example, Steller referred to the sea cow as monogamous and provided descriptions of their mating behaviour (Box 2.10).

In a modern analysis of the evolution of sirenian mating systems, Anderson (2002) assumed the sea cow was indeed monogamous as Steller described, and postulated plausible conditions of the species' dispersion and ecology on this basis. Nonetheless, Steller (1751) also provided the following description for the sea otter: 'They preserve their conjugal affection most constantly, and the male does not serve more than one female. They live together both on sea and on land. The one-year-old cubs . . . live with their parents until they set up housekeeping on their own score.' Modern knowledge of sea otter social behaviour and ecology (e.g. Riedman and Estes 1990) suggests that Steller's observations on these subjects in sea

---

Box 2.10. On reproductive behaviour

'In the spring they come together in the human fashion, and especially about evening in a smooth sea. But before they come together they practice many amorous preludes. The female swims gently to and fro in the water, the male following her. The female eludes him with many twists and turns until she herself, impatient of longer delay, as if tired and under compulsion, throws herself upon her back, when the male, rushing upon her, pays the tribute of his passion, and they rush into each other's embrace.'

'Most commonly whole families live together in one community, the male with one grown female and their tender little offspring. They appear to me to be monogamous. The young are born at any time of year, but most frequently in autumn, as I judged from the new-born little ones that I saw about that time. From this fact, as I noticed that they copulated by preference in the early spring, I concluded that the foetus remained more than a year in the womb. From the shortness of the [uterine] cornua (*ex cornuum brevitate*), and from the fact that there are only two mammae, I infer that they have but one calf, and I have never seen more than one with the mother at a time.'

Excerpts from Steller's (1751) observations on form and function of the sea cow (from the Miller and Miller translation of *De Bestiis Marinis*)

otters were somewhat fanciful and anthropomorphic, as they may well have been for his sea cows. If Steller was accurate and his sea cow was indeed monogamous, then Anderson (2002) argues that such a mating system was facilitated because they probably only occupied an area within about 100 m of shore (based on modern bathymetry); had a compressed breeding season; were limited in group size by destructive foraging; females occupied exclusive matriarchal ranges; and bi-parental protection of offspring was favoured because of physically harsh conditions. Certainly it is plausible that such a singular animal may have had an unusual mating system for a sirenian, but such a scenario can never be fully verified given the information that is currently available.

Anderson (1995a) has pointed out that the accounts by Steller and Waxell document that sea otters were highly abundant and easy to kill on Bering Island. Initially, sea otters provided much fresh meat, despite its poor quality, and because their furs were highly prized commercially, they were also taken for pelts in large numbers by the stranded seamen. Soon after the crew's return to Kamchatka, word spread about fur-bearers on the Commander Islands and hunters further removed large numbers of sea otters. Sea otters are a 'keystone species', and their removal from kelp ecosystems causes a well-documented trophic cascade (Estes and Palmisano 1974; Estes and Duggins 1995). They feed on invertebrates that graze on kelp (such as sea urchins), and normally suppress populations of these grazers. In the absence of sea otters, urchins and other invertebrate grazers become superabundant and quickly cause drastic biomass reductions in the standing crop of kelp (Estes and Duggins 1995). Thus the high harvest of sea otters at the Commander Islands may also have acted as an indirect pressure on the sea cow food supply, and a reduction in kelp perhaps hastened their demise. Anderson (1995a) further suggested that the final constriction of Steller's sea cow to the uninhabited Commander Islands may also have been in part due to sea otter removal by aboriginal people (in addition to overkill by hunting of sea cows themselves). Humans occupied many areas of the past range of the sea cow for thousands of years, and in the Aleutians archaeological evidence shows that they caused periodic declines in sea otter populations with accompanying shifts in characteristics of the invertebrate grazers (Simenstad et al. 1978). However, limited and fragmentary evidence also suggests that sea cows may have co-existed with native people for at least hundreds of years in the western Aleutian Islands – oral history recorded in the late 1800s suggests they were hunted at Attu in the 1700s, primarily by women (Domning et al. 2007).

There is no information on predators of Steller's sea cow. Domning (1978a) suggested that sharks and killer whales may have taken them, and pointed out that the buoyancy of the sea cows may have made it difficult for killer whales to kill them by drowning, whereas the rocky kelp forests may have helped them avoid sharks. Steller's descriptions suggest that the young were kept in positions guarded from predators by adults (Box 2.11).

The likely numerical and demographic processes involved in the decline and extinction of Steller's sea cow were subjected to a population simulation model by Turvey and Risley (2006). Life history inputs were extrapolated from estimates for dugongs (Chapter 6), and included a life expectancy of 90 years and litter size of one. Variable ages of first conception (12–20 years), interbirth intervals (5–10 years) and population sizes at carrying capacity (1000–2000) were used in sensitivity analysis. Resulting maximum annual population growth rates were low at 1.35%, allowing a sustainable take of only 17 individuals each year, about one-seventh of the number taken as estimated by Stejneger (1887). Turvey and Risley (2006) also modelled the impact of three alternative hunting regimes on the fate of an initial population of 1500 Steller's sea cows: that estimated by Stejneger; wasteful hunting but with no provisioning for outbound voyages; and hunting without wastage but including provisioning for long journeys. Under the conditions suggested by Stejneger, the median date for extinction was 1756 (95% confidence limit (CL) 1754–1757); without provisioning of outbound voyages extinction would have occurred in 1778 (95% CL 1770–1786); and without wastage but with provisioning the end would have been reached in 1817 (95% CL 1799–1841). The calculated date of extinction under the most likely hunting conditions (1756) was slightly earlier than that recorded historically. Turvey and Risley (2006) noted that this discrepancy may result from uncertainty in the initial population size. A projected extinction date of 1760, with a 95% confidence limit that includes the date of extinction estimated by Stejneger's historical analysis, is reached when the initial population size is estimated to be 2000

Box 2.11. On care of young

'They keep the young and the half-grown before them while they feed, but they are careful to surround them on the flank and rear and always to keep them in the middle of the herd.'

Excerpt from Steller's (1751) observations on form and function of the sea cow (from the Miller and Miller translation of *De Bestiis Marinis*) animals (as speculated by Domning 1978a). Turvey and Risley (2006) also note that hunting alone may have been sufficient to have caused extinction; the loss of this giant sirenian at the Commander Islands would have taken place without any contribution that hypothetically may have occurred with the concomitant decline of sea otters and any resulting reductions in kelp biomass caused by increased grazing by the sea otter's invertebrate prey (as suggested by Anderson 1995a).

Steller and Waxell's observations that the sea cow did not submerge are supported, with reservations, by anatomical analysis (Domning 1978a). Considering this possible strict limitation to diving, these animals could only have fed on surface canopy kelps in the North Pacific Ocean. Kelps secrete secondary defensive chemicals for deterrence of herbivory (see Chapter 4), but surface canopy kelps have much lower concentrations of defensive chemicals than kelps that grow at greater depth (Steinberg 1985; Estes *et al.* 1989; Steinberg *et al.* 1995), and surface kelps are also more subject to the destructive forces of ocean waves. The effects of grazing on surface kelps by sea cows may have been overridden by mechanical damage, and thus there was no evolution of chemical resistance to sea cow herbivory by surface kelps, unlike the case with chemical deterrence of invertebrate grazers by epibenthic kelps in many parts of the world (Estes *et al.* 1989). The role of sea cows in structuring kelp forest communities may have been stronger during earlier times in areas farther to the south of California, where the forest is dominated by a different kelp species (*Macrocystis pyrifera*). There it is possible that surface grazing by sea cows during the calm summer period of kelp growth allowed greater light penetration to permit growth of epibenthic macroalgae, resulting in a different community structure than exists today. The production of reproductive tissues of two surface kelp species near the plant base just above the holdfast may be an adaptation in response to kelp grazing by sea cows (Estes *et al.* 1989). Unfortunately, the opportunity for direct study of ecological interactions between Steller's sea cow and its food supply has been lost with its extinction.

There has been a long historical record of attempts to secure skeletal specimens of Steller's sea cow once it was recognised to be extinct, as chronicled by Mattioli and Domning (2006). This material is largely all that remains of this giant end point of sirenian evolution, and is now scattered in about 51 institutions across 16 countries. It consists of 62 skulls, 27 mostly composite skeletons and over 550 bones (Mattioli and Domning 2006). These samples are available to qualified researchers, subject to institutional policies. Their modern study has included determination of calcium isotope ratios in bone (Clementz *et al.* 2003a); carbon isotope ratios in bone bioapatite and bone collagen in comparison with dietary sources

---

Box 2.12. On the head

'In comparison with the huge mass of the rest of its body the head is small, short, and closely connected with the body . . .. The nose is situated in the farthest tip of the head . . .. The eyes are situated exactly half way between the end of the snout and the ears in a line parallel with the top of the nostrils, or just a very little higher. They are very small in proportion to so huge a body, being no larger than a sheep's eyes. They are not provided with shutters, or lids, or any other external apparatus, but protrude from the skin through a round opening, scarcely a half inch in diameter . . .. When the lachrymal sac is cut a great amount of sticky mucus is found in its cavity . . .. The ears outside open only with a small hole . . .. There is not the slightest trace of an external ear, and the holes can be seen only by examining very closely; for the opening of the ears can not be distinguished from the rest of the pores, and would scarcely admit the quill of a chicken's feather.'

Excerpts from Steller's (1751) observations on form and function of the sea cow
(from the Miller and Miller translation of *De Bestiis Marinis*)

(Clementz *et al.* 2007); oxygen isotope ratios in bone (Lécuyer *et al.* 1996); and bone structure and histology (Kaiser 1974; Marmontel 1993; Ricqlès and de Buffrénil 1995). The relationship of brain to body size as expressed by encephalisation quotients was analysed based on measurement of some of these archived skulls, and the adjusted relative brain size was found to be among the smallest of all the mammals, probably as a result of natural selection for large body size (O'Shea and Reep 1990). Steller also noted that the head was very small relative to the size of the body (Box 2.12).

Genetic analysis of old DNA using the mitochondrial cytochrome b gene (Ozawa *et al.* 1997) has tended to confirm Domning's estimated time of divergence between dugongs and the sea cow lineage, based on paleonto-logical data (Chapter 3). The DNA was obtained from preparations of a scapula of a Steller's sea cow from Bering Island. The bone surface was ground clean, the bone pulverised and decalcified and remaining cells lysed, followed by extraction, concentration and amplification of DNA. Sequence divergence was estimated based on 1005 base pair sequences. The molecular clock approach to the analysis of substitution rates yielded a date of 22 million years ago for the divergence of the *Hydrodamalis* lineage from the *Dugong* lineage. This result is more consistent with the paleonto-logical evidence (which suggests an earlier divergence) than a previous immunological analysis of proteins from a skull of Steller's sea cow. The immunological assays from the latter studies suggested a divergence of only about 4–10 million years ago (Rainey *et al.* 1984; Lowenstein and

Scheuenstuhl 1991). Perhaps future analyses from museum specimens of this unusual sirenian will provide additional insights into its biology.

CONCLUSIONS

Steller's sea cow was the first human-caused extinction of any marine mammal. It represented a major end point in an unusual evolutionary trajectory of the Sirenia into cold marine waters, and a loss at the level of subfamily in mammalian diversity and classification (Chapter 3). Since then, at least two other species of marine mammals have become extinct, both much smaller species inhabiting warmer waters. The last Caribbean monk seal (*Monachus tropicalis*) was seen in 1952 (Kenyon 1977; Kovacs 2008). The Yangtze River dolphin or baiji (*Lipotes vexillifer*) was last documented in 2002 and could not be found by a comprehensive visual and acoustic survey in 2006 (Turvey *et al.* 2007; Pyenson 2009); its loss represents the first modern extinction at the level of a mammalian family (Lipotidae). Given that all marine mammals have life history characteristics that make them vulnerable to extinction (long-lived, slow breeding; Chapter 6) further extinctions of other species are likely without successful directed conservation.

The extinction of Steller's sea cow is proof that a marine mammal with a once-extensive distribution is not immune from extinction; it is a stark historical reminder of the vulnerability of many of today's isolated populations of sirenians, particularly populations occurring in the waters surrounding islands. Most of the known local extinctions of the dugong (Marsh 2008) and the West Indian manatee (Deutsch *et al.* 2008) have occurred around islands, and more such extinctions are likely (Chapter 8). Examples of modern island populations of sirenians at risk include the dugong in Okinawa (Chapters 1 and 8), Palau (Brownell *et al.* 1981; Marsh *et al.* 1995; Chapter 8) and the Antillean manatee in Puerto Rico, where anthropogenic mortality is high and genetic diversity low (Mignucci *et al.* 2000; Hunter *et al.* in press; Chapter 8). Purvis *et al.* (2000) point out that because of the low number of extant species, the Sirenia are one of three orders of mammals at highest risk of extinction. The degree of morphological and phylogenetic uniqueness at the order level seems to us a much more exceptional potential loss to global biodiversity than disappearance at the level of species, however tragic any such loss may be. Should the remaining species of sirenians follow Steller's sea cow to extinction, it will represent destruction of a singular evolutionary lineage that began over 50 million years ago with the diversification of mammals into the unique adaptive zone of purely aquatic herbivory (Chapter 3).

Affinities, origins and diversity of the Sirenia through time

INTRODUCTION

A full appreciation of the ecology and conservation of the Sirenia requires an understanding of their evolutionary history. Modern manatees and dugongs provide only a limited view of a much deeper biological continuum that extends back over 50 million years. There were numerous branchings along this continuum: dozens of species of sirenians existed through time. They ranged in size from little sea cows perhaps 150 kg in body mass, to the largest mammal other than the great whales to exist in historic time – Steller's sea cow – at a plausible body mass of over 10 000 kg (Chapter 2). Early sirenians walked on land with sturdy hind limbs, but fed on aquatic vegetation. Later forms were fully aquatic, with a variety of foraging strategies: some ate delicate seagrass leaves, some had large and powerful tusks that dug or cut through tough seagrass rhizomes, some may have specialised on molluscs, and others had no teeth at all and fed on soft kelps higher in the water column. These sirenians prospered or became extinct according to shifting climatic, geologic, oceanographic and biological conditions. Unlike the very rapidly changing environmental conditions of today that are products of human population and technological growth, varying conditions of the past acted more slowly, allowing ancient sirenians to adapt and evolve altered modes of life.

In this chapter we summarise this evolutionary history. We begin with two essential questions that have long fascinated scholars of the Sirenia: what were their ancestors like, and who are their closest living relatives? First, we briefly review the history of morphological and palaeontological studies of these questions, and how the hypothesised affinities have been reflected in the placement of the Sirenia in various mammalian classification schemes. Then we summarise the explosion of molecular and genetic data that have forced some radical reinterpretations of the evolutionary history of mammals within the past decade, concentrating on how these new data relate to the Sirenia and their closest affinities.

Following the overview of proposed relationships and affinities, we highlight findings from the fossil record that point to the likely circumstances surrounding the emergence and evolution of the Sirenia. These include a fascinating 'missing link' from the Eocene that walked on four limbs. We also give an overview of the fossil sirenians. To place the diversity of fossil sirenians in perspective, we review in detail patterns of sirenian evolution in two major regions: the western Atlantic–Caribbean, and the North Pacific Ocean. These regions have yielded a compelling series of fossils that illustrate a far wider diversity of co-existing and serially replaced sirenian morphotypes than exist today.

PROPOSED AFFINITIES AND PLACEMENTS OF THE SIRENIA BASED ON MORPHOLOGY AND PALAEONTOLOGY

The Order Sirenia has long been recognised to have affinities with other mammals, particularly elephants, another highly specialised group of major conservation concern. The overarching goal of modern higher-level classifications of the mammals has been to accurately reflect their evolution. As such, new refinements in classifications are scholarly attempts to pose hypotheses about origins and affinities. Efforts to reconstruct the puzzles of higher-order phylogenetic relationships within and among the mammals have been, and will remain, a long-term work in progress. McKenna and Bell (1997), for example, provided a hierarchical classification that utilised 12 categorical levels above the rank of order and below the class Mammalia. Such detailed results are beyond the scope of this book; conclusions regarding higher mammalian evolutionary relationships will likely remain in flux for many years.

A century prior to Darwin and the recognition of evolution as a framework for classification, Linnaeus (1758) grouped the living mammals into eight orders. He placed manatees and elephants together under one order: the Bruta (literally, 'Brutes'). He also included sloths, anteaters and pangolins in the Bruta, linking the five groups primarily on the basis of an absence of incisors and their subjectively peculiar modes of locomotion. The name 'Sirenia' was first formally applied when Illiger (1811) recognised manatees, dugongs and Steller's sea cow as a family within his Order Natantia (which also included one other 'family', today recognised as the

Order Cetacea). De Blainville (1836) created a distinct Order 'Gravigrades' that consisted of two groups, the sirenians (Sirenei) and elephants (Proboscidei), perhaps revealing an insight into their common ancestry ahead of his time (see below). Later classifications of mammals also grouped the sirenians and elephants together in various combinations. However, classification schemes in the 1800s generally placed sirenians closest to cetaceans based on adaptations to the aquatic environment. Gregory (1910) and Simpson (1932) summarised the history of classification schemes of mammals as they pertain to sirenians, which in past centuries had proposed alignments that included fishes, walruses and amphibious ungulates.

The morphological similarities thought to support a relationship between the sirenians and proboscideans were addressed by Gregory (1910; Table 3.1), who also took fossils into account. In reviewing past work (especially Gill 1872), Gregory (1910) connected these two orders

Table 3.1 *Morphological similarities between the Sirenia and Proboscidea (elephants) as noted by various authorities. Synapomorphies are considered true shared derived characteristics. See text and Supplementary Material Appendix 3.1 for summaries of molecular evidence linking these two and other groups of mammals. See also Chapters 4 and 6 for additional information on similarities between the internal morphology of sirenians and elephants.*

Proposed morphological links between Sirenia and Proboscidea	Reference
Living species	
Pectoral mammae; thick skin; abdominal testes; bifid apex of the heart; long gastrointestinal tract; characteristics of bones of the snout; absence of a lacrimal foramen; aspects of placentation; bilophodont molars with a tendency to form an additional lobe from the posterior part of the cingulum; horizontal tooth replacement (but with very different patterns and only in the Trichechidae among sirenians).	Andrews (1906); Gregory (1910)
Fossil species	
Anterior placement of the orbits; shape and placement of the squamosal bones; general appearances of the palate, occipital region, scapula, pelvis and cranial endocasts.	Gill (1872); Gregory (1910)
Synapomorphies	
Bifid apex of the heart; pectoral mammae; shape of the molars; position of the anterior border of the orbit; lateral expansion of the zygomatic process of the squamosal bone; reduction of the mastoid process of the temporal bone.	Tassy and Shoshani (1988)

and, to a lesser extent, also linked early proboscideans with the Order Hyracoidea (hyraxes; see 'Paenungulata and Tethytheria' below). Some similarities, however, were also acknowledged to vary within the Sirenia and Proboscidea, or were shared similarities also seen in other groups and could have evolved independently. More recently, Tassy and Shoshani (1988) considered six characters as shared derived characters (synapomorphies) of the Sirenia and Proboscidea (Table 3.1).

PAENUNGULATA AND TETHYTHERIA

Simpson (1945) provided a milestone in the classification of mammals above the level of order. He formally designated a Superorder Paenungulata that included the Orders Sirenia, Proboscidea and Hyracoidea (as well as four extinct orders). He did not elaborate on the morphological similarities among these groups, but built on the findings of earlier workers (e.g. Gill 1872; Gregory 1910). The affinities of these three orders subsequently were the subject of numerous analyses that continue to the present.

The clearest and least debated of the proposed relationships based on morphology and palaeontology has been the link between the proboscideans and the sirenians. McKenna (1975) reflected this in a major rearrangement of mammalian classification. He joined the Orders Sirenia and Proboscidea with the extinct Order Desmostylia under the Tethytheria (a new mirorder,[1] so named because early members were thought to inhabit the coasts of the ancient Tethys Sea). Desmostylians were amphibious marine herbivores that occurred along the North Pacific coasts during the Oligocene–Miocene (Domning et al. 1986; Domning 2001a; see Table 3.2 for definitions of geological time divisions). They were contemporary with sirenians (see below), perhaps like ground sloths in appearance and terrestrial locomotion, but swam like polar bears (Domning 2002). Desmostylians most likely rested on land and fed on plants in intertidal and subtidal zones, as well as on freshwater and estuarine vegetation (Domning et al. 1986; Clementz et al. 2003b). Simpson (1945) and others considered desmostylians to be sirenians, but Reinhart (1953, 1959) and later investigators recognised them as distinct but closely related. Also in contrast to Simpson's (1945) classification, McKenna's (1975) scheme aligned hyraxes more closely with perissodactyls (odd-toed, hoofed mammals like horses and rhinoceroses) and extinct condylarths than with the Tethytheria. McKenna and Bell (1997) more radically refined this arrangement. They realigned the

Table 3.2 *The geologic time scale (modified after US Geological Survey Geologic Names Committee 2007) as pertinent to the evolution of the Sirenia.*

Era	Period	Epoch	Approximate age of boundary	Major events impacting the Sirenia
Cenozoic	Quaternary	Holocene	11.5 ka	Human hunting, boating, accidental netting, habitat disruption, climate change (Chapter 7)
		Pleistocene	1.8 Ma	Glaciation and sea-level changes
	Tertiary	Pliocene	5.3 Ma	Closure of the Central American Seaway (mid-Pliocene), continued Andean orogeny
		Miocene	23.0 Ma	Uplifting of the Andes mountains in South America (late Miocene)
		Oligocene	33.9 Ma	Central American Seaway remains open, some cooling as global oceanic circulation patterns expand
		Eocene	55.6 Ma	Most continents separated, but major Tethyan Seaway region remains present. Shorelines of this tropical area become the site of origin for the Sirenia, with early sirenians already dispersed to the Caribbean by mid-epoch. Higher seagrass diversity. Tropical conditions exist at high latitudes.
		Paleocene	65.5 Ma	Appearance of many small forms of terrestrial mammals
Mesozoic	Cretaceous		145.5 Ma	Spreading of flowering plants. Dinosaurs flourish, then become extinct at the end of the period. Continued breakup and drifting of continents
	Jurassic		199.6 Ma	Breaking up of Pangaea begins. 'Age of Dinosaurs'
	Triassic		251.0 Ma	The world continent of Pangaea is the main landmass, engulfing the large Tethys Sea
Paleozoic			542 Ma	Beginning of high diversity of multi-celled organisms, formation of the earliest fossils such as trilobites. Mass extinctions at the end of the era. Closed with formation of Pangaea.

ka = 10^3 years before present; Ma = 10^6 years before present.

hyraxes somewhat closer to sirenians than to perissodactyls, and proposed a new Order Uranotheria to embrace three suborders: the Tethytheria, Hyracoidea and the extinct Embrithopoda (large, mainly Oligocene herbivores). They divided the Tethytheria into the two infraorders,[2] Sirenia and Behemota; the Behemota were further subdivided into the Proboscidea and Desmostylia (two parvorders).

McKenna (1975) and McKenna and Bell (1997) did not specify morphological details for the placement of the Sirenia. However, Domning et al. (1986) provided data and a cladistic analysis in support of the Tethytheria concept.[3] They defined and measured 92 morphological characters of known fossil sirenians, proboscideans and desmostylians to create a cladogram representing the hypothesised relationships. Domning et al. (1986) interpreted the results as demonstrating an early branching of the Sirenia. They concluded that the Proboscidea and Desmostylia shared a more recent common ancestor than either did with the Sirenia, although this was debated (e.g. Tassy and Shoshani 1988). A new phylogenetic analysis by Gheerbrant (2009) supports a Sirenia–Desmostylia clade as a sister group to the Proboscidea within the Tethytheria. The morphological evidence linking the Hyracoidea with the Sirenia and Proboscidea has been more difficult to interpret. In contrast to the long-recurring hypothesis that the hyraxes are an earlier group allied with these other two orders (e.g. Gregory 1910; Simpson 1945; Novacek et al. 1988), some experts (e.g. Fischer 1986; Prothero et al. 1988) suggested that hyraxes were indeed more closely allied to perissodactyls. More recent morphological analyses have weakened support for this hypothesis, and again more closely align the Hyracoidea with the Tethytheria (Shoshani 1986; see also Novacek et al. 1988), consistent with the Paenungulata concept of Simpson (1945).

A thoughtful synthesis of some of the above classification schemes for the paenungulates and related taxa was proposed by Gheerbrant et al. (2005). This scheme recognises the group Altungulata (ranked as a mirorder by McKenna and Bell 1997) to include both the Orders Perissodactyla and the Paenungulata. Within the Paenungulata, Gheerbrant et al. (2005) defined one major division as the Order Hyracoidea and another as the Tethytheria. The tethytheres included the amphibious fossil group Anthracobunia (unspecified rank) and the Order Desmostylia, Order Embrithopoda, Order Proboscidea and Order Sirenia. This proposed classification (Table 3.3) has appeal, but we encourage consulting the literature for the latest thinking on this topic. Indeed, modern genetic analyses offer additional hypotheses about sirenian affinities as outlined below.

Table 3.3. *Position of the Sirenia and associated orders within the morphologically based hierarchical classification of the mammals according to Gheerbrant* et al. *(2005).
The Anthracobunia are omitted. Approximately equivalent formalised rankings given by McKenna and Bell (1997) are also listed. See text and associated references for other arrangements of these and associated groups.*

Rank in classification	Group (Gheerbrant et al. (2005))	Ranking (McKenna and Bell (1997))
Class	Mammalia	
Subclass rank unspecified	Altungulata	Mirorder Altungulata
Supraordinal rank unspecified	Paenungulata	Order Uranotheria
Order	Hyracoidea	Suborder Hyracoidea within order Uranotheria
Supraordinal rank unspecified	Tethytheria	Suborder Tethytheria
Order	Desmostylia	Parvorder Desmostylia within Infraorder Behemota within Suborder Tethytheria
Order	Embrithopoda	Suborder Embrithopoda within Order Uranotheria
Order	Sirenia	Infraorder Sirenia within Suborder Tethytheria
Order	Proboscidea	Parvorder Proboscidea within Infraorder Behemota within Suborder Tethytheria

AFFINITIES OF THE SIRENIA BASED ON MOLECULAR AND GENETIC ANALYSES: THE AFROTHERIA

Molecular genetic analysis places the Sirenia as part of a superficially dissimilar array of living mammals with deep biogeographic roots in the ancient Afro-Arabian landmass: Superorder Afrotheria. This assemblage is a seemingly odd assortment: sirenians, golden moles, the aardvark (*Orycteropus afer*), sengis (also called elephant shrews), tenrecs, hyraxes and elephants (Figure 3.1). This molecularly based hypothesis initially raised much debate (e.g. Novacek 2001). There are gaps and contradictions in the fossil record inconsistent with the affinities of the Afrotheria indicated by the molecular data. These putative relationships as they pertain to the Sirenia are discussed from a palaeontological perspective by Gheerbrant et al. (2005). Several viewpoints exist on molecularly based taxonomic arrangements within the Afrotheria, and we can only partially summarise the extensive literature on this topic here (see Supplementary Material

Figure 3.1. The Order Sirenia belongs to the Afrotheria, an ancient clade of molecularly defined but morphologically diverse mammals with origins on the African continent. In addition to the Sirenia, other modern representatives include: (a) elephants (Order Proboscidea); (b) golden moles (Order Afrosoricida, Family Chrysochloridae); (c) the aardvark (Order Tubulidentata); (d) tenrecs (Order Afrosoricida, Family Tenrecidae); (e) hyraxes (Order Hyracoidea); and (f) elephant shrews (Order Macroscelidia). Photos courtesy of: (a) Cynthia Moss, Amboseli Trust for Elephants; (b) Galen Rathbun, California Academy of Sciences; (c) Andrew Taylor; (d) © P. J. Stephenson; (e) Hendrik Hoeck; and (f) Galen Rathbun.

Appendix 3.1 for more details). Recent reviews should be consulted regarding these new molecular-based mammalian phylogenies (e.g. Springer and Murphy 2007; Tabuce *et al.* 2008), and new information is expected to arise. Nonetheless, the basic framework of the Afrotheria, however revolutionary it was in its genesis over a decade ago, now seems to be an acceptable group in which to place the Sirenia (Supplementary Material Appendix 3.1). Below we summarise some of the remarkable molecular work that has newly illuminated the affinities of sirenians and their allies.

Preliminary molecular and biochemical studies of sirenian affinities

Indirect biochemical and molecular information was applied to explore relationships among the Sirenia and other mammals, beginning largely in the 1980s, prior to the advent of direct genetic analyses. These earlier studies shed the first light on some of the relationships later revealed by genetics, but also reaffirmed some inferences based on morphology. De Jong et al. (1981) analysed amino acid sequences of an eye lens protein from living representatives of 15 orders of mammals, including the Amazonian manatee. They reaffirmed the likely monophyly of the paenungulates, and suggested that paenungulates were one of the earliest offshoots in the evolution of placental mammals. However, unlike morphological studies, results placed the aardvark (Order Tubulidentata) close to the sirenians, hyraxes and elephants. In concurrence with these results, the immunological distance assays of antisera to purified serum albumins by Rainey et al. (1984) indicated that the Sirenia, Proboscidea, Tubulidentata and Hyracoidea were likely a monophyletic group (with the hyrax showing greatest differences), which they referred to as 'an "Old African" unit among the placental mammals'. Miyamoto and Goodman's (1986) analysis of amino acid sequences from polypeptide proteins also suggested that the four orders were closely related and that the aardvark was a paenungulate. Kleinschmidt et al. (1986) analysed amino acid sequences in haemoglobin proteins of various mammals (not including aardvarks) and also concluded that the sirenians, proboscideans and hyraxes formed a monophyletic group that was among the earliest divisions of the placental mammals. They too placed sirenians and proboscideans closer to one another than to hyraxes.

Shoshani (1986) combined morphological and early biochemical approaches. The morphological analysis confirmed close affinities of the Sirenia, Desmostylia, Proboscidea and Hyracoidea along with the embrithopods and dinoceratans, compatible with Simpson's (1945) Paenungulata. Biochemical techniques verified the mutual affinities of the three paenungulate orders, but again linked these with the aardvark, a linkage that failed to emerge from morphological characters. Wyss et al. (1987) also utilised both morphological and amino acid sequence data and aligned the sirenians, hyracoids and proboscideans together. Biochemical evidence linking these three groups with aardvarks was weaker (Wyss et al. 1987).

Genetic and cytogenetic evidence of sirenian affinities within the Superorder Afrotheria

In addition to the Sirenia, there are five other orders of living mammals in the molecularly based Superorder Afrotheria (Figure 3.1). Order Afrosoricida (Stanhope *et al.* 1998a) includes the tenrecs (Family Tenrecidae) of Madagascar and central Africa and about 21 species of golden moles (Family Chrysochloridae) found in central and southern Africa (Bronner and Jenkins 2005). Golden moles are small, burrowing, insectivorous mammals that are not true moles, but convergent with them. The insectivorous and omnivorous tenrecs consist of about 30 species of diverse morphology and habitats, including the hedgehog tenrecs, rice tenrecs, shrew tenrecs and otter shrews. The sengis or elephant shrews (Order Macroscelidea) occur widely in Africa; there are about 16 species of these small- to medium-sized (up to 700 g; Rovero *et al.* 2008), cursorial, long-snouted, insectivorous mammals. The order Tubulidentata has one living species, the large (40–60 kg) termite- and ant-eating aardvark of Africa. The Order Hyracoidea comprises four species of hyraxes, herbivores ranging in size from 1.8 kg to 4.5 kg (Jones 1978; Olds and Shoshani 1982) of Africa and Arabia. The three species of elephants from Africa and Asia form the Order Proboscidea.

Extensive genetic analysis provides the support for the unification of this diverse assemblage under the Afrotheria. This evidence has mounted quickly during the last 10–15 years, and encompasses a wide foundation of molecular and cytogenetic approaches (but based only on sampling extant members). The Superorder Afrotheria was formally proposed for this clade by Stanhope *et al.* (1998a,b) based on a major analysis involving complete sequences of four mitochondrial genes and additional sequences from four nuclear genes, sampled from many representatives of the above families (including the dugong and the West Indian manatee) and other mammals. The alignments of the Afrotheria were clear and well supported based on all genetic data sets, separately and combined. Within the Afrotheria, the Sirenia aligned closely with elephants and hyraxes in all analyses (although relative proximity varied among genes), affirming the relationships of the paenungulates hypothesised by earlier morphological work. The Tubulidentata aligned near the paenungulates, but closer to other forms, a finding repeated in most subsequent molecular analyses (Supplementary Material Appendix 3.1).

The designation of the Afrotheria was revolutionary because it used purely molecular genetic data from living forms to designate higher-level mammalian taxonomic relationships without invoking morphology or palaeontology. Other research has attempted to define new morphological similarities among afrotherians (e.g. Asher *et al.* 2009), such as delayed appearance of permanent dentition (Asher and Lehmann 2008) and placentation (Carter *et al.* 2004). However, the number of molecular genetic studies that support the validity of the Afrotheria is much more impressive (Supplementary Material Appendix 3.1). In general, all affirm the Afrotherian lineage, a paenungulate grouping within the Afrotheria and the placement of the Sirenia within the paenungulates. The greatest area of disagreement, as in the morphological research, involves the alignment of the hyraxes with other paenungulates.

The novelty of the purely molecular-based classification of the Afrotheria prompted evaluation of the relative weight of morphological and genetic data in defining the group. Asher *et al.* (2003) and Springer *et al.* (2007) concluded that molecular data were superior to morphological and fossil data in defining the Afrotheria and Paenungulata. However, within-paenungulate affinities (such as the closeness of sirenians to either hyraxes or elephants) varied with methodology. Seiffert (2007) analysed afrotherian phylogeny combining morphology and genetic information and grouped the Sirenia and Proboscidea together under the Tethytheria, and joined tethytheres with the Hyracoidea to form the Paenungulata, but left open questions about the strength of the potential monophyly of various pairs of relationships among the Sirenia, Proboscidea and Hyracoidea. Other molecular genetic investigations also show varying strengths of relationships between pairs within these three paenungulate groups, and consider them major ambiguities in the molecular classification of the Afrotheria (Murphy *et al.* 2004; Asher 2007; see Supplementary Material Appendix 3.1). The relationships and degree of divergence of the hyraxes with other paenungulates based on molecular data remain enigmatic, as found by past morphological analyses.

Chromosome painting – which hybridises human chromosomes with those of other mammals and then maps loci in common – also shows strong support for the Afrotheria and the Paenungulata (Kellogg *et al.* 2007; Pardini *et al.* 2007). The chromosome painting technique reveals changes that are large and infrequent in comparison to changes in nucleotide sequences; in the parlance of molecular systematics these are referred to as 'rare genomic changes'. Reciprocal cross-species chromosome painting studies were carried out on Florida manatees, the rock hyrax (*Procavia capensis*) and the African elephant (*Loxodonta africana*), using aardvarks and humans as outgroups (Pardini *et al.* 2007). Based on these comparisons,

Pardini *et al.* (2007) were able to reconstruct a suggested ancestral paenungulate karyotype ($2n = 58$) prior to the divergence of the three orders. Using estimated chromosomal rates of evolution, they concluded that paenungulates were monophyletic within the Afrotheria (Pardini *et al.* 2007).

DIVERGENCE AND CHARACTERS OF THE EARLIEST SIRENIA AND THEIR CLOSEST RELATIVES

The timing of the radiation of ancient afrotherian progenitors into the first paenungulates is somewhat conjectural, with various models suggesting different times (e.g. Murphy *et al.* 2004; Springer *et al.* 2005; Springer and Murphy 2007; Wible *et al.* 2007). Some molecular clock analyses suggest that the paenungulates emerged about 62–65 million years ago (Springer *et al.* 2003; Murphy *et al.* 2004; Pardini *et al.* 2007), and that the Sirenia diverged perhaps 2–3 million years after the earliest paenungulates (Springer *et al.* 2005; Springer and Murphy 2007). This rapid diversification probably created some of the molecular ambiguity in defining whether sirenians are closer to hyraxes or elephants (Amrine and Springer 1999; Nishihara *et al.* 2005). Cytogenetic analyses suggest that the rate of chromosomal change was low within the likely short time periods (estimated at 4–7 million years) between the successive radiations of these three paenungulate lineages (Pardini *et al.* 2007). In the fossil record the earliest proboscideans and hyracoids are Early Eocene, whereas the first known desmostylians appear later, in the Early Oligocene (Gheerbrant *et al.* 2005). Morphological characteristics also tie the early sirenians, proboscideans and desmostylians to each other (Savage *et al.* 1994; see 'Affinities of the Sirenia based on molecular and genetic analyses: the Afrotheria' above).

The unity of the living dugongs and manatees was formally recognised by Illiger (1811), who, as noted above, first applied the name Sirenia to them. Unique morphological features of the Sirenia are discussed by Gheerbrant *et al.* (2005). Including the most primitive fossil forms, the derived characters shared among all members of the order include pachyosteosclerosis[4] (histological evidence for this extends to the Eocene; de Buffrénil *et al.* 2008), premaxillary–frontal bone contact, an enlarged rostrum and mesorostral fossa, lack of paranasal sinuses, and the long and deep mandibular symphysis with parasagittally aligned incisor–canine toothrows (Domning 2001a). Early sirenians had five premolars, unlike the early ungulates, which possessed four (e.g. Domning *et al.* 1982). The significance of the presence of five premolars has been enigmatic in relation to hypothesised relationships of the early sirenians with other mammals (McKenna 1975; Savage *et al.* 1994).

The first fossil sirenians are from the latest Early Eocene or earliest Middle Eocene, about 50 million years ago (Savage *et al.* 1994; Table 3.2). It has been speculated that the evolution of sirenians from fully terrestrial mammals took place in swampy or marshy habitats and involved foraging on floating or emergent aquatic plants (Savage *et al.* 1994). The Eocene shores of the Tethys Seaway were the likely centre of origin. For many millions of years the Tethys Ocean was a large body of water bordering the contiguous landmasses of Laurasia and Gondwana. As the continents formed and drifted the ancient Tethys lost its early shape, but its remnant form in the Eocene can be considered to be a long band of continuous water from what is now the north-western Indian Ocean through the Mediterranean Sea, with open access to the Caribbean (Figure 3.2). It is important to note that during the Eocene the climate was much warmer than today, with tropical biotas at latitudes now at the Arctic Circle. Oceanic circulation patterns were very different, with most currents travelling around the equator rather than across latitudes as in the present (Ivany *et al.* 2003). The Eocene Tethys Seaway, as the centre of origin, set the stage for the early distribution of the Sirenia. As noted by Domning *et al.* (1982, pp. 60, 64),

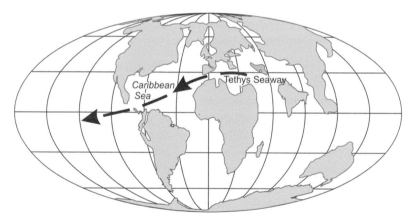

Figure 3.2. Approximate positions of the major landmasses in the Middle Eocene. The Tethys Seaway (along the shores of which the Afrotheria and early sirenians arose at an earlier time) was open at both ends, allowing dispersal of sirenians to the Caribbean via warm circumtropical currents (arrows). The Caribbean was also closer to the Tethys Sea in the Early Eocene than later in Earth's history, and warm waters extended above latitudes of the landmasses now known as Europe. Map supplied by BP Archive, redrawn by Adella Edwards from Smith *et al.* (1994), with permission.

from almost the beginning of their recorded history, the sea cows have had what can justly be called a 'pan-Tethyan' distribution, being found throughout the length of the Tethyan Seaway, which formed the heart of the Paleogene marine tropics ... the Paleocene–Eocene continuity of the Caribbean and Old World Tethyan marine tropics ... would lead us to expect a relatively homogeneous sirenian fauna.

THE FOSSIL SIRENIA

In this section we discuss the fossil history of the Sirenia, including their classification, evolutionary relationships, palaeogeography and inferences about their ecology. More detailed information on this topic can be found in works by Simpson (1932), Reinhart (1959) and Domning (1978a). In particular, the exhaustive bibliography and index prepared by Domning (1996, 2010) and its classification and synonymy should be consulted as a key to past works and their interpretation. Much of the contemporary understanding of the fossil sirenians is based on Daryl Domning's labours, and we draw heavily on his work for this overview.

Over 30 years ago, Domning (1978a) noted that,

> The Sirenia are a morphologically close-knit group, conservative in their evolution, limited in their diversity, subtle in their differences. The narrowness of their adaptive zone has made their history at once simple and complex: simple by its relative lack of opportunities for adaptive radiation, but complex by the parallelism it has induced in their adaptations.

Over the past few decades the mounting sirenian fossil record has revealed more ecological differences and a higher diversity of radiations that now warrant greater appreciation. New species and genera of sirenians are being discovered nearly every year (Domning 2001a). Fossil sirenians appear on every continent except Antarctica (Gheerbrant *et al.* 2005). However, there is a paucity of fossil records for large areas of the world of importance to modern sirenians, including the Indian Ocean and South Pacific, home to the dugong (see Chapters 1 and 8) and the centre of modern seagrass diversity (Bajpai and Domning 1997; Domning 2001a). Thus a complete account of sirenian fossil history will remain unfulfilled for many years.

The extinct sirenians are grouped into four families: the Prorastomidae, the Protosirenidae, the Dugongidae and the Trichechidae (Table 3.4; Figure 3.3). These four families have been generally accepted for decades (some since the 1800s; Simpson 1945; Domning 1996). Relationships of the various species of sirenians to one another have been proposed and updated based on cladistic

Table 3.4 *The families of fossil Sirenia. References cited emphasise more recent literature and reviews. Domning (1996) should be consulted for a more comprehensive listing of additional sources. See Supplementary Material Appendix 3.2 for listing of fossil species within families and Table 3.2 for details of the geologic time scale.*

Family	Time of occurrence	Distribution	References
Prorastomidae	Early Middle to Late Eocene	West Indies, North America	Domning (2001b,c); Gheerbrant *et al.* (2005); Savage *et al.* (1994)
Protosirenidae	Early Middle to Late Eocene	Western Atlantic, Mediterranean, Indian Ocean regions	Domning (2001a); Domning and Gingerich (1994); Gheerbrant *et al.* (2005)
Dugongidae	Middle Eocene to Recent	Mediterranean, Europe, North Africa, western Atlantic–Caribbean, Indian and Pacific Oceans	Domning (1994, 2001a,b); Gheerbrant *et al.* (2005)
Trichechidae	Oligocene to Recent	Europe, South America, western Atlantic–Caribbean, North America, west Africa	Domning (1982a, 1994, 2001a,b); Gheerbrant *et al.* (2005)

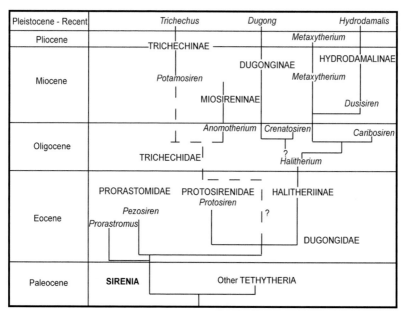

Figure 3.3. Evolutionary tree for the families and subfamilies of sirenians. Modified from Gheerbrant *et al.* (2005), with permission.

analysis of morphological features with resulting phylograms to describe likely relationships (Domning 1994, 2001a). The morphological analyses are generally congruent with stratigraphic findings. These proposed relationships continue to be refined and updated as new discoveries are made (e.g. Bajpai and Domning 1997; Domning and Aguilera 2008), and further taxonomic revisions are to be expected.

The Prorastomidae and *Pezosiren*: the quadrupedal sirenians

The Family Prorastomidae (Cope 1889) includes the most primitive sirenians and was first known from Jamaica; thus by the start of the Middle Eocene sirenians had dispersed from the Eurasian–African margins of the former Tethys Sea to the Caribbean (Gheerbrant *et al.* 2005). For nearly 150 years only one named species of prorastomid was known to science, *Prorastomus sirenoides* (Owen 1855; see Supplementary Material Appendix Table 3.2.1). It was presumably quadrupedal and amphibious, judging by its close relative *Pezosiren* (see below). Based on its narrow and undeflected rostrum, *P. sirenoides* was probably a selective browser on floating and emergent aquatic plants and may have fed on seagrasses only to a minor degree (Savage *et al.* 1994; Domning 2001b). These creatures probably used fluviatile or estuarine habitats rather than marine or terrestrial ecosystems (Savage *et al.* 1994). *P. sirenoides* is the oldest and most primitive known sirenian, but was probably not an ancestor to other known members of the order (Savage *et al.* 1994). Prorastomids occurred in the West Indies and North America from late Early or early Middle to Late Eocene (Gheerbrant *et al.* 2005). There are other unnamed Eocene prorastomids from this region, and possibly from Israel (Goodwin *et al.* 1998). *P. sirenoides* is known only from fragmentary skeletal elements, and one skull and mandible (Savage *et al.* 1994).

The lack of more complete specimens of prorastomids changed during the 1990s. Fossil remains found in early Middle Eocene (about 50 million years ago) marine deposits at Seven Rivers, Jamaica, were crowned 'one of the most marked examples of morphological evolution in the vertebrate fossil record' (Domning 2001c, p. 625). This was a true missing link in the transition from terrestrial ancestors of the Sirenia to the subsequent fully aquatic forms. Slightly younger than *P. sirenoides*, the new find was named *Pezosiren portelli* by Domning (2001c) (*pezos* is Greek for walking; Roger Portell discovered the site). This beast was about the size of a pig (slightly over 2 m long), had four legs and was fully capable of terrestrial locomotion (Figure 3.4). The skeleton of *P. portelli* is well represented (only lacking the atlas, bones of the feet, caudal vertebrae and a few other bones). The pelvic girdle was long and narrow but well developed, and limb bones were similar

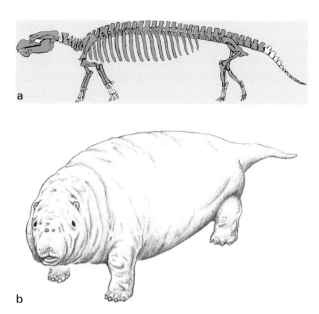

Figure 3.4. (a) The articulated fossil skeleton of *Pezosiren portelli*, an early Middle Eocene sirenian in the Family Prorastomidae (missing bones in white). This sirenian was fully quadrupedal and amphibious (illustration from Domning (2001c), with permission). (b) Artist's model of possible appearance of *P. portelli*. Redrawn by Gareth Wild from image at locolobo.org.

to those of primitive ungulates. The tail vertebrae were unlike those of modern sirenians, in which propulsion is by up-and-down motion of the fluke. Instead, their structure indicates that aquatic locomotion was through spinal undulation coupled with paddling by backward and upward thrusts of the hind limbs, similar to otters. The thoracic vertebrae provided attachments for strong neck ligaments to support the head out of water. The skull was decidedly sirenian (see Domning 2001c for osteological details). Few teeth were recovered, but the sockets indicated a dentition comparable to those of other Eocene sirenians. The ribs were pachyosteosclerotic to provide ballast, as is typical for sirenians as far back as this epoch (de Buffrénil *et al.* 2008). *P. portelli* likely spent much of its life in the water feeding on aquatic vegetation. Its phylogenetic relationship to *P. sirenoides* is unclear (Domning 2001c).

The Protosirenidae
The Family Protosirenidae (see Supplementary Material Appendix Table 3.2.1) is known from Middle to Late Eocene sites bordering the western Atlantic Ocean, Mediterranean Sea and Indian Ocean (Domning *et al.*

1982; Domning 2001a; Gheerbrant *et al.* 2005; Bajpai *et al.* 2006, 2009). The Protosirenidae was given family rank by Sickenberg (1934), a decision subsequently affirmed by Simpson (1945), Domning (1994, 1996) and others. The protosirenids were mainly aquatic, with short forelimbs and reduced hind limbs indicating an amphibious life with restricted terrestrial locomotion, analogous to modern pinnipeds (Domning and Gingerich 1994). Protosirenids begin to show typical sirenian cranial traits such as a down-turned rostrum indicating bottom feeding, and a broadened mandibular symphysis indicating underwater 'grazing' (Domning and Gingerich 1994; Domning 2001a). One species (*Protosiren fraasi*) was described over 100 years ago as from the early Middle Eocene of Egypt. Another from Egypt, *P. smithae*, is a slightly younger (late Middle Eocene) and larger form than *P. fraasi* (Domning and Gingerich 1994). Likely a descendant of *P. fraasi*, it retained hind limbs and had larger incisor tusks and slightly greater rostral deflection. Isotope composition of teeth of *P. smithae* and other, unnamed species of *Protosiren* are consistent with a diet of seagrasses (MacFadden *et al.* 2004; Clementz *et al.* 2006, 2009).

The protosirenid *Ashokia antiqua* was recently described from the Middle Eocene of India based on cranial material (Bajpai *et al.* 2009). Additional species of *Protosiren* are reportedly distinctive (see Supplementary Material Appendix 3.2.1), but represented only by postcranial elements: *P. eothene* from the early Middle Eocene of Pakistan may be the oldest and smallest protosirenid (Zalmout *et al.* 2003); *P. sattaensis* from the Middle Eocene of central Pakistan is intermediate between *P. fraasi* and *P. smithae* (Gingerich *et al.* 1995). Specimens from additional unnamed species of *Protosiren* are known from Eocene deposits in India (Bajpai *et al.* 2006), Florida and North Carolina (Domning *et al.* 1982). According to Domning and Gingerich (1994) and Domning (2001a), the forms from the south-eastern United States likely merit new generic designations.

The Dugongidae

Dugongids appeared in the Middle Eocene of the Mediterranean and became widespread by the Early Oligocene, by which time the prorastomids and protosirenids had disappeared (Domning 2001a; Gheerbrant *et al.* 2005). Early dugongids show progressive reduction in the bones of the pelvic girdle and hindlimb (internal and non-functional) in comparison with prorastomids and protosirenids (Domning 2000). The extinct dugongids constituted by far the most diverse group of sirenians (in contrast to today, with the dugong the only remaining species). The multiple fossil forms are now placed into three subfamilies: the Halitheriinae,

Dugonginae and Hydrodamalinae (see Supplementary Material Appendix
Tables 3.2.2 and 3.2.3; Figure 3.3).

Halitheriinae
This subfamily is paraphyletic and probably includes the conservative
dugongid stem (Domning 1978a, 2001a). The earliest dugongids were
Middle and Late Eocene halitheriines in genera such as *Eotheroides*,
Eosiren and *Prototherium* (see Supplementary Material Appendix
Table 3.2.2). Fossils from these three genera are primarily from European
and North African areas of the former Tethys Sea (*Eotheroides* is also known
from Madagascar; Samonds *et al.* 2009). *Eotheroides aegyptiacum* is the most
primitive and is likely a sister group to other dugongids; *Eosiren* may have
been euryhaline rather than exclusively marine (Domning 1994; Domning
et al. 1994). Like other early sirenians, halitheriines had five premolars
(Domning *et al.* 1982), and probably were the earliest to lose functional
hind limbs (Domning 2001a). Most members of the Halitheriinae had
small, subconical tusks and probably specialised on seagrass leaves and
small rhizomes, also reflected in the isotope values of tooth enamel
(Clementz *et al.* 2009). The early dugongids from the Eocene need taxonomic revision (Domning *et al.* 1994; Domning 1994, 2001a).
   Fossil halitheriines have their longest geologic record in Europe.
Halitherium schinzii probably branched from *Eosiren* in the Late Eocene
(Domning *et al.* 1994). Another closely related species, *Halitherium taulan-
nense*, is known from the Late Eocene of France (Sagne 2001a). The sub-
family includes several species in the genus *Metaxytherium* (probably
derived from *Halitherium*), widely distributed in the Miocene and
Pliocene on both sides of the Atlantic, and also reaching the Pacific
(Domning 2001b; see Supplementary Material Appendix Table 3.2.2).
Metaxytherium likely arose in the North Atlantic, and is known from the
Late Oligocene of the south-eastern United States (Domning 2001a). Most
members of the *Halitherium–Metaxytherium* lineage had strongly down-
turned snouts and small- to mid-sized tusks, and were probably generalised
consumers of the leaves and rhizomes of smaller seagrasses. European
species of *Metaxytherium* may have stemmed from *Halitherium schinzii*
and *H. christolii* (Domning and Thomas 1987; Domning and Pervesler
2001) and appear to have been the only European sirenians to survive
from the Paleogene (the period of the Paleocene–Oligocene epochs) to the
Neogene (the Miocene and thereafter). The species of *Metaxytherium* in
Europe show a stratigraphically and cladistically validated lineage from
Metaxytherium krahuletzi to *M. medium* (which was very similar to

M. floridanum), to *M. serresii*, and finally to *M. subapenninum*. These European *Metaxytherium* showed little morphological change from the Late Oligocene to the Late Miocene (Domning and Thomas 1987), and thus the lineage was characterised by Domning (2001b) as the only example of lengthy evolutionary stasis in the sirenian fossil record. This group tended to increase in body size during this period; the size increase in *Metaxytherium* correlates with cooling climatic conditions (favouring larger body mass because of thermal advantages, as occurred in *Hydrodamalis* in the North Pacific, see 'Hydrodamalinae' below; Bianucci *et al.* 2008). The earlier *M. krahuletzi* may have been about 3.25 m long, whereas the later *M. subapenninum* may have reached 4.0 m (Bianucci *et al.* 2008).

An intriguing reversal in body size towards dwarfism took place in Late Miocene and Early Pliocene Mediterranean *Metaxytherium*, and by the Late Pliocene the European forms had evolved much larger tusks (Domning and Thomas 1987; Bianucci *et al.* 2008). These changes were likely in response to the Messinian Salinity Crisis of the Late Miocene (5–6 million years ago), a time of repeated sea level, salinity and temperature fluctuations. During this period the Mediterranean Sea went at least partially dry and was probably broken into isolated basins with deteriorated habitat quality for sirenians (Bianucci *et al.* 2008). *M. serresii* of this time was about 2.3–2.4 m long, and its smaller size was likely due to isolation or nutritional stress brought on by shifting species composition of seagrass beds (Bianucci *et al.* 2008). As a result of the Messinian Salinity Crisis, a change in the dominant seagrass species from the large *Posidonia* to plants with more modest-sized rhizomes may have favoured the intermediate tusk size found in *M. serresii*. The eventual return of *Posidonia* to dominance in the Mediterranean afterwards may have led to evolution of larger tusks, as seen in *M. subapenninum* (Bianucci *et al.* 2008). Stable isotope data from tooth enamel corroborate the changing environmental conditions faced by halitheriines persisting on marine seagrasses in this region (Clementz *et al.* 2009).

In addition to European halitheriines, *Metaxytherium crataegense* was found in the south-eastern United States and the eastern North Pacific, where it apparently gave rise to the Hydrodamalinae (Domning 1978a, 2001b; see 'Hydrodamalinae' below), through the intermediate *Metaxytherium arctodites* (Aranda-Manteca *et al.* 1994). In the Caribbean region *M. crataegense* probably gave rise to *M. floridanum*, a species with small tusks and a strongly deflected rostrum (indicating bottom feeding), which died out by the end of the Miocene (Domning 1988, 2001b). Earlier, the Caribbean was also home to *Caribosiren turneri* of the Oligocene, a small halitheriine with a strongly down-turned snout (but apparently no tusks) that fed on seagrass leaves and perhaps smaller rhizomes (Domning 2001b).

Dugonginae

The Subfamily Dugonginae (see Supplementary Material Appendix Table 3.2.3) arose from Oligocene halitheriines of the south-eastern United States, and diversified in the Caribbean and western Atlantic (Domning 1997a, 2001b,c). They dispersed to other tropical waters of the world in the Late Oligocene and Early Miocene (Bajpai and Domning 1997; see the latter reference and Domning and Aguilera (2008) for cladistic analyses of the dugongines). Dugongines tended to evolve large blade-like tusks, presumably for digging rhizomes of larger seagrasses (Domning 2001a; Domning and Beatty 2007). A preference for rhizomes is also consistent with carbon isotope values in tooth enamel (Clementz *et al.* 2009). The most primitive known dugongine was *Crenatosiren olseni*, a small sea cow with medium-sized tusks that existed in the south-eastern United States during the Late Oligocene (Domning 1997a, 2001a). During the Early Miocene a basal clade comprising the smallest known dugongines probably arose from the *Crenatosiren* lineage. Its members have been grouped in the genus *Nanosiren*, found in the western Atlantic and Caribbean of North and South America and the eastern Pacific from Early Miocene to Early Pliocene times (Domning and Aguilera 2008). These little sea cows were about 2 m in length and 150 kg in body mass (perhaps smaller than Eocene prorastomids), and were probably at the lower limits of sirenian body size, constrained by the energetics of living in water (Domning and Aguilera 2008). Small size allowed access to very shallow water unavailable to larger contemporary halitheriines such as *Metaxytherium* or other co-existing dugongines such as *Corystosiren* and *Xenosiren* (see below). *Nanosiren* possessed small, conical tusks and had strongly deflected rostra, suggesting a diet of small seagrasses (Domning and Aguilera 2008).

Other dugongines included the highly derived *Corystosiren varguezi* and *Rytiodus capgrandi* with their large, flattened, blade-like, self-sharpening-tusks (Figure 3.5), known from the Early Miocene to Pliocene of south-eastern North America and the Caribbean (*Corystosiren*), France, Libya and possibly South America (*Rytiodus*) (Toledo and Domning 1991; Domning 2001b). *Corystosiren* had a massive, solid skull roof of unknown function (Domning 1990, 2001a). *Dioplotherium* is another dugongine genus in a sister group to the *Corystosiren–Rytiodus* clade. *Dioplotherium* included the more primitive *D. manigaulti* from the Early Miocene (Domning 1989a), and its derived descendant *D. allisoni*, known from the

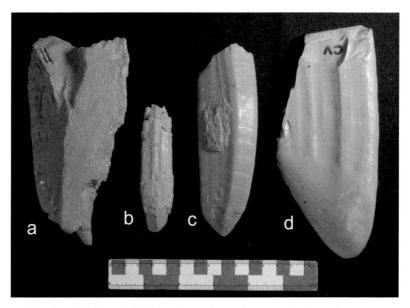

Figure 3.5. Replicas of fossil sirenian tusks used in seagrass-harvesting experiments by Domning and Beatty (2007). (a) *Metaxytherium floridanum*, left tusk in premaxilla, medial view. (b) *Crenatosiren olseni*, right tusk, medial view. (c) *Dioplotherium manigaulti*, left tusk, medial view. (d) *Corystosiren varguezi*, right tusk, lateral view showing broad wear surface, partly restored. The tusks are self-sharpening because of the variable thickness of the enamel on different sides of the tusk. Scale bar = 15 cm. Reproduced from Domning and Beatty (2007), with permission.

eastern North Pacific Ocean and possibly Brazil (Domning 1978a; Toledo and Domning 1991). The latter species had greater snout deflection and more slender, self-sharpening tusks than in *Corystosiren* and *Rytiodus*. It also had an anterior gape due to greater mandibular than rostral deflection, perhaps related to mechanical cutting of rhizomes. *D. allisoni* died out in the eastern Pacific before the Late Miocene as a result of changing environmental conditions and possible competition with desmostylians for bottom plants (Domning 1978a; see 'Sirenian diversity through time: ecology of co-existing species and serial replacements' below). *Xenosiren yucateca*, known from the Yucatan Peninsula of Mexico in the Late Miocene or Early Pliocene, may have been a direct descendant of *Dioplotherium* (Domning 1989b, 2001b), and probably dug forcefully with its large, self-sharpening tusks (Domning 2001b).

The *Dugong* lineage was unique in evolving root hypsodonty, with reduced but ever-growing, open-rooted cheek teeth with flat dentine surfaces. The fossil record of the modern *Dugong dugon* is unknown, but the skull of a very close relative (perhaps a congener) is known from the Late Pliocene of Florida (Domning 2001a). This fossil is the latest dugongine known from the western Atlantic Ocean, and suggests a subsequent dispersal to the Indo-Pacific that resulted in the modern dugong (Domning 2001a). More recently, phylogeographic analysis based on mitochondrial DNA (mtDNA) suggests that modern dugong populations were genetically structured by a geographic land barrier across the Torres Strait region during recent glacial maxima (McDonald 2005; D. Blair, personal communication 2010). A formerly single Australian dugong population became separated by this land bridge for much of the time between 115 000 and 7000 years ago (but with some intermittent periods of high water and suitable seagrass habitat). They have been free to mix again for the last 7000 years, but such mixing has not been complete and geographic population structuring remains evident (McDonald 2005; D. Blair, personal communication 2010).

Hydrodamalinae

The hydrodamalines (see Supplementary Material Appendix Table 3.2.3) undertook a decidedly different evolutionary path, leading to invasion of cool waters, gigantism and a turn away from feeding on bottom plants that ultimately led to a diet of surface kelps (Chapter 2). (It is likely that the kelps radiated before the Middle Miocene and probably originated much earlier; Domning 1989c.) Domning (1978a) interpreted the hydrodamalines as a single evolutionary lineage, but other evidence from Japan has led to a hypothesis of splitting within the hydrodamalines (Domning and Furusawa 1994; Domning 2001a). The earliest hydrodamaline, *Dusisiren reinharti* of the Early Miocene of Baja California, likely evolved from a *Metaxytherium* similar to *M. arctodites* (Domning 1978a; Domning and Furusawa 1994). *Dusisiren reinharti* then evolved into *D. jordani* of Middle to Late Miocene deposits in California, followed by *D. dewana* of the Late Miocene of Japan (Domning 1978a; Takahashi *et al.* 1986) (and possibly also *D. takasatensis* in Japan; Kobayashi *et al.* 1995). *D. dewana* has been described as a 'perfect morphological and chronological link' between *Dusisiren* and *Hydrodamalis* (Domning 2001a).

The *Hydrodamalis* lineage survived into the Pliocene, Pleistocene and Recent eras by adapting to cold water, exposed coastlines and feeding on plants close to the surface, especially kelps (see details in 'Sirenian diversity through time: ecology of co-existing species and serial replacements' below). It consists of *H. cuestae* of the Late Miocene to Late Pliocene of

California and Baja California; *H. spissa* of the Early Pliocene of Japan; and *H. gigas* of the Pleistocene to Recent, distributed from California across the Aleutian Islands to the Commander Islands off the Kamchatka Peninsula, and thence south to Japan (Domning and Furusawa 1994). *Hydrodamalis gigas* fed exclusively at or just below the surface and had no teeth (see Chapter 2).

The Trichechidae

Another branching within the Sirenia led to the mostly New World Family Trichechidae (Figure 3.3; see Supplementary Material Appendix Table 3.2.4), best known for the manatees. The nature of this branching remains uncertain. Domning (1982a) first hypothesised that trichechids may have been derived from protosirenids that reached South America in the Eocene by waif dispersal from the Tethyan area. This hypothesis has been challenged by his more recent cladistic analysis of fossil morphology (Domning 1994), indicating that trichechids evolved from a Late Eocene halitheriine dugongid. However, an analysis by Sagne (2001b) supports the earlier hypothesis of trichechid descent from protosirenids. In either case, the branching that resulted in the Trichechidae probably took place in the Late Eocene or Early Oligocene, leading to two subfamilies, the Miosireninae and the Trichechinae (Domning 1994, 1996; Figure 3.3).

Miosireninae

The miosirenines may represent a peculiar adventure in the evolutionary history of the Sirenia. *Anomotherium langewieschei* of the Late Oligocene and *Miosiren kocki* of the Early Miocene were North Atlantic forms known from European fossil beds. These northern species had thickened palates, thought to be adaptations for feeding on shellfish (Sickenberg 1934; Savage 1976; Domning 2001a), although the likely co-occurrence of sea-grasses in the habitat of *A. langewieschei* has also led to the suggestion that it was primarily a seagrass feeder (Diedrich 2008).

Trichechinae

The earliest recognised trichechine was the Miocene *Potamosiren magdale-nensis* from Colombia. It lacked the indeterminate replacement of cheek teeth characteristic of later trichechines (Domning 1982a, 1997b; Domning and Hayek 1984). Other notable traits of all living trichechids include six cervical vertebrae, an elongated acromion process of the scapula and an inflated zygomatic process of the squamosal bone. Fossil *Potamosiren* bones that could reveal these traits have not been unearthed, allowing a counterargument that *P. magdalenensis* might be an unspecialised offshoot in another lineage (e.g. the Protosirenidae) rather than a trichechid (Domning 1982a). *Potamosiren* was followed in time by *Ribodon limbatus* of Late Miocene (5–6 million years old) deposits in Argentina. *Ribodon* had the new pattern of tooth replacement that played a major role in the subsequent evolution and distribution of the Sirenia. *Ribodon limbatus* fossils clearly show continuously erupting, supernumerary cheek teeth, although teeth were larger with distinctly different, less lophodont cusp patterns than in *Trichechus*, and with fewer teeth present at any one time (Domning 1982a). Fossil *Ribodon* sp. also occurred later in south-eastern North America (probable Pliocene deposits; Domning 1982a).

Specimens assigned to *Trichechus* appeared during the Pliocene–Pleistocene transition in the western Amazon region (Domning 1982a). They were similar to North American manatees of the Pleistocene, and more closely resemble modern *T. manatus* than *T. inunguis*. Pleistocene fossils referable to the West Indian manatee appeared throughout Florida and elsewhere in the south-eastern United States (Domning 1982a, 2005; see 'Sirenian diversity through time: ecology of co-existing species and serial replacements' below). Quantitative analysis of divergence of cranial traits suggests that all three living species of manatees probably separated from one another over a short interval at about the same time (Domning and Hayek 1986). No subspecies of living Amazonian and West African manatees can be distinguished, whereas morphometric data segregate living West Indian manatees into two subspecies: the Florida manatee and the Antillean manatee (Domning and Hayek 1986; see below regarding genetically based interpretations of evolution in *Trichechus*). A fossil subspecies (*T. manatus bakerorum*) is also recognised (see 'Sirenian diversity through time: ecology of co-existing species and serial replacements' below).

The evolution of the tooth replacement pattern in trichechids was accompanied by an increasing complexity of tooth crown patterns. The appearance of these features is analogous to the evolution of high-crowned-cheek teeth in plains ungulates that feed on grasses with a high silica content, suggesting that a radical change in abrasiveness of the diet must have begun in South American sirenians at about the time *Ribodon* appeared in the Late Miocene (Domning 1982a). Uplifting of the Andes Mountains occurred in the Miocene through Late Pliocene–Early Pleistocene, with a concomitant runoff of water, nutrients and silt. This fertile runoff likely favoured a proliferation of freshwater grasses and other aquatic plants. In addition to eating seagrasses in coastal areas today, modern manatees feed on a wide variety of aquatic plants (Chapter 4), including many true grasses (Poaceae, formerly known as Gramineae) that have siliceous phytoliths that produce heavy wear on teeth. The Amazonian manatee in particular feeds heavily on true grasses that form large floating meadows (Chapter 4). Domning (1982a) argued that terrestrial grasses probably appeared and colonised freshwater aquatic systems in South America in the Miocene, including the Magdalena, Orinoco and La Plata rivers and areas along their coastal outflows to the Caribbean and Atlantic. Early trichechids, including *Ribodon*, occurred in these coastal and fluvial areas, and it is likely that these populations first evolved the unique trichechid pattern of tooth replacement in response to the dietary availability of true grasses.

In the Middle Miocene (when *Potamosiren* was present) the Amazon Basin had an outlet to the Pacific Ocean, but drainage divides existed between the basin and the Atlantic and Caribbean. Domning (1982a) thought it likely that coastal manatees entered the Amazon Basin through the outlet to the Pacific Ocean or through intermittent connections with Atlantic or Caribbean drainages. Due to further Andean uplift, by the Early Pliocene the Pacific outlet had closed, forming a large enclosed Amazon Basin in which *Trichechus* or similar forms were isolated from coastal populations. During the Pliocene these isolated manatees further diverged adaptively for feeding on freshwater grasses. This isolation ended in the Late Pliocene–Early Pleistocene, when the drainage barrier between the inland Amazon waters and the Atlantic Ocean was breached. Manatees within the Amazon Basin may have been in genetic isolation and evolved more rapidly in response to the proliferation of grasses and floating meadows than coastal manatees. This model allows allopatric speciation of *T. inunguis* from a *T. manatus*-like progenitor. Consistent with the allopatric speciation hypothesis, Domning (1982a; Domning and Hayek 1986) analysed the morphology of modern trichechids and concluded that whereas *T. manatus* and *T. senegalensis* are united by certain derived characters to the exclusion of Amazonian manatees, they are on the whole more primitive than *T. inunguis* (the latter species has other derived characters, in addition to smaller and more complex molars, a lack of nails on the flippers, reduced number of dorsal vertebrae, a thickened supraoccipital bone and a higher diploid chromosome number). According to this scenario, while *T. inunguis* was evolving in the Amazon Basin, Pliocene–Pleistocene populations of coastal manatees expanded their distribution through the Caribbean into coastal North America, where they remained little changed up until the present.

Somewhat different scenarios favouring the evolution of supernumerary molars in trichechids in response to flourishing grasses can also be envisioned. Domning (1982a) noted that supernumerary molars could have originated in the isolated freshwater Amazon Basin (rather than in coastal areas) in a *Potamosiren*-like progenitor. This scenario would require descendants to exit the basin and reach the coast in the Late Miocene (through re-established connections with the ocean) to form the *Ribodon* lineage that replaced coastal *Potamosiren*. Evolution towards a *T. manatus*-like form could have continued within the basin with an eventual second dispersal along the coasts later in the Pliocene, resulting in replacement of coastal *Ribodon* and widespread dispersal along the coasts of the Americas. *T. inunguis* then would have arisen with continued evolution within the basin. Yet another possibility could be that *Ribodon* gave rise to *T. inunguis* and *T. manatus* separately (Domning 1982a). Genetic evidence suggests that other scenarios for the origin of present-day manatees may eventually prove more viable (Garcia-Rodriguez *et al.* 1998; Cantanhede *et al.* 2005; Vianna *et al.* 2006; see 'Sirenian diversity through time: ecology of co-existing species and serial replacements' below). In any case, the evolution of indeterminate replacement of the cheek teeth in manatees provided a competitive advantage that likely sounded a death knell for New World dugongids (see 'Sirenian diversity through time: ecology of co-existing-species and serial replacements' below).

The three living species of manatees occupy different areas, as discussed in Chapters 1, 5 and 8. The euryhaline *T. manatus* shows little overlap in distribution with the more specialised freshwater *T. inunguis*. The two species are considered parapatric at the mouth of the Amazon (but genetic evidence suggests some contact and possible hybridisation; see 'Sirenian diversity through time: ecology of co-existing species and serial replacements' below); *T. inunguis* occurs only in the freshwater Amazon Basin, whereas *T. manatus* occupies coastal lagoons and larger inland waterways where it is a generalist feeder on both seagrasses and freshwater plants (Chapter 4). The West African manatee (*T. senegalensis*) shows very little morphological difference from *T. manatus* and occupies a similar range of habitat types. It is thought to be the result of Late Pliocene or Pleistocene dispersal of a *T. manatus*-like manatee from South America. Palaeoclimatic conditions may have created ocean currents more favourable to such dispersal from South America than exist today, and far-flung extended travels of individual *T. manatus* are known from modern times (Domning 1982a, 2005; Reid 1995). Much evidence shows the New World as the centre of origin for *Trichechus*, and morphologically (Domning and Hayek 1986) and genetically (Vianna *et al.* 2006) the West Indian and West African manatees are each other's closest relatives. The two species even share identical or closely related endoparasites (Sprent 1983; Beck and Forrester 1988). Molecular clock calculations yield an estimated time of divergence of these two species of about 309 000 years ago (Vianna *et al.* 2006).

The West Indian manatee has the best-known Pleistocene and Recent fossil history and biogeography of any sirenian. All Pleistocene manatees from the United States examined thus far are largely indistinguishable from West Indian manatees, and most closely resemble the Antillean subspecies. However, about 125 000 years ago (Late Pleistocene) a morphologically differentiated subspecies, *T. manatus bakerorum*, occurred from Florida to North Carolina (Domning 2005). This subspecies existed during an inter-glacial period with a climate at least as warm as today (Domning 2005). Pleistocene changes in climate and sea level, and associated factors, probably resulted in a series of evolutionary tendencies within the West Indian manatee at the northern margins of its distribution. In modern times the Florida subspecies seems to be geographically separated (but not completely isolated, see Chapter 8) from other populations to the south by the deep Straits of Florida with its strong currents. To the north and west, the Florida manatee is separated from Antillean manatees of Mexico–south Texas by cool winters on the northern Gulf of Mexico coast (Domning and Hayek 1986). The northern extent of the range of the Florida subspecies is limited in winter to the southern part of the Florida peninsula unless other warm-water sources are available (Moore 1951a; Deutsch *et al.* 2003; Domning 2005).

Fossil manatees can be found much farther north, providing evidence that during the Pleistocene and Holocene, water temperatures showed alternating warmer and cooler periods. Domning (2005) hypothesised the following scenarios to help explain the occurrence of morphologically sep-arable fossils of West Indian manatees in relation to changing climates. (1) Distributions shifted northward when climate warmed, and endemic isolates such as the present *T. manatus latirostris*, otherwise limited to warm winter ranges in the lower peninsula of Florida, would mix with more tropical forms such as the Antillean manatee. During warm times manatees also travelled up the Mississippi River on the northern Gulf Coast, as evidenced by fossil *T. manatus* in Arkansas and Ohio (Williams and Domning 2004). (2) During exceptionally cool periods their warm, lower Florida peninsula refugia would likely disappear altogether (see below). (3) Subsequent warming would allow reinvasion of Florida by Antillean forms. The ability of the *T. manatus bakerorum* subspecies to maintain its integrity as far north as North Carolina during the warm interglacial period may have been due to persistence of the cool water barrier in the northern

Gulf of Mexico, even while warmer waters extended as far as North Carolina on the Atlantic coast, consistent with paleoceanographic evidence (Domning 2005).

As a curious footnote to the documented Pleistocene history of *T. manatus* in North America, there is a persistent erroneous claim about the indigenous nature of Florida manatees that sometimes appears in the popular press and in statements by members of special interest groups. This is the claim that manatees in modern Florida are an exotic species introduced by people. This notion is contrary to all the fossil, biogeographic and molecular genetic evidence in the scientific record, but is sometimes used as an argument to counter regulations for manatee conservation. The idea is simply not true.

Molecular genetics suggests additional complexities in manatee evolution

Genetic studies affirm some of the above scenarios concerning the evolutionary biogeography of *Trichechus*, but also suggest new possibilities and reinterpretations. The characterisation and distribution of 16 haplotypes based on 410 base pair fragments of the mtDNA control region (D-loop) gene were carried out on 86 samples of West Indian manatees from multiple areas within the species' distribution, ranging from Florida to Brazil (Garcia-Rodriguez *et al.* 1998). Sixteen individual *T. inunguis* from the Amazon Basin were also included. No polymorphisms existed in West Indian manatee populations in Florida (23 individuals sampled), Mexico or Brazil (six individuals sampled in each). The lack of haplotype diversity in the Florida samples and the sharing of that haplotype with manatees in the West Indies may indicate a recolonisation of Florida by waif dispersal from the Antilles after the Wisconsinan glaciation (Garcia-Rodriguez *et al.* 1998; Domning 2005), consistent with Domning's (2005) supposition that *T. manatus bakerorum* could not survive in Florida during the Wisconsinan glacial period at the end of the Pleistocene. Alternatively, low haplotype diversity could be explained by a recent population 'bottleneck' affecting the Florida population (Garcia-Rodriguez *et al.* 1998).

Strong geographic differences in haplotype distributions of modern manatees exist, with differentiation grouping into three clusters (Garcia-Rodriguez *et al.* 1998). One cluster is a Florida–West Indies grouping (but also including some samples from Colombia); the second cluster holds samples from an area extending from Mexico to the rivers of northern South America bordering the Caribbean (a Caribbean mainland group); and the third is from north-eastern South America (along the Atlantic coast from Guyana to Brazil). A lack of significant interchange among these groups due to stretches of unsuitable coastal habitat seems likely, although there is evidence of minor interchange. (This tripartite grouping of West Indian manatees based on mtDNA differs from the morphologically based cladistic assignment of modern *T. manatus* into two subspecies by Domning and Hayek (1986). However, it is consistent with the prior subspecies in that the new divisions are within the *T. manatus manatus* morphological grouping, which can be considered paraphyletic to *T. manatus latirostris*, the more peripheral isolate derivative of the West Indies genetic group (D. Domning, personal communication 2010).)

The genetic divisions among the above three mtDNA groupings are described as being as deep as the species-level divisions between *T. manatus* and *T. inunguis* from the central Amazon of Brazil, and comparable to the depths of divisions among some genera of whales (Garcia-Rodriguez *et al.* 1998). Provisional molecular clock calculations suggest that the three divisions of West Indian manatees could have split from each other at somewhat less than one million up to seven million years ago, consistent with Domning's (1982a) palaeontological estimate that *T. manatus* or a *T. manatus*-like form first appeared in the mid-Pliocene. Cantanhede *et al.* (2005) reported a similar estimate of genetic divergence between West Indian and Amazonian manatees that corresponds to a split about four million years ago.

Surprisingly, hybridisation between *T. manatus* and *T. inunguis* has been described, first based on three samples from Guyana (Garcia-Rodriguez *et al.* 1998). Although hybridisation could have been an accidental result of placing of the two species together for weed control in Guyana during the 1960s (Allsopp 1969), analysis of additional samples has verified the presence of hybrids in the wild in French Guiana and in Brazil at the mouth of the Amazon River (Cantanhede *et al.* 2005; Vianna *et al.* 2006). Hybrid and likely backcrossed individuals were also confirmed using cytogenetic karyotype analysis and sequencing of two microsatellite loci (Vianna *et al.* 2006). Indeed, some of the hybrids were morphologically intermediate between the two species (general appearance like *T. manatus* but lacking some nails on the pectoral flippers and with light spots on the abdomen), and had chromosome numbers intermediate between *T. manatus* (2n = 48) and *T. inunguis* (2n = 56) (Vianna *et al.* 2006).

Haplotype distribution has also been investigated in Amazonian manatees. Cantanhede *et al.* (2005) used the mtDNA control region (361 base pairs) of *T. inunguis*, and employed in part the approach Garcia-Rodriguez *et al.* (1998) developed for *T. manatus*. Their work revealed that the two species share two haplotypes, and that the Florida–West Indies cluster of *T. manatus* may be more closely related to *T. inunguis* than it is to either of the other two groupings of *T. manatus*. (However, a competing model of monophyly within *T. manatus* could not be overruled.) Overall, the genetic evidence suggests four lineages of *Trichechus* in the New World: one inhabits the freshwater Amazon ecosystem, whereas the other three inhabit coastal systems and associated inland waterways (Cantanhede *et al.* 2005). Genetically, the three coastal lineages appear to be as different from one another as they are from the freshwater lineage, leaving open the questions of whether there are four or two species of New World manatees, or a single species with four divisions (Cantanhede *et al.* 2005). This is important not only for understanding the diversity and evolution of the Trichechidae, but also for conservation and management. It certainly calls for additional rigorous analyses on larger numbers of samples from more locations using more extensive sampling of the genome. Inclusion of the cytochrome b sequence data by Vianna *et al.* (2006), for example, has supported a hypothesis that *T. inunguis* is cladistically more basal than *T. manatus* and *T. senegalensis*, and the survivor of 'an ancient lineage adapted to fresh water' (Vianna *et al.* 2006, p. 444; see 'Sirenian diversity through time: ecology of co-existing species and serial replacements' below). *T. inunguis* from Brazil do not show geographically based structuring (Cantanhede *et al.* 2005; Vianna *et al.* 2006).

Vianna *et al.* (2006) used the 410 base pair fragments of the mtDNA control region employed by Garcia-Rodriguez *et al.* (1998), but in addition also analysed a 615 base pair region of the cytochrome b mtDNA gene and two microsatellite loci. They sampled 330 individuals of all three species of *Trichechus*, some from the same regions sampled by Garcia-Rodriguez *et al.* (1998) and Cantanhede *et al.* (2005). They found 20 haplotypes of the mtDNA control region genes in West Indian manatees, as well as four other haplotypes more typical of Amazonian manatees. They affirmed the occurrence of a single haplotype in the Florida population, and found that manatees morphologically referred to as *T. manatus* from French Guiana had haplotypes that belonged to *T. inunguis*. This work revealed further complexities in the degrees of isolation of different geographic populations of *T. manatus*. Findings of this more extensive analysis agreed in broad outline with those of Garcia-Rodriguez *et al.* (1998) and Cantanhede *et al.* (2005), but the inclusion of more samples suggested instead that two major 'evolutionarily significant units' exist: the north-eastern South American group and a heterogeneous assemblage from the other areas. Vianna *et al.* (2006) suggest that discontinuities between the north-eastern South American and other populations of *T. manatus* may be the result of a barrier to gene flow formed by what are now the Caribbean islands during the Pleistocene, when sea levels were lower. They agreed with Cantanhede *et al.* (2005) that, genetically, *T. inunguis* appeared to be most closely related to a geographic cluster of *T. manatus* similar to the Florida–West Indies cluster of Garcia-Rodriguez *et al.* (1998). This relationship also bolsters the hypothesis that *T. manatus* is a paraphyletic species (and *T. manatus manatus* a paraphyletic subspecies). The discontinuity between genetic data and past palaeontological hypotheses was further highlighted by the results of the mtDNA cytochrome b sequencing (Vianna *et al.* 2006), which places *T. inunguis* as the basal species in the *Trichechus* clade. However, their molecular clock and coalescent time calculations suggest that the evolutionary lineage leading to the modern *T. inunguis* acquired its present array of mtDNA lineages later than *T. manatus*.

SIRENIAN DIVERSITY THROUGH TIME: ECOLOGY OF CO-EXISTING SPECIES AND SERIAL REPLACEMENTS

Premises about the ecological and evolutionary capacities of sirenians

Domning (1976, 2001b) defined a series of premises about what he termed the 'aquatic megaherbivore adaptive zone' that provide a background on the forces and constraints faced by mammals that occupy this zone. We refer to these as 'Domning's Postulates' (Box 3.1). There are also four other points that Domning raises about conditions within this adaptive zone, and the range and limits of sirenian adaptations within which natural selection can work in allowing resource exploitation and sometimes co-existence.

(1) The postulate that rostral deflection corresponds to the degree of bottom feeding can be relaxed if a sirenian specialises by digging at a single spot because a head-down vertical posture can be employed, rather than maintaining a horizontal position to allow the body to move forward during feeding.
(2) The minimum depths at which feeding can take place are limited by factors related to predation and stranding, whereas maximum depth is limited by anatomical and physiological constraints of lung structure and ballast.
(3) Food plants can vary in size, accessibility of the leaves and especially rhizomes, and in their toughness and abrasiveness. Reliance on more

Box 3.1.   Domning's Postulates on ecological and morphological
characteristics of sirenians that mould their evolutionary potential (Domning
1976, 2001b)

(1) Sirenians are shallow-water, nearshore animals.
(2) Sirenians are typically tropical or subtropical.
(3) Sirenians are moderately vagile.
(4) Sirenians are opportunistic feeders that typically prefer angiosperms.
(5) Seagrasses are shallow-water, nearshore plants typically preferring
    protected tropical waters.
(6) Sirenian bodies should be horizontal for efficient forward locomotion.
(7) Specialisation on plants by growth habit is most economically achieved by
    altering the position of the mouth opening while maintaining the body's
    axis horizontal.
(8) The degree of rostral deflection (down-turning of the snout) correlates
    with the degree of specialisation as bottom feeders.

deeply buried, thicker, and more fibrous or lignified rhizomes would
require greater anatomical specialisations in size and shape of tusks for
cutting and uprooting, and reinforcement of skull parts involved in
associated mechanical actions.

(4) Specialisation on tougher, more fibrous plant parts may require larger
body size to allow a longer gut and more efficient digestion, and dental
specialisations for better mastication of tougher and more abrasive plant
parts and substrates.

Domning (2001b) assumed that without human influence sirenian
populations would be limited by food supply, and that competition for
food would be a major factor in structuring sirenian communities. The
speculated feeding niches he describes on the basis of anatomical adaptations define ways in which those adaptations would allow a particular sea
cow taxon to be competitively superior and more efficient than co-
existing sirenians. However, Domning emphasises that, more broadly,
the diets of all sirenians can include leaves of seagrasses, and sea cows
that lack tusks are still capable of excavating small rhizomes (Domning
2001b) as demonstrated by three of the four extant species (Chapter 4).
Domning's scenarios are based on the understanding that in the tropical
and subtropical marine environment sirenians are: obligate seagrass
feeders; the only potential consumers of seagrass rhizomes; and the most
important consumers of seagrass leaves (Chapter 4).

Domning's Postulates and additional factors are reflected in morphological adaptations of co-existing fossil sirenians. These adaptations likely indicate the axes along which food resources were partitioned. These axes include body size, rostral width, rostral deflection, characteristics of cheek teeth, and presence, size and morphology of tusks. To test hypotheses inferred from tusk adaptations, Domning and Beatty (2007) explored the influence of tusk morphology on extraction of naturally occurring seagrass rhizomes with plastic replicas of tusks of the Middle Miocene halitheriine *Metaxytherium floridanum*, the Oligocene dugongines *Crenatosiren olseni* and *Dioplotherium manigaulti*, and the Pliocene dugongine *Corystosiren varguezi* (Figure 3.5). Excavation using model tusks of these species indeed differed in the number of strokes needed to uproot sample quadrats of seagrasses; in the mass of rhizomes excavated in the species of seagrass with the largest, deepest rhizomes (Domning and Beatty 2007) the large blade-like tusks of *Corystosiren* excavated the largest mass. Cranial architecture suggested that dugongines with blade-like tusks used a downward and backward movement in digging rhizomes, and comparison of geometric aspects of the tusk tips of *Corystosiren* with those of *Dugong* demonstrated that *Corystosiren* had tusk shapes that were optimal for maximising cutting with minimal friction with food (Domning and Beatty 2007).

Co-existing species and serial replacements in the Caribbean Sea and western Atlantic Ocean

The Caribbean Sea and western Atlantic Ocean hosted the greatest known diversity of sirenians. This region is where the sirenian fossil record began (although earlier roots are Old World), starting with the Eocene prorastomids and leading up to the Recent manatees. Thus, over the last 50 million years, it has been home to all four families of sirenians, often with overlapping multiple species assemblages, and it is the only area on Earth continuously occupied by the Sirenia since the Eocene. Indeed, up to six species of dugongids may have been sympatric in the region during the Late Oligocene (see 'Oligocene sirenians in the Caribbean Sea and western Atlantic Ocean'). Given this intriguing and rich history, Domning (2001b) pondered likely ecological circumstances that may have permitted the co-existence or caused the serial replacements of these sirenians, and offered hypotheses regarding the co-evolution of sirenians and aquatic plants in the region. In this section we follow a temporal, epoch-by-epoch sequence to highlight Domning's (2001b) scenarios. Sirenian history in this region reminds us that the current situation, marked by essentially no spatial overlap or competition for food among modern sirenians, is much more simplified than in the past.

Eocene sirenians in the Caribbean Sea and western Atlantic Ocean
During the Eocene, the western Atlantic Ocean and Caribbean Sea were marked by the presence of prorastomids, protosirenids and early dugongids, including distinct but unnamed taxa. In addition to the known species (see Supplementary Material Appendix Tables 3.2.1 and 3.2.2), in the Middle Eocene the south-eastern United States alone was home to at least one or two other prorastomids, one or two protosirenids, and likely a primitive dugongid that are not named (Domning *et al.* 1982; Savage *et al.* 1994). Some of these and other unidentified sirenians also were found elsewhere in the western Atlantic during this time (Domning *et al.* 1982; Domning 2001b). The prorastomids were amphibious, as also to a lesser degree were the protosirenids (but probably spending less time on land). *Protosiren* had a broader and more down-turned snout than the prorastomids, suggestive of unselective aquatic grazing and a concentration on bottom feeding (Domning and Gingerich 1994; Domning 2001b). Prorastomids and early dugongids had largely straight and/or narrow snouts, suggesting more selective browsing. Both may have used fresh water to some extent, where a richer and more diverse aquatic flora probably allowed subdivision through selective feeding (Domning 2001b). Thus, early partitioning of the sirenian adaptive zone probably fell along habitat lines (marine, estuarine and fluvial), selective browsing versus grazing, and perhaps position of food in the water column (near-surface versus bottom). In any case, these Middle Eocene sirenians were the groups available to later move into what Domning (2001b, p. 40) thought of as the 'relatively vacant' aquatic megaherbivore adaptive zone. The dugongids developed enhanced adaptations for swimming and for feeding on seagrasses and were more successful, eventually becoming the most diversified family of sirenians, whereas prorastomids and protosirenids, with their less extensive adaptations for fully aquatic herbivory, became extinct by the end of the Eocene.

Oligocene sirenians in the Caribbean Sea and western Atlantic Ocean
The diversification of the dugongids in the Caribbean and western Atlantic during the Oligocene also included unidentified and yet-to-be-named-species that added to sirenian diversity during this epoch (Domning 2001b). Named forms include two species of *Halitherium*, *H. antillense* and *H. alleni* (although the status of *H. antillense* is dubious, the fauna likely

included these or similar halitheriines such as species of *Metaxytherium*); the halitheriine *Caribosiren turneri*; two dugongines, *Crenatosiren olseni* and *Dioplotherium manigaulti*; and perhaps a member of the clade that produced *Corystosiren* and *Rytiodus* (known from the region soon afterwards in the Miocene). Domning (2001b) characterised the *Halitherium–Metaxytherium* forms as having strongly down-turned snouts and small tusks that suggest they were generalised feeders on seagrass blades and on rhizomes of smaller seagrasses. *Caribosiren turneri* lacked tusks and had probably the most highly deflected rostrum of any sirenian, and thus was, by analogy with the dugong, likely a specialist on small seagrasses or the leaves of larger seagrasses. *Crenatosiren olseni* had larger tusks than the halitheriines and likely fed on mid-sized rhizomes, whereas the large tusks of the larger-bodied *Dioplotherium manigaulti* suggest feeding on large rhizomes.

Early members of the *Corystosiren* lineage in the Oligocene likely had the largest tusks and thus could have specialised on robust rhizomes of the very largest seagrass species (Domning 2001b). (At the time of Domning's analysis, the existence of very large seagrasses in the early Caribbean was not documented, but a new Eocene species (*Thalassites parkavonensis*) from Florida has since been described (Benzecry and Brack-Hanes 2008).) The likely specialisation of some species on large rhizomes is also indicated by tusk shape and structure: the tusks of *Dioplotherium manigaulti* and *Corystosiren* were not only large, but were also flattened and blade-like. Further, the tusks had enamel only on one side in *Corystosiren* and in later species of *Dioplotherium*, making them self-sharpening and presumably more efficient at rhizome cutting. In the Late Oligocene the group of co-existing dugongids also could have been joined by early trichechids in South America, but this inference has not yet been confirmed by fossils (Domning 2001b).

Miocene sirenians in the Caribbean Sea and western Atlantic Ocean
The fossil Sirenia from the Early to Middle Miocene in South American marine areas (Toledo and Domning 1991; Domning 2001b) included halitheriines (*Metaxytherium*-like) and at least two dugongines (*Dioplotherium allisoni*-like and *Rytiodus*-like sea cows). The larger number of Early to Middle Miocene sirenians known from the south-eastern United States and Caribbean includes the halitheriines *Metaxytherium crataegense*, a *Dioplotherium*, an unnamed primitive *Corystosiren*, a likely dugongine from Puerto Rico and small dugongines in the *Nanosiren* lineage (Domning 2001b; Domning and Aguilera 2008). In the Middle to Late Miocene, *Metaxytherium floridanum*, a descendant of *M. crataegense*,

occurred in Florida. The earliest known trichechid, *Potamosiren magdalenensis*, appears in the Middle Miocene of Colombia (Domning 1997b). *M. crataegense* and its slightly larger descendant *M. floridanum*, with strongly down-turned snouts and small tusks, presumably continued to be generalised feeders on seagrass blades and small rhizomes. *M. floridanum* showed a minor increase in rostral deflection and body size, suggesting to Domning (2001b) an increase in niche breadth in response to competitive release from the extinction of *Crenatosiren* or other sirenians. The dugongines, *Dioplotherium allisoni*, *Rytiodus* and *Corystosiren*, with their large, self-sharpening, cutting tusks likely fed on intermediate to large rhizomes. In addition to these species, the small dugongines in the *Nanosiren* lineage appeared in the Miocene and likely fed on small seagrasses in shallow waters that could not be accessed by their contemporaries (Domning 2001b; Domning and Aguillera 2008). Domning (2001b) suggested that *Nanosiren* may have competed with and possibly replaced the halitheriine *Caribosiren*. There were also trichechids present in this epoch, with *Potamosiren magdalenensis* known from the Middle Miocene, but most likely these early manatees occupied freshwater and estuarine habitats and thus did not yet overlap extensively with the marine sirenians (Domning 2001b).

Pliocene sirenians in the Caribbean Sea and western Atlantic Ocean
The Pliocene epoch was a time of much change in the western Atlantic and Caribbean. The halitheriines were no longer there, but dugongines present at the beginning of the Pliocene included *Xenosiren yucateca* and *Corystosiren varguezi*, and the diminutive *Nanosiren garciae* (Domning 1989b, 1990, 2001b; Domning and Aguillera 2008). However, only one species, a *Dugong*-like form, is known in the region by the Late Pliocene, and that lineage disappeared entirely by the end of the epoch. The trichechids, represented by *Ribodon* in the Early Pliocene and by *Trichechus* later in the epoch, greatly expanded their geographic distribution and entered more marine habitats. From the end of the Pliocene to the Recent, trichechids were the only sirenians known from across this wide region.

What occurred during the Pliocene that might account for these declines in diversity in the western Atlantic and Caribbean sirenian fauna? Domning (2001b) suggested the following scenario. Seagrasses were established by the end of the Cretaceous, and nearly all of the Eocene fossil seagrasses are referable to living genera and species and had similar growth forms. The seagrass flora in the Caribbean, however, was more diverse during at least the early part of the Tertiary; along with species still found today, the area also included genera now growing only in the Old World (Domning 2001b). Some of these are bulkier plants with large fibrous rhizomes, providing a wider range of food sizes for megaherbivores. In the mid-Pliocene the Central American Seaway that provided interchange with the Pacific Ocean closed due to the rising of the Isthmus of Panama (Iturralde-Vinent 2003; Figure 3.6). Closure resulted in a major disruption to ocean circulation systems, altered marine communities of the Caribbean and western Atlantic and led to the extinction of many invertebrates and vertebrates, including other seagrass-associated species (Allmon *et al.* 1996; Jackson *et al.* 1996). Domning (2001b) suggested that this major extinction event (possibly also precipitated in part by climatic cooling) reduced seagrass diversity, thereby restricting the number of feeding niches available to sirenians. Furthermore, the former presence of sirenians specialising on the large rhizomatous species of seagrasses (that later disappeared), probably acted to increase patches of higher plant diversity through their digging. Thus sirenian feeding activities may have opened niches for smaller successional species of plants, perhaps disrupting the establishment of more uniform stands of climax vegetation. These larger sirenians would have been keystone species that permitted the co-existence of other sea cows specialising on different plants. Removal of keystone sea cows may have caused cascading effects across the relatively simple sirenian community. The Pliocene decline in dugongid diversity was also hastened by the simultaneous spread of the trichechids, with their more efficient dentition. The latter may have been particularly advantageous with higher silt loads from mountain building. Rostral deflection in trichechids appears to have increased once dugongids disappeared from the fossil record (Domning 1982a).

Co-existing species and serial replacements in the North Pacific Ocean
The evolution of the Sirenia in the North Pacific followed a seemingly simple but unique pathway. It began with dispersal of dugongids from the Caribbean to the Pacific through the Central American Seaway in the Miocene; it ended with the gigantism and other remarkable adaptations for feeding on kelps in cold water attained by Steller's sea cow. As described in Chapter 2, the latter species was hunted to extinction in historic times, ending the presence of all sirenian lineages in the temperate Northern and eastern Pacific. Once again, Daryl Domning and colleagues (e.g. Domning 1976, 1978a; Aranda-Manteca *et al.* 1994; Domning and Furusawa 1994) have provided much of the documentation and interpretation of this evolutionary pathway.

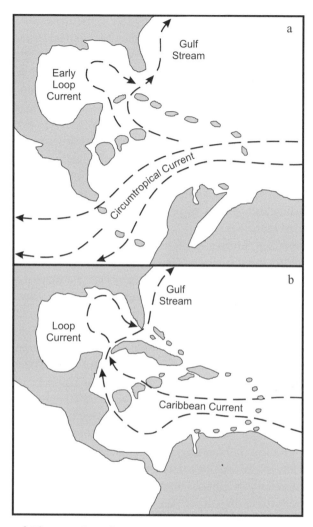

Figure 3.6. The open Central American Seaway was a dominant feature throughout much of the Tertiary, allowing diversification of species associated with seagrass, including sirenians in the Caribbean. The Central American Seaway also allowed dispersal of dugongids into the North Pacific in the Miocene. Closure of the seaway due to volcanic activity and rising of the Isthmus of Panama in the Pliocene disrupted tropical ocean circulation patterns, reduced seagrass diversity and hastened the decline of sirenian diversity in this region. (a) Relative position of the Americas and ocean circulation patterns in the Late Oligocene to Middle Miocene. (b) Oceanic circulation patterns and lack of access to the Pacific from the Caribbean region from the Pliocene onwards. After Itturalde-Vinent (2003), with permission.

Solid fossil evidence shows that three stocks of sirenians appeared in the North Pacific in the Early to Middle Miocene, exemplified by the halitheriine *Metaxytherium arctodites*, the dugongine *Dioplotherium allisoni* and the hydrodamaline *Dusisiren reinharti*. *D. allisoni*, also known from Early Miocene deposits in the western Atlantic (Toledo and Domning 1991), was a bottom feeder that likely specialised on seagrass rhizomes, whereas *M. arctodites* may have been a more generalised seagrass feeder. Both became extinct in the Pacific by the Late Miocene, whereas *D. reinharti* marked the beginning of the Hydrodamalinae. The hydrodamaline lineage successively evolved morphologically through a series of sea cows (a conclusion also well-supported stratigraphically) that survived by adapting to climatic cooling and feeding on plants growing at or near the surface on rocky, exposed coastlines.

The contrasting fortunes of these lineages were tied to a series of geographic, climatic and resulting ecological conditions that prevailed from the Middle Miocene onwards. *D. allisoni* may have been a specialist on smaller seagrasses than those eaten by the sympatric desmostylian herbivore *Desmostylus*; *D. allisoni* possessed a strongly deflected rostrum (about 70°), self-sharpening tusks and other anatomical features associated with feeding on seagrass rhizomes (Domning 1978a). Found on the California coast, it and the earliest members of the other two Early to Middle Miocene stocks (*Metaxytherium* and *Dusisiren*) probably reached this region by waif dispersal from the Central American Seaway, as some of their seagrass food plants may have done. Waif dispersal to California is likely because much of the west coast of Mexico lacked favourable habitat and was too rocky and exposed to promote extensive areas of seagrasses. This broad region of some 2500 km of coastline was thus a barrier to genetic interchange with Caribbean sirenian populations (Domning 1978a), a possibility that became fully moot with the closing of the Central American Seaway in the Pliocene (Figure 3.6). Coastal California was tropical and subtropical in climate throughout most of the Miocene, and the coastline was much different than today, with numerous embayments, estuaries and protected shorelines favourable to seagrass meadows. However, two factors changed these conditions. Climatic cooling began in the Late Miocene and Pliocene, and coastal mountain orogeny also began in the Late Miocene, causing a draining of shallow-water habitats extending to the later Pliocene and Pleistocene. These changes caused coastal California to become unfavourable for seagrasses. *D. allisoni* became extinct after the late Middle Miocene, an extinction perhaps hastened further by competition with desmostylians (Domning 1978a). The more generalist feeder *Metaxytherium arctodites* possibly represents an isolated California taxon derived from another species that crossed through the Central American Seaway: *M. crataegense*, known from the Early to Middle Miocene of the south-eastern Pacific in Peru, as well as the Caribbean (Muizon and Domning 1985). *M. arctodites* also co-existed with *Dusisiren* in coastal California and Baja California Sur. With strong rostral deflection but small tusks, *M. arctodites* probably had a feeding strategy intermediate between *Dioplotherium* and *Dusisiren*, and seems to have depended primarily on seagrass leaves and rhizomes of smaller seagrasses (Aranda-Manteca *et al.* 1994). *M. arctodites* probably represents a sister group to the hydrodamalines, and also is known only from the Middle Miocene (Aranda-Manteca *et al.* 1994). Thus two out of the three co-existing lineages disappeared with changing ecological conditions that reduced seagrass abundance and diversity in the eastern North Pacific Ocean.

The hydrodamaline lineage, in contrast, was pushed in its unique direction by these same conditions. *D. reinharti* of the Early Miocene may have been a seagrass feeder, but it likely had only small or vestigial tusks and did not specialise on bottom plants. It seemingly had a less deeply deflected rostrum (perhaps about 40°) than *D. allisoni*, which facilitated feeding on plants higher in the water column. Climatic cooling, the increase in rocky coastlines and a reduction in shallow habitats in coastal California beginning in the Late Miocene favoured algae and kelps over seagrasses. Unlike *Dioplotherium*, the hydrodamaline trajectory showed the capacity to evolve in response to these changes. *D. reinharti* was replaced by an intermediate form that then led to *Dusisiren jordani* of the Middle to Late Miocene of California, about 10–12 million years ago. By this stage *Dusisiren* had increased in size to about 4.5 m (compared with a likely 3.5 m in *D. reinharti*) and showed a reduction in number and size of adult teeth and other morphological tendencies that became amplified in descendants, including *Hydrodamalis* (Domning 1978a). Large size and reduced dentition are seen as new adaptations for subsisting on kelp in cooling waters. *Dusisiren dewana* was the next species in the lineage, about 4.3 m long with a rostral deflection of 30°–40°, further reductions in dentition and a reduced, claw-like forelimb resembling that of *Hydrodamalis*, as described by Steller (Chapter 2). *D. dewana* is known from the Late Miocene in both California and Japan (Takahashi *et al.* 1986); another species, *D. takasatensis*, may also have occurred in Japan (Kobayashi *et al.* 1995).

The first recognisable species of *Hydrodamalis* was the Late Miocene to Late Pliocene *H. cuestae* from southern California and Baja California (Domning 1978a; Domning and Furusawa 1994). This sea cow had low rostral deflection, no teeth in adults but vestigial teeth in the young, lacked phalanges in the flippers, had a braincase that was similar to that of *Dusisiren* but had a more rectangular rostral masticating surface. Domning (1978a) estimated it to be over nine metres long, more than twice the size of *D. jordani*, representing a very large increase in sirenian body size over just a few million years. Evidence that hydrodamalines inhabited rocky, high-energy coastlines exists in the fractured ribs and other pathological conditions seen in fossil skeletons of both *Dusisiren* and *Hydrodamalis* (Domning 1978a). *Hydrodamalis gigas* appeared in the fossil record in the Late Pleistocene of California and the Aleutian Islands as a descendant from *H. cuestae*. It also spread northward across to the western Pacific and south to Japan, remaining in coastal California as recently as 20 000 years ago (Domning 1978a; Domning and Furusawa 1994). The existence of another nominal species, *Hydrodamalis spissa*, in the Early Pliocene of Japan remains open to interpretation pending additional discoveries. It may be an individual variant of *H. cuestae* or could represent a separate lineage to *H. gigas*, relegating *H. cuestae* as a side branch with no further descendants (Domning and Furusawa 1994). At any rate, the culmination of sirenian adaptations for feeding on kelps in cold water was reached by *H. gigas*. These adaptations were described by Steller (1751, 1925) and Domning (1978a), and are summarised in Chapter 2. They include large body size, absence of teeth, a likely inability to submerge, loss of phalanges and other modifications of forelimbs (to serve as holdfasts, for propulsion and for uprooting algae). Because *Hydrodamalis gigas* persisted until its elimination by humans in historic times, it also allowed the written documentation of its soft anatomy and behaviour by Steller (1751, 1925; Chapter 2), topics that are only subjects of speculation for other extinct lineages of the Sirenia.

CONCLUSIONS

Extant dugongs and manatees provide only a glimpse of the wonderful diversity of sirenians that came before them. The group has an old and complicated origin that has fascinated scholars for centuries. Molecular evidence for some of their seemingly bizarre affinities provides a major example of how traditional evolutionary findings can be subject to radical reinterpretations, as well as reinforcement, by modern research technology. The fossil record of sirenians is substantial, and includes some unusual mammals at the boundary of the major ecological transition from a terrestrial existence to the adaptive zone of purely aquatic megaherbivory.

Subsequent radiations within that adaptive zone are well documented and show diversification with changing environmental conditions, with some adaptations allowing co-existence with other sirenians in space and time. Eventually, most forms were replaced by others, until today when we are left with three Atlantic species and one in the Pacific (Figure 1.2). These four species have essentially no overlap in distribution (Chapters 1, 5 and 8), but can show major differences in feeding ecology (Figure 1.3; Chapter 4). However, with the exception of the separate lineage that invaded the cold North Pacific and was extirpated by an exploitative civilisation (Chapter 2), both the extinct and the contemporary Sirenia were all confined to tropical or subtropical coastal waters (Chapter 5).

The warm, tropical, coastal environment within which most sirenian lineages existed over long periods of time remains in place. However, the pace, scale and nature of change in that environment over the last century are more radical than some of those that took place over hundreds of thousands to millions of years of evolutionary history. Today's sirenians inhabit shores and rivers of countries that are suffering from poverty and scarce resources, leading to opportunistic and intentional killing for food across tropical parts of the globe (Chapters 7 and 8). Human population growth, habitat deterioration, changes in fishing technology and even recreational pursuits are also accelerating the potential for extinction of the remaining species of sea cows (Chapter 7). As discussed in Chapter 9, conservation solutions are urgently needed to prevent the curtain from closing on 50 million years of sirenian evolution and adaptation.

NOTES

1. Mirorder is the taxonomic category above superorder; see also Table 3.3.
2. Infraorder and parvorder rank below order in the system of taxonomic classification. Infraorder is a higher rank than parvorder. See also Table 3.3.
3. Cladistics is a form of biological systematics that classifies organisms into hierarchical monophyletic groups by focusing on shared derived characters (synapomorphies).
4. Pachyosteosclerosis is a thickening of the bones of the ribs and other elements combined with replacement of trabecular (spongy) bone with compact bone. The resultant 'internal weight belt' is thought to aid bottom feeding and positioning in the water column.

Feeding biology

The feeding biology of sirenians, the only group of fully aquatic herbivorous mammals, is arguably their most defining characteristic and the basis for their popular name – sea cows. In this chapter, we describe the feeding biology of wild populations of the four species of extant sirenians and the methods used to obtain this information. We then explore the significance of this information for sirenian conservation. We have not considered the feeding biology of captive sirenians, which is discussed by various authors, especially Best (1981) and Marshall *et al.* (1998, 2000, 2003), or the feeding biology of Steller's sea cow and other extinct sirenians, which are discussed in Chapters 2 and 3.

OPTIMAL FORAGING THEORY

MacArthur and Pianka (1966) and Emlen (1966) first articulated the theory of optimal foraging, which states that natural selection favours animals whose behavioural strategies maximise their net energy intake per unit time spent foraging. Foraging time includes both the time spent searching for food and the time spent handling food. The theory was devised to attempt to explain why, out of the wide range of foods available, animals often restrict themselves to a few preferred types. The theory is based on assumptions that may not apply to wild animals and is contentious (e.g. Pyke 1984; Pierce and Ollason 1987), but provides a useful heuristic framework for considering the trade-offs that sirenians have to make as they go about their daily lives.

For large herbivores, such as sirenians, the effects of food quality on nutritional status can outweigh those of food quantity (Owen-Smith and Novellie 1982; Owen-Smith 1990). Even though the quantity of food available may seem far from limiting, plant material is typically of relatively low quality: high in fibre and low in protein. The nutrient content of the plant may vary with its phenological stage. Plants may also possess chemical defences. Consequently, much of the plant material that is apparently available as food for sirenians may not be of sufficient quality to be part of their food resources.

Optimal foraging theory predicts that an animal must balance two competing strategies: spending longer (i.e. using more energy) searching for highly 'profitable' food items, or devoting less time (i.e. using less energy) to more abundant but less profitable foods. We might expect that as large herbivorous mammals, manatees and dugongs would be trying to maximise their intake of higher-quality foods (relatively higher concentrations of protein and energy) while minimising their intake of lower-quality foods (high in fibre and secondary compounds). We would also expect sirenians to try to minimise the time spent feeding and the risk of predation. These trade-offs are important determinants of habitat choice (Chapter 5) and are therefore highly relevant to the development of effective conservation strategies.

FEEDING STYLE

Scientists studying large, terrestrial, herbivorous mammals distinguish between grazing and browsing, a distinction based primarily on plant type (e.g. Janis 2008). Grazing refers to eating grass: monocotyledonous (single seed leafed) plants of the grass family (Poaceae, formerly Gramineae) (Kellogg 2001). Eating woody or non-woody dicotyledonous plants is referred to as browsing. Sirenian researchers have tended to use these terms interchangeably, even though the definitions used by terrestrial ecologists are not strictly applicable when dugongs and manatees feed on seagrasses, their most important marine food plants. Seagrasses are not true grasses, or even a monophyletic group. Rather, 'seagrass' is a functional term referring to marine, vascular, flowering plants of the families Cymodoceaceae, Zosteraceae and Hydrocharitaceae that share numerous morphological and physiological characteristics (Larkum and den Hartog 1989). Although seagrasses are monocotyledons and superficially similar to terrestrial grasses in some respects, they are quite distinct taxonomically and are adapted to living in the sea. Thus the terms 'grazing' and 'browsing' are not really suitable to describe the feeding behaviours of dugongs and manatees, so we have chosen not to use them here.

Nonetheless, both dugongs and manatees also employ two different feeding modes. Following Wirsing et al. (2007a), we refer to them as 'excavating' and 'cropping'. We define excavating ('rooting' sensu Anderson 1981a; 'furrow grazing' sensu Preen 1992) as feeding on both the above-ground and below-ground parts of plants by disturbing the sediment; and cropping ('surface grazing' *sensu* Preen 1992) as feeding on the above-ground parts only. Sirenians mostly use excavating when feeding on seagrasses with accessible rhizomes, presumably because of the nutritional advantages in eating both the above- and below-ground parts of these plants (as explained in 'Food quality' below). Cropping occurs when feeding on other plants. Because seagrasses are modular plants, even when a sirenian feeds on both the above- and below-ground portions of a plant, the whole food organism is rarely killed.

FOODS

Methodology

The methods used to document the foods eaten by wild sirenians have been reviewed by Beck and Clementz (2012) and are referenced in the sections below, which document the diets of each of the four extant species. In this overview, we summarise each of these techniques and consider their inherent strengths and limitations.

Direct field observations of feeding sirenians by scientists, indigenous peoples and fishers provide unequivocal dietary information that has been very important for determining the foods of West Indian manatees and dugongs occurring in clear water, but are less informative for all four species in turbid waters. Dugongs and manatees feeding on whole seagrass plants generate clouds of sediment and feeding plumes, which are important signs of bottom feeding activity. Feeding scars in seagrass beds have also been used as indicators of excavating by dugongs (Figure 4.1) and West Indian manatees. With the exception of manatees feeding on emergent terrestrial grasses and floating vegetation, signs of cropping are more difficult to detect and are a less reliable index of the food habits of sirenians than signs of excavating. In addition, it is very difficult to obtain unbiased information on the relative importance of different foods in the diets of sirenians based on direct observation alone, because some individuals and some forms of feeding are much easier to observe than others.

It is also important to remember that even large numbers of digesta samples are unlikely to be from a random sample of animals. Nonetheless, comparing the histology of samples of sirenian digesta with a reference collection of potential food plants arguably provides the best information on the foods eaten by wild manatees and dugongs, but samples from a large number of carcasses can be restricted to areas where anthropogenic mortality and human populations are high.

Figure 4.1. Dugong feeding trail in a seagrass meadow near Phuket, Thailand. Len McKenzie photograph, reproduced with permission.

Faecal samples are generally easier to collect but less reliable than samples from the mouth and stomach because components of the diet are differentially digested as they pass through the alimentary canal, potentially providing biased information on their relative importance. Plants recognisable in faeces are likely to be the species that are least digested rather than those most frequently eaten. Quantitative micro-histological analysis of digesta is very time-consuming. Near infra-red reflectance spectroscopy (André and Lawler 2003) is a rapid and relatively inexpensive method of identifying and quantifying the species present in digesta, but requires the chemical signatures of a reference collection of food plants to be calibrated.

All these methods provide snapshots of the foods recently eaten by individual sirenians. Information on the average dietary intake of individuals over a longer period of time (months) can be obtained by measuring the ratio of stable isotopes in various tissues. This method has been used to investigate the relative importance of freshwater and marine plants to Florida manatees (Ames *et al.* 1996; Reich and Worthy 2006; Alves-Stanley and Worthy 2009; Alves-Stanley *et al.* 2010). However, isotope ratios do not provide fine-grained, precise information on the species or plant parts in the diet. Such information is essential to understanding food selection and interactions with the food supply that may be of greatest relevance to conservation.

Analysis of fatty acid constituents in blubber (so-called fatty acid signature analysis) has been used to study foraging and diet in a number of marine mammals (see review by Budge *et al.* 2006). Although the technique works well in at least some species, Ames (2002) and Ames *et al.* (2002) noted that fatty acids among the aquatic plants are quite similar. Thus, the utility of fatty acid signature analysis for assessing the diet of sirenians is questionable.

Dugongs

Because dugongs are strictly marine and restricted to the tropics and subtropics, they have limited access to floating vegetation (other than seagrass, algae and mangrove leaves in drift) or to terrestrial vegetation other than mangroves. There are no records of dugongs feeding on floating leaves in the wild, although they do so in captivity. The circumstances surrounding the only record of a dugong feeding on mangrove leaves suggest that it was feeding at high tide when the mangrove was submerged (Johnstone and Hudson 1981). Thus, unlike manatees, which feed throughout the water column, wild dugongs are bottom feeders (Figure 4.2a), as expected from their anatomy (see 'Food acquisition' below).

No sex or body size differences in the diet of dugongs have been detected (Johnstone and Hudson 1981), apart from the fact that young calves suckle from their mothers for up to 18 months (possibly more) after birth (see Chapter 6). Nonetheless, dugong calves start eating seagrass (including the rhizomes) within a few weeks of birth (Marsh *et al.* 1982). Only adult male dugongs and a few very old females have erupted tusks (Marsh 1980). Although erupted tusks wear during feeding, tusks do not seem to affect diet (Domning and Beatty 2007) in contrast to the situation in extinct dugongids (Chapter 3).

Seagrasses are undoubtedly the most important component of the dugong's diet (Figure 4.3; Tables 4.1–4.2; Supplementary Material Appendix Table 4.1), as they have been for most sirenians throughout their evolutionary history (see Chapter 3). Valentine and Heck (1999) review the evidence for animals eating seagrass and conclude that seagrass herbivory is still widespread in the world's oceans and carried out by sea urchins, gastropods, crustaceans, fish and water fowl, as well as by marine megaherbivores: the green turtle (*Chelonia mydas*), the dugong and the West Indian and West African manatees. Green turtles occur throughout the range of the dugong and the roles of dugongs and green turtles in modern seagrass

Figure 4.2. Recent sirenians and their food plants. (a) Dugong feeding by excavating in a low biomass seagrass bed in New Caledonia. Note the scars on the back inflicted by the tusks of an adult male dugong and the sediment clouds produced by bottom feeding (photograph by Luc Faucompré). (b) Three Amazonian manatees among freshwater vegetation (photograph by Anselmo d'Affonseca). (c) Florida manatee cropping the introduced aquatic water weed *Hydrilla* (photograph by USGS); (d) Florida manatee cropping shore vegetation (photograph by USGS). Note the difference in food biomass in (c) and (d) compared with (a). All photographs reproduced with permission.

ecosystems are often compared (discussed in 'Competition with green turtles' below).

Seagrasses are anchored in the sediment by their roots and rhizomes. The below-ground biomass (roots and rhizomes) is typically greater than the above-ground biomass (leaves) in the seagrass species eaten by dugongs (de Iongh *et al.* 1995, 2007; Lanyon and Marsh 1995a; Aragones and Marsh 2000; Masini *et al.* 2001; Table 4.3). The ratio of above-ground biomass to total biomass often varies seasonally (Preen 1992; de Iongh *et al.* 1995, 2007; Lanyon and Marsh 1995a) with a summer/wet season peak in several seagrass species. This difference is important as the nitrogen concentration in the leaves is typically greater than that in the rhizomes and roots, whereas the converse is generally true for starch (see 'Food quality' below).

As explained in 'Food acquisition and processing' below, it is difficult for a dugong to feed on individual species of structurally small seagrass in a mixed-species seagrass meadow. Under such circumstances dugongs

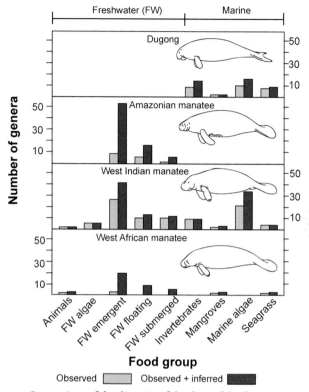

Figure 4.3. Comparison of the diversity of the diets of the four extant species of sirenians. The data on which this figure is based are found in Supplementary Material Appendix Tables 4.1–4.4. FW = fresh water. Drawing by Shane Blowes, reproduced with permission.

typically feed on several species of seagrass in a single dive (Johnstone and Hudson 1981) and often incidentally consume algae and invertebrates as well. Quantitative analysis of dugong stomach and mouth contents indicates that dugongs consume significant quantities of both seagrass leaves and rhizomes (Johnstone and Hudson 1981; Marsh et al. 1982; André et al. 2005), making it inappropriate to refer to the dugong as a rhizovore (but see Anderson 1998).

The range of the dugong (Chapters 1 and 8) is broadly coincident with the distribution of seagrasses in the tropical and subtropical Indo-West Pacific, but not all the seagrasses in this area are accessible to dugongs. Some seagrass meadows occur beyond the dugong's known diving range and may be permanently inaccessible. For example, in the Great Barrier

Reef region, Lee Long *et al.* (1996) recorded seagrasses to depths of 58 m. The deepest reported dugong dive is 36.5 m (Sheppard *et al.* 2006) and the deepest recorded dugong feeding trail is 33 m (Lee Long *et al.* 1996). Other seagrass meadows are inaccessible on a seasonal or daily basis. At the high latitude limits of the dugong's range (such as the Arabian/Persian Gulf or Shark Bay in Western Australia), some seagrass meadows are not used in winter, presumably because the water is too cold (Anderson 1986, 1998; Marsh *et al.* 1994; Preen *et al.* 1997; Masini *et al.* 2001; Gales *et al* 2004; Holley *et al.* 2006; Chapter 5).

Tides restrict dugong foraging on intertidal seagrass meadows on a daily basis (Anderson and Birtles 1978; Nietschmann and Nietschmann 1981; Sheppard *et al.* 2009; Figures 4.1 and 5.17a). In areas where the tidal range is large, such as tropical Australia, the area of intertidal seagrass meadows to which dugongs have limited access is vast. For example, Roelofs *et al.* (2005) estimated a total area of 70 000 hectares (700 km²) of intertidal seagrass meadows in the Gulf of Carpentaria, northern Australia.

As summarised in Figure 4.3 and the Supplementary Material Appendix Table 4.1, dugongs feed on nine of the ten genera and probably on most of the approximately 26 species of seagrass that occur within their range (Green and Short 2003), with at least two likely exceptions. Dugongs have not been confirmed to eat the robust, fibrous, temperate seagrasses *Posidonia australis* or *P. coriacea*, the ranges of which overlap the high latitude limits of the dugong's range in Western Australia.

Only four studies have quantitatively analysed the stomach or mouth contents of more than ten dugongs. Comparisons are compromised by the different methods used to record the data as well as the sampling constraints discussed above. Nonetheless, comparisons indicate that the relative importance of seagrass genera differs among locations (Table 4.1), and confirm that the relative importance of seagrass genera changes at the same location during times of seagrass loss from plant dieback and extreme weather events (Table 4.2).

Dugongs expand their diet opportunistically in times of nutrient shortages, when seagrass beds are seriously depleted. Such shortages may occur as a result of the natural or human-induced loss of seagrass, or be due to seasonal or tidal restrictions on access to entire plants or parts of plants. During 1972, a year after local seagrass beds were severely damaged by a cyclone, the brown algae *Sargassum* sp. was the predominant component of the stomach contents of 11 of the 49 dugongs drowned in nets near Townsville, North Queensland, Australia (Spain and Heinsohn 1973; Marsh *et al.* 1982), even though dugongs generally appear to have limited

Table 4.1. *Percentage of individual dugongs for which each seagrass genus was identified in the stomach or mouth contents based on three studies[1] from northern Australia–southern Papua New Guinea*

	Mabuiag region Torres Strait (Nietschmann 1984)	Daru region Torres Strait (Johnstone and Hudson 1981[2])	Various locations in north Queensland (Marsh et al. 1982)
No. dugongs sampled	16	102	95
Sampling site	Stomach	Mouth	Stomach
Date	1977	1978–1979	1968–1978
Seagrass			
Cymodocea	19	32	61
Enhalus	94	18	8
Halodule	6	67	95
Halophila	63	80	89
Syringodium	0	9	2
Thalassia	94	73	39
Thalassodendron	50	0	0
Zostera	0	0	3
Plants other than seagrass			
Algae n/i[3]	0	7	51
Avicennia	0	1	0

Dugongs usually had been eating more than one species of seagrass and more than one species from some genera.
[1] Results from André et al. (2005) not included as their results are not conducive to this comparison.
[2] Approximate values.
[3] n/i = not identified.

capacity to masticate and digest algae (Annandale 1905; Lipkin 1975; Marsh et al. 1982; but see Whiting 2002a and 2008). The relative importance of different seagrass genera in the stomach contents of the dugongs changed during the year after the 1971 Townsville cyclone (Marsh et al. 1982), and Spain and Heinsohn (1973) noted a considerable amount of silt and sand in the stomach contents at that time.

Preen and Marsh (1995) found algae, dead seagrass rhizomes and anoxic sediment in the stomachs of dugongs that died in the Hervey Bay region in south-east Queensland after the loss of more than 1000 km² of seagrass caused by two floods and a cyclone in 1992. Comparison of the diets of dugongs killed by indigenous hunters at Mabuiag Island in Torres Strait in 1977 (a time of seagrass dieback) and 1997–1999 (when the

Table 4.2. *Dugong diets change over time even at the same location, as demonstrated by comparing the average percentage by volume of items in the stomach contents of dugongs caught by hunters at Mabuiag Island in Torres Strait in 1977 (a time of seagrass dieback) and 1997–1999 (when the seagrass was healthy)*

	Nietschmann (1984)[1]	André et al. (2005)[2]
No. dugongs	16	128
Date	1977	1997–1999
Cymodocea	1	13
Enhalus	29	Low amounts
Halodule	1	Low amounts
Halophila	8	Low amounts
Syringodium	0	12
Thalassia	39	71
Thalassodendron	4	Low amounts
Algae n/i[3]	Not recorded	Low amounts

[1] Microhistological examination of stomach contents; note, percentages do not sum to 100% as some material was unidentified.
[2] Near infra-red reflectance spectroscopy examination of stomach contents.
[3] n/i = not identified.

seagrass beds were healthy; Table 4.2) indicates that the dugongs ate relatively more of the fibrous species *Enhalus acoroides* when seagrass was scarce. Remaining in an area but changing their diet (and thereby risking starvation) is one of two functional responses of dugongs in a time of food shortage; the other is to move from the affected area.

Dugongs may also change their diet on a seasonal basis. Such changes are most apparent at the high latitude limits of their range, where access to some seagrass meadows is limited by the water being too cold for dugongs at some times of the year (Chapter 5). Anderson (1998) observed dugongs feeding on *Halodule uninervis* in Shark Bay in summer, and inferred from their behaviour that they fed on *Amphibolis antarctica* (Anderson 1982) and *Halophila spinulosa* (Anderson 1994) at other locations in Shark Bay in winter; the last two species are relatively low in nutrients (see 'Food quality' below). *Zostera capricorni*[1] grows in two morphological variants, a broad-leafed form and a narrow-leafed form. This variation in leaf width, which also occurs in some other seagrasses (see *Halodule uninervis* below), is due to plasticity in growth form stimulated by local conditions (Waycott et al. 2004); the narrow-leafed forms are generally considered to be more desirable as food for dugongs as they are less fibrous (see 'Food quality' below). Although *Zostera capricorni* is not usually eaten in Moreton Bay, Queensland, Preen (1992) observed some dugongs feeding on the narrow-leafed morph of this species in winter, possibly because it grows near the passage leading from the bay to warmer, oceanic water. Preen (1992) also observed dugongs feeding on the narrow-leafed morph of *Zostera capricorni* in spring, when it was fruiting. Seasonal changes in diet are not limited to the subtropics; Johnstone and Hudson (1981) noted that the dugongs with algae in their mouths were sampled in winter in tropical Torres Strait.

In some environments, dugongs apparently subsist largely on algae. Whiting (2002a, 2008) documented dugongs closely associated for long periods with algal-covered, rocky reefs in the Northern Territory in tropical Australia. Bjorndal (1980) speculates that, for green turtles, individual selectivity for an algal or seagrass diet may reflect the specificity of the internal microflora. It is not known whether the gut microflora of dugongs eating algae for long periods differs from that of those eating mainly seagrasses.

During times of apparent nutrient shortage, dugongs also deliberately eat invertebrates, at least at the subtropical limits of their range such as Shark Bay and Moreton Bay in Australia (see Figure 8.7 for locations). Anderson (1989) reports that dugongs created circular craters as they foraged for thin-shelled burrowing mussels (*Botula vagina*) and possibly sea pens (*Virgularia*) in Shark Bay during winter. On the basis of direct observations of feeding trails and analysis of digesta, Preen (1995a) demonstrated that invertebrates such as ascidians (especially *Sycozoa pulchra*) and unidentified chaetopterid worms are an important part of the diet of dugongs in Moreton Bay, especially in early spring. As yet there is no scientific evidence that dugongs deliberately feed on invertebrates in the tropics, although dugongs certainly consume invertebrates incidentally when feeding on whole seagrass plants (Preen 1992; Figure 4.3; Supplementary Material Appendix Table 4.1). Preen (1995a) postulated that dugongs feed on invertebrates to meet their nitrogen requirement in times of seasonal shortage in the subtropics and speculated that dugongs may be less omnivorous in tropical areas where the growth of seagrasses and hence the availability of nitrogen is less seasonally variable. In addition, subtropical seagrass meadows are generally less diverse than tropical meadows (Green and Short 2003). Accordingly, dugongs have fewer choices among seagrass species in the subtropics.

Diet selection may be defined as an animal's choice of food from an array of available food items. Individuals of many species of terrestrial herbivores can be kept in cages, enabling the biomass and nutrient content of both the food offered and the food consumed to be measured with precision. We thus have a very good understanding of the basis for diet selection of many mammalian herbivores, particularly farm animals. Sirenians are very expensive to keep in captivity and the logistics of diet selection experiments are usually prohibitive. Only one 'cafeteria experiment' has been performed on the dugong, and it provides limited insights. This experiment consisted of only two trials on a single dugong that had been in captivity for 17 years (de Iongh 1996).

It is very difficult to confirm diet selection in the wild, where diet selectivity is usually measured by comparing the relative abundance of food in an animal's diet with the relative abundance of the foods available. This approach has limited applicability for herbivores as food quality is likely more important than food quantity (see 'Optimal foraging theory' above). The situation is further complicated when food availability changes with tide and season as for the dugong. Some researchers who have analysed dugong digesta or studied dugong behaviour have inferred that dugongs selectively target pioneer, low biomass genera of seagrasses, especially *Halodule* and *Halophila*, in preference to more fibrous climax genera, such as *Zostera*. Unfortunately, most of these claims are not supported by data on the biomass or nutrient concentrations of the available seagrasses. Nonetheless, there is some evidence that these claims are correct at certain locations. Preen (1992) ranked the dugong's preference for seagrasses in Moreton Bay, Queensland, in summer and autumn (when the abundance of seagrass is greatest) on the basis of: (1) frequency of encounter with feeding dugongs in different seagrass communities; (2) the relative abundance of different seagrasses in areas accessible to dugongs; and (3) the signs left by feeding dugongs. His ranking was: (1) *Halophila ovalis* (most preferred); (2) *Halodule uninervis* narrow-leaf morph; (3) *Syringodium isoetifolium*; (4) *Halodule uninervis* broad-leaf morph; and (5) *Zostera capricorni* broad-leaf morph.

This hierarchy of seagrasses as preferred food for dugongs is not clear and consistent across all locations or times. *Zostera capricorni* appeared to be the main seagrass eaten by dugongs in the inshore waters of a significant dugong habitat, Shoalwater Bay, Queensland, in 1975, although there was some evidence that they avoided dense old stands (Anderson and Birtles 1978). *Amphibolis antarctica* is almost certainly the seagrass eaten most often by dugongs in Shark Bay, the area that supports the second-largest known dugong population (indeed, the second-largest sirenian population) in the world (Marsh *et al.* 2002). *Thalassia hemprichii*, a climax species (van Tussenbroek *et al.* 2006), appears to be an important food of dugongs in Torres Strait (Tables 4.1 and 4.2), the area which supports the world's largest dugong population (Marsh *et al.* 2002). *Thalassia hemprichii* was found in the mouths of 73% of the dugongs from the Daru region of Torres Strait, a higher proportion than *Halophila ovalis* (67%) or *Halodule uninervis* (52%) (Johnstone and Hudson 1981). On the basis of stomach contents analysis, André *et al.* (2005) concluded that the dugongs caught by hunters in 1997–1999 in the Orman Reef area of Torres Strait (Table 4.2) were mainly eating *Thalassia hemprichii, Cymodocea* sp. and *Syringodium isoetifolium*, the seagrasses which dominated the area in biomass. *Halophila ovalis* and *H. spinulosa* were consumed only incidentally, despite being present at densities comparable to those in Moreton Bay, where Preen (1992) considered them preferred foods. *Enhalus acoroides* was present in abundance in Torres Strait, but was not targeted by dugongs in 1997–1999, in contrast to the time of seagrass dieback in the mid-1970s (Table 4.2). André *et al.*'s (2005) data suggest that diet selection by dugongs is not solely based on abundance, but that animals avoid seagrasses that are very high in fibre, presumably because they have difficulty masticating fibrous seagrasses (see 'Mastication' below).

Sheppard *et al.* (2010) conducted the most comprehensive study of factors influencing dugong diet selectivity. Their analysis was limited to a $24\,km^2$, intensively used seagrass habitat in subtropical Hervey Bay, south-eastern Queensland, in winter. Sheppard *et al.* tracked seven male dugongs at a fine spatial scale (<10 m) using GPS transmitters. They modelled resource selection within the habitat by comparing the dugongs' use of space (which they assumed was an index of their consumption of seagrass) with the species composition, biomass and nutrient characteristics of the seagrass landscape, which had been mapped in detail (Sheppard *et al.* 2007). Patterns of association between dugongs and the four seagrass species present, all of which were low biomass species, were complex and indefinite. Dugongs were associated with *Halodule uninervis* and *Halophila spinulosa* only on day-time low tides when the animals' habitat choices were limited by water depth and possibly vessel traffic. The dugongs were associated with *Halophila ovalis* only at intermediate tides at night. In general, the dugongs tended to avoid areas with a high density of *Halophila spinulosa* and *Zostera capricorni*. Sheppard *et al.*'s results are consistent with some of Preen's (1992) findings from Moreton Bay, but demonstrate that the selection of seagrass by dugongs is probably influenced by many factors other than the relative prevalence of seagrass species per se (as discussed below).

Herbivores make decisions about where to forage and for how long in response to both the biomass (which in turn influences the intake rate) and nutritional quality of their food plants (Searle *et al.* 2005). de Iongh *et al.* (1995, 2007), Anderson (1998) and Masini *et al.* (2001) concluded that dugongs preferentially feed on *Halodule uninervis* to maximise energy intake because its rhizomes are rich in starch (see 'Food quality' below). While not eliminating this possibility, Sheppard *et al.* (2010) demonstrated that the situation is more complex at their study site in Hervey Bay. The association of dugongs with seagrasses high in starch was positive during both day and night high tides when the animals could access the intertidal areas where the seagrass biomass was generally low. Nonetheless, the researchers found that the tracked dugongs were consistently associated with seagrass patches where nitrogen concentrations were relatively high, except during the day at low tides, when their choice was restricted. The dugongs then associated with seagrass patches of high biomass. Sheppard *et al.*'s (2010) finding that dugongs consistently seek foods high in nitrogen supports Preen's (1995a) explanation for their feeding on invertebrates at the high latitude limits of their range. We discuss why nitrogen might be a limiting resource for dugongs in 'Food quality' below.

Within their relatively specialised habitats of intertidal and subtidal tropical and subtropical seagrass meadows (Figures 4.1, 4.2a and 5.17a), dugongs are clearly capable of exploiting a relatively wide diet, including macro-invertebrates and algae (Figure 4.3). This capacity probably explains their large range. Thus we suggest that dugongs should be referred to as 'seagrass community specialists' rather than 'seagrass specialists' or 'rhizovores'. Unlike most extinct members of the Subfamily Dugoninae (Chapter 3), modern dugongs do not have to compete with other sirenians (or most other marine megaherbivores other than green turtles – discussed under 'Competition with green turtles' below). This situation has probably allowed modern dugongs to utilise a broader and more variable feeding niche within seagrass ecosystems than was occupied by some of the other marine sirenians of the past (Chapter 3).

Amazonian manatee

The Amazonian manatee is widely distributed in the freshwater rivers and lakes of the Amazon Basin below about 300 m in elevation. This region supports an amazingly diverse flora of aquatic and semiaquatic plants, including 'floating meadows' of true grasses. Amazonian manatees favour calm, shallow waters away from human settlements and where the productivity of aquatic and semiaquatic plants is high (Marmontel 2008; Figures 4.2b, 5.16). The availability of food plants varies with the annual 10–15 m changes in water levels resulting from seasonal rains. The inter-annual variation in the plant communities is also large (Rosas 1994; Junk and Piedade 1997; Arraut et al. 2010; Chapter 5). When water levels inundate the flood plains, Amazonian manatees have access to a wide variety of submerged, floating and emergent freshwater vegetation (Figure 4.3). As plant productivity increases while the river levels are rising, the new growth is nutritionally valuable and manatees have access to an abundance of high-quality food. When the water levels recede, the floating plants become dormant, or die and decompose, and the flood plains revert to a terrestrial plant community that is inaccessible to manatees. As is explained in Chapter 5, for the dry season Amazonian manatees migrate to the rias (long, narrow lakes formed by the partial submergence of river valleys), where food is limited and they are forced to fast for long periods when they may only eat algae, fallen leaves and other organic material, including rotting stalks and mud (Best 1981, 1983; Arraut et al. 2010). In at least some regions during the cyclic dry seasons, low water levels can limit access to food plants and force manatees to fast for 3–4 months (Best 1983); during prolonged dry seasons they may have to fast for up to seven months. This seasonal availability of food drives the Amazonian manatee's reproductive cycle (Chapter 6).

Because the river waters are typically very turbid, Amazonian manatees are rarely seen feeding in the wild and information on their feeding habits comes from the anecdotal reports of fishers, feeding signs on partially eaten plants and the examination of digesta. In the most comprehensive study to date, Guterres et al. (2008) used the first two methods to identify 63 species of plants potentially eaten by Amazonian manatees (Figure 4.3), especially in the high biodiversity várzea (white water) systems. As predicted by Domning (1978a) on the basis of their cranial anatomy (see 'Food acquisition' below), Amazonian manatees feed on a wide variety of aquatic and semiaquatic macrophytes, including emergent and floating plants (Supplementary Material Appendix Table 4.2). There are relatively few submerged plants available to Amazonian manatees because the low light levels restrict the growth of submerged vegetation in the rivers (Junk and Piedade 1997).

Colares and Colares (2002) identified the plant material in samples of stomach contents and faeces from Amazonian manatees at two study sites in the central Amazon. They identified eight species of plants, a maximum of three species per sample, from the digesta of manatees during high water when food was plentiful and the animals could be selective. The genera eaten during high water included *Eichhornia*, *Echinochloa*, *Hymenachne*, *Panicum*, *Paspalum*, *Pistia*, *Polygonum* and *Salvinia*. During low water when food was scarce, the manatees' diet was more opportunistic and diverse, expanding to 21 species (including seven of the eight eaten during high water), and up to seven species were identified per sample. Emergents comprised almost two-thirds of the species identified in the digesta; floating plants comprised about 30% and submerged plants the remainder. Terrestrial grasses were found in 96% of the samples analysed. Amazonian manatees have not been recorded eating animal material.

Amazonian manatees have access to a more diverse diet than dugongs. Their habitats support greater plant diversity and they have the capacity to feed throughout the water column. In addition, their diet is strongly influenced by the seasonal and highly variable climate of the Amazon Basin. Thus the Amazonian manatee's access to food changes seasonally, a situation analogous to that of the dugong in the subtropics, but contrasting with dugongs throughout most of their range where food access varies during the day according to the tidal cycle.

West Indian manatee

The West Indian manatee has a wide distribution that spans some 24 countries and territories in North, Central and South America, and the Antilles (Chapter 8). In Florida and South and Central America, manatees are frequently found in rivers and estuaries and along shallow, protected coastlines which support seagrass meadows. In at least some parts of the Greater Antilles, manatees appear to be more consistently found in marine environments than in the rest of their range (Powell *et al.* 1981; Lefebvre *et al.* 1989, 2001).

Thus the West Indian manatee exploits a wide variety of habitats (Figure 5.14), including: (1) Western Atlantic seagrass habitats analogous to the marine habitats used by dugongs in the Indo-West Pacific; (2) freshwater habitats similar to those occupied by Amazonian manatees in the Amazon Basin; (3) clear freshwater habitats where submerged vegetation is more available than in the Amazon Basin; and (4) estuarine habitats that occur between the manatees' marine and freshwater environments (Figure 1.3; Chapter 5). Domning (1982a) recognised the relationship between the wide feeding niche of the West Indian manatee and its cranial morphology (see 'Food acquisition' below).

West Indian manatees feed on submerged, floating and emergent plants depending on their availability (Hartman 1979), which, in turn, depends on the habitat and the depth to which light can penetrate. Nonetheless, and unlike the dugong, the West Indian manatees' use of marine habitats may be restricted by their need to access fresh water (Chapter 5). Manatees also appear to be restricted to feeding on seagrass meadows growing at depths of about 1–5 metres, and mostly feed in less than two metres (Hartman 1979; Lefebvre *et al.* 2000), in contrast to dugongs, the feeding trails of which have been recorded at water depths to 33 m (Figure 1.3). Throughout most of the range of the West Indian manatee (Chapter 8), the tidal range is generally small (<1 m), and in contrast to the dugong, access to food is generally not tidally constrained except in some mangrove and marsh habitats.

Like the Amazonian manatee, the West Indian manatee's food supply varies seasonally in flooded wetlands in areas with a pronounced wet season, such as the mangrove swamps of the Amapá coast of Brazil (Lefebvre *et al.* 1989). The wet season also influences manatee habitat use, and presumably its food, in the Greater Antilles, where manatees reportedly frequent river mouths during the rainy season (Lefebvre *et al.* 1989). At least in Florida, some foods and foraging areas are seasonally unavailable; the growth of aquatic vegetation is terminated or much reduced in winter, and seagrasses can suffer extensive leaf kill and algae die-off (Hartman 1979). Additionally, some summer foraging areas, particularly in northern Florida and south-eastern Georgia, are also unavailable to manatees in winter because they are too far from warm water aggregation sites (Deutsch *et al.* 2003). Manatees use seagrass beds offshore from the Florida Everglades more in summer than in winter, when they are drawn to the warmer waters of inshore and fluvial habitats with the passage of cold fronts and related thermal stress (Stith *et al.* 2006; Chapter 5).

There are numerous observations of manatees feeding in the wild, particularly in Florida (Figures 4.2c and 4.2d). Daniel (Woody) Hartman (1979) provides the most comprehensive and enduring account. He describes the Florida manatee as an indiscriminate and opportunistic feeder that eats a wide variety of plants both aquatic and terrestrial, fibrous and non-fibrous, vascular and non-vascular, preferring submerged, floating and emergent vegetation in that order. In waters clear enough to support submerged vegetation, manatees feed almost exclusively underwater, largely ignoring natant and emergent vegetation. In turbid waters, where submerged vegetation is unavailable, manatees feed on floating vegetation. Especially in waters without aquatic vascular flora, manatees resort to feeding on freshwater or marine algae and bank growth. Zoodsma (1991) observed manatees feeding on the cordgrass *Spartina alterniflora*, moving along the marsh edges and taking the grass at or just above the water's surface.

Most of Hartman's (1979) conclusions about manatee feeding were based on his keen natural history observations. His deductions are supported by subsequent field and laboratory studies using a range of techniques, including direct observations of wild animals or their feeding signs, inferences from the behaviour of wild animals and examination of stomach contents. Supplementary Material Appendix Table 4.3 lists the foods that various authors have noted that West Indian manatees eat in various parts of their range. As summarised in Figure 4.3, these foods include four genera of seagrasses, 34 or more genera of non-epiphytic marine algae (most of which are probably eaten incidentally), two genera of mangroves, 11 genera of freshwater submerged plants, 12+ genera of freshwater floating plants, 40+ genera of terrestrial emergent plants, five genera of freshwater algae, and invertebrate and vertebrate animals. The list is unlikely to be complete and does not include the complex mixture of algae, cyanobacteria, heterotrophic microbes and detritus that is attached to aquatic vegetation in fresh or salt water. West Indian manatees are certainly not fussy eaters!

Hartman (1979) considered that algae do not figure significantly in the diets of Florida manatees unless vascular vegetation is in short supply, such as in Tampa Bay during the winter of 1981 (Lewis *et al.* 1984). Like dugongs, Florida manatees mainly eat algae incidentally, but the gastrointestinal tracts of 84 manatees examined by Ledder (1986) included those from five animals that had consumed algae exclusively in the days prior to their deaths. All of these manatees were recovered from the west coast of Florida and included males and females, adults and juveniles. Hartman (1979) occasionally observed Florida manatees foraging along the bottom, nibbling on the surface of the substrate and chewing detritus. Ledder (1986) concluded that the low amounts but high incidence of mangrove material in the manatee digesta she examined is evidence that manatees consume detritus in seagrass beds, although whether such consumption was incidental or targeted is uncertain.

Florida manatees eat most of the aquatic weeds that have invaded US waterways since the late nineteenth century, as evidenced by the overlap between the weeds listed by the Center for Aquatic and Invasive Plants at the University of Florida (http://plants.ifas.ufl.edu) and the manatee foods listed in the Supplementary Material Appendix Table 4.3. In particular, *Hydrilla verticillata* and water hyacinth (*Eichhornia crassipes*) became major components of manatee diets in some locations (Hartman 1979). The biomass of these weeds is typically much higher than the endemic foods of manatees in many freshwater areas in Florida and elsewhere (Table 4.3; Figure 4.2c), a factor that may have contributed to the increase in the population of Florida manatees observed in the last decades of the twentieth century (see Chapters 6 and 8).

Manatees can be remarkably opportunistic in their diets, particularly when using waterways with impoverished vascular flora. On the east coast of Florida, manatees feed on the algae growing on jetties, pilings, floats, docks, mooring lines and the hulls of boats (Hartman 1979). O'Shea (1986) observed manatees feeding on the mast of the live oak, *Quercus virginiana*. The acorns were lying on the bottom of a thermal refuge, the Blue Spring Run in Volusia County, Florida during autumn and early winter. In contrast to wild dugongs, which have never been reported feeding on seagrass drift, Hartman (1979) described manatees congregating to feed on floating mats of *Syringodium filiforme* and *Halodule wrightii* blown inshore by the wind. He also told of manatees moving up canals to flood-control dams to feed on water hyacinth and other vegetation pouring through the gates. Manatees have even been seen eating lawn grass trimmings that have been cast upon surface water and coalesced in mats.

The opportunistic feeding of manatees is not restricted to plant material; they also eat invertebrate and vertebrate animals and even their own faeces. The Florida manatees killed in a mass die-off in late winter and early spring 1982 had eaten thousands of tunicates of the genera *Molgula* and *Styela*; the ascidians were hypothesised to have built up dinoflagellate (red tide) toxins that may have been lethal to the manatees (O'Shea *et al.* 1991; Chapter 7). Powell (1978) interviewed fishers in northern Jamaica who reported manatees systematically visiting their gill nets to deflesh small, scaled fish (carangids and scombrids) caught in the nets. Tyson (in Smith 1993) observed manatees eating the carcasses of flounder, redfish and grouper from waters adjacent to fish-cleaning tables in south-eastern Florida and saw a manatee surfacing with a live flounder. Bonde (in Smith 1993) photographed a manatee feeding on a dead rat that floated out from under a dock. Although coprophagy is not considered a significant contribution to the diet of West Indian manatees, wild Florida manatees have been observed eating manatee faeces (Hartman 1979; Best 1981). Coprophagy may be more prevalent in winter when manatee movements are restricted to thermal refugia where food supplies have been depleted.

Information on seasonal variation in the food of West Indian manatees is largely unquantified. However, Ledder (1986) used stomach contents analysis to quantify seasonal differences in the diet of manatees in south Florida that fed in both fresh and salt water. She found that manatees consumed more subsurface parts of plants (particularly seagrasses) in winter than in spring or autumn. This result is not surprising given that Florida seagrasses in some parts of the state suffer extensive leaf kill in winter (Hartman 1979), but may also be indicative of manatees needing a diet with a high calorie content when exposed to cooler temperatures (Ledder 1986). Alves-Stanley *et al.* (2010) could find no differences in the diets of Florida manatees on the basis of collection month using stable isotopes in skin. However, the samples were mainly collected in winter. Alves-Stanley *et al.* (2010) detected seasonal differences in similar samples from Belize. These differences could be the result of seasonal changes in manatee feeding, habitat use or the stable isotope ratios in their food plants, but should be regarded with caution because the sample sizes are small. Castelblanco-Martinez *et al.* (2009) found no seasonal differences in the diet of manatees in Chetumal Bay, Mexico on the basis of analysis of contents of the digestive tract and faeces.

Ledder (1986) also tested whether the diets of manatees from south Florida differed by age and sex. The most common plant species in the diet were equally represented in males and females. The seagrass *Syringodium* and the freshwater grass *Panicum* made a greater contribution to adult diets; algae contributed significantly more to the diets of juveniles. However, this conclusion was confounded with dietary differences resulting from the geographic origin of carcasses and needs to be substantiated. Alves-Stanley *et al.* (2010) could find no age- or sex-based differences in the diets of Florida or Antillean manatees using stable isotopes in skin. Castelblanco-Martinez *et al.* (2009) also found no age or sex differences in the diet composition of manatees in Chetumal Bay, Mexico on the basis of analysis of contents of the digestive tract and faeces.

Somewhat surprisingly, there have been few rigorous studies of the West Indian manatee's food preferences relative to availability in the wild. There are numerous anecdotal observations of manatees allegedly preferring or avoiding specific plants, but the conclusions are location specific. For example, Hartman (1979) found that manatees ate submerged plants – mainly the submerged introduced waterweed *Hydrilla verticillata* (Figure 4.2c) – at Crystal River, where floating vegetation held little attraction and emergent plants were ignored. In contrast, at the St Johns River, Hartman (1979) regularly observed manatees feeding on water hyacinth (*Eichhornia crassipes*) and alligator weed (*Alternanthera philoxeroides*), but they showed no interest in water lettuce (*Pistia stratiotes*). Powell and Waldron (1981) hypothesised that eelgrass (*Vallisneria americana*) was the preferred food of manatees in the St Johns River after observing manatees passing luxuriant rafts of *E. crassipes* to feed on small growths of eelgrass in extremely shallow water (< 0.4 m) by manoeuvring themselves with their flippers and cropping the eelgrass at its base. In the marine environment, Lefebvre *et al.* (2000) provided some evidence that manatees feeding in

Pelican Cove, a small embayment in eastern Puerto Rico, targeted *Halodule wrightii* in preference to *Thalassia testudinum* despite the former's more restricted distribution.

Bengtson (1981) conducted an experimental feeding study to test the food preferences of wild Florida manatees. He first presented manatees with 23 potential food plants on a feeding tray. The manatees ate most of the aquatic and shore-type species and avoided woody terrestrial plants. However, two herbaceous aquatic plants were shunned: *Nuphar* sp. (spatterdock) and *Hydrocotyle umbellate* (water pennywort); both contain high concentrations of toxic secondary compounds, a result consistent with Reynolds' (1981a) observations that Florida manatees may avoid certain algae that contain toxins. Bengtson then presented the manatees with a choice of four known food species. In 68.2% of trials the manatees ate the plant first encountered when they approached the tray and then moved from one end of the tray to the other as they finished eating at each site. This result may have been compromised by the amounts of vegetation supplied, preventing the manatees from feeding to satiation on a subset of the food plants.

The relative importance of marine, brackish and freshwater species to Florida manatees undoubtedly varies with location. From her study of the digesta of 84 manatee carcasses from south Florida, Ledder (1986) concluded that 39% had fed primarily on marine plants, 44% on freshwater plants and 17% on brackish water plants. Most individuals had 3–7 species of plants in their alimentary canal at the time of sampling. Florida presents a mosaic of habitat choices for manatees, and feeding habitat usage can be expected to be variable. In some areas of Florida, manatees feed entirely on freshwater vegetation year-round (such as the upper St Johns River; Bengtson 1981), whereas in other areas they utilise habitats that are largely estuarine or marine nearly all year (Deutsch *et al.* 2003).

As explained above, stable isotope analysis is increasingly being used to quantify the relative importance of freshwater and marine foods to manatees at a population level (Ames *et al.* 1996; Reich and Worthy 2006; Alves-Stanley and Worthy 2009; Alves-Stanley *et al.* 2010). Alves-Stanley and Worthy (2009) studied newly captive Florida manatees that were transitioning from a diet of aquatic forage to lettuce to measure the mean half-life for the turnover of the stable isotopes of carbon and nitrogen in manatee epidermis. The results were $\delta^{15}N$: 53 days for manatees rescued from coastal regions, 59 days for those rescued from riverine regions; $\delta^{13}C$: 27 days (coastal regions), 58 days (riverine). In the most comprehensive study to date, Alves-Stanley *et al.* (2010) analysed δ^{15}N and δ^{13}C in skin samples from 118 free-ranging Florida manatees and 91 wild Antillean manatees. They established that seagrasses were included in the diets of Florida manatees from all regions sampled, although dependence on seagrasses varied with location of capture. Seagrasses were a critical component of the diet of Antillean manatees in Belize and even more so in Puerto Rico; the last result is consistent with observations of their habitat use (Powell *et al.* 1981). These results support the hypothesis of Lefebvre *et al.* (2000), that manatees in Florida are less specialised grazers than manatees in Puerto Rico, where seagrasses are the major available food resource.

West Indian manatees are clearly habitat generalists (Figures 1.3 and 4.3) capable of opportunistically exploiting a remarkable array of natural and introduced foods across marine, estuarine and freshwater habitats. The costs and benefits of this feeding niche are discussed below in 'Comparative overview and implications for conservation'.

West African manatee

As summarised in Chapters 1, 5 and 8, the West African manatee occurs in most of the sheltered coastal marine waters, brackish estuaries and adjacent rivers along the coast of West Africa from southern Mauritania (16° N) to the Loge, Dande, Bengo and Cuanza rivers of Angola (18° S) (Powell and Kouadio 2008). Powell (1996) described their optimal coastal habitats as: (1) coastal lagoons with abundant growth of mangrove or emergent herbaceous growth (Figure 5.15a); (2) estuarine areas of larger rivers with abundant mangroves in the lower reaches and lined with grasses upriver; and (3) shallow protected coastal waters with fringing mangroves or large marine plants. In riverine habitats that have large fluctuations in water levels and flow rates, West African manatees prefer areas that provide them with access to deep pools or lakes during the dry season and that flood into swamps or forests during the wet season (Figure 5.15b). In general, the habitat requirements of *Trichechus sengalensis* seem similar to those of *T. manatus* (Powell and Rathbun 1984; Chapter 5): sheltered water with access to food and fresh water, a conclusion consistent with the geologically recent separation between the two species (Chapter 3). However, West African manatees probably use marine habitats less than their cross-Atlantic cousins because the West African continental shelf is narrow with high-energy beaches, a combination which restricts suitable marine habitats (Powell 1996).

In comparison to the other sirenians, especially the West Indian manatee, the diet of the West African manatee is poorly known and the diversity of plants eaten is almost certainly seriously underestimated. Their known food plants include representatives of all the plant biotopes eaten by the West Indian manatee with the exception of freshwater algae (Supplementary Material Appendix Table 4.4). These omissions are inconclusive given the paucity of data. West African manatees are known to eat at least 40 genera of plants, including two genera of seagrass, two genera of mangroves, five genera of submerged freshwater plants, nine genera of floating freshwater plants and some 20 genera of emergent freshwater/ terrestrial plants (Figure 4.3). In Senegal (J. Powell, unpublished) and Sierra Leone (Reeves et al. 1988), manatees are known to eat small fish captured in fishermen's nets. In Senegal and the Gambia, shell remains of freshwater clams have also been found in their stomachs. Villagers also report manatees eating fine clay, algae and detritus in the dry season (Powell 1996). Like other sirenians, West African manatees seek roots rich in carbohydrates, allegedly by scratching the shoreline soil with their flippers (Dodman et al. 2008). Hunters use tubers and mangrove fruits as bait to catch manatees.

West African manatees can be extremely opportunistic in their dietary choices. Powell (1996) presents anecdotal information suggesting that this species:

(1) eats the fallen fruit of trees growing on the banks of rivers;
(2) feeds in flooded forests;
(3) removes the flesh from small fish from fishermen's gill nets and from freshwater bivalves;
(4) follows hippopotami around, allegedly to eat their dung;
(5) eats cassava peels and other food scraps discarded by villagers;
(6) eats aquatic weeds such as *Eichhornia crassipes*, *Azolla africana*, *Salvinia nymphellula*, *Lemna aequinoctialis* and *Pistia stratiotes*;
(7) consumes rice plants in rice paddies where they are considered a serious pest by farmers (Reeves et al. 1988; Silva and Araujo 2001); and
(8) roots along the bottom during the dry season, eating algae, fallen leaves and sticks.

We predict that when more information becomes available, the feeding niche of the West African manatee will prove to be as wide as that of the West Indian manatee.

On the basis of their cranial morphology, Domning (1982a) deduced that West African manatees are adapted to feed on emergent or overhanging river bank plants growing extremely high in the water column. The rivers in most of West and Central Africa are very turbid, limiting the growth of submerged macrophytes (Powell 1996). Powell (1996) presents anecdotal information that indicates that West African manatees (like West Indian manatees) will partially come onto land to feed if the soil is moist. Nonetheless, they eat at least two genera of seagrasses, suggesting that they are also effective bottom feeders. The lack of knowledge of the dietary requirements of this species is a serious impediment to a comprehensive understanding of its habitat needs.

FOOD ATTRIBUTES

Plant biomass and productivity

Most of the research on the abundance of the food plants eaten by wild manatees and dugongs has focused on seagrasses. Table 4.3 summarises the data on the undisturbed plant biomass and total plant productivity of seagrasses in areas where Florida manatees and dugongs have been observed feeding. The below-ground (below-substrate) biomass of seagrass is generally much greater than the above-ground (above-substrate) biomass. The ranges of both the above-ground biomass ($5-208 \, g \, ww \, m^{-2}$) and the below-ground biomass ($8-1330 \, g \, ww \, m^{-2}$) are considerable.

The aquatic weed *Hydrilla verticillata* (Figure 4.2c), an important rooted food plant of Florida manatees in fresh water, formed extensive, detached, floating mats at Crystal River in the last decades of the twentieth century (Hartman 1979). Its biomass exceeded that of most seagrass meadows where sirenians feed (compare Figures 4.2a and 4.2c); the biomass of another aquatic food plant that floats entirely on the water surface, the water hyacinth *Eichhornia crassipes*, can be more than an order of magnitude greater again. The productivity of *E. crassipes* also vastly exceeds that of even very productive seagrasses such as *Halophila ovalis*. The fact that both marine and freshwater plants are important to Florida manatees at the levels of populations, subpopulations and individuals (Ledder 1986; Reich and Worthy 2006; Alves-Stanley *et al.* 2010; Figure 4.3; Supplementary Material Appendix Table 4.3) suggests that, like terrestrial herbivores, manatees select foods on the basis of quality as well as biomass.

Food quality

Herbivores are faced with food that varies in quality and quantity. As pointed out by Aragones *et al.* (2006; 2012a), a definitive set of determinants of food quality for sirenians is not yet available. The obstacles to assessing food quality more precisely require long-term experiments using captive animals, with diets varying in the components of interest, to measure relationships

Table 4.3: *Plant biomass and total plant productivity for various food plants of dugongs and manatees. The seagrass values are for known sirenian feeding sites. The values for freshwater weeds are for cultivated plants; the values for wild plants are likely to be somewhat lower*

Species	Plant biomass (g) wet weight m⁻²		Total plant productivity (g) dry weight m⁻² d⁻¹	Location	Sources
	Above-ground	Below-ground			
Seagrasses					
Halophila ovalis	5	8		Moreton Bay, Queensland, Australia	Preen (1992)
Halodule uninervis	10–40	11–32	7.5–15.6	Koh Bae Na, Thailand	Nakaoka and Aioi (1999)
	13–17[1]	138–179[1]		Moreton Bay, Queensland, Australia	Preen (1992)
	80–188	529–1330		Nang Embayment, Indonesia	de Iongh *et al.* (1995)
	63–144[1]	360–585[1]	0.1–1.6	Shark Bay, Western Australia	Masini *et al.* (2001)
Halodule wrightii	94[1]	298[1]		Indian River lagoon, Florida, USA	Lefebvre *et al.* (2000)
Syringodium filiforme	412[1]	914[1]		Indian River lagoon, Florida, USA	Lefebvre *et al.* (2000)
Syringodium isoetifolium	96–208	238–349		Moreton Bay, Queensland, Australia	Preen (1992)

Table 4.3. (*cont.*)

Species	Plant biomass (g) wet weight m^{-2}		Total plant productivity (g) dry weight m^{-2} d^{-1}	Location	Sources
	Above-ground	Below-ground			
Freshwater plants[2]					
Azolla			2.9		Debusk and Ryther (1987)
Cyperus			≤22.9		Debusk and Ryther (1987)
Eichhornia	≤33 625		24.2		Ryther et al. (1978); Debusk and Ryther (1987); William T. Haller, Center for Aquatic and Invasive Plants, University of Florida, personal communication
Hydrilla	≤2690		4.2		
Lemna			4.5		Ryther et al. (1978)
Pistia			14.2		Debusk and Ryther (1987)
Typha			≤52.6		Debusk and Ryther (1987)

[1] Assuming dry weight:wet weight ratio of 1:9.
[2] All freshwater plants listed here were grown in cultivation ponds under controlled conditions.

between diet and digestive efficiency, body condition and health, and other measures of performance. However, it is possible to make some predictions in spite of this lack of experimental data (as discussed below). Despite their aquatic lifestyles, dugongs and manatees face dietary challenges similar to those of other large herbivorous mammals, and the vast literature on terrestrial species informs the interpretation of the sparser data on sirenians.

Forage nutrients are conveniently divided into the components making up the structure of the plants (cell wall) and the nutrients contained inside the cell wall (cell contents). The cell content components studied in the plants eaten by sirenians have generally been protein, total nitrogen, water-soluble carbohydrate and starch. Nitrogen is a proxy for protein, whereas water-soluble carbohydrate is the most rapidly digestible part of the non-structural carbohydrate in plant tissues, and starch is the most important of the storage carbohydrates in plants (Aragones et al. 2006; 2012a).

The components of plant cell walls are cellulose and hemicellulose and related polymers, principally lignin (Parra 1978). Herbivorous mammals do not produce the enzymes required to digest cellulose and other fibrous components of their diets (Van Soest 1994). Hence some portion of their digestive tract must be modified to contain the symbiotic organisms capable of hydrolysing the substances resistant to the digestive enzymes secreted by the host animal. The main fermentation chamber in a herbivorous mammal can be in the foregut or the hindgut. Sirenians are hindgut fermenters (see 'Digestion' below).

Various biologists have measured the nutrients in some food plants of sirenians, especially the seagrasses that are known to be eaten by dugongs. The early assessments tended to be based on unreliable measures of food quality, such as caloric value (e.g. Birch 1975). In recent years, researchers have used the more established chemical measures developed for domestic herbivorous mammals (for a discussion of methods, see Van Soest 1994). Near infra-red reflectance spectroscopy, a widely accepted method for the determination of the chemical attributes of organic materials (Lawler et al. 2006), has also been used.

Nitrogen and protein

The structural building blocks of animal tissues are proteins, whereas in plants they are carbohydrates. Animal tissues are usually about 10% nitrogen, yet plant tissues are often as low as 1–2% nitrogen (Bentley and Johnson 1991). Consequently, nitrogen is often the dietary component in short supply for herbivorous mammals, particularly young, growing animals and females in the late stages of pregnancy and lactation (Frape 2004). Sirenians have fewer opportunities to capture the nitrogen synthesised by gut microbes than foregut fermenters like ruminants, because as hindgut fermenters most of their gut microbes are in the colon, relatively late in the gut passage (Van Soest 1994). So we might expect nitrogen to be a major limiting nutrient for sirenians, as also suggested by Lanyon (1991).

Although nitrogen has not been confirmed as the major criterion for food selection by sirenians, some data support this hypothesis. As discussed above, Sheppard *et al.* (2010) reported that the dugongs they tracked (which were all males, rather than pregnant or lactating females or rapidly growing young calves with greater nitrogen needs) were consistently associated with foods with relatively high nitrogen concentrations, except at low tide when the availability of such food plants was limited. Both small and large groups of dugongs feed repeatedly on areas of seagrass (de Iongh *et al.* 1995, 2007; Preen 1995b), suggesting that they take advantage of the increased nitrogen concentrations in new growth, despite the concomitant decrease in starch (energy) (see 'Sirenian–plant interactions' below). The importance of *Thalassia* leaves (which are rich in nitrogen) to dugongs in Torres Strait, the region which supports more sirenians than anywhere else in the world, and the tendency of sirenians to feed opportunistically on animals (see Supplementary Material Appendix Tables 4.1, 4.3 and 4.4) are additional arguments in favour of nitrogen being important for sirenian nutrition.

Some postgastric-fermenting mammals practise coprophagy to capture microbial proteins and vitamins (Van Soest 1994). As discussed above, West Indian manatees have sometimes been observed eating their faeces both in the wild and in captivity (Hartman 1979; Best 1981) and dugongs have been observed eating faeces in captivity (Lanyon and Marsh 1995b). Like other large herbivores, sirenians are not considered to be obligate coprophagers, although coprophagy may be an important source of gut microbes for young and sick animals (see 'Digestion' below). Rather, sirenians probably use microbial protein originating in the large intestine to supply essential amino acids lacking in the diet (Parra 1978), although this possibility has not been formally investigated.

The available data (Murray *et al.* 1977; Best 1981; Tucker and DeBusk 1981; Anderson 1986; Silverberg 1988; Duarte 1992; Aragones 1996; Sheppard *et al.* 2007, 2008) indicate that whole-plant nitrogen levels for seagrasses, the aquatic plants mostly eaten by sirenians, and macro-algae are mostly 1–4% of the total nutrients, within the range for horse forage (Frape 2004) (assuming a nitrogen to crude protein conversion factor of 5.7; Sosulski and Imafidon 1990). The nitrogen concentration in seagrass leaves is often more than twice that of the rhizomes (Lanyon 1991; Aragones 1996; Sheppard *et al.* 2007, 2008), although *Halophila spinulosa* and *Syringodium isoetifolium* can be exceptions to this pattern. The nitrogen values for the leaves of various species of *Thalassia* (1.6–3% dry matter [DM]; van Tussenbroek *et al.* 2006) and *Halodule* (>1.9% DM; Lanyon 1991; Provancha and Hall 1991; Aragones 1996; Sheppard *et al.* 2007) are typically relatively high; *Amphibolis antarctica* leaves are relatively low in nitrogen (~1–1.2% DM; Walker and McComb 1988). Thus the concentrations of nitrogen in seagrasses are variable, but not dissimilar to those of some terrestrial grasses. As explained in 'Secondary compounds', this generalisation could be negated by the fact that the values reported here represent total nitrogen, rather than digestible nitrogen.

Water-soluble carbohydrates and starch
Despite the arguments for the importance of nitrogen, de Iongh *et al.* (1995, 2007), Anderson (1998) and Masini *et al.* (2001) claim that dugongs favour foods rich in starch and water-soluble carbohydrates on the basis of their alleged preference for the rhizomes of *Halodule uninervis*. These claims ignore two salient facts: (1) relative to most other seagrasses, the leaves of *Halodule uninervis* are generally also rich in nitrogen (see below); and (2) the stomach contents of dugongs eating *Halodule* sp. contain a substantial proportion of leaves as well as rhizomes (Marsh *et al.* 1982). Nonetheless, it is likely that access to energy is more important to dugongs and manatees at some times of year and at some latitudes than others, making the energy from the starch in rhizomes relatively more important. The fact that seagrass leaves are generally richer in nitrogen and fibre, whereas the roots and rhizomes are richer in non-fibrous carbohydrate, confirms the advantage to sirenians of eating the leaves, roots and rhizomes of seagrass when they are accessible, rather than the leaves only.

Aragones (1996) measured starch and water-soluble carbohydrate in the above- and below-ground components of eight species of seagrass. Concentrations were always higher in the below-ground components with the exception of *Halophila spinulosa*, for which the starch concentration was higher in the leaves. Sheppard *et al.* (2007, 2008) showed that these patterns are not always consistent across species or locations. It is also likely that these concentrations vary seasonally, but such variation has not been investigated to our knowledge. Lanyon (1991), Aragones (1996) and Sheppard *et al.* (2007) all found that, relative to other species of seagrass, the roots and rhizomes of *Halodule uninervis* are consistently rich in non-structural carbohydrates, particularly starch: the total concentration of water-soluble carbohydrate and starch averages ~14% in rhizomes of this species (Aragones 1996). The rhizomes of *Syringodium isoetifolium* can also be rich in these components (average ~10%;

Aragones 1996), but the concentration of starch declines with depth (Sheppard *et al.* 2008). Sheppard *et al.* (2007) also found that total starch was higher in intertidal than subtidal *Zostera capricorni*. Silverberg (1988) recorded soluble carbohydrate concentrations in some freshwater plants in spring that are substantially higher even than the values recorded for *Halodule uninervis* (e.g. *Myriophyllum spicatum* up to 45%, *Ruppia maritima* up to 38%).

Structural carbohydrate and fibre

As discussed below, hindgut fermentation has been confirmed as an important source of energy in sirenians. Murray *et al.* (1977) measured high values for the products of fibre fermentation (volatile fatty acids – VFAs) in the caecum and colon of a dugong, and Murray (1981) noted that the relative concentrations of the principal VFAs produced were typical of those of terrestrial ruminants feeding on highly fibrous diets. Goto *et al.* (2004a) used microorganisms obtained from the faeces of two captive dugongs to demonstrate degradation of *Zostera marina* in vitro. Thus we would expect the amount of structural carbohydrate or fibre in the diet of sirenians to be important (Van Soest 1994).

The standard method developed by Robertson and Van Soest (1981) to analyse the fibre content of plant foods first measures neutral detergent fibre (NDF) (essentially the cell wall and usually equated with this fraction). The neutral detergent residue is then subjected to acid detergent extraction. The resultant acid detergent fibre (ADF) comprises lignin, cellulose and cutin, the values of each of which can be sequentially estimated by further extraction. ADF is widely used as a quick estimate of the fibre content in feeds (Van Soest 1994). Hemicellulose (plus some protein attached to cell walls) is usually determined by the difference between ADF and NDF. Lignin is the most significant factor limiting the availability of nutrients in plant cell walls to herbivores because it is generally indigestible. Lignin gives rigidity to cell walls and is often used as a natural marker in digestive physiology (see 'Digestion' below).

Fibre levels in the freshwater plants eaten by sirenians have generally been measured as cellulose expressed as a percentage of DM, whereas those in seagrasses are generally given as a percentage of NDF and ADF, making comparisons difficult. Best (1981) listed the cellulose values of 14 freshwater food plants of manatees: values range from 14% DM in *Vallisneria spicatum* to 37% DM in *Potamogeton crisais*. Lanyon (1991) measured cellulose in four species of seagrass: values ranged from means of 13% DM in the roots and rhizomes of *Halophila ovalis* to 22% DM in the leaves of *Zostera capricorni*. These values are generally less than the fibre levels in the forage plants of terrestrial herbivores (Best 1981), presumably reflecting the reduced need for structural reinforcement for plants growing in water as Lanyon and Sanson (2006a) point out. The mean NDF concentrations in seagrass leaves range from means of 32% DM in *Halophila ovalis* to 63% DM in *Zostera capricorni* and are often higher than the corresponding values for rhizomes (Lanyon 1991; Aragones 1996; Sheppard *et al.* 2007, 2008). The differences among seagrass species are not always consistent, presumably reflecting the ages of the plants sampled. Typical NDF values for terrestrial grasses are in the 50–70% range, whereas browse values tend to be in the region of 30–50% (Van Soest 1982). Seagrass is clearly not more fibrous than terrestrial forage.

Secondary compounds

From a herbivore's perspective, the nutritional quality of a plant depends on its nutrient content and the ability of the animal to extract the nutrients. Plant secondary metabolites may reduce the availability of some chemical components of plants to herbivores and may constrain both reproductive fitness and habitat use (e.g. Foley and Moore 2005; DeGabiel *et al.* 2009). The role of these compounds in the foods eaten by dugongs and manatees has received limited attention.

As noted above, Bengtson (1981) and Reynolds (1981a) observed that Florida manatees may avoid species with high concentrations of toxic secondary compounds. Best (1981) noted that the tannin concentrations of most species of manatee food plants that had been analysed were relatively low (2–3% DM) but that levels were high in the freshwater macrophytes *Cabomba caroliniana* (15.6% DM) and *Potamogeton crisais* (7.2% DM). Lanyon (1991) determined that the rhizomes of *Cymodocea serrulata* had higher condensed tannin values (6.3% DM) than the leaves or rhizomes of the other three species of seagrass she tested. The significance of this result to the nutritional value of *Cymodocea serrulata* to dugongs has not been investigated.

DeGabriel *et al.* (2008) developed a simple, integrative assay to quantify nutritional quality of food plants for herbivores and provide some index of the detrimental effects of tannins. Using this technique with sirenian food plants promises important insights into the availability of nitrogen in the foods of sirenians.

Minerals

Few researchers have investigated the mineral content of sirenian seagrass foods, assuming that mineral values are unlikely to be limiting (but see Anderson 1986; Provancha and Hall 1991). In contrast, Central American freshwater plants are known to be low in calcium and sodium, reflecting the geochemical deficiencies of the region (Best 1981). Brian Beatty (personal communication, 2009) has photographed two Florida manatees eating lime-stone, presumably to overcome a calcium deficiency in their diet. One of this book's authors (O'Shea) has also repeatedly observed Florida manatees eating clay-like deposits exposed along a short section of the submerged bank at Blue Spring on the St Johns River, presumably for the mineral content. The need to balance their mineral intake may be one reason that West Indian manatees consume low biomass seagrasses as well as higher biomass fresh-water aquatic plants. Birch (1975) measured the chloride and sodium con-centrations in seagrass and found that they were, respectively, about 10 and 20 times the corresponding levels in terrestrial pasture grasses. Nonetheless, as Ronald *et al.* (1978) pointed out, the mineral content of aquatic weeds growing in sewage or industrial wastewater can be exceptionally high.

Water content

Best (1981) noted that although the nutrient concentrations of the plants eaten by sirenians may be reasonably similar to those of the plants eaten by terrestrial herbivores when expressed as a percentage of dry matter, they are very different when expressed as a percentage of wet weight. Both marine and freshwater plants have a much higher proportion of water (> 90%) than terrestrial grasses and other forage (75–80%). Best (1981) assumed that sirenians would need to eat relatively large quantities of aquatic plants to achieve a reasonable nutrient intake.

In vitro dry matter digestibility

In vitro dry matter digestibility (IVDMD) attempts to simulate the digestive processes of herbivorous mammals using the enzymes pepsin and cellulase, and potentially integrates the individual assays for the other dietary compo-nents. Thus Aragones *et al.* (2006) consider IVDMD to be the most informa-tive measure of the quality of the foods of dugongs. Aragones (1996) measured the IVDMD of the leaves, roots/rhizomes and whole plants of ten seagrass species eaten by dugongs, including two species of *Cymodocea*, five species of *Halophila*, *Halodule uninervis*, *Syringodium isoetifolium* and *Zostera capricorni*. The mean IVDMD estimates were mostly extremely high (≥80% potentially digestible) and comparable to the values estimated for dugongs in vivo. The IVDMD values obtained by Sheppard *et al.* (2007, 2008) were generally somewhat lower than those of Aragones (1996), but still relatively high. The high IVDMD of *Zostera capricorni* is noteworthy. Both Preen (1995a) and Lanyon and Sanson (2006a) commented on the relatively large fragments of this plant in the faeces of dugongs and have assumed that it is not as efficiently digested as other seagrasses. The discrepancy is probably due to the dugong's difficulty in masticating *Zostera capricorni*, making it relatively less digestible in vivo than suggested by Aragones' (1996) in vitro measurements on material that had been ground by machine.

Sheppard *et al.* (2007) found that, because of the relatively low intraspecific differences in the nutrient composition of the seagrasses at Burrum Heads in Queensland, nutrients were concentrated or dispersed according to seagrass biomass. The only detectable influence of site on nutrient distribution was tidal exposure. Intertidal *Halophila ovalis* possessed a higher IVDMD than subtidal plants of the same species and total starch was also higher in intertidal than subtidal *Zostera capricorni*.

The nutritional basis of food selection by sirenians is clearly an important topic for further research. However, contrary to some of the literature (e.g. de Iongh *et al.* 2007), there is no basis for the claim that the forage eaten by sirenians is of poorer quality than that eaten by many wild, large, terrestrial, herbivorous mammals.

FOOD ACQUISITION AND PROCESSING

Dugongs feed by excavating or cropping, depending on seagrass morphology and the nature of the sediment. Manatees are likely to be similar. When feeding on structurally small seagrasses of the genera *Cymodocea*, *Halophila*, *Halodule*, *Syringodium* and *Zostera capricorni* (narrow), dugongs excavate the plant community as they swim forward, carving characteristic feeding trails in the sea bottom (Figure 4.1) and creating clouds of sediment (e.g. Figures 4.2a, 5.4; Anderson and Birtles 1978; Heinsohn *et al.* 1977; Preen 1992; Anderson 1998; Marshall *et al.* 2003). These feeding trails are usually 10–25 cm wide (roughly the width of a dugong's facial disk), serpentine (Figure 4.1), between 30 cm and several metres long and up to about 6 cm deep (Heinsohn *et al.* 1977; Anderson and Birtles 1978; Preen 1992); dugong feeding scars can also be circular (Aragones 1994; A. Preen, personal communication 1999) or elliptical (Nakanishi *et al.* 2008). Circular and elliptical scars presumably result from the sirenian digging at a single spot rather than moving forward during feeding (see 'Domning's Postulates', Chapter 3). Preen (1992) found no correlation between the mean length of feeding trails and the density of seagrass shoots at four sites where shoot density ranged from 261–2950 shoots per square metre, but noted that where *Syringodium isoetifolium* occurred at the exceptionally high density of ~8700 shoots per square metre, the trails were only 30–50 cm long. Dugong feeding trails are used both by scientists (Supplementary Material Appendix Table 4.1) and indigenous hunters

(Nietschmann 1984; see Frontispiece and Chapter 7) as evidence of dugong feeding activity. Feeding trails are presumably more reliable as an index of sirenian feeding activity on species that are excavated rather than cropped. Dugongs and Florida manatees can remove substantial proportions of the above- and below-ground biomass of some seagrasses (see Table 4.7).

When feeding on larger, more robust seagrasses or in compacted sediments, dugongs feed by cropping seagrass leaves. In Shark Bay, Western Australia, Anderson (1982) reported that dugongs fed on *Amphibolis antarctica* by 'stripping' the leaves off the stems. In Thailand, dugongs feeding on *Enhalus acoroides* cut the seagrass leaves at a consistent above-ground height, and marks consistent with dugong teeth were seen on the leaves (Nakanishi *et al.* 2008). Examination of the stomach contents of dugongs feeding on *Thalassia hemprichii and E. acoroides* indicates that their rhizomes are not generally eaten (Erftemeijer *et al.* 1993; André *et al.* 2005; Domning and Beatty 2007), presumably because they are either too fibrous or extend too deep in the sediment (e.g. 6–12 cm) for a feeding dugong to disturb (Domning and Beatty 2007).

West Indian manatees also feed by cropping (Figures 4.2c and 4.2d) or excavating, depending on the growth form of the food source and the part of the plant being consumed. They prefer to feed underwater, and when feeding on floating vegetation such as water hyacinth, they may drag the plants beneath the surface, manipulate them with their lips, bristles and flippers and delicately crop each leaf, usually rejecting the stalks and roots. Alternatively, they may eat entire plants (Hartman 1979). When feeding on alligator weed (*Alternanthera philoxeroides*), manatees usually remain at the surface with heads awash eating the stems, leaves and roots of the plants indiscriminately. However, when feeding on dense 'jungles' of *Hydrilla verticillata* (Figure 4.2c) or *Myriophyllum spicatum*, submerged plants that reach 2–6 m to the surface, Florida manatees feed seemingly haphazardly, indiscriminately consuming stems, leaves and flowers. Manatees also feed on rafts of *Hydrilla* that have been uprooted by their activity (Hartman 1979). In Panama, manatees commonly eat true grasses growing along the river banks as well as floating vegetation and mangrove leaves overhanging the surface (Mou Sue *et al.* 1990). During the dry season in eastern Venezuela, the main source of food has been reported to be mangrove leaves (O'Shea *et al.* 1988). When feeding on emergent vegetation, such as the salt marsh grass *Spartina alterniflora* (Gramineae), Florida manatees bite the shoots, shearing them off cleanly (Baugh *et al.* 1989; Zoodsma 1991). Florida manatees typically feed on *Spartina* at high tide, which allows them to feed with only the tops of their heads and snouts above the water, although at times one-half to two-thirds of the anterior dorsal surface may be exposed. Reynolds (1977) observed a large adult manatee haul over half its body out of the water to feed on *Panicum hemitomon*, despite this plant growing down to the water, making such hauling-out unnecessary. The anecdotal information reported by Powell (1996) suggests that the feeding behaviour of the West African manatee is similar.

Hartman (1979) considered that Florida manatees ate only the leaves of seagrasses. However, Ledder (1986) found seagrass rhizomes in the digesta of manatees from southern Florida, a finding confirmed by the field observations of Lefebvre *et al.* (2000). Lefebvre *et al.* (2000) described manatee feeding scars in *Halodule wrightii, Syringodium filiforme and Thalassia testudinum* as elliptical patches (4 m × 6 m) with an average area of 27 m² and depth of 9 cm (cf. 6 cm for dugongs; Preen 1992). Manatees removed 80–95% of the above-ground biomass and 50–67% of the roots and rhizomes. Nonetheless, manatees rarely eat *T. testudinum* rhizomes (Marsh *et al.* 1999). Thus, like dugongs, Florida manatees excavate and eat both the above- and below-ground components of some genera of seagrasses.

The chewing sounds of manatees (Reynolds 1977; Bengtson 1983; Etheridge *et al.* 1985; Zoodsma 1991) and dugongs (Tsutsumi *et al.* 2006; Hodgson *et al.* 2007) are clearly audible. Reynolds (1977) suggests that these feeding sounds may, through social facilitation, stimulate conspecifics to begin eating.

Food acquisition

Several aspects of the morphology of the skull and mouthparts of manatees and dugongs influence their trophic ecology and their capacity to acquire different foods. Domning (1978a, 1982a) and Domning and Hayek (1986) measured manatee and dugong skulls and quantified the marked differences between species in the deflection of the rostrum relative to the palatal plane (Figure 4.4). Domning (Box 3.1) suggested that the degree of snout deflection influences where in the water column a species of sirenian feeds most efficiently and its capacity to capture submerged, floating and emergent vegetation (Chapter 3). Dugongs have the most deflected snouts (~70°) of the extant sirenians. This high rostral deflection enables them to place their disk against the substrate while feeding (Figures 4.2a and 4.5), presumably an energetically advantageous position for a benthic forager (see Chapter 3). Amazonian (25°–41° deflection) and West African (15°–40°) manatees have the least deflected snouts, presumably an adaptation for feeding on natant and emergent vegetation. The snout deflection of West Indian manatees (Figure 4.4) is intermediate between these extremes (29°–52°), reflecting their generalist foraging

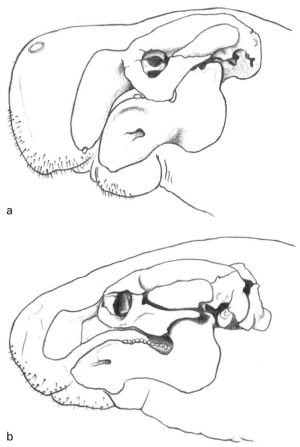

a

b

Figure 4.4. Diagrammatic representation of the differences between the skulls of (a) a dugong and (b) a Florida manatee, illustrating the differences in rostral deflection and opening of the mouth. Drawing by Gareth Wild, reproduced with permission.

niche. The neck musculature of the Amazonian manatee also facilitates surface feeding (Domning 1978c).

Vibrissae or whiskers are specialised hairs, which most mammals use only to pick up tactile cues. Like other mammals, manatees and dugongs use their vibrissae to help them explore their environments (Marshall *et al.* 1998, 2000, 2003; Chapter 5). Uniquely, sirenians also use their elaborate facial musculature, lips and vibrissae to capture their food (Marshall *et al.* 1998; Reep *et al.* 1998). Chris Marshall and his co-workers devised ingenious plexiglas feeding platforms to take close-up videotapes of the perioral regions of captive manatees and dugongs during feeding bouts (Marshall

et al. 1998, 2003; Figure 4.5), demonstrating that both dugongs and manatees use their oral bristles and lips to introduce vegetation into the mouth
through a series of cyclical manoeuvres.

The feeding behaviour of all three manatee species is very similar. They
use some of their bristle fields (Figure 5.2) as prehensile tools to grasp
vegetation and introduce it into the front of the mouth, in a variety of
behaviours which allow them to maximise their potential as generalist feeders.
The pinching and grasping actions of manatees' lips and bristles also enable
them to excavate rhizomes and remove sediment from seagrasses.

In contrast, dugongs use their lips and bristles to collect and introduce
seagrasses into the sides of the mouth, thereby decreasing the transport
distance within the mouth, which is longer in dugongs than manatees
because of the dugong's ventrally deflected snout (Figure 4.4). Although
unable to study how dugongs excavate seagrass rhizomes from the sediment in captivity, Marshall *et al.* (2003) hypothesised that dugongs used

Figure 4.5. Dugong feeding on plexiglass platform at Toba Aquarium, Japan during
experiments conducted by Chris Marshall to determine the anatomical basis for
dugong food acquisition. Dugongs quickly learned to feed on leaves of *Zostera
marina* pushed into holes drilled through the plexiglass at approximately 10 cm
intervals. The entire apparatus was held in place on the inside of the viewing window
with suction cups. Reproduced from Marshall *et al.* (2003), with permission.

their bristles in a biological disassembly line to excavate seagrass rhizomes and other benthic organisms, clean them of sediment and pass them into the mouth in a methodical, rhythmic and efficient manner. This mechanism is remarkably good at cleaning the sediment from the food; several researchers have commented on how little sand is typically found in dugong digesta (e.g. Heinsohn and Birch 1972; Erftemeijer *et al.* 1993), except when they are suffering from food shortage (Spain and Heinsohn 1973; Preen and Marsh 1995).

As explained in Chapter 3, large upper incisor tusks are a characteristic of most living and fossil sea cows of the Subfamily Dugonginae. Domning and his co-workers have speculated that these tusks were an important aid for fossil dugongids that fed on seagrass rhizomes. Do they play the same role in the feeding behaviour of modern dugongs, given that apart from occasional old female dugongs, adult male dugongs are the only extant sirenians to have erupted tusks? Domning and Beatty (2007) addressed this question by comparing the stomach contents of dugongs with and without erupted tusks and by examining the geometry and micro-wear of worn tusks. They concluded that male dugongs with erupted tusks do not consume more rhizomes than females without erupted tusks and that tusks do not play a significant role in feeding in modern dugongs. Nonetheless, each tusk is worn laterally into a blade-like shape, presumably as the excavating dugong moves forward (Figure 4.6).

Both captive and wild Florida manatees have frequently been observed using their flippers to manipulate foods (Hartman 1979), and captive Amazonian manatees at the Instituto Nacional de Pesquisas da Amazônia (INPA) facility in Manaus also use their flippers to introduce food into their mouths. In comparison, very little is known about flipper use in feeding by the West African manatee, although there are reports that they scratch the shore with their flippers to unearth roots rich in carbohydrate, as discussed above (Dodman *et al.* 2008). Despite various claims, there is no evidence that the dugong uses its flippers to manipulate food. Like West Indian manatees, both captive and wild dugongs have been observed using their flippers to walk along the bottom or support themselves while feeding (Jonklaas 1961), and dugong flippers bear conspicuous calluses, presumably resulting from contact with the bottom during feeding. Marshall *et al.* (2003) described captive dugongs and manatees using their flippers in self-care activities to remove debris from their mouthparts. Thus, although both dugongs and manatees use their flippers for a variety of activities associated with feeding, only manatees actually use their flippers to 'handle' their food.

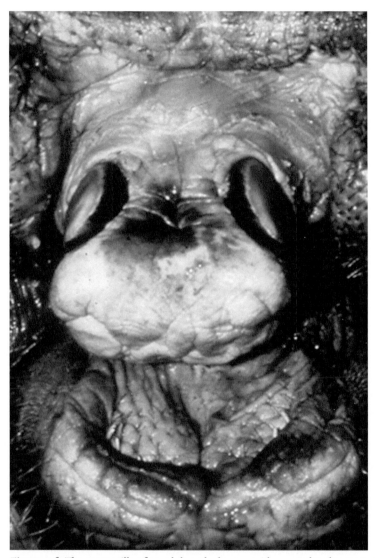

Figure 4.6. The premaxilla of an adult male dugong with erupted and worn tusks on either side. Extract from a photograph in Heinsohn and Marsh (1984), with permission.

Mastication

Herbivores derive energy from plant material through the enzymatic breakdown of cell contents and the microbial fermentation of cell walls (Keys *et al.* 1969; Van Soest 1994). As explained in 'Digestion' below, sirenians are hindgut fermenters like horses and elephants, and rely on symbiotic organisms in the caecum and colon to ferment the fibrous portion of their diet. In herbivores, the rates of fermentation and enzymatic digestion are affected by the size of the food particles presented to the micro-flora (Clauss and Hummel 2005). Thus, for a herbivorous mammal to access the nutrients in its food optimally, plant material must be broken down mechanically into small fragments; for example, a horse with sound teeth generally reduces hay and grass particles to < 1.6 mm long (Frape 2004). Teeth play the major role in this process in most mammals, and hindgut fermenters have typically placed a premium on effective dentition (Lanyon and Sanson 2006a,b). Mechanical breakdown by teeth also releases the plant cell contents and enables them to be absorbed directly in the anterior alimentary canal in an unfermented form. Hindgut fermenters such as the horse tend to rely much more on their teeth than do ruminants (Frape 2004). The abrasive properties of plants and the actions of chewing cause considerable tooth wear, potentially jeopardising masticator effectiveness and ultimately reducing the availability of nutrients, the rate of fermentation and general health (Logan and Sanson 2002; Frape 2004).

Dugongs and manatees have solved the problem of tooth wear very differently from most other mammals and from each other, presumably because of their different tooth-wear challenges. The three species of manatees all eat a wide range of aquatic plants, including terrestrial grasses (Poaceae, formerly Graminae; see Figure 4.3 and Supplementary Material Appendix Tables 4.2–4.4). Terrestrial grasses have abrasive particles of silica in their leaves and stems (McNaughton *et al.* 1985). In contrast, seagrasses, the staple food of the dugong and part of the diets of West Indian and West African manatees (Figure 4.3; Supplementary Material Appendix Tables 4.1, 4.3, 4.4), are not only less abrasive but are also much easier to rupture than terrestrial grasses (Lanyon and Sanson 2006a). Nonetheless, some abrasive particles of substrate may also be taken in and cause wear on the teeth when sirenians feed on seagrasses.

Manatees have a tooth replacement system that enables them to masticate abrasive food plants effectively (Domning 1982a). The 5–8 cheek teeth in each jaw quadrant (Figure 4.7) are replaced horizontally in an apparently limitless series of supernumerary molars throughout much or all of each animal's life (Domning and Hayek 1984). The Amazonian manatee probably has more silica in its diet than any other sirenian because of its

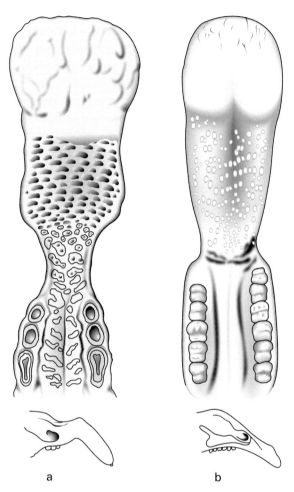

Figure 4.7. Drawing of upper mouthparts of (a) adult dugong and (b) adult Florida manatee, showing horny plate (top of figure), molar teeth and palate. Note that the manatee has five teeth in the right upper jaw (left in the figure) and six teeth in the left upper jaw, presumably because it had recently lost the right anterior tooth. Lower drawings show that the maxilla of the dugong (a) is much more downwardly deflected than that of the Florida manatee (b). This difference in deflection makes dugongs virtually obligate bottom feeders in contrast to manatees, which feed throughout the water column. Redrawn from Marsh *et al.* (1999) by Gareth Wild with permission.

relatively higher consumption of true grasses. Its teeth are smaller, have more complex patterns of lophs than the teeth of the other species and are best adapted to an abrasive diet (Domning 1982a; Domning and Hayek 1986). In contrast, dugongs, which presumably have the least abrasive diet of any extant sirenian, have simple, peg-like molars with degenerate enamelled crowns that wear quickly, exposing the much softer and less wear-resistant dentine (Figure 4.7). The dugong's last two molars are open rooted and grow throughout life. These ever-growing cylinders of dentine constitute the entire cheek dentition of most old adult dugongs (Marsh 1980). The dugong's flat-crowned teeth are probably not very efficient at grinding plants, and Lanyon and Sanson (2006b) believe they play a relatively unimportant role in the mechanical breakdown of the dugong's seagrass food.

Surprisingly, dugongs can break down some seagrasses as efficiently as West Indian manatees. Both sirenians reduce seagrass to fragments almost as small as those produced by horses chewing hay (Frape 2004). Marsh et al. (1999) measured the size of fragments of seagrasses in the stomach contents of dugongs and West Indian manatees eating the same genera of seagrasses and showed that, despite its simple dentition, the dugong is at least as effective at masticating the leaves and rhizomes of Halodule and the leaves of Thalassia as the West Indian manatee (neither species eats Thalassia rhizomes, as explained above). However, this finding does not mean that the teeth of dugongs are as effective as those of manatees in masticating seagrasses. The dentition of both dugongs and manatees is only a part of their masticatory apparatus; their entire oral cavity functions to break down their food.

Similar to Steller's sea cow, which had no teeth (Chapter 2; Figure 2.6), both dugongs and manatees have well-developed horny pads or plates at the front of their mouths (Murie 1872; Gohar 1957; Figure 4.7). In addition, the palate of the dugong (but not of the Florida manatee; Figure 4.7) is modified into regions of horny papillae and folds (Marsh and Eisentraut 1984) that may assist in mastication. The relatively small sirenian tongue is probably also important in positioning food in the mouth (Yamasaki et al. 1980; Levin and Pfeiffer 2002). Thus the sizes of the fragments in sirenian stomachs reflect the relative effectiveness of their entire masticatory apparatus rather than the teeth per se.

Comparison of the effectiveness of the masticatory apparatus of manatees and dugongs in breaking down plants is limited to the seagrasses of the genera Halodule and Thalassia. Dugongs are less effective at masticating fibrous seagrasses (e.g. *Enhalus acoroides, Zostera capricorni*) than low-fibre seagrasses such as species of *Halodule* and *Halophila* (Marsh *et al.* 1982; Preen 1992; Lanyon and Sanson 2006a). Indeed, Nakanishi *et al.* (2008) reported that when dugongs feed by cropping *Enhalus* the hard leaf edges were left. The macroalgal fragments in dugong stomachs are also often large (Lipkin 1975; Marsh *et al.* 1982). Presumably the dugong's masticatory apparatus would not break down terrestrial grasses very efficiently. In contrast, the combination of a limitless supply of molars plus horny plates (Figure 4.7) apparently allows manatees to exploit a much wider variety of food plants than dugongs (Figure 4.3; in Supplementary Material Appendix, compare Tables 4.1 with Tables 4.2–4.4). These differences between the masticatory apparatus of manatees and dugongs may be most important under lengthy periods of food scarcity, such as at times of seagrass dieback (Marsh and Kwan 2008) or during seasonal food shortages in the Amazon (Best 1984). Under such conditions, manatee dentition would be expected to confer a competitive advantage, allowing manatees to consume a wider variety of food plants including coarse and fibrous species. This difference may be one reason why the large temporal fluctuations in fecundity recorded for dugongs have not been noted in manatees (see Chapter 6).

Digestion

As explained above, herbivorous mammals do not produce the enzymes required to digest cellulose and other fibrous components of their diets (Van Soest 1994). Hence a region of their digestive tract must be modified to contain the symbiotic bacteria, protozoa and fungi capable of hydrolysing cellulose, hemicellulose and other substances resistant to their digestive enzymes. The principal organic end products of this fermentative digestion are volatile fatty acids (VFAs), which herbivores absorb and use as energy sources (Parra 1978; Van Soest 1994). Fermentation also leads to heat production.

The large intestine, and especially the colon, is the major fermentation chamber of large, hindgut-fermenting mammals like sirenians (Van Soest 1994; Clauss *et al.* 2007). Numerous species of bacteria have been cultured from the dugong hindgut (Goto 2004a; Tsukinowa *et al.* 2008). When compared with most other colon fermenters, the gastrointestinal tract of sirenians is remarkable, and the differences between sirenian species is relatively minor (Reynolds and Rommel 1996). The structural adaptations common to all sirenians include a discrete accessory digestive gland associated with the stomach, duodenal diverticulae and an exceptionally long and relatively narrow colon (Figure 4.8). This combination is unprecedented. These features enable sirenians to process their food in a way that

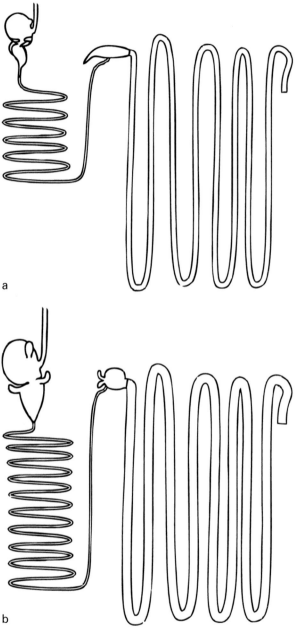

a

b

Figure 4.8. Diagrammatic representation of the alimentary canal of (a) the dugong and (b) the Florida manatee. The small intestine is relatively longer in the Florida manatee than the dugong and the large intestine relatively shorter. Redrawn from Reynolds and Rommel (1996) by Gareth Wild, with permission.

is different from most other colon digesters, which tend to consume bulky food at the expense of efficient fermentative action (Parra 1978; Van Soest 1994).

The sirenian stomach is a relatively simple sac, remarkable only for the prominent accessory digestive gland known as the cardiac gland (Figure 4.9; Langer 1984; see also Box 2.5), which contains most or all of the hydrochloric acid and enzyme-producing cells, depending on the sirenian species. Several other mammals also have cardiac glands associated with their stomachs (Reynolds and Rommel 1996), including herbivores (koalas, wombats and beavers) and insectivores (the grass-hopper mouse and pangolin). Most authors speculate that cardiac glands function to secrete copious amounts of mucus to protect the acid- and enzyme-secreting cells from the abrasive effects of ingested food. Marsh *et al.* (1977) and Reynolds and Rommel (1996) also suggest that the gland may produce extra mucus to lubricate the digesta, a plausible explanation for sirenians given the large amounts of mucus that typically coat their

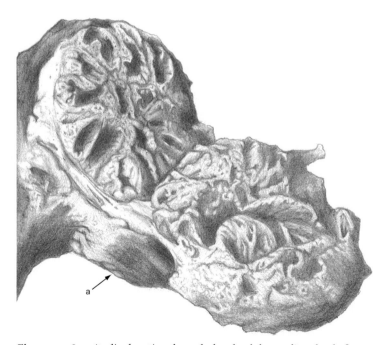

Figure 4.9. Longitudinal section through the glandular cardiac gland of a juvenile female dugong, showing the complex mucosal folds. a = aperture. The diameter of the cardiac gland is about 15 cm. Redrawn by Gareth Wild from a photograph in Marsh *et al.* (1977), with permission.

stomach contents, presumably minimising the absorption of nutrients (Reynolds and Rommel 1996). The digesta in the stomachs of manatees and dugongs is always remarkably dry, indicating that the stomach is also an important site for absorbing the water that must be ingested when they feed on aquatic plants (Marsh et al. 1977; Reynolds and Rommel 1996). Isolating the acid- and enzyme-secreting cells in the cardiac gland would increase the surface area of stomach epithelium available for absorbing water. A large quantity of water in the stomach would presumably dilute (and buffer in the case of sea water), the acid and enzymes secreted by the chief and parietal cells in the cardiac gland. Water colder than body temperature would presumably also reduce the capacity of these enzymes to operate optimally. Murray et al. (1977) found that the concentrations of VFAs in the dugong's stomach were very low, demonstrating that it is not an important fermentation site. Lanyon and Sanson (2006a) measured the particle size of the digesta at various places along the alimentary tract of three dugongs and found that about 50% of the post-oral breakdown occurred in the stomach, presumably mostly as a result of the muscular contractions of the stomach wall. Taken together, these features suggest that the sirenian stomach is a site of water resorption and particle size reduction rather than nutrient absorption.

Duodenal diverticulae are uncommon in mammals. They increase the surface area of the proximal duodenum, may enable a large quantity of food to pass from the stomach to the duodenum at once, and may also slow the passage of the digesta (Reynolds and Krause 1982; Langer 1984). Dugongs that had not fed for a week after being stranded by a storm surge still had full stomachs, suggesting that the passage of food from the stomach is stimulated by subsequent feeding (Marsh et al. 1986).

The small intestine of the Florida manatee is about the same length as its large intestine. In contrast, the dugong's small intestine is only half as long as the hindgut (Figure 4.8); the reason for this difference is unclear. Lanyon and Sanson (2006a) demonstrated that a further 25% of the post-oral reduction in the particle size of digesta occurs in the small intestine of the dugong. Burn (1986) estimated that 16% of the digestion of cellulose in the Florida manatee occurs in the duodenum, but concluded that this organ does not play a significant role in the digestion of organic matter, nitrogen or fat. Murray et al. (1977) concluded that digestion in the dugong small intestine is negligible. These findings are supported by the histological difference between the Florida manatee's small intestine and that of most other mammals; the manatee small intestine has fewer absorptive cells than a typical mammal (Reynolds and Rommel

1996). The small intestine of Florida manatees and dugongs typically contains little digesta, suggesting a rapid passage rate (Burn 1986; Reynolds and Rommel 1996; H. Marsh unpublished). We conclude that the small intestine probably does not play a significant role in dugong digestion but is important in reducing the size of particles of digesta; the small intestine may provide a more significant role in manatee digestion.

Clauss and Hummel (2005) point out that there are two ways to accelerate passage through a tubular system: shortening the tube and/or increasing its diameter. Conversely, passage will be slowed by increasing the length and/or decreasing the diameter of a tube. Sirenians have adopted this second strategy and the sirenian colon is relatively narrower and longer than that of most other colon-digesting mammals: the colon of a large manatee can be > 20 m long (6–7 body lengths); that of a large dugong up to 25 m (> 8 body lengths) (Spain and Heinsohn 1975; Figure 4.8; Table 4.4; see also Box 2.5). In addition, unlike most other large non-ruminant herbivores (Van Soest 1994), the large intestine of sirenians is not sacculated. Sacculation is a morphological adaptation to increase the surface area of the colon. Sirenians use a different strategy – they achieve a high ratio of colon surface area to volume of digesta relative to other large non-ruminant

Table 4.4. *Length of the alimentary canal and mean digesta retention time for sirenians compared with some other large hindgut fermenting mammals*

Species	Average length of alimentary canal (hindgut) (m)	Sources	Mean retention time (hr)	Sources
Dugong	30	Spain and Heinsohn (1975)	146–166	Lanyon and Marsh (1995b)
Florida manatee	40	Reynolds (1980)	146	Lomolino and Ewel (1984)
			147	Larkin et al. (2007)
Indian elephant	18	Owen in Miall and Greenwood (1878)	20.3	See Clauss et al. (2007)
Horse	21	Riegel and Hakola (1996)	28.5	See Clauss et al. (2007)

mammals by having an extremely long narrow colon, a feature which presumably contributes to their extraordinary digestive efficiency (as discussed below).

Fermentation of the more refractory components of a plant diet is a slow process requiring retention of food. Presumably as a result of the long colon, the digesta retention time of sirenians is unusually long (Table 4.4). Clauss et al. (2007) reviewed the mean ingesta retention times of 93 herbivorous mammals, including caecum and colon hindgut fermenters and ruminant and non-ruminant foregut fermenters. Only the three-toed sloth, Bradypus tridactylus (a non-ruminant foregut fermenter with a low metabolic rate), has been recorded as having a retention time comparable to that of dugongs and manatees.

Lanyon and Sanson (2006a) demonstrated that a further 28% of post-oral reduction in the particle size of digesta occurred in the caecum and large intestine of the dugong. The hindgut is the major site of fibre digestion and VFA production in both dugongs (Murray et al. 1977) and Florida manatees (Burn 1986).

The caecum (Snipes 1984) and colon are also the primary sites of protein digestion in the Florida manatee (Burn 1986) and dugong (Murray et al. 1977), and of fat-soluble material in the Florida manatee (Burn 1986); fat digestion has not been studied in the dugong. This situation contrasts with that in terrestrial hindgut-fermenting mammals where enzymatic digestion of the cell contents tends to occur in the foregut, whereas the cell wall fraction is broken down in the hindgut (Parra 1978). Murray et al. (1977) concluded that the contribution of the hindgut to the nutrition of the dugong was considerable, a conclusion that is almost certainly true for all sirenians.

Despite the length of the large intestine of sirenians, the total weight of digesta as a proportion of body weight (approximately 8% in the Florida manatee) is low compared with other colon digesters such as the horse (16%) or elephant (17%) (Clauss et al. 2007). These ratios may be distorted by sirenians being relatively heavier for their length than terrestrial herbivores as a result of their adaptations to aquatic environments, including a heavy skeleton and blubber layers. Nonetheless, the comparison suggests that the total weight of digesta in sirenians is not relatively higher than that of large terrestrial mammals, a conclusion supported by the relationship between body mass and absolute dry matter intake for herbivorous mammals (Clauss et al. 2007).

Digestive efficiency is inversely correlated with gut passage rate (Clauss et al. 2007), so it is not surprising that the digestive efficiency of sirenians is very high relative to that of other herbivorous mammals, especially other hindgut digesters. Two main methods have been used to estimate the digestive efficiency of dugongs and manatees: (1) tracking the apparent digestibility of various food components in various places in the alimentary canals of fresh carcasses using the change in the ratio of the component of interest to the lignin in the cell walls of the food plants, which acts as a naturally occurring indigestible marker; and (2) comparing the composition of ingesta and faeces in live captive animals assuming a gut passage rate derived from studies of captive animals. These methods have been used to measure slightly different parameters and have been used for different food plants, but give essentially similar results. Whereas their capacity to digest crude protein (nitrogen) is unremarkable (Table 4.5), manatees and dugongs, like green turtles (Bjorndal 1980), digest a very high proportion of the fibre in their diet compared with other colon fermenters like the horse (Table 4.5). This high digestive efficiency may explain the exceptionally high values for VFAs found by Murray et al. (1977) in the caecum and colon of a dugong. Despite these high values, Murray (1981) estimated that the VFAs would not contribute as much to the energy metabolism of a dugong as to terrestrial ruminants, which may be why sirenians also seek plants high in starch and water-soluble carbohydrates.

Food consumption

The most reliable estimates of food consumption in a wild sirenian are those of Kay Etheridge and her co-workers, who developed an ingenious method of estimating the intake of aquatic plants by Florida manatees of different body sizes (building on the technique developed by Bengtson 1983). Etheridge et al. (1985) listened to captive manatees chewing using a hydrophone, counted the number of chews required for manatees of known body weight to consume known amounts of Hydrilla verticillata and developed regression equations to predict chew counts from body weight. They then monitored the chew rates and time budgets of three radio-tracked wild animals and estimated that adult manatees eating H. verticillata can eat approximately 7% of their body weight (wet weight of food) in five hours of chewing time. The corresponding figure for a juvenile was 9.5% of body weight; for a calf 15.7%. The most reliable estimates of food consumption for dugongs are for captive animals and suggest that, contrary to the preliminary figures of Best (1981), dugongs consume about the same amount of food per unit body weight as manatees (see Table 4.6).

Assuming that H. verticillata is 8% dry matter (see Best 1981), the relative dry matter intake of ~29 g kg$^{0.75}$ consumed by adult wild Florida

Table 4.5. Digestive efficiencies of sirenians as measured by apparent digestibilities of various dietary components compared with another colon digesting herbivore, the horse. The term 'apparent digestibility' is used because the digesta contain metabolic excretions such as microbes, gut secretions and sloughed cells from the gut lining that result in an underestimation of the actual digestibility of some dietary components

Sirenian species (sample size)	Food plant	Apparent digestibility %						Sources
		Dry matter	Crude protein (N)	Crude fibre/NDF	ADF	Cellulose	Dietary energy	
Seagrasses								
D. dugon[1] (1)	Halophila/Halodule		70	84	82			Murray et al. (1977)
D. dugon[1] (2)	Zostera marina	71–95					70–96	Aketa et al. (2003)
		>90						Goto et al. (2004b)
Freshwater vegetation								
T. inunguis[2] (4)	Cabomba	51	–9[3]	78	71	81	46	Atkinson and Best (unpublished)
T. inunguis[2] (?)		55	–1[3]	78			29	Best (1981)
T. inunguis[2] (?)	Pistia	68	60	77			49	Best (1981)
T. manatus[2] (1)	Eichhornia	83	78				80	Lolomino and Ewel (1984)
Terrestrial grasses								
T. inunguis[2] (4)	Brachiaria	58	59	57	58	63	56	Atkinson and Best (unpublished)
	Panicum	44	45				34	Best (1981)
Unknown but variable								
T. manatus[2] (7)			50–75	64–89		64–89		Burn (1986)
Lettuce								
T. manatus[2] (1)	Latuca	91	94				89	Lolomino and Ewel (1984)
Meadow hay and pelleted concentrate								
Horse[2] (3 and 4)		58–62	65–69	38–40	41–46		62–64	Miraglia et al. (1999)

[1] Carcass study using lignin as an internal indigestible marker.

[2] Captive animal study comparing ingesta and faeces.

[3] Negative nitrogen digestibility attributed to high tannin levels in Cabomba.

Table 4.6. *The most reliable estimates of daily food intake of captive and wild sirenians (wet weight) as a percentage of body weight. Best (1981) presents a more extensive list but includes data that may not be reliable*

Sirenian species	Status	Food plant	Calf (%)	Juvenile (%)	Adult (%)	N/S[2] (%)	Sources
Dugong dugon	Captive	*Zostera marina*	14.3[1]		7.0		Goto et al. (2004b)
Trichechus inunguis	Captive					8.0	Best (1981)
T. manatus	Wild	*Eichhornia, Pistia, Vallisneria*				6.6	Bengtson (1983)
		Hydrilla	15.7	9.6	7.1		Etheridge et al. (1985)[3]

[1] One-year-old calf.
[2] N/S = age group not specified.
[3] % body weight/five hours (not daily).

manatees (based on Etheridge *et al.* 1985) is low relative to other large hindgut-fermenting mammals (Clauss *et al.* 2007), presumably reflecting the manatee's relatively low metabolic rate (Irvine 1983) or relatively heavy skeleton and blubber layer. Clauss *et al.*'s (2007) estimate for juvenile dugongs ($37\,\mathrm{g\,kg}^{0.75}$) was rather higher than the corresponding value for a juvenile manatee ($31\,\mathrm{g\,kg}^{0.75}$) (calculation based on data from Etheridge *et al.* 1985), possibly reflecting the inaccurate estimates of dugong food consumption acknowledged by Lanyon and Marsh (1995b), or perhaps a relatively higher metabolic rate for the dugong (see Lanyon *et al.* 2006a).

The amount of time an individual sirenian spends feeding presumably reflects its life history stage, whether it is feeding by cropping leaves only or excavating to eat whole plants, and the biomass of its food. Bengtson (1983) reported that manatees feed for 4.2–6.9 hours per day, depending on the season. Etheridge *et al.* (1985) obtained accurate information for one animal that fed for a mean of about 4.5 hours per day. The only estimate for wild dugongs is Chilvers *et al.*'s (2004) crude calculation of 16 hours per 24-hour period based on the diving profiles of 15 wild dugongs fitted with time–depth recorders. Whether these species differences are real or a methodological artefact is unknown; dugongs feeding on low biomass seagrasses would be expected to have to spend much more time feeding than manatees feeding on aquatic water weeds of considerably higher biomass (compare Figures 4.2a and 4.2c, d; Table 4.3). Acoustic methods for monitoring sirenian feeding are promising (Etheridge *et al.* 1985; Tsutsumi *et al.* 2006; Hodgson *et al.* 2007) and should enable more accurate estimations of the time sirenians spend feeding.

SIRENIAN–PLANT INTERACTIONS

There is a large body of literature documenting the responses of plant communities to grazing by large terrestrial herbivores (e.g. Olff and Ritchie 1998; Bakker *et al.* 2006). The ecosystem effects caused by the feeding of extant sirenians are much less well understood and have mainly been investigated for dugongs and Florida manatees in seagrass ecosystems. There are almost no similar data available regarding the effects of any species of manatees on other aquatic plant communities apart from some work investigating their potential role in aquatic weed clearance (see 'Manatees and aquatic weed clearance').

As is discussed in Chapter 8, the present sizes of most populations of all extant sirenian species are believed to be significantly smaller than they were in the early nineteenth century. In addition, prehistoric sirenians were probably not only more abundant than modern sirenians, they also exhibited higher species diversity and morphological variation between and among species. For example, it is likely that several sympatric lineages of dugongids inhabited the West Atlantic–Caribbean region approximately 25 million years ago (Domning 2001b; see Chapter 3). Such diversity implies greater and more complex interactions with seagrass and other aquatic plant communities than occur today (Box 4.1).

Box 4.1.

'For most of the past 50 Ma, Caribbean seagrass communities have had to withstand heavy, sustained grazing pressure from several sympatric lineages of large mammalian herbivores. This fact is almost totally absent from both these communities today (wherein manatees are scarce or absent in most areas) and from the thinking of the aquatic botanists and marine ecologists who study such communities. Consequently the long-established tenet that seagrass ecosystems are largely detritus based ... must be revised to recognize that the modern situation is anomalous and that the "normal" pattern throughout most of tropical seagrass history has been that much (probably most) of the primary productivity has been channeled through the guts of herbivores, particularly sirenians.'

(Domning 2001b)

Seagrass scientists increasingly acknowledge the importance of herbivory in tropical seagrass ecosystems (Valentine and Duffy 2006) and that the local effects of dugongs and manatees on subtropical and tropical seagrass ecosystems can be significant. Supanwanid *et al.* (2001) review the methods for studying the effects of large herbivores on seagrasses. These effects have been studied empirically using several approaches: (1) observations of known feeding sites within meadows, typically including comparisons of (a) undisturbed sites with sites where sirenians have fed and/or (b) the same site over time (e.g. Wake 1975; Anderson and Birtles 1978; Packard 1984; de Iongh *et al.* 1995, 2007; Preen 1995a,b; Peterken and Conacher 1997; Nakaoka and Aioi 1999; Lefebvre *et al.* 2000; McMahon 2005); (2) comparisons among meadows with different intensities of sirenian feeding in the same general area (e.g. McMahon 2005); (3) field experiments in which sirenians were excluded from locations within seagrass beds (e.g. Lefebvre *et al.* 1988; Provancha and Hall 1991; Preen 1995b; Masini *et al.* 2001); and (4) field experiments in which the two types of sirenian feeding (whole plant versus leaves only) have been simulated in seagrass beds (e.g. de Iongh *et al.* 1995, 2007; Supanwanid 1996; Aragones and Marsh 2000; McMahon 2005; Aragones *et al.* 2006).

Effects on the food plant community

The effect of sirenian feeding on their food plants can be measured at several levels and by using different response variables. The levels include the individual feeding scar, the area disturbed per day by an individual animal and the effect of a large group of animals on an individual plant community. The response variables include: microbial processes, above- and below-ground plant biomass, plant species composition, plant nutrients, invertebrate community composition and detritus, plus the time taken by each of these variables to return to the pre-disturbed condition. We deal with each of these effects below.

Microbial processes and detritus

Perry and Dennison (1996, 1999) showed that the rates of microbial nutrient cycling in seagrass sediments in subtropical Moreton Bay, Queensland, increased after intensive excavating by dugongs. The nitrogen fixation rates they measured in seagrass sediments disturbed by dugong feeding were the highest ever measured for a seagrass system. They concluded that excavating dugongs aerate the substrate and mix some of the detritus with sediment, producing a substrate for bacterial nitrogen fixation that apparently increased seagrass productivity.

Herbivory also alters the relative abundance of detrital (dead organic) matter in a seagrass meadow. Aragones and Marsh (2000) showed experimentally that intensive dugong excavating reduces detritus, presumably because most of the plant material is eaten rather than left to die and decay. In Moreton Bay, McMahon (2005) also found more detritus on seagrass in an area where dugongs rarely fed than in two areas where they excavated intensively. Heinsohn et al. (1977) observed that dugongs avoided dense, old stands of *Zostera capricorni*; Preen (1992) noted that dugongs avoided seagrass with the high epiphyte loads that can be accrued over long periods without physical disturbance. Collectively these findings suggest that excavating dugongs increase variation in the age structure of seagrass communities, thereby reducing the build up of detritus.

Plant biomass and productivity

Comparison between feeding scars and the adjacent vegetation demonstrates that excavating manatees and dugongs remove a high but variable proportion of the seagrass biomass from feeding patches (Table 4.7).

As outlined in Table 4.6 and in 'Food consumption', adult sirenians consume about 7% of their body weight per day; calves and juveniles consume relatively more – but absolutely less – food. Thus, even an individual sirenian can disturb a considerable area of seagrass in a single day, especially in areas with low biomass. Preen (1992) assumed that an individual dugong consumed 3.22 kg dry weight per day and estimated that each dugong disturbed about 400 m² of seagrass each day in Moreton Bay, Queensland, where the median biomass was 12.3 g dry matter per square metre. In the Great Barrier Reef region, also in Queensland, where the biomass of the seagrasses on which dugongs feed ranges from 5.8 to 10.4 g dry matter per square metre, Marsh et al. (2005) estimated that each dugong would disturb about 300–800 m² of seagrass each day, assuming a daily consumption of 3.16–4.52 kg dry matter. Provancha and Hall (1991) calculated that an adult Florida manatee consuming about 5.9 kg dry weight per day would crop almost 46 m² of *Syringodium filiforme* leaves per day, reducing the blade height from > 25 cm to 5–10 cm.

Although the numbers and densities of sirenians have been reduced over most of their ranges, they still occur at high densities at some sites, and their local impact on seagrass biomass can be very significant. Preen (1995b) estimated that a group of some 460+ dugongs consumed more than 151 000 tonnes (wet weight) of *Halophila ovalis* from 41 hectares in Moreton Bay, Queensland, in under 17 days, reducing shoot density by

Table 4.7. *Percentage of seagrass removed by feeding sirenians*

Seagrass species (location)	% Biomass removed			Sources
	Above-ground	Below-ground	Whole plant	
Florida manatee				
Halodule wrightii, Syringodium filiforme (Florida)	80–95 (short shoots)	50–67 (roots and rhizomes)		Lefebvre *et al.* (2000)
Dugong				
Zostera capricorni (Shoalwater Bay, Queensland)			58.6–78.2%	Anderson and Birtles (1978)
Mixed species meadow (Moreton Bay, Queensland)	85.6 (shoots) 90.8 (above ground biomass)	58.5 (rhizome) 25.1 (root)		Preen (1992)
Halodule uninervis (Indonesia)	93	75		de Iongh *et al.* (1995)
Halophila ovalis (Thailand)	94	61		Suzuki, in Nakaoka and Aioi (1999)

about 95%. McMahon (2005) monitored dugong feeding trails along a fixed 50 m transect in a *Halophila ovalis* meadow also in Moreton Bay, where herds of 50–150 dugongs were regularly observed feeding. She observed dugong feeding trails in 11 out of 13 months, and estimated removal of a monthly average of 23% of seagrass at the site. The highest percentage removed was in late winter (August), when the dugongs removed 65% of seagrass from the meadow. Masini *et al.* (2001) estimated that during the summer, an aggregation of dugongs feeding in a *Halodule–Penicillus* meadow in Shark Bay, Western Australia caused a loss (ingestion plus drift) of over 50% of production.

The seasonal aggregations of Florida manatees in warm-water areas during winter and spring can also significantly impact nearby vegetation (Lefebvre and Powell 1990; Packard 1984; Provancha and Hall 1991). Packard (1984) studied the effect of manatees feeding in Jupiter Sound, Florida, in winter and estimated that about 40% of the seagrass beds (179 000 m²) were disturbed, and that biomass was 46–67% lower in meadows with feeding scars compared with undisturbed meadows. Provancha and Hall (1991) estimated that a large concentration of up to

200 manatees consumed about 30% of the standing stock of seagrass blades in a meadow dominated by *Syringodium filiforme* in the north Banana River in Florida over about 60 days before moving away, thereby reducing the percentage cover of sites open to manatees by 90% compared with experimental areas from which manatees were excluded.

McMahon (2005) studied seagrasses at three meadows in Moreton Bay, Queensland, each of which represented a different level of intensity of natural dugong feeding on seagrass leaves, roots and rhizomes. Unfortunately, the levels of feeding were not replicated across meadows, making it impossible to unequivocally attribute the differences between meadows to differences in feeding intensity. McMahon found that *Halophila ovalis* growing in a meadow where dugongs intensively removed leaves, roots and rhizomes were twice as productive in summer and 1.5 times as productive in winter as plants of the same species growing in a meadow that was not disturbed by dugongs. This difference in productivity (50–100%) between plants that are eaten and those that are not has not been replicated experimentally. The effect of sirenian feeding on plant productivity clearly requires further investigation, especially in ecosystems other than seagrass communities. Whitham *et al.* (1991) reviewed the literature on the compensatory responses of terrestrial plants to herbivory and the physiological mechanisms that may be involved. They concluded that there is a continuum of compensatory responses ranging from negative to positive; we expect that the same is true for the plant species eaten by sirenians.

Plant community structure and composition

Field experiments demonstrate that feeding sirenians can have significant effects on the community structure and dynamics of multi-species seagrass meadows. Provancha and Hall (1991) constructed exclosures on a seagrass flat dominated by *Syringodium filiforme* to assess the impact of large concentrations of Florida manatees. There was a significant relative decrease in *S. filiforme* outside the exclosures as a result of heavy cropping by the manatees and a slight increase in density in the pioneer species *Halodule wrightii*. Preen (1995b) conducted a simulated excavating and exclosure experiment in a community in Moreton Bay, Queensland, containing approximately equal proportions of *Zostera capricorni*, *Halophila spinulosa* and *Halophila ovalis*. Low-intensity excavating did not alter the relative abundance of *Z. capricorni* or *H. spinulosa* and may have reduced the relative abundance of *H. ovalis*. In contrast, intensive excavating increased the relative abundance of *H. ovalis* at the expense of *Z. capricorni* and *H. spinulosa*. Aragones and Marsh (2000) simulated dugong feeding in a meadow at Ellie Point, in tropical Queensland, and found that both light and intensive excavating changed the species composition of the experimental plots in favour of *H. ovalis* at the expense of *Z. capricorni* and *Cymodocea rotundata*. All of these experiments demonstrate how sirenian feeding can alter the composition of seagrass communities by favouring the growth of pioneer genera of seagrass such as *Halophila* and *Halodule*.

Food quality

The response of seagrasses to the short-term disturbance caused by feeding manatees and dugongs can also change the chemical composition of the plant material. Aragones *et al.* (2006) studied the responses of five species of seagrass to experimental removal, in a manner that simulated the two dugong feeding modes: (1) excavating seagrass leaves, roots and rhizomes; and (2) cropping leaves only. *Halophila* and *Halodule* were the main genera that showed interesting changes in nutritional qualities. The whole-plant nitrogen concentrations of *H. ovalis* and *H. uninervis* increased by 35% and 25%, respectively, even after nearly a year of recovery from intensive excavation of whole plants. These gains were tempered by a concomitant reduction in starch and increase in fibre concentrations. In the short term, the nitrogen concentrations increased while the fibre concentrations decreased. The proportion of new foliage with relatively less structural material increased after the above- and below-ground parts of the seagrass were removed, leading to increased nitrogen concentrations, whereas mobilisation of energy reserves for rebuilding the above-ground biomass led to reduced starch concentrations. As explained above, the lack of a definitive set of determinants for sirenian food quality is a barrier to understanding the significance of these changes in the nutrient composition of seagrasses eaten by sirenians.

Benthic animals

There have been few studies of the direct and indirect effects of sirenian feeding on benthic animals in seagrass communities. Direct effects include the consumption of invertebrates by dugongs and manatees (Supplementary Material Appendix Tables 4.1, 4.3 and 4.4) and the disturbance of the sediment caused by digging up seagrasses. In addition, because the feeding scars remain for weeks or months while the seagrass recovers, it is possible that this disturbance may induce secondary indirect effects on the benthic community by altering habitat structure.

Preen (1995a) studied dugongs feeding on a stalked benthic colonial ascidian (*Sycozoa pulchra*) that was growing at densities of >3500 colonies per m² in subtropical Moreton Bay in spring. Dugongs removed 93% of the ascidian colonies from the feeding trails. Presumably dugong feeding was a major local impact on the ascidian community, although the effects were not studied. Nakaoka *et al.* (2002) and Skilleter *et al.* (2007) studied the community structure and abundance of benthic animals in *Halophila* dominated meadows in Thailand and Moreton Bay, Queensland respectively. The densities of most (but not all) groups of invertebrates were generally higher in intact vegetation than in actual (both studies) or simulated (Moreton Bay only) dugong feeding trails. The magnitude of the differences in density varied among functional groups and taxa and different groups responded differently to the effects of physical disturbance on the sediment and seagrass removal *per se*. Differences in invertebrate community composition were detected in both studies. These differences persisted for at least weeks after the creation of the trails. Longer-term studies are required to understand how sirenian feeding affects benthic communities in a range of seagrass communities.

Recovery of plant communities

Even the most intense and sustained removal of seagrass leaves, roots and rhizomes by sirenians leaves relatively small spaces (< 1 m) between surviving tufts of seagrass (Packard 1984; Preen 1995b), allowing recovery to occur by clonal growth, augmented by sexual reproduction in at least some species (e.g. *Zostera capricorni*; Peterken and Conacher 1997). McMahon (2005) conducted the most comprehensive study of the factors enabling *Halophila ovalis* to cope with dugong feeding disturbance. She concluded that genetically diverse meadows, fast growth, flexible clonal reproduction, regular and flexible sexual reproduction and breeding systems, and a persistent and abundant seed bank were all important to rapid recovery.

Recovery from feeding can be examined at the scale of a feeding trail, patch (group of feeding trails) or meadow, and can be measured in terms of cover, biomass or shoot density. The time scales of recovery of seagrass beds from sirenian feeding depend on many factors, including the intensity, nature and timing of the impact, its location within the meadow, the nature of the seagrass community, latitude and availability of light. *Halophila ovalis* can recover very rapidly from the removal of leaves, roots and rhizomes by dugongs. McMahon (2005) concluded that this rapid recovery rate most likely results from rapid clonal growth. Recovery times of a month or less have been recorded in both the tropics and subtropics, for example by Nakaoka and Aioi (1999) in Thailand and McMahon (2005) in Moreton Bay, Queensland. Seagrasses other than the species of *Halophila* generally take longer to recover. de Iongh *et al.* (1995) estimated that the biomass of *Halodule uninervis* took five months to recover from dugongs removing leaves, roots and rhizomes in Indonesia in the wet season and did not recover at all during the dry season. Aragones and Marsh (2000) showed experimentally that recovery times ranged from months for *Zostera/Cymodocea* and *Halophila ovalis* at one location in tropical north Queensland to more than a year in a monospecific meadow of *Halodule uninervis* at another site in the same general area.

Lefebvre *et al.* (2000) demonstrated that short shoots of both *Halodule wrightii* and *Syringodium filiforme* showed significant recovery within one growing season after Florida manatee feeding and noted that the locations of heavily disturbed sites have remained stable over ten years. Manatee feeding was heavy for only about three months of the year (in winter) in their study area, with much less intense use during the warm season growth period after the manatees migrated northward.

The speed and nature of seagrass recovery from sirenian feeding clearly varies with time of year and location (Supanwanid 1996; Aragones and Marsh 2000; McMahon 2005). As Aragones *et al.* (2006) point out, the variable times taken for seagrasses to recover from sirenian feeding indicate that the appropriate timing for dugongs and manatees to revisit sites is also likely to vary greatly in both time and space. Thus, it is not surprising that there is considerable variability in the feeding patterns of sirenians and the return times to particular locations (e.g. Provancha and Hall 1991; Preen 1995b; de Iongh *et al.* 1995, 2007; Anderson 1998; Lefebvre *et al.* 2000; Masini *et al.* 2001; Hodgson 2004; McMahon 2005). There are so many variables associated with environmental and anthropogenic changes to seagrass beds that reliable predictions of seagrass recovery and sirenian return times may be impossible, making reliable estimations of carrying capacity difficult (see Chapter 5).

Feeding optimisation?

The studies of the interactions between dugongs and seagrasses clearly demonstrate that feeding sirenians can have significant effects on the biomass, community structure and chemical composition of their food plants. Preen (1992, 1995b) coined the term 'cultivation grazing' for the apparently beneficial effects (from the dugongs' perspective) that dugongs feeding in large herds had on the seagrass community. de Iongh *et al.* (1995,

2007) showed that this effect was not limited to dugongs feeding in large herds and even small groups of dugongs repeatedly return to the same areas. de Iongh (1996) speculated that the resultant increase in food quality (higher IVDMD) that he measured compensated for the lower intake per bite resulting from the reduced standing crop caused by frequent feeding.

Lefebvre and Powell (1990) also concluded from aerial photographs that Florida manatees fed in the same patches of *Halodule wrightii* in a seagrass meadow in successive winters; whether these were the same animals was not confirmed. Bengtson (1981) and Deutsch *et al.* (2003) tracked individual Florida manatees to the same feeding locations in successive years (Figures 5.6 and 5.7), and there was some evidence that calves learned of these favoured feeding sites from their mothers (Chapter 5). Bjorndal (1980) also observed green turtles maintaining plots of young leaves of *Thalassia testudinum* by constant recropping.

It is a fundamental tenet of modern evolutionary theory that selection operates to increase the inclusive fitness of the individual rather than the group. Consequently, the concept of resource management such as 'cultivation grazing' by terrestrial vertebrates is contentious because it tends to invoke group selectionist arguments. Gordon and Lindsay (1990) conclude that herbivorous mammals are very unlikely to actively manage their food resources unless individuals exhibit long-term territoriality or use exclusive home ranges. Sirenians have not adopted either of these social systems (Chapter 5). Thus, the use of the term 'cultivation grazing' *sensu* Preen (1995b) is criticised by some ecologists. Nonetheless, like some large terrestrial herbivores and green turtles, sirenians apparently return to 'traditional' feeding sites (see 'Evidence for matrilineal use of space' in Chapter 5), thereby maintaining 'grazing' or 'browsing lawns' *sensu* McNaughton (1984) and Skarpe and Hester (2008), or at least local feeding patches of relatively young food plants of improved food quality (particularly through increased nitrogen and lower fibre content). How much time a foraging herbivore should spend in a patch of food to maximise its nutritional return is a central question in classical foraging theory (Searle *et al.* 2005). This question has not yet been investigated for sirenians. Nonetheless, the maintenance of grazing patches or lawns is probably one of the reasons that sirenians sometimes aggregate in groups (discussed further in Chapter 5).

Perry and Dennison (1999) attributed the higher shoot nitrogen they observed in meadows of *Zostera capricorni* and *Halophila ovalis* where dugongs had removed seagrass leaves, roots and rhizomes to microbial nitrogen fixation stimulated by feeding dugongs. McMahon (2005) disagreed with this conclusion because the nitrogen isotope signature in the disturbed plants was close to that of dugong faeces, suggesting that the plants were using nutrients in dugong faeces. In terrestrial systems, herbivory may provide positive feedback to the plants via local scale inputs of faecal and urinary material (Augustine *et al.* 2003). Aragones *et al.* (2006) postulated that the feedback loop might be different in the sirenian–seagrass ecosystem because the waste material generated by feeding sirenians was likely to be moved away by water currents. They suggested that positive feedback would occur via the enhancement of the detrital cycle caused by the activity of nitrogen-fixing bacteria, as documented by Perry and Dennison (1999). McMahon's results suggest that dugong faeces also promote seagrass growth. The feedback loops in sirenian–seagrass ecosystems clearly warrant further investigation.

Competition with green turtles

All three species of manatees are typically the only large herbivores in the subtropical and tropical freshwater systems in their ranges. In contrast, dugongs, West Indian manatees and West African manatees share seagrass meadows with green turtles, sometimes at locally high densities. André *et al.* (2005) compared the stomach contents of sympatric dugongs and green turtles caught by an indigenous fishery on the Orman Reefs in Torres Strait between Australia and Papua New Guinea. They found that dugongs fed exclusively on seagrasses (mainly the leaves of *Thalassia hemprichii*, and the leaves and rhizomes of *Cymodocea* spp. and *Syringodium isoetifolium*), whereas turtles consumed both seagrasses (especially the leaves of *Thalassia* and the fibrous *Enhalus acoroides* – usually avoided by dugongs) and macro-algae (mainly *Hypnea* spp., *Laurencia* and *Caulerpa* spp.). *Thalassia*, the most abundant seagrass in the area, was the only overlap in the diets of the two species. However, the overlap would presumably have been greater in the same area at a time of seagrass dieback in the 1970s when dugongs consumed relatively more *Enhalus* (Table 4.2). André *et al.* (2005) concluded that a comprehensive study of food partitioning between dugongs and green turtles would require a detailed and concurrent study of the food resources and the animals' movements. We are unaware of any study of diet partitioning between West Indian manatees and green turtles in the Caribbean, where green turtles also maintain seagrass patches at early succession growth stages, apparently with resultant nutritional benefits (Bjorndal 1980; Thayer *et al.* 1984; Zieman *et al.* 1984).

Manatees and aquatic weed clearance

Aquatic weeds are a major international problem and in the 1960s there was widespread enthusiasm for using the West Indian manatee as a control agent (e.g. Sguros 1966). Manatees have continued to be used to clear weed-choked canals in Guyana (formerly British Guinea) for many years (Allsopp 1960, 1969), and various trials suggested that manatees could keep restricted areas clear of weeds provided that the water was warm year-round and the manatees could be confined to the specific area that needed clearing (see Etheridge *et al.* 1985 for references). Etheridge *et al.* (1985) estimated the consumption rates of wild manatees eating *Hydrilla* and compared them with the biomass of *Hydrilla* in King's Bay, Florida, an over-wintering site for a large number of manatees. They demonstrated that the number of manatees required to winter in King's Bay to consume the standing biomass of *Hydrilla* would be ten times more than the number using King's Bay in 1985. When they included plant productivity in their estimates, it became clear that manatees would be a very inefficient control agent and that waterbirds such as coots (*Fulica americana*) might be more efficient because of their higher metabolic rate and greater reproductive output. Manatees are not presently considered as useful control agents for aquatic weeds because their numbers are too few, they are not well-suited for transport between sites that need weed control and they are protected (Langeland 1996).

COMPARATIVE OVERVIEW AND IMPLICATIONS FOR CONSERVATION

This chapter demonstrates that the three species of manatees and the dugong are all highly adapted to exploit a relatively wide range of the macrophytes that occur in their respective habitats (Figures 1.3, 4.3). This capacity to exploit different plants is particularly important in times of food plant shortage, which can occur either seasonally or in response to extreme weather events. All extant sirenians seem to prefer vascular vegetation to algae, and probably consume some animal protein opportunistically. The dietary differences among sirenian species reflect their habitats (Chapter 5) and evolutionary origins (Chapter 3). The largest difference in feeding niche (Figure 4.3) is between the dugong, which is strictly marine, and the Amazonian manatee, which occurs only in fresh water. West Indian and West African manatees eat vascular plants growing in both marine and freshwater habitats and their feeding niches overlap with those of both the Amazonian manatee and the dugong, even though the extant sirenians do not occur sympatrically except at the mouth of the Amazon River (Chapter 8).

The dugong is a subtropical and tropical seagrass community specialist, which we consider to be less dependent on pioneer genera of seagrass such as *Halodule* and *Halophila* than claimed by some other researchers (e.g. de Iongh *et al.* 1995, 2007; Anderson 1998; Aragones and Marsh 2000; Masini *et al.* 2001; Aragones *et al.* 2006, 2012a). Dugong food choices are not restricted by Latin names! Nonetheless, as Lanyon and Sanson (2006a) point out, dugongs, unlike manatees, seem only able to masticate relatively non-fibrous species of food plants effectively. Seagrass communities are also important habitats for the West Indian manatee and probably for the West African manatee. Seagrasses are limited to coastal waters and are highly vulnerable to human impacts because the coastal zone supports a high proportion of the world's human population, a proportion that is increasing. In their global assessments of the status of seagrasses, Green and Short (2003) and Waycott *et al.* (2009) concluded that the global decline in seagrass habitats is dramatic and accelerating. Waycott *et al.* (2009) estimated that 29% of the known area has been lost since 1879. However, they admit that there are significant gaps in their assessment, especially for West Africa and the tropical Indo-Pacific region, areas of significance to the West African manatee and the dugong, respectively. The threats to seagrass habitats include over-exploitation, physical modification, nutrient and sediment pollution, introduction of non-native species and global climate change (Waycott *et al.* 2009; Chapter 7). Climate change represents a relatively new risk to seagrass habitats, the impact of which is uncertain (Waycott *et al.* 2007); Brouns (1994) expects it to be positive, although he does not allow for the limits to expansion caused by human occupation of the coastal zone. These threats are considered further in Chapter 7.

Habitat loss is also a threat to Amazonian manatees (Chapters 7 and 8). Deforestation and contamination of the food supply by mercury, oil or pesticides are potential threats (Rosas 1994; Chapters 7 and 8). Climate change threatens to extend the dry season in the Amazon Basin (Malhi *et al.* 2008; Chapter 7), which will exacerbate the impacts of dry season food shortages on manatees. Like the other two manatee species, Amazonian manatees have the capacity to eat aquatic water weeds that can occur at very high biomass (compare Table 4.3 with Supplementary Material Appendix Table 4.2). Nonetheless, the seasonal fluctuation in river levels in the Amazon Basin (Best 1983) reduces their food access. As habitat generalists, the West Indian and presumably the West African manatees are able to exploit a wider range of food than dugongs or Amazonian manatees. This capacity has costs and benefits. Thus we consider that the reproductive rates of Amazonian manatees and dugongs are more likely to be reduced as a result of food resource limitations than those of West Indian or West African manatees. The plasticity of their feeding ecology allows West

Indian and West African manatees to exploit habitats that have been highly modified by humans, increasing the risks associated with the overlap between people and manatees, particularly in regions where the human population is increasing rapidly (Chapter 7). The consequences of this overlap between people and sirenians are most clearly seen for the Florida manatee, but potentially apply to the other species and subspecies of manatees as well, when this overlap brings increased hunting pressures and likelihoods of drowning in artisanal fisheries (Chapter 7).

All sirenians spend a considerable proportion of their day feeding. When their feeding is interrupted by human-related activities, such as vessel traffic (Chapter 7), sirenians may find it difficult to feed for long enough to satisfy their nutritional needs. This problem is likely to be greatest for sirenians when access to their food supply is limited, such as: (1) Florida manatees aggregating in warm-water refuges in winter; (2) dugongs and possibly West African manatees feeding on low biomass seagrasses in habitats where access to food is limited by the tidal cycle; and (3) Amazonian manatees in the dry season.

The studies of the interactions between dugongs and seagrasses demonstrate that feeding sirenians can deliver important ecosystem services by having significant effects on the biomass, community structure and chemical composition of their food plants through maintaining local patches of improved food quality, particularly enhanced nitrogen content. We infer that regular access to such feeding patches is important, especially in view of the evidence from dugongs and Amazonian manatees that variation in their food supply affects their reproduction (Chapter 6). High rates of reproduction currently observed in Florida manatees in some regions may even relate to the relatively recent proliferation of invasive aquatic plants from other continents. The suspected widespread reduction in the sizes of most dugong and manatee populations over the last 200 years has presumably changed their food plant communities, especially in areas not affected by some other form of physical disturbance (e.g. tropical storms). At present, it is impossible to assess the ecosystem significance of the decline in populations of dugongs and manatees in most areas and the resultant impact on ecosystem processes. The potential for important sirenian habitats to reach a threshold state beyond which the quality of available food sources prevents dugongs and manatees from recolonising these areas is a vital matter for sirenian conservation, as mentioned by Aragones *et al.* (2012a). We hope that the conservation initiatives discussed in Chapter 9 are sufficient to prevent this concern from becoming a reality.

SUGGESTIONS FOR FUTURE RESEARCH

Not unexpectedly, the research needs of each of the four sirenians are different and location-specific. The priority for Amazonian and West African manatees must be a more comprehensive documentation of their diet in the wild. The nutritional basis for food selection relative to food availability in the wild is poorly understood for all species, and will be an increasingly important question in the face of climate change. We need to understand how sirenians select their food at different stages of their life cycle, and how such selection changes seasonally and in response to extreme weather events and other causes of habitat loss in different habitats.

We have a very limited understanding of the interaction between sirenians and their food communities, especially in freshwater habitats and for invertebrates in marine habitats. Such information is critical to understanding the role of manatees and dugongs in ecosystem processes, and is fundamental to effective conservation of both sirenians and their habitats. We do not know why individual manatees and dugongs choose to feed in different feed patches, or the stimulus for them to move on to another patch.

The digestive anatomy and physiology of sirenians are clearly very different from those of other large mammalian herbivores. A coordinated, multidisciplinary approach to studying sirenian digestive anatomy and physiology in the context of their metabolic rate should provide fundamental insights into their digestive physiology, and extend our knowledge of large mammalian herbivores more generally.

NOTE

1. Also known as *Zostera muelleri capricorni* (Jacobs *et al.* 2006).

Behaviour and habitat use

INTRODUCTION

Knowledge of the behavioural ecology and habitat use of sirenians is critical for their conservation and management. In this chapter we summarise what is known about how sirenians perceive their environments; their swimming and diving behaviours; their activity and time budgets; their movements and migrations; and their anti-predator behaviour. These individual behaviours are clearly critical to understanding issues pertinent to the management of manatees and dugongs. Use of tactile senses to satisfy curiosity, for example, may explain their proclivity for entanglement and injury in lines and rope. Time spent at the surface can influence the susceptibility of dugongs and manatees to predators from below or to boat strikes or hunters from above. Activity and time budgets show the strong influence of seasonality on susceptibility to cold stress at the latitudinal margins of their distribution and to hunting during dry seasons in seasonally flooded environments. The scale of migrations and movements that sirenians employ can be highly informative for planning of reserves and protection of habitat corridors.

We also review knowledge of the social behaviour of sirenians, information that is fundamental for conservation. Do they occur in stable social groups that must have their integrity maintained? Are territories required for successful breeding? What factors may limit the potential diversity of sirenian mating systems? How do they communicate? Are patterns in the use of space transmitted through generations? How are young cared for and protected?

In the last section of this chapter we summarise habitat use by sirenians. What does current knowledge tell us about the habitat components that seem to be most highly associated with sirenian abundance and distribution? What among these factors might be manipulated for management? How flexible are sirenians in their use of habitat? Do they require wilderness? Knowledge of both the behaviour and habitat use of sirenians provides insight into their overall flexibility in responding to conservation actions.

INDIVIDUAL BEHAVIOUR

Sensation and perception

Audition

The auditory system of sirenians has been studied mostly in manatees. Field observations suggest they have 'exceptional acoustic sensitivity' (Hartman 1979). This sensitivity is well known to both manatee and dugong hunters; in some cultures the tiny stapes (the innermost middle ear bones) are removed and treated as magical objects (O'Shea *et al.* 1988; Chapter 7). Although most of this ability involves perception of sounds in water (sound waves travel much faster in water than in air), sirenians are capable of detecting some sounds in the air above the surface (Hartman 1979). Sound also provides an important communication medium for sirenian social behaviour (see 'Vocal communication' below). Anatomical specialisations of sirenian ears have been described and reviewed by Robineau (1969) and Ketten *et al.* (1992). A fused contact between the periotic and squamosal bones and the enlarged, spongy and oil-filled zygomatic process of the squamosal bone suggest that this region of the skull may also play a role in sound reception, perhaps as a low-frequency resonator, and in determining the direction from which a sound comes (Ketten *et al.* 1992; Ames *et al.* 2002).

Two of the areas of research that have assessed the hearing capacities of sirenians are discussed here. (The third approach is anatomical and outside the scope of this book; see Ketten *et al.* 1992.) One method involves recording electrical-evoked potentials detected outside the skull in response to controlled sound stimuli. The second technique is more powerful but laborious, and involves developing a behavioural audiogram by training captive subjects to perform behavioural actions in response to presentation of sounds at various frequencies and intensities. Evoked potentials have been recorded for Amazonian manatees (Bullock *et al.* 1980; Klishin *et al.* 1990; Popov and Supin 1990) and Florida manatees (Bullock *et al.* 1982; Mann *et al.* 2005). The evoked potential studies demonstrated variable sensitivity over a range of frequencies from about 200 Hz to 35–40 kHz in Florida manatees and Amazonian manatees, with greatest sensitivity in the lower range at 1–1.5 kHz in Florida manatees and 3.0 kHz in Amazonian manatees, dropping substantially above 8 kHz (one Amazonian manatee tested by Popov and Supin (1990) had a broader range of sensitivity). Frequencies of greatest sensitivity generally correspond to the lower-frequency components of communication sounds. In one study, Amazonian manatees showed greatest sensitivity to a sound source placed over the zygomatic process (Bullock *et al.* 1980), but in another experiment sensitivity occurred over a broader region of the head (Klishin *et al.* 1990). In Florida manatees, sensitivity over the zygomatic process was slightly less than over the small external ear opening, with a broad region of the side of the head showing responses rather than any one specific portal (Bullock *et al.* 1982). Based on evoked potential studies, manatees have high temporal resolution ability compared with terrestrial mammals (but less than dolphins), suggesting good ability to localise sound sources. Their sensitivity to higher frequencies may also enhance their ability to localise sound sources using inter-aural intensity differences, especially the higher harmonics in communication sounds (Mann *et al.* 2005).

Reliable behavioural audiograms of two captive Florida manatees identified the frequency range of best hearing to be from 6 kHz to 20 kHz, with peak sensitivity at 16–18 kHz (Gerstein *et al.* 1999; Gerstein 2002). The manatees had a general hearing capability over a frequency range of 0.4–46 kHz, similar to the findings of the evoked potential studies; but frequencies outside the 6–20 kHz range required marked increases in sound pressure levels for detection, and hearing sensitivity dropped sharply below 2 kHz. At low frequencies (0.015–0.2 kHz), sound detection was speculated to be by tactile senses. In some other species of animals the frequencies of greatest hearing sensitivity correspond with the frequencies of communication sounds (Owings and Morton 1998), but this does not seem to be the case for Florida manatees, based on behavioural audiograms (see below and 'Vocal communication'). The peak sensitivities of the behavioural audiograms (Figure 5.1) also do not correspond with frequencies that produce the greatest auditory evoked potentials, as described above. The greater sensitivity to higher frequencies observed in the audiogram research were thought to be an adaptation to avoid the complications in manatees perceiving sound reflections propagated from the water–air interface. Such complications occur with lower frequencies at the shallow depths typical of many manatee habitats; sounds at lower frequencies can be strongly attenuated or cancelled by such surface reflections (Gerstein *et al.* 1999; Gerstein 2002).

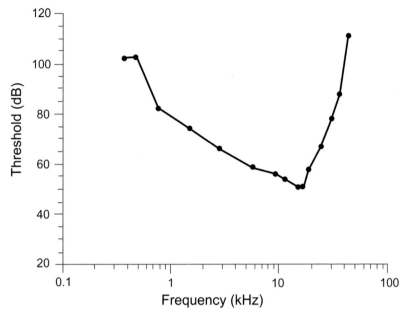

Figure 5.1. Behavioural audiogram for two captive Florida manatees. Results of the testing were consistent from one manatee to the other (redrawn by Shane Blowes after Gerstein (2002), with permission). The sound pressure levels that evoke trained behavioural responses were least (hearing most sensitive) at 16–18 kHz, but fell rapidly above 20 kHz and below about 800 Hz. See narrative, Gerstein (2002), Gerstein *et al.* (1999) and Colbert *et al.* (2009) for details.

Behavioural testing of responses to sounds made in water was also used to test the sound localisation ability of two other captive Florida manatees (Colbert *et al.* 2009). In this research, the subjects were presented with signals at two tonal frequencies (of one duration) and three broadband signals (of four durations). Their response rates were calculated at four angles relative to 0° (straight ahead): 45°, 90°, 270° and 315°. The response rates were well above chance for the broadband signals (0.2–20 kHz, 6–20 kHz and 0.2–2 kHz) at all angles, and above chance but less accurate for the tonal signals (4 kHz, corresponding to mid-range fundamental frequencies of communication sounds; and 16 kHz, corresponding to the mid-range best hearing frequencies in the above audiogram studies). Greater localisation ability was seen at the 4 kHz tones than at the 16 kHz tones. These manatees showed good localisation abilities at frequencies inclusive of those of the engine noises of recreational boats and at frequencies corresponding to those used in manatee communication (Colbert *et al.* 2009).

Tactile senses

The tactile sensory modality is highly developed in sirenians; Sarko and Reep (2007) described them as 'tactile specialists'. Modified hairs are used as sensory organs that differ in function between the orofacial region (where the hairs are most densely distributed) and the rest of the body. In addition to a higher density of hairs, the orofacial areas of sirenians have about 2000 highly modified hairs or perioral bristles distributed in six fields (Reep *et al.* 1998; Marshall *et al.* 2003; Figure 5.2). Unique among mammals, these bristles function along with modified facial musculature as a prehensile system to acquire, manipulate and ingest vegetation (Chapter 4). Based on experiments with a captive Antillean manatee, the ability of the bristles to resolve fine textural differences was found to be similar to that of the trunk of elephants (Bachteler and Dehnhardt 1999). Behavioural tests with captive Florida manatees suggested the ability to detect particle displacements of less than a micron (Bauer *et al.* 2010). The brain has a large and complex structural dedication to corresponding somatosensory regions, and sirenians possess

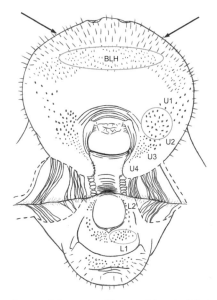

Figure 5.2. Frontal view of the perioral hair and bristle follicle fields of a Florida manatee (cheek muscles cut at dashed lines). Arrows indicate the orofacial ridge. The orofacial areas of sirenians have about 2000 highly modified hairs or perioral bristles distributed in six fields. Facial follicle areas include bristle-like hairs (BLH), upper perioral fields (U1–U4) and lower perioral follicle fields (L1–L2). The U2 and L1 fields are highly specialised for grasping. Re-drawn after Reep *et al.* (2001) and Sarko *et al.* (2007b), with permission from publisher S. Karger AB Basel.

unique cell clusters in the cortex of the brain, presumed to be for specialised processing of tactile information (Marshall *et al.* 2007; Sarko and Reep 2007; Sarko *et al.* 2007a). The hairs and bristles on the face also function in tactile exploration of the environment, and have a sensitive microanatomy similar to that of the vibrissae of other mammals, with highly innervated follicle–sinus complexes that differ in characteristics among facial regions and possess some unique features of innervation (Reep *et al.* 2001; Sarko *et al.* 2007b).

Hairs distributed over the rest of the bodies of manatees and dugongs (Figures 5.3 and 5.4) are anatomically similar to the sensitive facial vibrissae in other mammals and also have unique features of innervation (Kamiya and Yamasaki 1981; Reep *et al.* 2002; Sarko *et al.* 2007a,b). These hairs are spaced about 20–40 cm apart across the skin in Florida manatees, decreasing in density from dorsum to ventrum (Figure 5.3). It is likely that these hairs

a

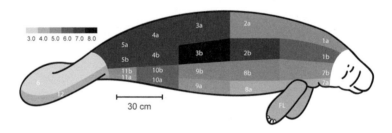

b

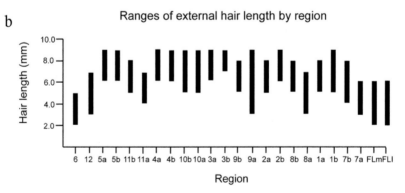

Figure 5.3. Distribution, density and lengths of tactile body hairs on a 300 kg, 257 cm long male Florida manatee. There are about 3000 hairs, 9–20 mm long, spaced about 20–40 cm apart, decreasing in density from dorsum to ventrum. Redrawn after Reep *et al.* (2002), with permission from publisher S. Karger AB Basel.

Figure 5.4. A foraging dugong with the facial disk above the substrate, raising a small sediment cloud. The small black dots (e.g. below the eye) mark locations of vibrissae which can be seen more clearly in the enlarged inset picture. Note also the faint tusk scrapes and scars along the upper dorsum. Photograph courtesy of Pierre Larue; reproduced with permission.

constitute a distributed tactile system capable of sensing the directionality of pressure waves. Because pressure waves travel well in water, this system may function analogously to the lateral line system in fish (suggested earlier by Hartman (1979) and Reynolds (1981b) based on anecdotal observation). This sensory system appears to be unique among mammals in providing a broad receptor field capable of detecting and localising water displacements by approaching animals and objects, perhaps providing information important to social behaviour (see below), as well as supplanting other sensory modalities at night and in turbid or shallow water (Reep *et al.* 2002). Reliance on close tactile inspection of objects rather than vision (see 'Vision' below) may place Florida manatees in harm's way when inspecting foreign objects that could cause injury or entanglement, such as crab trap lines, fishing lines and floodgates (Bauer *et al.* 2010).

Vision

Sirenians use vision, but the efficiency of this sensory modality can be limited by environmental conditions (night, turbid water). Dugongs approach divers and boats and inspect them visually (Anderson 1982). Hartman (1979) describes how Florida manatees typically investigate auditory stimuli by approaching and looking at the sources, often head-on, suggesting binocular vision (ocular fields show slight overlap in the Amazonian manatee; Piggins *et al.* 1983). Hartman observed that manatees apparently perceive objects visually at distances of 35 m in clear water, and inferred from reflections of artificial lights by the presumed *tapetum lucidum* that vision was good under low light conditions (Hartman 1979). (However, it was later established that a *tapetum lucidum* is lacking in manatees; Piggins *et al.* 1983; West *et al.* 1991.) It is likely that near objects (<1 m) are perceived with monocular cues (Piggins *et al.* 1983). Anecdotal observations (Anderson 1979; Hartman 1979) suggested that dugongs and Florida manatees see in air when their heads are raised above water, and this has been confirmed by visual discrimination tests on captive West Indian manatees (Griebel and Schmid 1996, 1997).

Experiments and modern anatomical and physiochemical studies on manatee eyes provide additional insights on vision (for a review of earlier anatomical studies of sirenians, see Piggins *et al.* (1983) and Griebel and Schmidt (1996, 1997)). Experimental brightness discrimination tests on captive West Indian manatees, using 30 shades of grey, showed a brightness discrimination ability about the same as two species of fur seals (*Arctocephalus*), but about one-third of that of humans (Griebel and Schmidt 1997). Colour discrimination experiments showed dichromatic colour vision as in diurnal mammals: West Indian manatees could discriminate blues and greens from grey, but not reds or blue-green (Griebel and Schmidt 1996). This result agrees with morphological inferences based on the presence of rods and two types of cone receptors in the retina of Florida manatees (Cohen *et al.* 1982). Molecular analysis of these receptors confirmed this dichromatism (other marine mammals generally have monochromatic vision) and found that the receptor genes have affinities with those of elephants (Newman and Robinson 2006; Chapter 3). The presence of both rods and cones in Florida manatees suggests the use of vision both at night, or dusk, and in daylight, consistent with their observed activity patterns (see 'Activity patterns and time budgets' below). In Amazonian manatees, however, the retina is dominated by rods, which is more consistent with the eyes of primarily nocturnal mammals (Piggins *et al.* 1983), perhaps reflecting their mostly turbid environments ('Habitat use and habitat components' below; Chapter 4). Accommodation of the lens (to change focus from near to far) may utilise unique vascular mechanisms in manatees (Natiello and Samuelson 2005).

There are some interesting anatomical differences between the structure of the eyes of manatees and dugongs. Corneal avascularity – the absence of blood vessels in the cornea – is required for optical clarity and optimal vision. Both Florida and Antillean manatees have vascularised corneas, in contrast to the avascular corneas of dugongs, dolphins and whales (Ambati *et al.* 2006). (The corneas of the other manatees have not been investigated.) The functional basis of the vascularised West Indian manatee cornea is intriguing (Ambati *et al.* 2006). Manatees are physiologically dependent on fresh water (see 'Salinity and access to fresh water' below), and corneal vascularisation could protect against, or perhaps result from, irritations due to this hypotonic environment.

Behavioural tests demonstrated marked individual differences in the visual acuity of two captive-born Florida manatees, suggesting natural variation within populations (Bauer *et al.* 2003). Tests of visual acuity, or ability to discriminate fine detail, in the better-performing manatee suggested a range intermediate between the Amazon river dolphin at the low end and some pinnipeds or the bottlenose dolphin at the higher end (Bauer *et al.* 2003). Manatees are likely either emmetropic or hypertropic (normal or far-sighted) like other marine mammals, and appear capable of using vision at intermediate to longer distances to orient towards larger objects; acuity does not improve at distances less than 1 m, where they appear to rely chiefly on their tactile senses for close inspection (Bauer *et al.* 2003). Their dichromatic colour vision and modest visual acuity suggests Florida manatees are best adapted for seeing patches of plants at intermediate to longer distances (Bauer *et al.* 2010).

Chemoreception

Like other marine mammals, the olfactory areas of the brain of sirenians are reduced; the relative size of the sirenian olfactory bulb is among the smallest in mammals, suggesting regression of an unused sensory system (Reep *et al.* 2007). Unlike many terrestrial mammals that rely on chemical signalling for mating and social behaviour (including other paenungulates), but similar to cetaceans, sirenians do not possess vomeronasal organs (Mackay-Sim *et al.* 1985). Sirenian tongues possess fewer tastebuds than those of terrestrial mammals, but more than those of other marine mammals (e.g. Levin and Pfeiffer 2002), suggesting that, in addition to detecting chemicals in food, sirenian tastebuds may function as receptors for chemical cues from the environment (e.g. fresh water), and perhaps signals from conspecifics for mating or other behaviour (e.g. hormones or pheromones), a situation hypothesised for Florida manatees (Hartman 1979; Larkin 2000; Levin and Pfeiffer 2002). Anecdotal observations suggest that manatees may use 'rubbing stones' for behaviour that seems analogous to scent marking (Rathbun and O'Shea 1984), but no form of chemical communication in sirenians has been confirmed, and experimental tests of chemoreceptive abilities have not been conducted.

Orientation and navigation

The apparently directed long-distance moves of hundreds of kilometres between sites made by some dugongs and manatees suggest a great capacity for orientation and navigation and a keen spatial memory (Deutsch *et al.* 2003; Sheppard *et al.* 2006; see 'Long-distance movement patterns and migrations' below). Even in local moves of a few kilometres, it is impressive to track a lone manatee as it travels down a black-water river without hesitation, across various bends and creek mouths only to turn into a pocket where it joins other individuals already in place. Sirenians also travel in the dark of night. However, the importance of the various sensory modalities in orientation and navigation is unknown. Sheppard *et al.* (2006) speculate that dugongs may use coastal geomorphology to navigate during long-distance moves via some combination of visual, magnetic, chemosensory or tactile cues, concordant with observed travels below the surface. Like manatees, dugongs possess an array of highly developed sinus hairs, especially on the muzzle (see 'Tactile senses' above; Kamiya and Yamasaki 1981; Marshall *et al.* 2003; Chapter 4), which may play a role in exploitation of geomorphic variations in the benthos as an orientation aid (Sheppard *et al.* 2006). Orientation and navigation of sirenians remains an unexplored frontier for research.

Swimming and diving

Locomotion

Locomotion in sirenians (especially Florida manatees) is described extensively by Hartman (1979). Dugongs swim by more advanced caudal oscillations of the tail, similar to cetaceans, whereas manatees swim by dorsal–ventral, sinusoidal undulations of the body and the flattened, spatulate tail (Kojeszewski and Fish 2007). Dugongs are more fusiform than manatees (Figure 6.3), and the peduncle provides greater thrust, with undulations at normal swimming speed timed at 0.56 per second

(Anderson 1982). Nonetheless, Florida manatees can swim great distances and seem to travel further than dugongs (Sheppard et al. 2006). One Florida manatee moved up to 4800 km at speeds of 20–35 km per day along the Atlantic coast of the United States (Reid 1995). Florida manatees have been reported to swim at typical speeds of 2–7 km per hour (18–25 km per hour in short fleeing sprints). Undulations are greatest at the tip of the tail (Hartman 1979; Kojeszewski and Fish 2007). Dugongs can also travel at fast speeds, cruising at up to 10 km per hour, with short sprints of up to about 20 km per hour (Jarman 1966; Marsh et al. 1981). Dugongs and manatees hold their flippers appressed to the body while cruising or travelling. They use their flippers independently or together for fine movements, such as sculling, approaching or retreating and facing objects, and for bottom-walking. Unlike dugongs, manatees also use their flippers in food handling (Chapter 4).

Resting adult manatees push off the bottom with their flippers to rise for breathing without use of the tail, descending to nearly the same spot (Hartman 1979). Florida manatees are neutrally buoyant and can rest at locations in the water column ranging from lying flat, with their bellies on the bottom, to hanging suspended at the surface (for further information on buoyancy control, see Domning and de Buffrénil (1991) and Rommel and Reynolds (2000)). They usually rest with their eyes closed but often open them while surfacing to breathe. Sleep during resting has been monitored electrophysiologically in a small captive Amazonian manatee. It exhibited slow-wave sleep and paradoxical sleep, both of which were interrupted at each respiration; the two forms of sleep were similar to those known for pinnipeds but unlike that of dolphins, which do not show paradoxical sleep (Mukhametov et al. 1992).

Respiration and diving
Respiration frequency is influenced by multiple factors, including activity level, body size, water temperature, water depth and reproductive status. Bottom-resting adult Florida manatees rise to breathe every 1–8 minutes (maximum 24 minutes; Reynolds 1981a), but breathe more frequently while travelling (Hartman 1979). Exhalations last about 1–3 seconds, with inhalations slightly shorter, depending on activity level (Hartman 1979). Manatee cows and their calves often breathe synchronously (Hartman 1979; Reynolds 1981b), and synchronous breathing also occurs in small groups of manatees and dugongs (Jarman 1966; Hartman 1979; Hodgson 2004). Whiting (2002b) directly observed dugongs in water 1–6 m deep in the tropical Northern Territory of Australia and recorded a maximum dive

time of about 11 minutes, a mean dive time of about 6 minutes (ten dives per hour), a mean surface time of 13 seconds and an average number of 1.6 breaths per surface interval. These dive times were much longer than those reported for dugongs at higher latitudes (Anderson and Birtles 1978; Anderson 1982, 1998; Hodgson 2004). Hodgson (2004) observed dugongs from above during 87 focal follows using a blimp-mounted video camera in Moreton Bay, Queensland, and measured 997 submergence intervals and 1110 surface intervals in water 0.9–4 m deep. Submergence intervals ranged from <1–317 s (mode 108 s); the mean surface time was $2 \pm SE 1$ s with a range of 1–5 s. The overall mean dive rate was 47 complete dive cycles per hour.

Direct observations of sirenians show that submergence times vary with activity and size class; larger, resting individuals have longer submergence times (e.g. Hartman 1979; Reynolds 1981a; Anderson 1982). Field observations also provide some detailed descriptions of variability in sirenian behaviour while breathing, including various postural adjustments to wave action and other environmental conditions (Anderson and Birtles 1978; Anderson 1982, 1998). Respiratory intervals of a small captive Amazonian manatee averaged 248 s when resting and 170 s when moving (Mukhametov *et al.* 1992). Hodgson (2004) video-recorded dive times and behaviours for 28 single individuals and 28 mothers with accompanying calves in clear, shallow water in Moreton Bay, Queensland, under calm conditions. The mean submergence time of all individuals was 75 s, very close to the value (73 s) recorded in Shoalwater Bay, Queensland, by Anderson and Birtles (1978). Hodgson (2004) reported that calves had significantly shorter submergence times (mean 72 s) than their mothers (mean 82 s). Submergence intervals were not affected by water depth (over the 0.9–4 m depth range), but were affected by concurrent behaviour. Submergence intervals were shortest (mean 68 s) when individuals conducted a single behaviour other than feeding or travelling; longest when they were undertaking a combination of behaviours (mean 92 s). The average time spent submerged while feeding (83 s) or travelling only (74 s) was intermediate between these two values.

Time–depth recorders suggest that dugongs and Florida manatees spend most of their time in shallow waters. Edwards *et al.* (2007) reported that Florida manatees at a power plant discharge in winter spent 19% of their time at depths <1 m, with one female and calf spending 82% of their time at <1 m. Fifteen dugongs tagged at widely separated regions of Australia spent nearly half of their activity at depths ≤1.5 m, and 72% of their time at depths ≤3.0 m from the surface (Chilvers *et al.* 2004). About 67% of the dives were interpreted as feeding dives, with average dive durations and depths of 2.7 minutes and 4.8 m, respectively, and maxima of 12.3 minutes and 20.5 m. Considerable individual variation was observed, but an average of 11.8 dives per hour were recorded across all dugongs sampled (Chilvers *et al.* 2004). The depths recorded suggest that the dugongs were highly dependent on the seagrasses growing in intertidal and shallow subtidal areas. However, dugongs may dive to 36 m (Sheppard *et al.* 2006), and dugong feeding trails have been recorded to depths of 33 m (Lee Long *et al.* 1996; Chapter 4).

The above data were interpreted without knowledge of the bathymetry of the locations where the dives were measured. Rie Hagihara (unpublished) used data from six dugongs fitted with time–depth recorders and GPS transmitters in Hervey Bay, Queensland, and found that dugongs use different parts of the water column depending on the local bathymetry. As expected, as water depth increased to about 5 m, the dugongs reduced the proportion of time they spent within 2.5 m of the surface. However, in water depths of more than 12 m, the proportion of time spent near the surface increased again, suggesting that dugongs have difficulty spending extended periods of time near the bottom in deeper water. The dugongs spent very little time in the middle of the water column irrespective of the water depth.

Activity patterns and time budgets

Most information suggests that sirenians do not have well-defined periods of circadian activity (e.g. Anderson and Birtles 1978; Hartman 1979). This lack of marked diel activity patterns is consistent with the absence of a pineal organ in the brain, which can act as a regulator of daily rhythms in temperate-zone mammals (Ralph *et al.* 1985). However, the need to conserve energy in response to seasonal changes in temperature or food availability can alter this lack of pattern, as can hunting pressure or boating activity (e.g. Jarman 1966; Reynolds 1981a; Bengtson 1981; Brownell *et al.* 1981; Rathbun *et al.* 1990; Miksis-Olds *et al.* 2007a,b). Dugongs also exploit tidal cycles to access areas unavailable at low water (Anderson and Birtles 1978; Preen 1992; Sheppard *et al.* 2009; Chapter 4).

R. Hagihara (unpublished) found some diel differences in the dugong's use of the water column. The six dugongs she studied spent more time close to the surface in the afternoon (1200–1800 hours) than in the morning or at night, suggesting that they were feeding more at these times than in the afternoon. Sheppard *et al.* (2009) analysed the GPS tracks of the same dugongs and found that they were closer to shore at night than during the day.

During warm seasons, Florida manatees show no circadian patterns in activity. In a study in the upper St Johns River they spent about 20–25% of the time feeding, an equal amount of time resting, about 10–15% of the day in 'cavorting' social behaviour (see 'Social behaviour' below) and about 30% (females) to 45% (males) of the time moving; the proportion of time spent feeding increased in late summer and autumn prior to the winter cold season (Bengtson 1981). The numbers of sightings of manatees engaged in various behaviours, as observed from survey aircraft in the lower St Johns River and Intracoastal Waterway of north-eastern Florida, also varied with season, with higher proportions observed resting in cold months and a greater proportion observed feeding, travelling or cavorting during warm months (Kinnaird 1985).

During winter months when water temperatures were cooler in the surrounding upper St Johns River, manatees at Blue Spring typically rested in the constant-temperature spring water (where food is unavailable). They left to feed by late afternoon when air temperatures were highest, fed during the night, returned several hours before noon the next day, and rested until late afternoon. Feeding periods were 3–15 hours in duration, and trips to feeding areas were direct. However, during prolonged cold weather manatees would remain in the spring and forego feeding for periods of up to a week (Bengtson 1981). Similar seasonal and diel patterns in use of springs at Crystal River by Florida manatees have also been well documented (Rathbun *et al.* 1990). Amazonian manatees also have seasonally inactive periods when water levels are low (Best 1983; Arraut *et al.* 2010; see 'Habitat use and habitat components' below).

Hodgson (2004) used her blimp-mounted video camera to quantify the time budgets of focal dugongs in Moreton Bay in winter. Dugongs spent most of their time feeding (41%), travelling (32%) and ascending to and descending from the surface (18%), and relatively little time resting, socialising or in other behaviours. Environmental variables accounted for little of the variability in the proportion of time dugongs spent in each behavioural category. Time budgets did not differ significantly between single individuals and mothers with calves. However, mothers spent significantly more time feeding than their calves, presumably because the calves had access to milk as well as seagrass. (Alternative methods of estimating the time sirenians spend feeding are considered in Chapter 4.)

Long-distance movement patterns and migrations

Sirenians show great variability in movement patterns and migration, depending on the study area and the influence of seasonal temperature or

Figure 5.5. A Florida manatee with a tethered transmitter, drinking from a freshwater hose in a tidal creek at Merritt Island National Wildlife Refuge, Florida. J. P. Reid photograph, reproduced with permission.

rainfall on regional ecosystems. Even within the same region, however, considerable polymorphism may be observed in movement patterns among individuals within a population (e.g. Deutsch *et al.* 2003; Sheppard *et al.* 2006). The ability to track sirenian movements has advanced greatly since the development of floating tethered transmitters (Figure 5.5), with the successive availability of radio, satellite and GPS-based tracking technologies (Rathbun *et al.* 1987; Reid and O'Shea 1989; Marsh and Rathbun 1990; Sheppard *et al.* 2006).

Dugong movements

Radio and satellite-monitored transmitters applied to six male dugongs on the north Queensland coast showed the existence of overlapping home ranges, a close association with inshore seagrass beds, occasional forays as far as 140 km travelled in two days, and some use of tidal creeks (Marsh and Rathbun 1990). This technology was extended by Sheppard *et al.* (2006), who applied satellite and GPS tracking technology to 70 dugongs in Queensland and the Northern Territory of Australia for periods of 15–551 days between 1986 and 2004. Travelling dugongs remained in coastal waters, generally within the ranges of seagrass beds. Mean maximum distances from shore were 12.8 ± SE 1.3 km from the coast. The scales of movement were heterogeneous. Twenty-six dugongs (37%) moved <15 km, 28 (40%) moved 15–100 km (mesoscale movements) and 14 (20%) moved 100–560 km (macroscale movements; Sheppard et al. 2006). Travel of the dugongs that moved >15 km (considered large-scale movements) was rapid and directed, averaging about 180 hours for a mean straight-line distance of 244 km. Mean daily distances travelled in these directed moves varied from about 6 km to 72 km. There was no evidence that dugongs stopped to feed during these directed movements, even when seagrass resources were available. Males, females and females with calves all made large-scale movements; dugongs that failed to move >15 km also encompassed these age and sex categories. Some movements were documented to be return movements, suggesting ranging rather than dispersal movements. Sheppard et al. (2006) proposed that dugongs have evolved to cope with unpredictable and patchy seagrass abundance (large-scale diebacks of seagrass are well documented (e.g. Preen and Marsh 1995; Preen et al. 1995; Marsh and Kwan 2008)) by making directed moves to alternative areas known to have seagrasses in the past. Use of these areas may be behaviours learned by calves from their mothers (see 'Evidence for matrilineal use of space' below).

Although dugongs tagged in the tropics made large-scale movements unrelated to temperature, in some areas dugongs moved in response to drops in water temperature to below about 17–18 °C (Sheppard et al. 2006; see 'Water temperature' below). In Moreton Bay, Queensland, near the high latitude extreme of the species' range on the east coast of Australia, dugongs made microscale movements of about 10 km to where the waters of the East Australian Current were >5 °C warmer than the 17–18 °C water in Moreton Bay. The dugongs regularly rode the currents in and out of the bay to move between cool-water feeding areas and warm-water resting areas. Another six dugongs made repeated return movements in winter across Hervey Bay, Queensland, to warmer ocean waters despite the apparent lack of seagrass and likely higher numbers of sharks in the oceanic location (Sheppard et al. 2006). Tracking studies also demonstrate that dugongs make seasonal movements within Shark Bay (the high latitude limit of the dugong's range in Western Australia) and have seasonally distinct core areas of activity that enable them to minimise exposure to low water temperatures during winter (Holley 2006).

Florida manatee movements
A multi-year (1986–1998) telemetry study of 78 Florida manatees along the Atlantic coast of the United States also revealed considerable complexity in movement patterns (Deutsch et al. 2003). Manatee movements had a strong

seasonal component, from north to south in autumn and early winter, and from south to north in spring. Individuals repeated their migratory patterns from one year to the next, and showed high fidelity to both warm season and winter destinations (Deutsch *et al.* 2003). Annual patterns of migrants fell into four classes: long-distance migrants (one-way distances of 575–831 km, with a maximum outlier of 2360 km in an adult male), medium-distance migrants (150–400 km), short-distance migrants (50–150 km) and year-round residents (<50 km in one-way movements). These patterns were consistent among individuals from year to year and, like dugong movements, did not relate to sex, age class, body size or female reproductive status. Three geographic foci were involved as migratory end points: (1) north-eastern Florida and south-eastern Georgia, summer habitat consisting largely of saltmarsh cordgrass available only at high tide; (2) the seagrass habitats of the Banana River and Indian River lagoons in Brevard County on the central coast, used in different ways as summer or winter habitat; and (3) seagrass habitats of south-eastern Florida used in winter, including Biscayne Bay near Miami (Deutsch *et al.* 2003). In general, travel between these areas was rapid, with only a few areas used as stopovers *en route*. Stopover sites often included access to fresh water as well as to submerged aquatic vegetation. Travel routes typically followed the dredged channel of the Intracoastal Waterway between the mainland and barrier islands, although some manatees also travelled in the ocean just beyond the breaking surf. Only 12% of the manatees tagged on the Atlantic coast were characterised as year-round residents.

Most seasonal movements southward were initiated abruptly in response to the arrival of substantial cold fronts (Deutsch *et al.* 2003). These directed moves began when water temperatures fell below at least 20 °C and continued as temperatures dropped further (93% of tagged manatees departed during periods when temperatures were falling). However, there was substantial variability among individuals in the precise temperature that sparked southward moves, with some manatees not migrating until temperatures of 15 °C were reached. Despite individual variation in temperatures that prompted migration, individuals were consistent from year to year in the temperatures at which they departed. Adult females with calves did not time migrations any differently than those without young, and no variation in initiation of moves was apparent by age class, size or sex. Manatees moved during the night as well as in day-time (see 'Activity patterns and time budgets' above for information on circadian rhythms). Short-distance migrants travelled at rates of 25 km per day, and medium-distance migrants at 36 km per day. Southward migrations averaged ten days, including two days at stopover points and eight days of travel time, but northward migrations averaged about 15 days, with five days at stopover points (Deutsch *et al.* 2003). About half the manatees had no stopover in autumn, and about one-third had no stopovers in spring. Migrating manatees travelled alone or in small groups, joining with larger numbers at stopover points. Those that moved the farthest north in spring generally delayed their migrations about five weeks behind those that moved from southern Florida to central Florida, typically stopping over at warm-water effluents of power plants. Large manatees arrived at northern destinations earlier than smaller ones. Some manatees initiated premature northward migrations during winter warming periods, only to turn back at the next major cold front. Manatees that made such moves during winter often travelled at faster rates (up to 58 km per day), presumably to reduce time spent in cool water, with one male travelling on the ocean side of the surf at a rate of 87 km per day (Deutsch *et al.* 2003).

The movement rates of male and female Florida manatees did not differ during migration or in winter, but on their warm season ranges adult males travelled more than females (3.8 km per day vs 2.0 km per day). The travel rates of subadult manatees were similar to those of adult females. Similar to findings in the St Johns River (Bengtson 1981, see 'Mating systems and mating behaviour' below), the increased warm season travel of males coincided with the period of heightened reproductive drive (Chapter 6). Most manatees maintained their migratory patterns (the four patterns as described above) from year to year, including fidelity to the same warm season and winter destinations (Figure 5.6). Warm season destinations showed high overlap from year to year, and were relatively small clusters of locations (Figure 5.6a), with the ninetieth percentile between cluster centres and all daily locations averaging 8.0 km, prompting Deutsch *et al.* (2003, p. 36) to note that 'clearly manatees were not nomadic in their movements during the warm season', contrary to earlier speculation (Hartman 1979). The manatees showed high site fidelity to their winter destinations (Figure 5.6b), with the ninetieth percentile averaging 6.4 km between cluster centres and all daily locations. Areas that were occupied at these destinations showed a high degree of overlap with other manatees, and were not exclusive or territorial.

Manatees from the Atlantic coast that were calves when their mothers were tracked were also tracked as subadults (Figure 5.7); the subadults adopted the migratory patterns of their mothers and showed strong philopatry to the warm season ranges that they had occupied as nursing calves. They also continued to use the same winter destinations. However, they did not travel with their mothers after weaning (Deutsch *et al.* 2003).

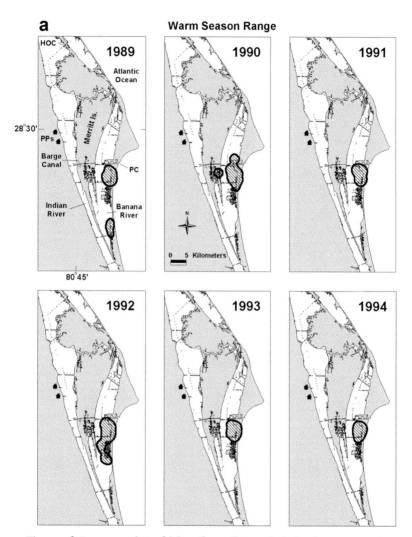

Figure 5.6. Inter-annual site fidelity of a satellite-tracked Florida manatee to her seasonal range based on Argos locations and visual observations. Hatched polygons denote 95% home ranges. Warm-water power plant effluents are marked with house symbols. (a) Warm season range of a female Florida manatee monitored for six consecutive years in the Banana River on the central Atlantic coast of Florida. (b) Winter range of the same manatee in southeastern Florida at two sites, respectively 275 km and 320 km south of the warm season range. Reproduced from Deutsch *et al.* (2003), with permission.

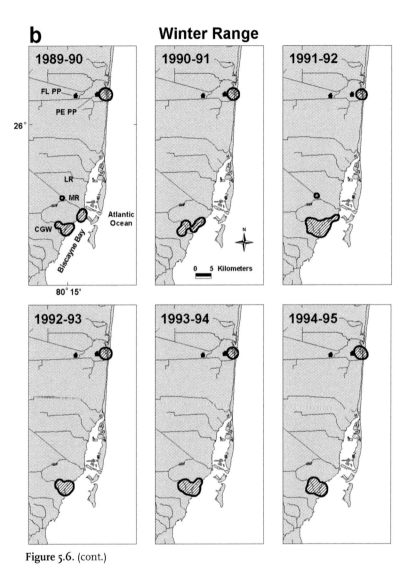

Figure 5.6. (cont.)

Movements of at least 33 wild manatees have been studied in far south-western Florida (encompassing the Everglades National Park) using GPS and satellite-monitored telemetry (Stith *et al.* 2006). In this region, the winter climate is generally very moderate. Most manatees did not make exceptional north–south migrations as on the Atlantic coast, but instead had regular offshore–inland movement patterns. The tracked manatees

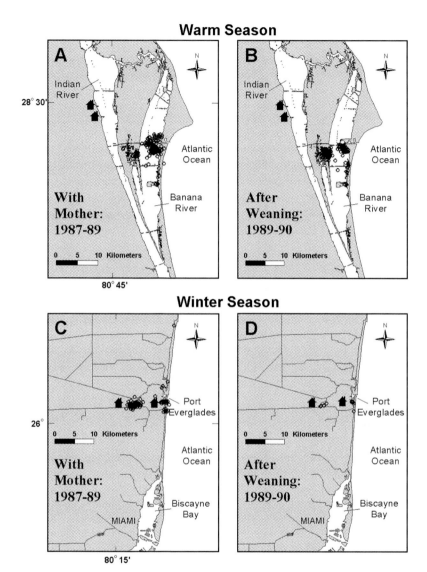

Figure 5.7. Warm season (central Atlantic coast) and winter (south-eastern Florida) ranges of a young Florida manatee both before and after weaning (after Deutsch *et al.* (2003)). Circles represent Argos locations, visual observations or triangulated positions of the calf or his mother. Warm-water power plant effluents are marked with house symbols. Reproduced from Deutsch *et al.* (2003) with permission.

made localised (*c.*15–35 km) but strongly seasonal shifts in distribution. They used shallow offshore areas with seagrasses in the Gulf of Mexico during summer and autumn. During winter they used inland zones and enclosed bays, where they occupied passive thermal refugia during cold snaps (Stith *et al.* 2010). No manatees used the offshore areas when water temperatures were less than 15 °C, and most made the shift closer to shore when temperatures in the Gulf of Mexico dropped below 18–20 °C (Stith *et al.* 2010). Manatees did not show any major change in individual ranging patterns in either shallow offshore or inland habitats, despite the passage of tropical storms and hurricanes (Langtimm *et al.* 2006). Like those tracked on the Atlantic coast, manatees generally used core areas that shifted with the season, but use of space showed some distinct patterns. In the wet months manatees spent more time at shallow offshore seagrass beds, and in dry seasons they used inland areas with access to fresh water more often (Stith *et al.* 2006). All tracked manatees moved from offshore seagrass beds to inland rivers, creeks and other sources of fresh water every 2–8 days, and travelled along distinct linear corridors (Stith *et al.* 2006).

In north-western Florida, manatees using the winter refugia in the Crystal River area generally did not migrate as far as some of the manatees on the Atlantic coast. In spring, they dispersed both north and south, but more than twice as many radio-tagged manatees moved north (Rathbun *et al.* 1990). Most distances travelled ranged up to about 125 km, with at least one individual on the north-western coast travelling about 175 km between Tampa Bay and the Suwannee River (Rathbun *et al.* 1990). Additional movement patterns of Florida manatees have been inferred from photo-identification records, and these sightings generally conform to patterns revealed in more detail by telemetry (Reid *et al.* 1991, 1995). An extreme movement included an adult female repeatedly seen at Crystal River during 1979–2006 sighted some 700 km away at Cuba in winter 2007 (Alvarez-Aleman *et al.* 2010).

Antillean, Amazonian and West African manatee movements

Antillean manatee movements are best documented for Puerto Rico. Manatees at Puerto Rico and neighbouring Vieques Island ranged up to 50 km along the coast, utilising inshore seagrass beds and periodically travelling to freshwater sources (Reid 2006). Heavy use was made of the de facto sanctuary provided by Roosevelt Roads Naval Station (now abandoned as a military facility), including shallow coves and bays protected from wind and wave action; all feeding sites located were at seagrass beds in shallow water, with a mean depth of 2.0 m (range 1–5 m; Lefebvre *et al.* 2000).

Manatees in the central Amazon of Brazil made strong seasonal movements each year, based on water levels and consequent access to foraging sites or refuge from predators (see 'Habitat use and habitat components' below; Arraut *et al.* 2010). These movements traversed a study area of about 120 km in length. Ongoing or recently completed tracking studies of manatees have taken place elsewhere in Central America, South America and Africa, but published accounts are generally not yet available. We look forward to wide dissemination of these results.

Anti-predator behaviour and habitat use

As long-lived, slow-breeding species (Chapter 6), sirenians may be expected to maximise survival by minimising risk from predation. Dugongs and manatees are somewhat protected by their large size, generally good auditory and tactile perception (see 'Sensation and perception' above), thick skin, stout rib cage and fusiform body shape with minimal constriction or appendages that predators can easily grasp. They can also minimise predation through behaviour. Although manatees and dugongs are generally considered to be relatively slow swimmers compared with most other marine mammals, they are capable of short bursts of speed to escape from perceived danger (see 'Locomotion' above). Both dugongs and manatees change their behaviour and orient towards and use deeper water for escape from predators or perceived predators (Nowacek *et al.* 2004; Heithaus *et al.* 2007; Wirsing *et al.* 2007a,b,c,d). However, the risks of predation in modern environments seem to differ between manatees and dugongs, as does the degree to which anti-predator behaviour and microhabitat use have been investigated.

Information concerning non-human predators on manatees is largely limited to anecdotal reports of takings of Amazonian manatees by jaguars (*Panthera onca*) and caimans (*Melanosuchus niger, Caiman crocodilus*; Arraut *et al.* 2010), likely predation on West African manatees by crocodilians and sharks (Reynolds and Odell 1991) and rare evidence of attempted attacks on West Indian manatees by sharks (Mou Sue *et al.* 1990) or alligators (*Alligator mississippiensis*; Wells *et al.* 1999). Three recent instances of shark attacks on living Florida manatees (i.e. antemortem attacks) in south-western Florida have been reported (D. Semeyn, unpublished). In contrast, the history of human predation on manatees is long, and Arraut *et al.* (2010) partially attribute seasonal movements of Amazonian manatees in the western Amazon to a strategy to reduce vulnerability to both animal and human predators. Manatees respond reflexively to threats by diving; this behaviour suggests that fear of terrestrial predators (including humans) has historically shaped manatee responses and reactions, as expressed by Barrett (1935), writing of Antillean manatees in Nicaragua: '[Manatees] are afraid of a leaf falling into the water, yet they pass their lives among sharks and crocodiles, and the reason for their terror of things above the surface can hardly be explained by hereditary instinct, for their safety has always been jeopardised by aquatic predatory enemies.' Although the quick flight taken by wild manatees in reaction to shoreline or above-water disturbances is commonly observed, there have been no specific studies of manatee behaviour in relation to predation. It is of note, however, that West Indian manatees of all ages and sexes give loud, long alarm calls while fleeing in such situations (O'Shea and Poche 2006; see 'Vocal communication' below).

Manatee calves may be more vulnerable to predators than larger individuals. In Florida, female manatees usually vocalise to call calves to their sides when disturbed, and typically place themselves between their calves and people swimming in the water (O'Shea and Poche 2006; T. J. O'Shea, personal observation), perhaps a generalised response of shielding calves from potential predators. Manatees will also typically list to one side while moving to minimise exposure of the ventrum to human swimmers. However, some wild manatees become habituated and tame to human swimmers, soliciting petting and stroking with no apparent fear (Hartman 1979; see 'Availability of quiet or sheltered areas' below).

The risks of non-human predation are presumably much greater for dugongs, because of their exposure to marine predators. Attacks on dugongs are documented for saltwater crocodiles (*Crocodylus porosus*), killer whales (*Orcinus orca*) and, particularly, tiger sharks (*Galeocerdo cuvier*). To our knowledge, the lone evidence of saltwater crocodiles eating dugongs is an aerial photograph taken off Cape York in Queensland, showing a large crocodile with a dugong in its jaws. It is unknown if the dugong was killed or scavenged. Anderson and Prince (1985) relate anecdotal accounts of dugongs attacked by groups of killer whales in Shark Bay in Western Australia, including an account of about ten killer whales attacking a group of dugongs, replete with blood in the water and the presence of wounded and dying dugongs.

Both living (Patterson 1939) and dead dugongs are savaged by sharks. The first written account comes from the explorer William Dampier at Shark Bay in Western Australia in 1699. Dampier (2005, p. 37) thought that remains found inside the shark were from a 'hippopotamus', but they were likely from an adult female dugong with two unerupted tusks:

> Of the sharks we caught a great many which our men eat very savourily . . . one which was 11 foot long. The space between its two eyes was 20 inches, and 18 inches from one corner of his mouth to the other. Its maw was like

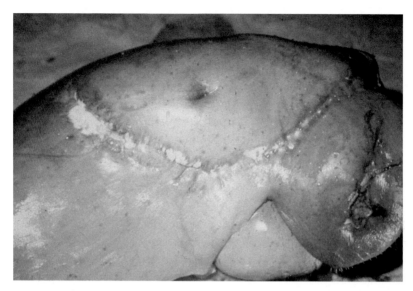

Figure 5.8. Head of a dugong killed by traditional hunters off Cape York, Australia showing the scar from a huge, healed shark bite. The fresh wound above the upper lip (right) is a hunting wound. H. Marsh photograph.

a leather sack ... in which we found the head and bones of a hippopotamus; the hairy lips of which were still sound and not putrefied, and the jaw was also firm, out of which we plucked a great many teeth, 2 of them 8 inches long and as big as a man's thumb, small at one end, and a little crooked; the rest not above half so long.

We cannot know whether Dampier's dugong was alive or dead when eaten by the shark. Nonetheless, some adult dugongs bear healed scars (Figure 5.8) indicating that they have survived attacks from large sharks; tiger sharks have been observed harassing, killing and consuming apparently healthy adult dugongs, and the abundance of large tiger sharks seems to vary with availability of dugongs as prey (Anderson 1981b, 1995b; Wirsing *et al.* 2007b,c). The most comprehensive information on the incidence of shark attack on dugongs comes from the Marine Wildlife Strandings and Mortality Program that operates along the urban coast of Queensland. The accompanying database maintained by Queensland's Department of Environment and Resource Management has records of shark bites on 70 stranded dugongs (95% CI 9%–14%) out of 628 carcasses examined between 1991 and 2008 (H. Marsh, unpublished). It is not known what proportion of these attacks occurred before death. (In contrast, evidence of shark bites is very rare in Florida manatees.).

Dugong remains (usually in small quantities and lacking bones) were commonly found in stomachs of tiger sharks sampled from Western Australia (Heithaus 2001; Simpfendorfer et al. 2001). It is not possible to determine what proportion of the shark bites on dugongs or the remains found in shark stomachs were taken antemortem or postmortem. However, the direct observations of sharks attacking dugongs and the likelihood that at least some remains found in stomachs were from dugongs taken alive provide good evidence that there is a real risk of shark predation on dugongs, and behavioural studies (below) bear this out.

Risk has two elements: the likelihood of something happening and the consequences if it happens (Norton et al. 1996). As pointed out by Wirsing et al. (2007d), animals can reduce risk from predation by: (1) reducing their risk of encountering a predator, such as by using a refuge; and (2) decreasing the consequences of the encounter by increasing the likelihood of escape without being killed or seriously wounded. The relative importance of these two strategies is likely to be situation-dependent. We consider the responses of dugongs to predation in this context, noting that attack on the body from below is more likely to be fatal than attack from above, because the dorsum and sides of dugongs are protected by the rib cage. Attacks on dugongs are also likely to be fatal if wounds are inflicted on the head or the tail.

Wirsing et al. (2007a,b,d) documented that dugongs alter their foraging behaviour under conditions of heightened risk from shark attack. In the Eastern Gulf of Shark Bay, Western Australia, dugongs were most abundant in summer, coincident with the peak in shark densities. However, the dugongs were distributed differently in winter and summer. They used shallow (2–4.5 m) habitats and deep (>6 m) habitats in proportion to the biomass of the seagrass *Amphibolis antarctica* in winter, when large tiger sharks were scarce, but used deep habitats (where seagrass biomass was lower) more than expected when tiger sharks were common. At a still finer spatial scale, the dugongs underused safe edge microhabitat when the sharks were scarce and overused such habitat (which offers escape to deeper water) when sharks were common (a crude index of food quality was lower in edge habitats). The researchers concluded that dugongs adaptively managed their risk by spending more time in impoverished but safer foraging patches when the likelihood of encountering sharks increased. However, dugongs are occasionally seen from the air in open areas feeding close to large sharks (H. Marsh, personal observation), suggesting that dugongs will tolerate the presence of large sharks when they have access to deep water (Preen 1992) or need to access warm water (Sheppard et al. 2006). Excavation behaviour probably reduces vigilance and raises clouds of sediment (Figures 4.2a, 5.4; Chapter 4), which obscures the ability to scan for predators (and may also attract them).

Dugongs must be vulnerable to predation while giving birth, and consequently seek refuge in areas such as very shallow sandbanks that are inaccessible to large sharks and killer whales. On at least two occasions, cows calving in very shallow or sheltered waters have been aground but awash, behaviour that has been interpreted as a predator-avoidance strategy (MacMillan 1955; Marsh *et al.* 1984c). Sandbanks would not be a refuge from crocodile attack, and in areas of crocodile abundance it is possible (but unknown) that dugongs may use different habitats for calving. Patterson (1939) describes a dugong cow defending her calf from an attacking shark by placing herself between the calf and shark, and working her way into shallow water until the shark became stranded. The cow was fatally mauled.

During long-distance travels, some satellite-tracked dugongs repeatedly dived to >25 m and spent low amounts of time at the surface, likely to reduce the risk of surprise shark attack from below (Sheppard *et al.* 2006). This strategy is similar to the sub-surface travelling behaviour of migrating green turtles, which is interpreted as a strategy to reduce the visibility of their silhouette against the surface to hunting sharks (Hays *et al.* 2001). Sub-surface travelling should also reduce the consequences of the encounter because attacks from above are less likely to result in serious wounds than those from below. Dugongs in New Caledonian waters travel through passes rather than through the adjacent reticulated reef complex, possibly to enhance their capacity to escape from large sharks (Garrigue *et al.* 2008). Within the reef complex aerial observers saw a dugong (provisional identification) being attacked by 10 tiger sharks and 20 other, smaller sharks. The dugong could not escape the barrier created by the reef and pursuit lasted only a few minutes before the sharks were successful (Garrigue *et al.* 2008).

In Moreton Bay, Queensland, Hodgson (2004) concluded that dugongs may prefer to rest on the edges of the sandbanks, where they are closer to the relative safety of deep water. Dugongs in New Caledonia also rest in deeper water, in this case outside the barrier reef, and may rest in herds. Garrigue *et al.* (2008) observed a herd of 45 dugongs, including ten cow–calf pairs, resting outside the barrier reef, where they could easily escape from predators. Wirsing *et al.* (2007b) also presented evidence that dugongs at their study site in the Eastern Gulf of Shark Bay tended to rest in deeper water (>6 m), where the opportunity for escaping from sharks was greater than in shallower areas (2.5–4.5 m).

Current information suggests that foraging tactics and the use of micro-habitats by dugongs to reduce risk from predators is situation-dependent, and conforms to the availability of options to reduce the likelihood of encountering a predator, as well as to decrease the consequences of the encounter. Although shark predation may not be high, the presence of sharks probably alters dugong behaviour and habitat use through fear (Wirsing et al. 2007b). The situation-specific trade-offs differ in at least three dimensions: (1) the topography of the site and the location of escape routes; (2) the nature of the dugong's primary activity (e.g. feeding, travelling, resting, calving, nursing); and (3) other habitat and ecological constraints (see also Heithaus et al. 2009).

SOCIAL BEHAVIOUR

Group sizes and composition

The data from sightings of sirenians during aerial surveys are consistent with a general lack of evidence for cohesive social structure, other than a single female with its nursing young. However, groups are nonetheless commonly observed in dugongs and West Indian manatees. 'Herds' of dugongs ranging from 12 to about 300 have been reported during aerial surveys, particularly at higher latitudes (e.g. Heinsohn et al. 1978; Marsh and Sinclair 1989a; Preen et al. 1997; Lanyon 2003; Preen 2004; Holley et al. 2006). The largest aggregation, estimated to contain some 670 dugongs in two main groups, was seen in the Arabian/Persian Gulf (in winter at a presumed warm-water source; Preen 2004) Large aggregations of hundreds of Florida manatees also form around warm water during winter cold spells (see Chapter 7 and below). Most observations suggest that groups of sirenians are transitional in composition, forming around resource concentrations such as seagrass beds, shelter from rough seas, warm-water sources in winter, or freshwater sources in the case of manatees (see below). Aerial surveys outside the winter season suggest a modal group size of one in dugongs and West Indian manatees (insufficient survey data are available for Amazonian or West African manatees). These estimates are likely biased low (Chapter 8) and are dependent on the observer's definition of a 'group'. However, the proportion of total sightings of manatees or dugongs found in groups of two or more on such surveys (even at times other than winter) is often greater than the number seen as single animals (Table 5.1).

These proportions can also vary seasonally based on local conditions. For example, Lanyon (2003) reported that most dugongs observed during aerial surveys of Moreton Bay, Queensland, in 1995 were seen in groups of two or more (Figure 5.9). These groups tended to be somewhat smaller and

Table 5.1 *Proportions of individuals seen that were accompanied by one or more other individuals, based on cumulative aerial sightings of dugongs or West Indian manatees.*
Reports from winter near the latitudinal limits of distribution are generally excluded. The original reports should be consulted for definitions of 'groups'. Group sizes have not been reported using standardised methodology across species. Additionally, the extended area survey methodology used often in manatee studies (Lefebvre et al. 1995) probably results in greater accuracy in the estimates of group size than the passing mode transect methodology typically used in dugong surveys (Marsh and Sinclair 1989a,b), which may underestimate the sizes of small groups.

Taxon	Location	Number seen	% seen with others	Reference
Dugong	Trang Province, Thailand	264	68.5	Hines *et al.* (2005a)
	Shark Bay, Western Australia (summer)	687	60.0	Holley *et al.* (2006)
	Great Barrier Reef, Queensland, Australia	761	55.5	Marsh and Saalfeld (1989, 1990)
Antillean manatee	Belize, Chetumal Bay, Mexico	644	64.1	Morales-Vela *et al.* (2000)
	Panama	70	81.4	Mou Sue *et al.* (1990)
Florida manatee	Southwest Florida	554	66.2	Irvine *et al.* (1982)

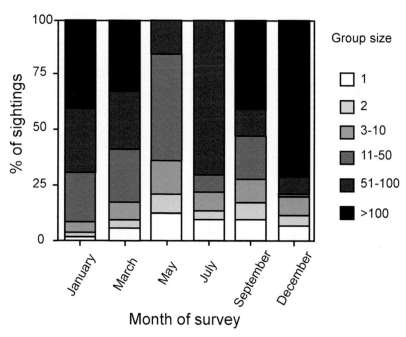

Figure 5.9. Sightings of dugongs in groups of various sizes in Moreton Bay, Queensland, Australia during six bimonthly aerial surveys in 1995. Re-drawn after Lanyon (2003) with permission from CSIRO Publishing, http://www.publish. csiro.au/nid/144/paper/WR98082.htm

Figure 5.10.

more dispersed in winter when dugongs leave the bay intermittently for warmer water (Preen 1992; Sheppard et al. 2006; see 'Dugong movements' above and 'Water temperature' below). (This situation contrasts with Florida manatees, which tend to aggregate at warm-water sites in winter; Chapter 7 and 'Water temperature' below.) Reynolds (1981b) observed Florida manatees from shore and a canoe in a closed system with about 60 individuals and also found most manatees in groups of two or more. Anecdotal and interview data also suggest the occurrence of West African manatees in small groups (mean 2.5; range from 1 to 15; mode 1) in the Bijagós Archipelago of Guinea-Bissau (Silva and Araújo 2001). Thus a substantial proportion of dugongs and West Indian manatees are generally seen in the company of other individuals. Such sightings may reflect local resource aggregations, but nonetheless point out the potential for social interactions.

Despite the tendency for sirenians to occur in groups, group stability tends to be low apart from the social bond between a cow and her calf. Reynolds (1981b) noted that the frequency with which any two Florida manatees other than females and offspring were seen together was low, indicating a lack of group stability (see also Bengtson 1981; Deutsch et al. 2003). Although dugongs arrived at the seagrass meadows in Shoalwater Bay, Queensland, in groups, group composition and integrity broke down during the day (Anderson and Birtles 1978). Anderson (1998), however, later suggested that female and subadult dugongs may be more gregarious than adult males, and that groups were more cohesive while rooting (excavating; see Chapter 4 for terminology), resting or responding to disturbance, with departure or recruiting of individuals into a group rare. This conclusion does not accord with the observations of Hodgson (2004) in Moreton Bay. Each focal dugong maintained the same nearest neighbour for an average of only one minute, despite the persistent records of large herds at this location (e.g. Heinsohn et al. 1978; Preen 1992; Lanyon 2003; Hodgson 2004).

Although large groups of dugongs have been recorded in many parts of the species' range (Figure 5.10), persistent reports come only from

---

Caption for Figure 5.10
Two types of dugong groups. (a) The largest group of dugongs ever recorded, estimated to contain some 670 animals, seen in the Arabian/Persian Gulf in winter at a presumed warm-water source. (b) A dense herd of about 60 feeding dugongs stirring up sediment near Islam Islets, Exmouth Gulf, Western Australia. (a) Tony Preen photograph; (b) Grant Pearson and Bob Prince photograph; reproduced with permission.

Moreton Bay. To determine the function of the Moreton Bay herds, Hodgson (2004) studied individual dugong behaviour in relation to group dynamics using her blimp-mounted video system. Dugongs in large, dense herds spent significantly more time feeding than dugongs in smaller or scattered groups, suggesting that the larger dense groups in Moreton Bay primarily facilitate feeding. Dugongs did not use large herds for resting, and mother–calf pairs were less likely to occur in large herds than single individuals, suggesting that dugongs do not shelter in herds when most vulnerable to shark attack. Group structure was fluid, as noted above. These results support Preen's (1992) hypothesis that large herds may function to maintain feeding patches of relatively young plants of improved food quality (Chapter 4). de Iongh et al. (1995) showed that this effect could also be achieved by small groups of dugongs. Why group sizes tend to be larger in Moreton Bay than most other areas, including locations where the absolute number of dugongs is much greater (such as Torres Strait; Marsh et al. 1997) is not known.

To our knowledge, there have been no detailed long-term studies of wild sirenians that targeted group composition, turnover or stability, and associations of individuals that form transitory groups away from thermal refuges. At least superficially, there may be some parallels in the formation of transitory groups of sirenians with the fission–fusion societies of other mammals, as exhibited by some species of cetaceans, primates, bats and other orders (e.g. Whitehead et al. 1991; Henzi et al. 1997; Willis and Brigham 2004). Fission–fusion social organisation can arise when there are variable trade-offs to the costs and benefits of group living (e.g. Conradt and Roper 2000; Couzin 2006).

Mating systems and mating behaviour

Mammalian mating systems can take several distinct but predictable forms. These forms depend on factors such as resource distribution and social structure, and how these relate to the separate interests of males and females in terms of lifetime reproductive success. In most species of mammals, the lengthy period of resource allocation devoted to pregnancy and lactation makes the relative investment by females in young much greater than that of males. Thus in most species of mammals, females may choose mates on the basis of the likely quality of the offspring they will produce, whereas individual males employ a strategy of inseminating as many females as possible.

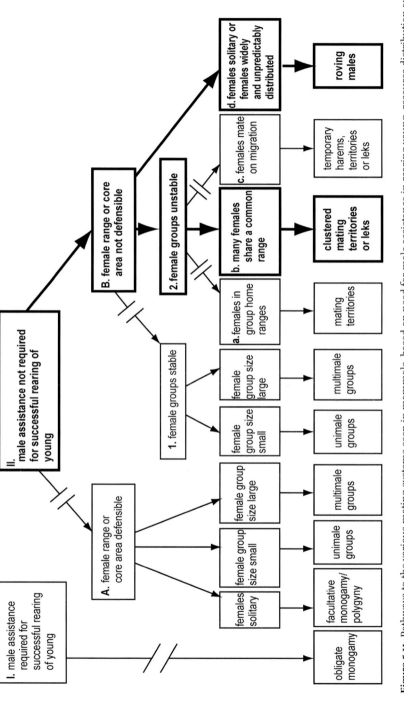

Figure 5.11. Pathways to the various mating systems seen in mammals, based on need for male assistance in rearing young, resource distribution and stability of female groups. Bold pathways (boxes and arrows) indicate the potential mating systems that are feasible for sirenians. Re-drawn by Shane Blowes after Clutton-Brock (1989), with permission.

Mammalian mating systems were extensively reviewed by Clutton-Brock (1989), who developed a graphical model of how different mating systems are moulded (Figure 5.11). It is useful to examine the mating systems of sirenians in relation to this model. At levels I and II in Figure 5.11, the outcome of pathways leading to monogamy or to other mating systems is dependent on the capability of males to assist in the rearing of young. This pathway seems unavailable to modern sirenians: males cannot suckle young and male provisioning of aquatic plants seems unlikely (young feed independently on plants beginning at an early age; Hartman 1979; Marsh *et al.* 1982; Chapter 4). Male defence of a stable territory where a female and young can obtain protection from predators and adequate nutrition for the full period of calf dependency also seems unlikely, given the lack of rich, concentrated clumps of high-quality forage available to aquatic herbivores. Only Steller's sea cow was claimed to be monogamous, but this may have been an eighteenth-century anthropomorphic projection (Chapter 2; see also Anderson 2002 for a different interpretation).

This leaves sirenians only pathways A or B from option II in Figure 5.11, where male assistance in rearing young is not needed. Given the wide-ranging movements of female sirenians (see 'Long-distance movement patterns and migrations'), their ranges or core areas seem indefensible by males, and there is no evidence that females form stable social groups that can be defended by males against other males. There is also no evidence that female mating is limited to migratory periods (Chapter 6). This situation leaves only two likely outcomes for sirenian mating systems: scramble promiscuity with roving males (Figure 5.11 pathway II–B–d) or the formation of leks (or small, clustered mating territories; Figure 5.11 pathway II–B–2–b). Observational accounts support male and female promiscuity in dugongs and West Indian manatees, with single oestrous females tended by roving males that coalesce into larger groups termed 'mating herds' (Figure 5.12a). However, there has also been an intriguing set of observations of dugongs in Shark Bay that suggest that a lek mating system may occur under some as yet poorly understood conditions (Anderson 1997, 2002).

Mating herds of dugongs have been observed at the latitudinal limits of the species' range in Queensland and Western Australia (some 24 000 km apart; see Figure 8.7; Preen 1989a; Holley 2006) and at least one location in between (H. Marsh, unpublished). These mating herds were similar in many respects to mating herds of West Indian manatees (see below).

Figure 5.12. (a) Mating herd of Florida manatees. As female manatees enter oestrus, they become the focal point of a mating herd in which a number of different males may participate over a period of days or weeks. The female sometimes moves into shallow water, presumably to escape her consort; here, a large female lies on her side with her ventrum away from the group of males comprising a mating herd in west central Florida. In this part of the state, in the summer months, mating herds are not uncommon along the beaches of the Gulf of Mexico, as shown here. (b) A pair of dugongs mating in Palau. (a) Manatee Research Program, Mote Marine Laboratory photograph; (b)Mandy Eptison photograph; both photographs reproduced with permission.

Preen (1989a) observed groups of up to 20 dugongs moving rapidly in a cluster, with bodies often in contact, structured around a single adult, presumed to be a female. Preen (1989a, p. 384) also observed 'fighting', where 'the dugongs remained in one spot and engaged in a continuous bout of violent clashes characterised by explosive splashes, tail thrashing, body rolls, and body lunges. Tails, heads, backs, and bellies were constantly thrust above the surface and other dugongs.' Mounting behaviour was observed twice, in which one group of three males, and in another case four males, clung to the cow, with one male embracing the female from below in a belly-to-belly position while other males clung to the sides: 'Each male embraced the female with his flippers and attempted to manoeuver his tail so as to bring his genital slit in opposition with the female genital slit and effect intromission.' Preen (1989a) notes that this behaviour seems more violent than in mating herds of manatees, and that the small tusks of males could inflict injury, as evidenced by numerous tusk scars seen on some dugongs (Figures 4.2a, 5.4). Similarly, Holley (2006) observed a large mating herd in South Cove of Shark Bay, Western Australia, comprising more than ten dugongs in a tight cluster, behaving in a similar manner to the mating herds described by Preen (1989a).

The mating behaviour of dugongs may vary with location. In contrast to observations of dugongs in mating herds, lone pairs of dugongs have been photographed mating in Palau (Figure 5.12b), and Anderson (1997) reported that dugongs in South Cove of Shark Bay formed leks in two consecutive summers. Up to 20 solitary dugongs occupied 'activity zones' that were stable and persistent for the summer, averaging about 24 ha of bottom area. These residents showed high fidelity to the sites, which were exclusive to a single dugong, and regularly patrolled them, vocalising frequently with multiple calls (Anderson and Barclay 1995; Anderson 1997). The dugongs also exhibited a series of postural displays on the sites, with neighbouring dugongs sometimes engaging in 'confrontations' and fights (Anderson 1997). Non-patrollers and 'ephemeral' dugongs also visited the cove, with some involved in pair matings or apparent matings with residents (described in greater detail by Anderson 1997). The activity zones appeared to have insufficient food to support even one dugong. Although it was not usually possible to determine the sex of individuals (Anderson 1997), this evidence for lekking in dugongs may not be incompatible with the observations of mating herds: distinctly different mating systems can be exhibited within a single species under different conditions (Clutton-Brock 1989). Lekking is certainly a theoretically possible outcome of the resource and social conditions that prevail for sirenians (Figure 5.11). Most of the ungulate species that form leks are grazers with females that live in unstable, mixed-sex herds that range over substantial areas (Clutton-Brock 1989), conditions similar to those of dugongs. Incidental observations suggest that mating behaviour is unusually flexible in moving herds, a situation that may apply in Shark Bay, where movements are temperature-related (Holley 2006; see 'Water temperature' below). In some species of mammals, leks can be replaced by roving-male strategies in low-density populations. Both leks and the scramble promiscuity characteristic of mating herds are considered default strategies for mating systems when resources are not defensible and females do not form stable groups (Bradbury 1977; Bradbury and Vehrencamp 1977). Additional study of dugong mating strategies, although difficult, would be worthwhile to shed further light on possible diversity in sirenian mating systems. The use of genetic techniques to check paternity at the South Cove site could be particularly informative.

Mating herds in manatees (Figure 5.12) have been described only for the West Indian species, primarily in Florida (Hartman 1979; Bengtson 1981; Rathbun et al. 1995; Wells et al. 1999). Male Florida manatees travel more widely and more often than females (Bengtson 1981; Deutsch et al. 2003), either solitarily or in loose groups that can persist for several days. These travels involve searches for oestrous females. In the upper St Johns River system, males patrol circuits of interconnecting water bodies every 6–10 days (Bengtson 1981). When an oestrous female is located, the numbers of males in these mating herds can grow substantially, at times involving vigorous 'cavorting' (see 'Cavorting, play and other social behaviours' below) with substantial rolling, pushing and mouthing of the body surfaces of the female and each other for hours at a time (well described in Hartman 1979). Females have been known to enter very shallow water in apparent attempts to discourage males (Hartman 1979; Reynolds 1981b; Figure 5.12). Although males clash in seemingly violent collisions and wrestling, serious injury among males has not been observed (Hartman 1979). Bengtson (1981) reported on three such mating herds of up to 15 males that pursued focal females for distances up to 160 km for periods of 9–12 days, with a maximum daily movement of over 30 km (mating herds can persist for periods ranging from a week to over a month; Hartman 1979). Mating herds spent about 28% of the time involved in cavorting, 33% moving, 24% feeding and 18% resting (Bengtson 1981). Following the dissolution of mating herds, males in the upper St Johns River resumed their patrolling circuits, presumably in search of additional oestrous females (Bengtson 1981). Movements of females in mating herds may be attempts to attract the greatest number of possible mates to choose from. Although the sometimes violent, vigorous interactions between males and focal females have been described as possible contests of stamina and strength among males (Rathbun and O'Shea 1984), subadult-sized males also join these groups (subadult-sized manatees may be capable of producing sperm; Hernandez *et al.* 1995). Only two observations of copulation of Florida manatees have been published, one involving a subadult male and two adults each copulating with an adult female (Hartman 1979). Multiple copulations infer sperm competition among male manatees (Reynolds *et al.* 2004; Chapter 6; Supplementary Material Appendix 6.1). In north-western Florida, an adult female was observed as the focus of a mating herd that persisted for three weeks, accompanied by at least 20 different males over this period, but with a changing composition of male identities (Rathbun *et al.* 1995). Other herds of presumed males were observed during the course of aerial surveys by Rathbun *et al.* (1995), with a range of 5–22 manatees, with an average of ten. These herds were least commonly observed during cool months, when reproductive activity of Florida manatees is lowest (Chapter 6). In comparison with Florida manatees, little has been reported on mating systems of Amazonian or West African manatees. Keith and Collins (2007) reported sighting a mating herd in Gabon, suggesting a mating system in common with West Indian manatees.

Cavorting, play and other social behaviours

Sirenian biologists frequently use the term 'cavorting' to describe group behaviour that resembles behaviour in mating herds but at a lower intensity. Cavorting has been most intensively observed in studies of Florida manatees, but likely occurs in all sirenians (described for dugongs by Anderson 1998). All ages and sexes can be involved in cavorting groups of manatees, but such groups most commonly contain males. Group size can vary from two to more than ten manatees. The primary sense used in cavorting is tactile: manatees mouth and nibble over one another's body surfaces using the perioral bristles and hair, clasp and embrace, thrash and roll as in mating herds, sometimes intermittently for several hours. About 10% of the daily time budget can be spent in cavorting during the warm season (Bengtson 1981).

At Blue Spring in Florida, even small suckling calves may occasionally leave their mother's side to join in cavorting groups, only to return, squealing, if the activity is too rough (T. J. O'Shea, personal observation). Manatee calves sometimes play with each other as if cavorting. Cavorting behaviour likely reflects the importance of practising and perfecting the kinds of behaviour that are all-important in sirenian mating systems. Florida manatee calves also have been observed to play alone by 'twisting, tumbling, and barrel-rolling through the water ... one calf completed several minutes of play by "rocketing" to the surface so that his chest and flippers broke the water' (Hartman 1979 , p. 110). Adult manatees also have been observed to play by 'body surfing' in a swift current, initiated by one, then joined by others in apparent social facilitation (Reynolds 1981b).

Manatees also 'kiss' at the water's surface, pressing muzzle-to-muzzle at respiration, perhaps exchanging information by chemoreception (Hartman 1979) or through the specialised tactile hairs abundant on the muzzles (see Figure 5.2 and 'Tactile senses' above). Florida manatees have been observed to attend to carcasses of their own species. In some cases these attentions may involve possible mating drives of males, but epimeletic (care-giving) behaviour of cows apparently attempting to keep dead calves at the surface also have been reported (Hartman 1979), as has epimeletic behaviour of captive Amazonian manatees towards another captive that had difficulty breathing (Rosas 1994).

Dugongs may show more aggressive behaviour than manatees. Seemingly mildly injurious scars from the small tusks of males are commonly observed on their backs (Figures 4.2a and 5.4), and the mating herds described above appear qualitatively more violent (Preen 1989a). Anderson and Birtles (1978) also describe 'rushing' behaviour, in which lone dugongs 'surged swiftly and powerfully' towards them with the head and anterior body above the surface in two rushes about 15–20 m in length; dugongs were also observed directing similar behaviours towards each other. Fights between dugongs in leks described by Anderson (1997) involved dugongs ramming each other at high speed and other behaviours, once including an impact forceful enough to throw the recipient 'halfway out of the water'. Aggressive behaviour of Florida manatees is apparently limited primarily to jostling in mating herds (Hartman 1979), and has not been described for other species and subspecies of manatees.

Mother–infant behaviour

Dugong females and calves remain close together throughout the period of calf dependency, calves rarely straying more than 1–3 body lengths from their mothers (Anderson 1981a,b, 1982, 1998). Contact is mediated through exchanges of vocalisations (Anderson and Barclay 1995; see below). During travelling, calves often swim above their mothers, which likely functions to reduce their visibility to predators from below. Anderson

(1998) suggested that small dugong calves may not forage on rhizomes, but may start grazing by three months of age (Heinsohn 1972). Nonetheless, there is evidence that dugongs start feeding very soon after birth. Marsh *et al.* (1982) observed a neonate with seagrass, including rhizomes, in its stomach. Anderson (1984) reported that dugong calves often suckle from the axillary teats in an inverted position (belly-up), unlike Florida manatees. He timed three suckling bouts as 1.6, 1.6 and 1.1 minutes. In contrast, Hodgson (2004) recorded dugongs suckling with their dorsum uppermost in Moreton Bay. She timed 15 suckling bouts, averaging 87 s (1.45 minutes; range 70–105 s).

One captive manatee observed giving birth raised the newborn calf to the surface on her back to assist it in breathing (Moore 1951b). Information about mother–infant behaviour in wild Florida manatees derives primarily from observations at Crystal River (Hartman 1979) and at Blue Spring (T. J. O'Shea, unpublished). Mother and calf Florida manatees typically stay in close contact (often within 1 m of each other during the first year), but the degree of separation shows individual variability. Calves more typically travel by the mother's side rather than over the back as in dugongs. Florida manatee young approach the female and nuzzle the side near an axilla, often exchanging vocalisations until the calf grasps the axillary teat. Cows almost always alternate successive nursing bouts between the right and left teats, and discourage calves from the most recently nursed side by not opening the axilla or moving away. Nursing bouts last about two minutes (also reported by Reynolds 1981b) and occur about once or twice each hour throughout the period of calf dependency, which typically lasts from one to two years (Chapter 6). Weaning may occur abruptly, but has not been intensively studied. Females often nurse their calves just prior to travelling. Small manatee calves begin to nibble vegetation at an early age, and typically follow the female's lead in timing of surfacing, resting, travelling and feeding. Synchronous breathing of cow–calf pairs varies with activity, and in one study occurred in 82–92% of respirations (Hartman 1979), and up to 40% of respirations while resting in another (Reynolds 1981b). Calves sometimes engage in play while the mother feeds or rests (see above). The auditory and tactile senses seem to dominate in mother–young communication, with calves frequently nuzzling, mouthing and kissing the female and both responding to each other's vocalisations (see 'Vocal communication' below). Females typically position themselves between their calves and sources of apparent danger. Similar behaviour was reported for dugongs by Patterson (1939).

Vocal communication

All living species of sirenians use underwater sound for communication. There have been numerous studies of sirenian communication sounds, ranging from early analysis of basic physical properties (e.g. Schevill and Watkins 1965; Evans and Herald 1970; Nair and Lal Mohan 1975; Reynolds 1981b), continuing through recent investigations of the potential for detecting the presence of sirenians by their sounds for applied purposes (e.g. Muanke and Niezrecki 2007; Gur and Niezrecki 2009; Ichikawa et al. 2006, 2009). Characteristics of individual call notes seem fairly similar among the species of sirenians. Frequency ranges are typically from 1–18 kHz, often with harmonics and non-harmonically related overtones (e.g. Anderson and Barclay 1995; Sousa-Lima et al. 2002; O'Shea and Poche 2006). Hartman (1979) described calls of Florida manatees as sounding to the human ear like 'squeals, chirp-squeaks, and screams' and as 'groans' or 'grunts'. Anderson and Barclay (1995) referred to dugong calls as 'chirp-squeaks, barks, and trills' with some sounds intermediate among these categories. The most commonly recorded vocalisations of sirenians seem to function as contact calls, with variations in acoustic properties signalling motivational states of the caller, such as alarm or distress. Anderson and Barclay (1995) hypothesised that the short, lower-frequency barks of dugongs may signal male aggression.

Durations of typical Florida manatee call notes are about 200–300 ms (milliseconds), but duration can vary with context. Single Florida manatee calls can be as long as 900 ms, with longer calls given in contexts of fear or alarm (O'Shea and Poche 2006). Adult and subadult Amazonian manatee calls averaged 224 ms, whereas those of calves averaged 173 ms (Sousa-Lima et al. 2002). Dugong vocalisations recorded in Shark Bay, Western Australia, are shorter than manatee calls (60 ms for chirp-squeaks, 30–120 ms for barks), except for trills, which were 100–2200 ms (Anderson and Barclay 1995). Calls of dugongs recorded off Thailand were grouped as short-duration (mean 126 ms) chirps or long-duration (mean 1737 ms) trills, with chirps observed much more often and trills more commonly at the middle or end of multiple-call sequences (Ichikawa et al. 2006; Okumura et al. 2006). Dugongs seem to issue sequences of multiple call notes more than manatees, and vocalise while feeding on rhizomes (which can obscure vision by raising sediment clouds (Figures 4.2a and 5.4); Anderson and Barclay 1995; Okumura et al. 2006).

Vocalisations are individually distinctive in captive Amazonian manatees and wild and captive Florida manatees, with distinctiveness based on single or combinations of properties such as fundamental frequency, frequency modulations, 'contours' seen in sound spectrograms and frequencies of harmonics with greatest energy or intensity (Sousa-Lima *et al.* 2002; Williams 2005; O'Shea and Poche 2006). Some vocalisations probably convey information through the use of acoustic features that are 'noisy' or non-linear, such as subharmonics and biphonations (Mann *et al.* 2006). Such variability in aspects of call structure also occurs in dugongs (Anderson and Barclay 1995).

Adults of both sexes produce vocalisations, but exchanges of communication calls are most common between cows and their nursing calves. Calves of Amazonian manatees and Florida manatees vocalise at much greater rates than adults (Sousa-Lima *et al.* 2002; O'Shea and Poche 2006). Antiphonal exchanges ('dueting') occur at rates as high as 20 calls per minute when wild Florida manatee mother–calf pairs rejoin after moving apart, with calves issuing about 2–5 times more calls than mothers (Hartman 1979; Reynolds 1981b; O'Shea and Poche 2006). Wild Florida manatee mothers and young show evidence of mutual recognition based on vocalisations (O'Shea and Poche 2006), and this phenomenon has also been reported for a captive Amazonian manatee mother and calf (Sousa-Lima *et al.* 2002). Some of the distinctive traits of Florida manatee vocalisations can persist within individuals for periods of up to six years, from suckling age until past sexual maturity (O'Shea and Poche 2006), with one wild female recognisable by researchers based on a unique 'trill' evident in some of her vocalisations from about six months to 22 years of age (Williams 2005). However, characteristics of individual vocalisations generally change as manatees mature, typically including downward shifts in call frequencies (Williams 2005).

Manatees other than cows and calves vocalise at rates that vary with activity and behavioural context. Rates are lowest during resting, intermediate while travelling, and highest at nursing and other social situations (Reynolds 1981b; Bengtson and Fitzgerald 1985; Williams 2005; O'Shea and Poche 2006; Miksis-Olds and Tyack 2009). Florida manatees may alter vocalisation parameters in response to environmental noise levels (Miksis-Olds and Tyack 2009). Dugongs seem to vocalise more often during dark early-morning hours (Ichikawa *et al.* 2006). No data are available on vocal communication in West African manatees, although recordings and sound spectrograms of calls of an isolated captive calf in Ivory Coast were similar to those of some Florida and Amazonian manatee calves (T. J. O'Shea, unpublished). Additional information about vocalisations of sirenians is available in the studies cited above, and in research reported for dugongs by

Marsh *et al.* (1978); Amazonian manatees by Sonoda and Takemura (1973); Antillean manatees by Sonoda and Takemura (1973), Alicea-Pou (2001) and Nowacek *et al.* (2003); and Florida manatees by Steel (1982).

Evidence for matrilineal use of space

Matrilinearly transmitted learned behaviour, commonly known as tradition, seems to play a large role in determining use of space and migratory habits of Florida manatees (Hartman 1979; Bengtson 1981; Deutsch *et al.* 2003) and possibly dugongs (Anderson 1981a; Sheppard *et al.* 2006). Although there is no evidence that sirenians show long-term social structure, a number of intriguing observations suggest that the use of space may follow matrilines that could enable both direct and indirect kin selection for seemingly altruistic behaviours. Evidence from telemetry research on the Atlantic coast of Florida shows that manatee calves learn migratory patterns and use of space at destinations from their mothers, and that manatees show high natal philopatry (Deutsch *et al.* 2003; Figure 5.7). These findings are supported by: (1) warm season tracking of subadults in the upper St Johns River (T. J. O'Shea, unpublished); (2) the manatees' high fidelity to wintering areas (e.g. Figure 5.6b; Deutsch *et al.* 2003); and (3) the lack of evidence for major dispersal of young observed in the longitudinal studies of photo-identified manatees described in Chapter 6 (e.g. Powell and Rathbun 1984; Reid *et al.* 1991).

Genetic studies of Florida manatee populations also support the notion that there may be a matrilineal spatial substructure to their populations (McClenaghan and O'Shea 1988; Bonde 2009). The occupancy of areas by matrilines can set the stage for natural selection for seemingly altruistic social behaviour, with manatees fitting the pattern of 'Type 1' natal philopatry (Waser and Jones 1983): parents tolerate offspring within their home ranges into adulthood, interacting only occasionally and foraging independently. One consequence of natal philopatry can be an increased probability that close kin are clustered geographically (Waser and Jones 1983). 'Even in the absence of any ability to recognise one's kin, natural selection can favor increased tolerance of a class of individuals if the probability of its members being close kin is high enough' (Waser and Jones 1983, p. 378).

Perhaps the greatest behavioural evidence for possible spatial matrilines in sirenians comes from observations of allomaternal care by female Florida manatees. Wild females have been repeatedly observed to 'adopt' and suckle orphaned calves, often for periods extending the full course of calf dependency (Bonde 2009). In other cases, wild female manatees can 're-adopt' their

Figure 5.13. A 3–4-year-old wild Florida manatee male that was 're-adopted' by its mother after her loss of a younger calf. She allowed it to suckle at the same rates and durations as if it was a normal young calf, for at least an entire winter. Photograph by T. J. O'Shea.

own offspring and direct kin selection is more obvious. For example, an adult female at Blue Spring weaned a male calf after nursing it for about two years. She became pregnant after weaning this calf but lost the newborn, and instead re-encountered her previous male calf (weaned about a year earlier) on the warm season range. She allowed it to suckle at the same rates and durations as a normal calf for at least an entire winter, even though it was adult-sized and 3–4 years old (Figure 5.13; T. J. O'Shea, unpublished).

HABITAT USE AND HABITAT COMPONENTS

This overview of sirenian habitat use is based on the habitat components (food, environmental conditions and related factors) that are currently known to be of general importance to dugongs and manatees. As is the case for most aspects of sirenian ecology, much more information exists with regard to habitat use by dugongs and West Indian (especially Florida) manatees than for West African or Amazonian manatees. Ortega-Argueta *et al.* (2012) discuss in detail various methods for investigating habitat use by sirenians. The utility of those methods varies depending on the habitat feature being assessed, and the importance of some habitat features remains uncertain.

A descriptive and useful definition of 'habitat' is 'an area with the combination of resources like food, cover, water, and environmental conditions (temperature, precipitation, presence or absence of predators and competitors) that promotes occupancy by individuals of a given species and allows those individuals to survive and reproduce' (Morrison *et al.* 2006). Habitats are used by animals at a number of different scales (e.g. Mayor *et al.* 2009), and distinctions exist among definitions of terms used in the study of habitats. We follow the suggestions of Hall *et al.* (1997) and define 'habitat use' as the way in which an animal uses a collection of physical and biological components in a habitat. The understanding of sirenian habitat use is in most cases not deep enough to use other terminology, such as 'habitat selection' (the process of decisions made by an animal about what habitat it uses at different scales), or 'habitat preferences' (the consequence of that process, resulting in the disproportionate use of some resources over others; Hall *et al.* 1997). Sheppard *et al.*'s (2010) research on dugongs' resource selection within a seagrass bed in Hervey Bay, Queensland, which is discussed in Chapter 4, is an exception to this overall lack of information on sirenian resource selection.

Multivariate habitat use studies

Only a few studies have been designed specifically to examine habitat use by sirenians. Logistic regression and general linear models were used to examine relationships between habitat variables and manatee use of mostly freshwater systems in adjacent areas of Nicaragua and Costa Rica (Jiménez 2005). Regional watercourses were sampled in 87 different units, defined based on habitat type. Each unit was characterised based on ten variables related to hydrological characteristics (water depth, visibility, temperature, current and width), vegetation (total abundance of aquatic vegetation, emergent vegetation and floating vegetation), forest cover on the shore and boat traffic. Manatee presence or absence was indicated by sightings of manatees or signs of manatee feeding. Results differed among sectors, but Antillean manatees in these coastal rivers generally tended to use wide, clear and warm waters with abundant emergent vegetation and forest cover; they were associated with slower currents, greater depths and lagoonal areas (Jiménez 2005).

In northern Chetumal Bay, Mexico, Antillean manatee distribution records from aerial survey and habitat metrics were entered into a geographical information system (GIS) database and analysed using Poisson and logistic regression (Olivera-Gomez and Mellink 2005). The area was divided into 309 cells, each 0.5 km by 0.4 km. Habitat variables for each cell included measures of depth, slope of the bottom, aquatic plant cover, distance to freshwater sources, shelter from wind and surf, and salinity. The proportion of cells used was affected strongly by distance to fresh water, and also varied positively with depth, plant cover and unevenness of the bottom. Overall frequency of occurrence was most strongly affected by distance to fresh water, depth and plant cover, with the first two variables of greatest importance (Olivera-Gomez and Mellink 2005).

The coastal seagrass meadows of eastern Queensland are important dugong habitats (Chapter 8). The seagrass species assemblages change across the latitudinal gradient of this vast region and over time (McKenzie et al. 1998; Rasheed 2000; Coles et al. 2003a), making habitat modelling challenging. Grech and Coles (2010) generated a GIS-based habitat suitability model for coastal seagrass habitats to approximately 15 m below mean sea level at the scale of the inshore waters of the Great Barrier Reef World Heritage Area (approximately 22 600 km^2). They used a Bayesian belief network to quantify the relationship between seagrass presence/ absence and eight environmental drivers: relative wave exposure, bathymetry, spatial extent of flood plumes, season, substrate, region, tidal range and sea surface temperature. The seagrass data resulted from 11 562 spot dives conducted during surveys by Rob Coles and his team from the Queensland Department of Primary Industries and Fisheries between 1984 and 2001. The analysis showed that, at this regional scale, the main drivers of seagrass presence were tidal range and relative wave exposure: low tidal ranges and low levels of wave exposure increased the likelihood of seagrass presence. Outputs of the model included probabilistic GIS surfaces of seagrass habitat suitability in both the wet and the dry seasons at a cell size of 4 km^2. This initiative has been coupled with the development of a spatially explicit model of dugong distribution and relative abundance at the same cell size based on data from dugong aerial surveys conducted from the mid-1980s to 2007, over an even larger area: the Queensland–New South Wales border through the Great Barrier Reef region, Torres Strait and the Gulf of Carpentaria (Grech and Marsh 2007; Grech et al. 2011a; Figure 9.1). These models are being used to inform marine conservation initiatives for dugongs and their seagrass habitats (Chapter 9).

At a finer spatial scale, Preen (1992) used hierarchical log-linear analyses and logistic regression to identify the environmental variables associated with the distribution and abundance of dugongs in Moreton Bay in south-east Queensland. The dugong data were generated from aerial surveys: the environmental variables tested were season, habitat type relative to availability, distance to deep (>2.5 m) water and water temperature in winter. During winter, dugongs were 4.9 (95% CI 1.88–12.81) times more likely to be in waters warmer than 19 °C than in areas where water temperatures were lower. Dugongs were also 1.74 (95% CI 1.24–12.45) times more likely to be found in areas within 1.5 km of deep water than in areas where deep water was further away. After correcting habitat preferences for water temperature and distance for deep water, it was clear that habitats dominated by species of *Halophila* were preferred overall (Chapter 4). However, there were seasonal differences in the dugong's habitat preferences. After the winter values were corrected for temperature, habitats dominated by *Zostera capricorni* (both narrow- and broad-leaf morphs; see Chapter 4) and channels and the area immediately outside a passage into the bay were preferred relative to the situation in the other seasons. The distribution of dugongs did not vary among years and was not affected by the presence of boats, even though incidental observations demonstrate that dugongs react to boats under certain circumstances (Chapter 7). Sheppard *et al.*'s (2010) modelling of resource selection by dugongs based on the nutrient value of their food is discussed in Chapter 4.

Forage

Large quantities of plants must be consumed to satisfy the nutritional needs of sirenians (Chapter 4). Dugongs are seagrass community specialists and seagrasses are also important components of the diets of West Indian and West African manatees (Chapter 4). Because seagrasses are light-limited, in marine and estuarine areas sirenians generally occupy relatively shallow waters of sufficient clarity or tidal exposure to support bottom-growing plants (Chapter 4). Thus dugongs tend to occur in shallow, sheltered bays, mangrove channels or in the lee of large inshore islands (Heinsohn *et al.* 1977). Dugongs also occur in deeper water offshore where the continental shelf is wide, shallow and protected by reefs. For example, Marsh and Saalfeld (1989) sighted dugongs ~58 km offshore in water 37 m deep inside the northern Great Barrier Reef, and Whiting (1999) and Whiting and Guinea (2003) reported dugongs (including calves) at Ashmore Reef on the Sahul Banks on the edge of the Western Australian continental shelf.

Unlike dugongs, manatees can also occupy water where they feed on emergent, unrooted floating or bank-growing vegetation (Chapter 4; Figures 5.14–5.16). Amazonian manatees are patchily distributed throughout the Amazon Basin: they seem to concentrate in areas of nutrient-rich, seasonally flooded tropical rainforest (Rosas 1994; Junk 1997), where annual production of aquatic and semi-aquatic plants is high. However, access to food varies greatly with season for Amazonian manatees

Figure 5.14. Antillean manatees utilise a variety of foraging habitats. (a) The Rio Manati in Bocas del Toro, Panama. The light *café con leche* colouration from sediment loads due to erosion contrasts with the clear tannin-stained water of a canal (R. K. Bonde photograph). (b) True grasses growing along the shoreline are the predominant food source in the Rio San San in Panama (R. K. Bonde photograph). (c) Seagrass beds around saltwater cays north-east of Belize City, Belize are regularly used by manatees (T. J. O'Shea photograph). (d) Antillean manatees in Rio San Juan, Monagas, Venezuela can be found where mangrove leaves at high tide are the only available forage (T. J. O'Shea photograph). All photographs reproduced with permission.

(Chapter 4). In the wet season, manatees in the western Amazon occupy *várzea* lakes (i.e. on floodplains of rivers with waters high in nutrients and silt) where they forage on a diversity of freshwater macrophytes (Arraut *et al.* 2010). Individual Amazonian manatees engage in long seasonal movements from flooded areas during the wet season to deep-water bodies such as *rias* (long, narrow lakes formed by partial submergence of river valleys) during the dry season (Kendall 2001; Arraut *et al.* 2010).

West Indian manatees show great variability in foraging habitats, as reflected in the variety of known dietary items from both freshwater and marine habitats (Chapter 4; Figure 5.14). The same individuals followed by telemetry in north-western Florida may feed on seagrasses in one area and on exotic freshwater vegetation in another on the same day (e.g. Rathbun *et al.* 1990). West African manatees also use a wide range of habitats (Figure 5.15), and have been reported in river swamps, coastal seagrass

Figure 5.15. West African manatees inhabit a variety of aquatic habitats. These include coastal lagoons (Lagune Tagba, Cote D'Ivoire) (a), as well as turbid rivers where only shoreline vegetation is available (such as these true grasses *Echinocloa* sp., lower vegetation), along the banks of the Bandama River, Cote D'Ivoire) (b). T. J. O'Shea photographs.

beds, mangrove forests and flooded agricultural lands (Reeves *et al.* 1988; Grigione 1996; Silva and Araújo 2001), reflecting their heterogeneous diet (Chapter 4).

Despite the extensive studies of sirenian diets (Chapter 4), certain types of information, important for management purposes, remain unknown. In both the United States and Australia, for example, federal and state agencies, as well as certain stakeholder groups, are keen to define the carrying capacity of manatees in Florida waters and dugongs in Great Barrier Reef waters. Such exercises should be conducted with great care for several reasons, not the least of which is the degree of uncertainty regarding the nutritional needs of manatees and dugongs, as discussed in Chapter 4. The challenge of estimating the dugong carrying capacity of the Great Barrier Reef World Heritage Area is discussed in detail in Marsh *et al.* (2005) and can be illustrated by considering the fallacy of calculating carrying capacity based on measures of the total submerged aquatic vegetation in an area. Without factoring in the *quality* of the food (see Chapter 4) and the effects that long-term consumption of low-quality food could have on health, reproduction and survival, any calculations of carrying capacity could represent dramatic overestimates. Long-term research and monitoring have shown that aquatic plant communities can change radically across years within the lifespans of individual sirenians. Seagrass diebacks in marine (and estuarine) habitats (e.g. Preen and Marsh 1995; Preen *et al.* 1995; Hall *et al.* 1999; Marsh and Kwan 2008; Chapters 4 and 6) and the changing nature of freshwater plant composition from native to exotic macrophytes (Hartman 1979; Powell and Rathbun 1984), to the current abundance of the algae *Lyngbya* at the Florida manatee refuge in Crystal River (C. Beck, personal communication, 2010; Hauxwell *et al.* 2004) pose challenging questions regarding the time scales and bounds of variation over which it is legitimate to project carrying capacity (see Marsh *et al.* 2005).

Salinity and access to fresh water

The extant sirenians span habitats with a wide range of salinities, but dugongs and manatees have very different tolerances of saline habitats. Dugongs are exclusively marine and do not require access to fresh water (interestingly, Steller (1751) noted that his sea cow, also a dugongid, seemed to prefer areas at the mouths of freshwater creeks; see Box 2.8). Dugongs can be found in areas with high salinity, such as parts of the Arabian/ Persian Gulf (Preen 2004), although in Shark Bay in Western Australia they apparently avoid hypersaline conditions (Holley *et al.* 2006). Manatees, in contrast, include one species that is almost entirely freshwater (the Amazonian manatee) whereas the other two are euryhaline. At least one of the euryhaline species, the West Indian manatee, seems to require periodic access to fresh water for drinking (Reich and Worthy 2006). The medullae of the kidneys of West Indian manatees are sufficiently developed to concentrate urine (Hill and Reynolds 1989; Maluf 1989), and individuals can persist in sea water while feeding on seagrasses with high salt content for days at a time, as borne out by natural history observations. Hartman (1979) observed manatees with marine barnacles and remoras (*Echeneis naucrates*) that returned to the freshwater springs at Crystal River. In addition, Reynolds and Ferguson (1984) observed barnacle-encrusted manatees at the Dry Tortugas, seemingly far from known sources of fresh water. Barnacles, however, are lost only gradually during forays into fresh water (Hartman 1979), and are not a reliable indicator of exclusive use of marine habitats by manatees.

The ability of manatees to spend several days at a time in saltwater but with a need for periodic access to fresh water was confirmed experimentally by Ortiz *et al.* (1998), who held captive manatees (both the Florida and Antillean subspecies) at different salinities and fed them foods with different water content. Manatees held for nine days in salt water and fed seagrasses were able to maintain electrolyte homeostasis, but lost body mass, increased plasma osmolality, reduced the time spent feeding and were thought to be susceptible to dehydration if deprived of fresh water for periods longer than nine days (Ortiz *et al.* 1998). This finding supports the observations through telemetry that all Florida manatees tagged near the Everglades move from offshore seagrass beds to inland freshwater sources every 2–8 days (Stith *et al.* 2006). Captive manatees held in salt water for several days and fed only seagrasses had lower water turnover than those that were not as water-deprived. These manatees were thought to be using fat stores to produce metabolic water from the oxidation of lipids and did not drink sea water (Ortiz *et al.* 1999). After 4–5 days without fresh water the manatees stopped eating and began to pass hard, dry faeces (G. Worthy, personal communication, 2010). These findings provide evidence that West Indian manatees have good tolerance for salt water, but physiologically must drink fresh water periodically. Other physiological mechanisms by which manatees adjust as they move between salt water and freshwater environments include variable production of anti-diuretic hormone (ADH) to help conserve body water when in a marine environment, and aldosterone to help retain sodium when in fresh water (Ortiz *et al.* 1998).

Maintaining the availability of fresh water to West Indian manatees is an important factor to consider in habitat management for their conservation.

Virtually all aerial surveys and tracking studies of West Indian manatees in North, Central and South America have noted their frequent travels to and higher relative abundance at freshwater sources (see reviews in Lefebvre et al. 1989, 2001). These studies substantiate Hartman's (1979) suggestion that West Indian manatees need periodic access to fresh water, and are consistent with numerous anecdotal observations, biogeographic patterns of relative abundance throughout the species' ranges (Lefebvre et al. 2001) and results of radio tracking studies (e.g. Rathbun et al. 1990; Deutsch et al. 2003). Olivera-Gómez and Mellink (2005) found that proximity to fresh-water sources and water depth were the two habitat features that best explained Antillean manatee distribution in Chetumal Bay, Mexico. It is well known in Florida that garden hoses or storm sewers that dump fresh water into inshore areas are frequent gathering points for manatees. A running hose in an area frequented by manatees provides a sufficient attractant that the drinking, unwary animals can be easily captured for tagging or health assessment (Figure 5.5). Occupancy of freshwater habitats also reduces the risk of predation by sharks (see 'Anti-predator behaviour and habitat use' above).

Access to fresh water does not seem important to dugongs. The struc-ture of the dugong kidney is non-lobular and is different grossly and micro-scopically from all other marine mammals, including manatees (Batrawi 1953, 1957; Pabst et al. 1999), and more closely resembles that of a dog (Beuchat 2002). The degree to which dugongs rely on drinking fresh water is likely to be minimal. However, Preen (2004) suggested that the large herds of dugongs he observed in the Arabian/Persian Gulf in winter may aggregate around underwater freshwater springs, presumably for warmth (see 'Water temperature' below). Whether dugongs under such circum-stances drink water of reduced salinity is not known.

Water temperature

As a likely adaptation for feeding on a relatively low-energy food resource (compared with foods of most other marine mammals), sirenians have a relatively low metabolic rate, a poor thermal conductance (resulting in heat loss to the environment) and a low scope for increasing metabolic rate to counter heat loss (Scholander and Irving 1941; Gallivan and Best 1980; Irvine 1983). Cool water temperature poses limits to the latitudinal distri-bution of extant sirenians. This has been best explored for West Indian manatees (e.g. Moore 1951a; Whitehead 1977; Irvine 1983), and is detailed for dugongs below. Reduction in surface area–volume ratios was probably the major force in the evolution of gigantism in Steller's sea cow (the major exception to this rule of intolerance of cool temperatures by sirenians; Chapters 2 and 3). Cold can cause mortality in sirenian populations at the margins of their distribution in the subtropical temperate zone (Chapter 7), and is a governing factor for seasonal migrations of Florida manatees and some dugong populations (see 'Long-distance movement patterns and migrations' above). The responses of Florida manatees to cool water temperatures is well known and is a focus of management concern (Chapter 7) because of the formation of aggregations of manatees at springs and industrial warm-water sources. Florida manatees also form cool weather aggregations at 'passive thermal refugia', where salinity properties allow formation of temperature-inverted haloclines that produce warmer, more saline bottom waters (Stith et al. 2010) (see 'Florida manatee movements' above). The influence of seasonal water temperatures on distribution of West African manatees has not been well studied, but cold currents at the Angola–Benguela Frontal Zone (von Bodungen et al. 2008) seem likely to impose the southern limit.

Dugongs make local or seasonal movements in response to water temperatures lower than about 17–18 °C (e.g. Sheppard et al. 2006; see above). Annual migrations to warmer sources of water typically occur in Florida manatees when water temperatures fall below about 20 °C (e.g. Hartman 1979; Bengtson 1981; Powell and Rathbun 1984; Deutsch et al. 2003), and Florida manatees alter their activity patterns in winter to avoid colder temperatures (see above). Provision of warm-water habitat for Florida manatees is a critical need (Laist and Reynolds 2005a, 2005b; Chapter 7).

Some populations of dugongs (e.g. in Okinawa, parts of Australia and the Arabian/Persian Gulf) may be subjected to cold weather, but there have been no modern studies of the thermal physiology of dugongs, and cold-related adverse effects have not been described. Nonetheless, some evidence indicates that most dugongs make movements to avoid sea temperatures less than about 18 °C. This evidence includes the summer-only sighting records of live dugongs along the New South Wales coast in Eastern Australia (Allen et al. 2004) and sighting records and aerial surveys at three of the high latitude limits to the dugong's range: Moreton Bay in Queensland (Marsh and Sinclair 1989a; Preen 1992; see 'Multivariate habitat use studies' above); Shark Bay in Western Australia (Anderson 1986, 1994; Marsh et al. 1994; Preen et al. 1997; Gales et al. 2004; Holley et al. 2006); and the Arabian/Persian Gulf (Preen 1989b). Anderson (1986) suggested that dugongs used eastern Shark Bay in summer and western Shark Bay in winter to avoid water temperatures less than ~19 °C.

Subsequent surveys (Anderson 1994; Marsh *et al.* 1994; Preen *et al.* 1997; Gales *et al.* 2004; Holley *et al.* 2006) have shown that the situation is spatially more complex than first thought, but confirm that most dugongs avoid water temperatures less than about 18 °C. Nonetheless, a few dugongs may use cooler waters: <4% of dugongs observed during the aerial survey of Shark Bay in 1989 were in water <18 °C (Marsh *et al.* 1994); the corresponding percentage for a 1994 survey was 14% (Preen *et al.* 1997). Wirsing *et al.* (2007b) also sighted some dugongs in Shark Bay in waters below 15 °C, and Lanyon *et al.* (2005a) reported seeing dugongs in Pumicestone Passage in the Moreton Bay region year round, despite water temperatures below 18 °C from June to August, and down to 15.4 °C in June.

Preen (1989b, 2004) found major differences in the winter and summer distributions of dugongs in the Arabian/Persian Gulf and attributed those differences to changes in water temperature. In summer the dugongs were dispersed, whereas in winter Preen observed a high proportion of dugongs aggregated in large herds, including adjacent groups of 578 and 97 animals (Figure 5.10a). The larger of these herds is the largest aggregation of dugongs observed anywhere (Preen 2004). As discussed above ('Salinity and access to fresh water'), Preen speculated that this aggregation formed around freshwater springs, the temperature of which may have been warmer than the surrounding water.

Like Florida manatees, dugongs can rapidly change their distribution in response to cold fronts. Marsh and Sinclair (1989a) conducted aerial surveys of dugongs in Moreton Bay during the Austral winter of 1985. Maximum June temperatures in the bay were the lowest recorded in seven years, and during the five-day survey period, most of the dugongs relocated to warmer oceanic waters outside the bay. Movements of individual dugongs associated with seasonal drops in water temperature also have been documented using satellite tracking (Marsh and Rathbun 1990; Preen 1992; Holley 2006; Sheppard *et al.* 2006). Preen (1992) tentatively concluded that the dugongs he tracked stayed in oceanic water outside Moreton Bay for up to several days at a time, making brief high-tide feeding sorties into the bay. Holley (2006) concluded that seasonal drops in temperature caused individual dugongs to move from inner to outer regions of Shark Bay. However, readings from GPS tag deployments and remotely sensed sea surface temperatures indicate that some individual dugongs can remain in waters between 15 °C and 18 °C for weeks at a time, and may actively search cooler waters for preferred forage (Holley 2006). Conversely, as reported above, Sheppard *et al.* (2006) tracked six dugongs repeatedly traversing Hervey Bay in winter, moving between their Burrum Heads seagrass feeding area where the water temperature was

17–18 °C to deeper and warmer oceanic waters (21–22 °C) at Sandy Cape, where seagrass has not been recorded.

Moving to warmer water is not without costs, although the trade-offs are likely to be complex, spatially variable and moderated by the animal's behaviour. Preen (1992) observed that Moreton Bay dugongs timed their movements in and out of the bay to take advantage of the tide and maintained activity centres close to the passage from the bay to the sea. This behaviour reduced the energetic demands of the 15–40 km round trip between the dugong's feeding areas and warm-water refuges. Nonetheless, when dugongs move to warmer waters, they may lose access to high-quality seagrass, or any seagrass at all (Preen 1992; Holley 2006; Sheppard et al. 2006). If seagrass is available it may be in deeper water (Anderson 1994; Holley 2006) and therefore require more energy to graze. Both Preen (1992) and Sheppard et al. (2006) claimed that moving to deeper water also increases a dugong's risk of shark predation, although the trade-offs probably depend on the topography of the site as well as its degree of use by large sharks (see 'Antipredator behaviour and habitat use' above).

Winter water temperatures at the high latitude extremes of the dugong's range may cause seasonal reductions in seagrass, increasing the stress on the animals (Preen 1989b). Eating cold food also is a challenge for endotherms such as sirenians because of the energetic cost of heating the food to body temperature (Irvine 1983; Wilson and Culik 1991). This situation is especially challenging for herbivores that have to eat a relatively higher proportion of their body weight than carnivores. Irvine (1983) suggests that Florida manatees can forage in water colder than 18 °C if they can later digest their food in warmer water (fermentative metabolism in the gut can act to increase body temperatures as well, and lack of feeding can increase susceptibility to cold stress; Rommel et al. 2003; Chapter 7). Preen (1992) speculated that some dugongs leave Moreton Bay for short periods in winter so they can digest their food in warmer oceanic waters. Taken together, the information on the distribution and movements of dugongs at the higher latitude limits of their range in Australia and the Arabian region suggest that the lower thermal threshold for dugongs is typically about 18 °C, similar to that of Florida manatees but, like Florida manatees, some individual dugongs tolerate temperatures at least 1.0–2.6 °C lower than 18 °C for extended periods.

The thermal tolerances of sirenians are presumably also delimited by an upper thermal threshold, although this aspect of their physiology has not been formally investigated. In some parts of their ranges, water temperatures in summer can become extremely high – in excess of 41 °C in some locations (e.g. Charlotte Harbor, frequented by Florida manatees in summer). Although their low metabolic rate helps reduce the amount of heat that must be dumped against this gradient, sirenians exposed to very warm water might be expected to alter their behaviour (e.g. spend their time in deep places with stratified layers of cooler water on the bottom).

Availability of quiet or sheltered areas

Anecdotal evidence suggests that female Florida manatees seek quiet areas in which to calve, and they may continue to use such areas for some period after the calf is born (e.g. O'Shea and Hartley 1995; Deutsch et al. 2003). This assumption was tested recently by Gannon et al. (2007), who used more than a decade of aerial survey data around Sarasota, Florida, to compare habitat use by female manatees with calves and that of manatees unaccompanied by calves. Indeed, outside of the winter and spring (when cold weather influenced habitat use more than any other factor), the females with calves tended to use protected areas (including locations officially protected by state or county regulations) near seagrass meadows. Gannon et al. (2007) suggested that such habitat selection could be influenced by a range of factors such as: (1) ready availability of food at a time when the mother's energetic needs were high due to lactation; (2) reduced noise and other disturbance; (3) protection from storms or high waves; and (4) protection from predators and watercraft collisions. Seclusion of newborn calves from the potentially violent activities of roving males has also been suggested as a factor favouring use of quiet areas by cows with neonates (O'Shea and Hartley 1995). Hodgson (2004) observed that mother–calf pairs were less likely to occur in large herds than single individuals in Moreton Bay; the reason for their doing this is unknown. Dugongs also seek quiet or protected areas in which to calve, likely for predator avoidance (Marsh et al. 1984c; Anderson 1998).

The relationship between dugong distribution and the disturbances associated with boat traffic in Moreton Bay was investigated by Preen (1992): he could find no discernible effect. However, vessel densities were generally low in the major dugong habitats. The relationship between noise and other disturbances associated with boat traffic has been assessed by Buckingham et al. (1999) and Miksis-Olds et al. (2007a,b) for manatees in Florida. Boating activity affects fine-scale manatee distribution patterns at the winter thermal refuge in Kings Bay, Crystal River, confirming the great importance of sanctuaries as areas where manatees can avoid excessive human activity (Buckingham et al. 1999). Without such safe areas, undue human activities could displace manatees from the warm-water resources they need to survive during cold weather in winter (Hartman 1979; Powell 1981; Kochman et al.

1985). Swimmers and divers at Crystal River often disregard regulations and seek contact with manatees, impacting manatee activity patterns of resting or nursing and increasing their use of no-entry sanctuaries (King and Heinen 2004). Manatees on the Atlantic coast of Florida also disproportionately use areas free from public boating and other human activity as sanctuaries during warm seasons (Provancha and Provancha 1988; Deutsch *et al.* 2003). These studies serve as examples of how threats to sirenians worldwide can include effects of increasing ecotourism on use of habitat.

Miksis-Olds *et al.* (2007b) considered the effects of boating activity on Florida manatees during warm seasons in Sarasota Bay. They used acoustic playback experiments with controls to simulate approaches by different types and speeds of vessels. Changes in manatees' behaviour, swim speed and respiration rate showed that differential responses occurred to the various acoustic stimuli; the most pronounced avoidance response was to personal watercraft sounds. In addition, Miksis-Olds *et al.* (2007a) examined ambient noise levels in habitats frequented by manatees, and correlated noise and manatee use. They found that in seagrass beds of equivalent species composition and density, manatees selected those that had lower ambient noise levels at frequencies below 1 kHz. In addition, they demonstrated that manatee use of seagrass beds was negatively correlated with high boating activity in the early morning hours.

The extent to which other sirenians may seek or need quiet locations to give birth or care for calves is not known. However, various researchers (e.g. B. Morales, personal communication, 2010) have suggested that Antillean manatees throughout the Caribbean seek habitats that are protected from waves and winds. In the Drowned Cayes area of Belize (Figure 5.14c), Antillean manatees are commonly observed resting in depressions in the substrate, locally referred to as 'manatee resting holes', many of which are in sheltered areas with slow currents. Bacchus *et al.* (2009) compared the physical and environmental attributes of 12 resting-hole sites with 20 non-resting-hole sites in the Drowned Cayes, using water depth, substrate type, vegetation, water velocity, salinity and water temperature, and concluded that manatees may select resting holes based on the tranquillity of the water and environment. Marmontel (2008) suggests that Amazonian manatees 'favor calm, shallow waters, away from human settlements'.

Seasonal rainfall

Many tropical areas are influenced by strong seasonality in rainfall patterns, which affects habitat use by sirenians (reviewed for manatees by Deutsch *et al.* 2003). Wet season–dry season differences in hydrology markedly alter water levels in rivers, lakes and coastal lagoons. During high waters of the rainy season, manatees have access to otherwise inaccessible tributaries, oxbows and flooded forests. This phenomenon is most well known for Amazonian manatees in the Amazon River and its tributaries in Brazil (Best 1983; Arraut *et al.* 2010; see 'Forage' above), Peru (Reeves *et al.* 1996) and Ecuador (Timm *et al.* 1986). In the lowlands of Ecuador, aquatic vegetation flourishes in lakes of the flooded black-water forests (pH 5.5), and Amazonian manatees and river dolphins literally swim among branches that, in the dry season, are high on the trunks of trees (Figure 5.16). In the dry season, the manatees persist in remaining pockets of water where they can be easy prey for hunters and predators. Large rainy and dry season differences in water levels and accessibility of vegetation are also widespread in habitats used by West Indian and West African manatees. Evidence for differential dry season–wet season use of rivers and associated waters by West Indian manatees has been reported in South America (O'Shea *et al.* 1988; Reynolds and Odell 1991; Montoya-Ospina *et al.* 2001; Castelblanco-Martinez *et al.* 2009) and Central America (Rathbun *et al.* 1983; Reynolds *et al.* 1995). Reports from these areas indicate that this seasonality in habitat use may influence susceptibility to mortality. West African manatees also shift patterns of usage of river systems in relation to seasonal rainfall to take advantage of increased access to vegetation (see Deutsch *et al.* 2003).

Tidal cycles
Both tidal height and time of day influence habitat use by dugongs (Anderson and Birtles 1978; Nietschmann and Nietschmann 1981; Nietschmann 1984; Sheppard *et al.* 2009; Chapter 4; Figure 4.1). Dugongs tend to be closer to shore or on top of reefs at high tide and at night, possibly reflecting two factors: (1) at high tide, shallow seagrass beds become more available for foraging as water depth increases (see Chapter 4); and (2) such distribution may reduce vulnerability to predation and to collisions with watercraft. Tidal cycles and water depth also influence manatee use of foraging sites (Hartman 1979; Lefebvre *et al.* 2000; Flamm *et al.* 2005; Olivera-Gomez and Mellink 2005), although the tidal range in Florida is much less than that in dugong habitat in Queensland (Chapter 4).

Rough water
Dugongs seek shelter in the lees of islands and reefs in rough weather (Nietschmann and Nietschmann 1981; Nietschmann 1984), where they may form large transient groups. For example, in Torres Strait during the

Figure 5.16. Flooded black-water forest landscapes are utilised by Amazonian manatees during high water seasons in the lowlands of Ecuador. (a) Laguna Grande, Cuyabeno Reserve. (b) Laguna Camanguena, Cuyabeno Reserve, where floating true grasses (*Paspalum* sp., foreground and at margin of the lagoon) and submerged macrophytes provide abundant forage. T. J. O'Shea photographs.

south-east tradewind season (May through September), gusting and strong winds and low tides during the day force dugongs from the large and shallow windward reefs; they move to leeside reefs and island margins during night-time high tides. This pattern is reversed during the north-west monsoon season (December to April), when high tides occur during the day and there is a 180° shift between windward and leeward. These changes are exploited by hunters. The most favourable but difficult hunting conditions occur when a north-west storm lashes the reefs, driving large numbers of dugongs into leeside island shelter.

Wilderness and anthropogenic change

Threats to sirenians from anthropogenic change are serious and growing (Chapter 7). These threats include greater pressures from fisheries, hunting, exposure to contaminants, boating activity, damming of rivers and global climate change. Some sirenians actively use habitats adjacent to the impact sites that are sanctuaries from such threats (e.g. Provancha and Provancha 1988; Deutsch et al. 2003). However, in many areas sirenians persist despite these threats, sometimes in apparent abundance. High-density human activities occur side by side with dugong habitat use in major ports in Queensland, such as Gladstone (Figure 5.17a) and Townsville, and Florida manatees use habitats in urbanised and industrial areas such as Jacksonville, Miami and Tampa Bay (Figure 5.17b). Unlike some large mammals of conservation concern, wilderness per se does not seem to be a necessary habitat component for sirenians. However, use of urbanised or industrialised areas where favourable habitat components (food, fresh water, warm water) are available can greatly increase the risk of exposure of sirenians to a host of threats (see Chapter 7) that may not exist in more wilderness-like areas. Activities of people that impact populations of dugongs and manatees must be managed more intensively in high human-use areas, to reduce the potential for reproductive isolation of sirenian populations that remain in the dwindling number of coastal wild places.

Uncertainties still exist with regard to habitat requirements of all species of sirenians, particularly the Antillean, Amazonian and West African manatees. However, as stated in Chapters 1 and 9, and by Reynolds et al. (2009), current scientific knowledge is likely sufficient to make conservation decisions to protect marine mammals like the sirenians, and the habitats on which they depend. What is needed is the vision and the will to make those decisions. Strategies for improving conservation of sirenians are elaborated in Chapter 9.

Figure 5.17. Sirenians will occupy heavily developed coastal environments and do not require wilderness per se. (a) Dugong feeding trails (see arrows) at low tide near a coal loader at Curtis Island in the port of Gladstone, Queensland (photograph by Fisheries Queensland). (b) A thermal refuge for Florida manatees at an industrial paper mill in north-eastern Florida (photograph by R. K. Bonde). Photographs reproduced with permission.

CONCLUSIONS

The sirenians have some unique patterns of behaviour and habitat use that tie directly to their long evolutionary history of obligate aquatic herbivory. Individual patterns of behaviour involving sensory perception and locomotion seem most 'hard-wired', and are geared towards life in the water. Dark at night and often murky by day, the aquatic environment requires visual senses adapted more to locating patches of vegetation at a distance than fine discrimination. Given that pressure waves travel fastest in water, sirenians rely primarily on acoustic and tactile senses for fine-scale information about their environment. Their patterns of swimming and diving are also largely not amenable to management; it is human surface activity that must be modified in order to accommodate sirenians (see Hodgson *et al.* 2007).

The good news is that within the confines of marine (dugongs) and euryhaline or freshwater environments (manatees), sirenians seem to be capable of some behavioural flexibility. They show a polymorphism in movement patterns that allows a wider base for management (e.g. rigid migratory herds do not need complete protection to prevent loss of populations); the mating system does not require territories or minimum numbers of females in stable groups; females rear young solitarily and likely transfer knowledge of spatial use of habitat to them by tradition. These habitats may already be the places with conditions that have afforded maternal survival to adulthood. Habitat use shows wide flexibility in the Florida manatee, which will shift use of space to make greater use of sanctuaries if they contain vital habitat components; it is uncertain how flexible dugongs and the other manatee taxa are in this regard. These traits set the stage for hopefulness for the conservation and management of sirenians if legal, cultural, economic and institutional barriers can be overcome (see Chapter 9).

SUGGESTIONS FOR FUTURE RESEARCH

Once again, the research needs of each of the sirenian species with respect to their behaviour and habitat use are different and location-specific. From a conservation perspective, a priority for all sirenians, especially the dugong, Antillean manatee, Amazonian manatee and West African manatee, is more local and regional telemetry and distribution studies using appropriate survey techniques to provide managers with some of the basic knowledge required to protect key habitats, including feeding and breeding areas and movement corridors.

Efforts to conserve most populations of sirenians would be improved by increased knowledge of their behavioural ecology. Understanding the environmental pressures that have influenced the behavioural strategies employed by a species enables researchers to predict whether these strategies are flexible enough to cope with novel circumstances. Behavioural studies typically require a long-term research commitment to obtain meaningful data. Conservationists and managers generally emphasise the need for studies that have direct application, while scientists interested in animal behaviour typically advocate fundamental research. However, often it is the fundamental research that provides new insights that are valuable for conservation (Arcese *et al.* 1997). Long-term studies of known individuals are needed to provide understanding of the social behaviour, movement patterns, habitat use, flexibility of sirenian behaviours under varying environmental pressures, and how these pressures affect animal health and population parameters. This understanding will aid managers in setting appropriate recovery targets. In this respect, behavioural studies are an essential tool, which when combined with studies of population trends (Chapter 8) and demographics (Chapter 6) will enhance our ability to inform strategies to protect sirenian populations. Longitudinal behavioural studies of known individuals coupled with genetics promise to provide rich insights, especially into sirenian mating systems and the matrilineal use of space.

As the only aquatic herbivorous mammals, extending the research on the sensory biology of sirenians promises rich theoretical and practical insights. To date, this research has largely been conducted on captive Florida manatees and should be extended to the other species and subspecies. Determining how sirenians orient and navigate in their typically murky environments remains an unexplored frontier for research.

Life history, reproductive biology and population dynamics

INTRODUCTION

As explained in Chapter 3, sirenians evolved from primitive terrestrial herbivores early in the Tertiary period. Like whales and dolphins (cetaceans) but unlike the pinnipeds (walruses, seals and sea lions), manatees and dugongs spend their entire lives in the water and do not return to land to give birth and suckle their young. Sirenians are believed to share a common origin with several superficially dissimilar mammals of African origin, grouped together at the superordinal level as the Afrotheria. Within the Afrotheria the sirenians are closely aligned with elephants and hyraxes, and the three have been linked together as the clade Paenungulata, a unique grouping recognised for most of the last century (see Chapter 3). Although features of the external form of sirenians reflect adaptations to their lives of swimming and diving, the phylogenetic history of dugongs and manatees is reflected in features of their reproductive biology, which in some respects is strikingly similar to that of elephants (as outlined under 'Reproductive cycles' below and in the Supplementary Material Appendix 6.1).

Understanding the life history and reproductive biology of sirenians is fundamental to the development of effective strategies for their conservation. For example, knowledge of when and where they breed is important for assessing the effectiveness of establishing sanctuaries or other protected areas to protect breeding habitat. In addition, knowledge of the mating system (Chapter 5), the average ages at which females start breeding, the litter size, average inter-birth interval, life span and the probabilities of reproduction and survival, and how these parameters are affected by age and changing environmental conditions, are essential for understanding the population dynamics of manatees and dugongs. Knowledge of population dynamics is a requisite for sound management policies and for estimating the levels of human-induced mortality that are likely to be sustainable. In this chapter, we characterise the biology of reproduction in the Sirenia, and describe the methodological and analytical approaches that have been taken over the years to determine their vital parameters of reproduction and survival. We summarise the results of these studies and then review how information about vital parameters has been incorporated into models of sirenian population dynamics and the implications of these models for future research and management.

METHODOLOGY

The life history and reproductive biology of sirenians are difficult to study. There are three basic approaches: (1) the analysis of reproductive samples, teeth and bones collected from carcasses; (2) monitoring the reproductive history of known wild individuals during their life span and using mark–recapture analysis to calculate vital rates (longitudinal studies); and (3) studying individuals in captivity. The study of genetic markers from tissue (usually skin) samples also has the potential to revolutionise longitudinal studies and to provide insights into mating patterns (Bonde 2009), but is not considered in detail here because few results using this technique have been published. Not all methods are appropriate for answering all questions about sirenian life history and reproductive biology. The optimal approach is to use as statistically robust a method as possible. Multiple sampling approaches may sometimes be useful in answering certain questions, because when more than one method leads to the same conclusion, the findings are reinforced. Alternatively, differences between the results obtained from studies using different techniques can yield important insights into problems caused by study populations being non-random samples of the overall population; such problems are known as sampling biases. However, investigators should always strive to use sampling approaches that have the strongest underpinnings in statistical theory.

Potential sampling biases

All these methods have their strengths and weaknesses, and it is particularly important to understand any sampling biases when interpreting the results. The potential biases associated with the various methods are discussed below. It is important to remember that even a large sample may be biased.

Carcass studies

A sample of carcasses is unlikely to be random. Some conditions may increase the chances of mortality of mother and/or offspring. For example, young naïve animals may be over-represented in a sample obtained though a carcass salvage programme. Alternatively, if a sample results from indigenous harvest (see Chapter 7), some life-history stages may be targeted; for example, indigenous dugong hunters in Australia allegedly prefer breeding females to resting females or males because the former are fatter (Nietschmann and Nietschmann 1981; Raven 1990; Johannes and MacFarlane 1991). The amount of information derived from a carcass depends on the season (animals decompose more quickly in summer) and the age of the animal (it is more difficult to determine the age of older individuals, as explained below); in addition, a pregnancy becomes easier to detect as it progresses, and early pregnancies are likely to be missed (Marmontel 1995).

Longitudinal studies

The sample included in a long-term observational study is also unlikely to be random. Individuals are more likely to be included in the sample if they bear distinctive marks and if they are easy to observe, photograph or capture. Thus, for Florida manatees the photo-identified sample tends to be biased in favour of distinctively marked and approachable individuals that frequent aggregation sites in clear water. Scientists are concerned that older individuals that have attained enough scars to meet the criteria for cataloguing are over-represented in the sample (C. Langtimm, personal communication, 2010), resulting in an upward bias in survival rates.

When sirenians are captured as part of a longitudinal study, there may be permit conditions that ban the catching of sensitive life stages, such as cows with calves. Reproductive rates based on observational studies may also be biased, particularly because of the difficulty of confirming the sex of wild sirenians without catching them or obtaining a skin sample to enable genetic methods of sex determination (Lanyon *et al.* 2009) or individual identification to be used. Consequently, the presence of a dependant calf is often used to identify females, emphasising the most reproductively active individuals (Koelsch 2001).

Captive animals

Captive sirenians are also a biased sample of wild populations. Many come into captivity because they are sick or injured or are the descendants of such animals; once rehabilitated, they are often well fed and live in the absence of predators or random variation in the environmental factors that influence life history events (environmental stochasticity). Captive sirenians have the advantage of being generally tractable to handling and are available for recapture, but the results of reproductive and life history studies are confounded by their captivity.

Differences in knowledge about the species

Knowledge of the life history and reproductive biology of the four extant species of sirenians is very uneven. The Florida manatee is the most comprehensively studied sirenian, and the understanding of its life history and reproductive biology is based on data obtained from all three sources outlined above: (1) the salvage and necropsy of several thousand carcasses collected since 1974 in coordination with the south-eastern United States Marine Mammal Stranding Network and numerous other agencies and organisations (Florida Fish and Wildlife Research Institute, unpublished), including published studies based on the histological and anatomical examination of 67 males (Hernandez *et al.* 1995) and 275 females (Marmontel 1995), all with estimated ages; (2) extensive long-term observational studies of the life history of over 1500 recognisable wild individuals (O'Shea and Hartley 1995; O'Shea and Langtimm 1995; Rathbun *et al.* 1995; Reid *et al.* 1995; Langtimm *et al.* 1998, 2004; Koelsch 2001; Kendall *et al.* 2004; Runge *et al.* 2004; Langtimm 2009; Beck and Clark 2012); and (3) observations of captive manatees including at least 28 calves born in captivity (e.g. Odell *et al.* 1995). In contrast, there are few detailed data about reproduction and life history for the Antillean subspecies. Although Amazonian manatees have been kept in captivity for many years (e.g. Best *et al.* 1982), data on the reproductive biology of wild individuals are sparse. There is virtually no published information on reproduction in the West African manatee.

Information on the reproductive biology of the dugong mostly comes from the analysis of material from carcasses of more than 1500 dugongs (Bertram and Bertram 1973; Marsh 1980, 1986, 1995; Nietschmann and Nietschmann 1981; Nietschmann 1984; Marsh *et al.* 1984a,b,c; Hudson 1986; Kwan 2002; Marsh and Kwan 2008), including some 400 individuals whose ages have been estimated. These data have been supplemented by relatively short-term behavioural observations (Anderson and Birtles 1978; Preen 1989a; Anderson 1997; Adulyanukosol *et al.* 2007). In addition, new information on dugong life history and reproductive biology is currently being collected from a longitudinal study of dugongs in Moreton Bay, Queensland (Lanyon *et al.* 2002, 2005b, 2006b, 2009). Dugongs have never bred in captivity.

AGE ESTIMATION

The ability to estimate the age of individual manatees and dugongs is fundamental to the development of estimates of the life history parameters needed for the assessment and management of their populations. Body length is an inadequate surrogate for age in sirenians because both dugongs and manatees live for many years after they grow to adult size, and may continue to grow slowly at variable rates for years after weaning (Marsh 1980; O'Shea and Reep 1990; Marmontel *et al.* 1996; Schwarz and Runge 2009). A valuable traditional approach to determining the life history parameters of sirenians is studying the reproductive organs of a large sample of individuals whose ages have been estimated. This method relies on access to a large sample of carcasses and is increasingly being replaced by maximum-likelihood-based statistical techniques using data obtained from longitudinal observational studies of known-age wild animals (e.g. Langtimm *et al.* 1998, 2004; Kendall *et al.* 2004). Nonetheless, even these studies are hampered by the paucity of known-aged individuals in the sample, and for the most part rely on a single age class for adults.

Methods

Age estimates of many animals can be obtained through anatomical study of growth layers in hard tissues removed from carcasses, such as the otoliths (structures in the inner ears) of fish. These tissues provide a record of the seasonal growth rhythms of individual animals. The ages of mammals are usually estimated from the number of growth layers in their bones or teeth (Klevezal 1980). This approach requires agreement about the definition of a growth layer and knowledge of the layer deposition rate. In 1978, a workshop on marine mammal age determination coined the term 'growth layer group' to describe a repeating layer or group of associated layers in hard tissue in the teeth or bones of marine mammals (Perrin and Myrick 1980). The time period represented by the growth layer group has to be determined separately for each species and tissue type through independent calibration. In sirenians, the growth layer group represents one year's growth, and there are typically two layers in each growth layer group, one representing a period of rapid growth, the other a period of slower growth (see Figures 6.1 and 6.2).

The most reliable way to determine the deposition rate of growth layer groups is to count the layers in tissues from the carcasses of animals of known age. Another method is to inject living individuals with small doses of tetracycline (Domning and Myrick 1980; Marmontel *et al.* 1996).

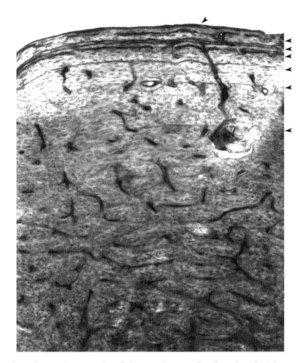

Figure 6.1. Photomicrograph of the ear bone of a female Florida manatee used in the validation of the age-estimation technique. The arrows point to the adhesion lines that correspond to the growth layer groups. The first arrow represents the first year zone. Reproduced from Marmontel (1995), with permission from USGS.

Tetracycline forms a fluorescent mark in the hard tissue laid down at the time of injection, and investigators can then count the number of growth layer groups formed between the time of injection and the time the animal dies. The work is laborious, because in order to confirm the deposition rate, a pool of known-age and tetracycline-injected marked individuals must be developed, with a system in place to recover their carcasses (preferably from the wild) after one to many years have elapsed. Manatee teeth are continually replaced throughout life and cannot be used for age determination (Chapter 4); bones are therefore the tissue of choice. Bones for age estimation must be set aside at necropsy, cut in rough sections, fixed in preservative, decalcified, cut into thin sections, stained and mounted for histological examination. Bones from both tetracycline-injected and known-age individual Florida manatees have been examined extensively and confirm a deposition rate of one growth layer group per year in the dome portion of the periotic bone (the bone surrounding the ear, see Figure 6.1) (Marmontel

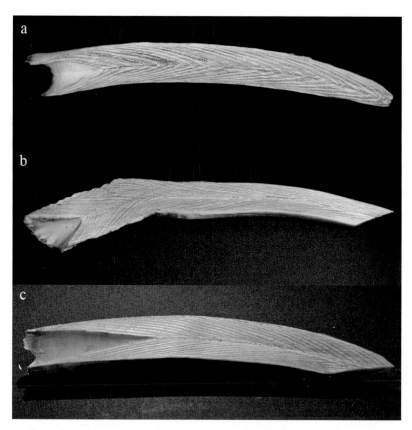

Figure 6.2. Dugong growth layer groups in longitudinally bisected dugong tusks prepared for age determination. One dark band plus one light band represents one year's growth. (a) The 14.5 cm-long female tusk (top) from Queensland is unerupted, indicating that this dugong was 23 years old when she died. (b) The 22 cm-long female tusk from Thailand provided by Kanjana Adulyanukosol is erupted and worn, and some growth layer groups have been lost. This dugong was more than 47 years old when she died. (c) The 13.5 cm-long male tusk from Queensland is erupted and worn and some growth layer groups have been lost. This dugong was more than 21 years old when he died. The tusks were photographed by Alastair Harry with assistance of Rie Hagihara, reproduced with permission.

et al. 1996). Other manatee bones do not provide such an accurate record. However, bone is a dynamic tissue that is continually being remodelled; over time bone can be subject to resorption and deposition that disrupts clear patterns in growth layer groups. In manatees greater than 10–15 years old, it may not always be possible for scientists to assign a precise age estimate. Remodelling of bone is more marked in females than males

(Marmontel *et al.* 1996), probably due to the demands for calcium during pregnancy and lactation.

Although most of the research on growth layer groups in manatee bones has involved Florida manatees, these layers have also been documented in bones of West Indian manatees from more tropical areas and in Amazonian manatees (Domning and Myrick 1980; Marmontel *et al.* 1996).

Unfortunately, it has not yet been possible to use the known-age or tetracycline approaches to calibrating growth layer groups in hard tissues of dugongs. However, the deposition rate of growth layer groups in dugong tusks (second incisors, which are present through the life of the animal) has been verified by the annual increment method (Marsh 1980). The seasonal deposition of the components of the growth layer group was confirmed by studying the nature of the layer being formed at the time each animal died. This approach shows that each growth layer group comprises a wide band laid down in spring though autumn and a narrow band formed in winter, apparently reflecting the seasonal growth pattern of the dugong's seagrass food (Marsh 1980). Dugong growth layer groups are usually remarkably clear and easy to count (Figure 6.2a), with less effort required in specimen preparation than for manatee bones. Unfortunately, the tusks erupt and wear in a few old females (Figure 6.2b) and in males after puberty (Figure 6.2c), and thus only minimum-age estimates are available for dugongs with erupted tusks. Tusk eruption in old females is quite different from male tusk eruption and usually results in the loss of very few growth layer groups.

Other anatomical methods of age estimation for sirenians are less developed. Kwan (2002) quantified a relationship between eye lens weight and counts of growth layer groups in dugongs. She concluded that the method had limited application because the resulting age estimates were imprecise. Counting growth layer groups in dugong ear bones, based on the technique used for Florida manatees, is likely to be a more robust method of estimating the ages of male dugongs with erupted tusks.

Molecular techniques may also hold some promise for age estimation, particularly if they can be applied to samples obtained from living individuals as well as to samples from carcasses. Telomeres are short, tandem, repeat sequences of DNA found at the ends of eukaryotic chromosomes, and function in stabilising chromosomal end integrity. In vivo studies of somatic tissue of some mammals have shown a correlation between telomere length and organismal age within a species (Nakagawa *et al.* 2004). This result suggested that telomere length could be used to improve age-related life history information in wild populations of sirenians where longitudinal studies of known individuals are lacking. Dunshea (2004) explored the use of telomeres to estimate the age of dugongs. Unfortunately, it seems unlikely that telomeres will enable determination of the absolute ages of individuals (G. Dunshea, personal communication, 2010). Telomere change over time is affected by genetic and environmental factors unrelated to chronological time. These processes vary spatially and temporally for animal populations.

LIFE SPAN

The oldest wild dugong whose age has been estimated is a female from Western Australia, estimated to be 73 years old (Marsh 1995). Despite the bias caused by bone resorption with age, some wild Florida manatees have had high growth layer group counts, including one individual with 59, suggesting that manatees can survive in the wild for about 60 years (Marmontel et al. 1996). Such estimates of life span are consistent with other knowledge of Florida manatee longevity in captivity and from field studies. Marmontel et al. (1996) listed the ages of four living captive Florida manatees that were 26–46 years old in 1994, three of which had been taken into captivity as adults and one that had been held in captivity since its birth in 1948. These manatees remained alive as of 2009 (age ranges 41–61 years). Two of the three oldest living wild manatees known from longitudinal field studies at Blue Spring in 1989–1991 (Marmontel et al. 1996) were known to still be alive as of 2009. These two were adults when first observed in 1970, indicating that they were more than 40 years old in 2009. The ages of several Antillean manatees were estimated at about 25 years, based on growth layers in the periotic dome (Marmontel 1993, 1995). Little is known about the longevity of African or Amazonian manatees. A male Amazonian manatee, held captive since it was a calf at the Steinhart Aquarium in San Francisco, died of a systemic fungal infection at 17 years of age (Morales et al. 1985).

Maximum longevity records provide biologically interesting information on potential life spans of sirenians and allow them to be categorised with other organisms with similar traits. However, such records are not a reliable index of average longevity or the overall age structure of a population. For example, Marsh (1995) estimated that <1.5% of the 331 wild dugong females she sampled were >60 years old, and Kwan (2002) recorded 2% of 92 female dugongs as >55 years old. In contrast, Marmontel (1995) sampled 511 female wild Florida manatees that died between 1976 and 1991. The average age of 143 sexually mature females was 12.6 years, but most of the carcasses were from young animals; only 3.7% of the carcasses were from animals aged over 25 years. Biases to ages of the carcass samples are likely as discussed above, but these findings suggest that, on average, manatees in Florida have shorter lives than dugongs in Australian waters.

Once manatees reach adult size, photo-identification-based survival rate estimates can be used to calculate mean life expectancies after first sighting (e.g. Cormack 1964). Over a comparable sampling period (1977–1991 at two areas, 1984–1991 at one) these estimates were 24 years after first sighting as an adult in the Crystal River area, 15 years at Blue Spring and 8 years for manatees along the Atlantic Ocean coast (O'Shea and Langtimm 1995). However, these results assume a stable age distribution and this approach has been superseded by the use of age-class (stage) specific survival rates in population models, as discussed below.

EXTERNAL SEXUAL DIMORPHISM

Sirenians show little sexual dimorphism. Although the asymptotic length of females may be slightly larger than that of males (Hartman 1979; Marsh 1980; Marmontel et al. 1992), the sex of sirenians cannot be inferred from body size. The distance between the genital opening and the anus is the most reliable index of sex (Figure 6.3). In females the genital opening is closer to the anus than to the umbilical scar, whereas in males the reverse is true (i.e. the genital opening is situated between the umbilical scar and the anus, but is closer to the former).

Dugongs and manatees have axillary mammary glands located under each flipper. These mammae vaguely resemble the breasts of human females and are the likely basis for the link between sirenians and the mermaid legend (see Chapter 1). The size of the mammae varies with reproductive status, being largest in lactating females. Dugong hunters from Torres Strait, between Australia and Papua New Guinea, claim that the length of the teats of a female dugong is proportional to the number of calves she has borne (Johannes and MacFarlane 1991).

As noted above, tusks erupt in male dugongs after puberty, and tusk eruption can be regarded as a secondary sexual characteristic (Marsh 1980; Marsh et al. 1984c). As in cetaceans, the testes of sirenians are permanently situated in the caudal portion of the abdominal body cavity.

Details of the internal reproductive anatomy of sirenians are presented in the Supplementary Material Appendix 6.1.

Figure 6.3. Ventral views of (a) male Florida manatee, (b) male dugong and (c) female Florida manatee. The figures show the differences in body form between dugongs and manatees and the differences in the distances between the male and female genital apertures (g), anus (a) and umbilicus (u). Dugong photograph by Karen Willshaw (www.karenwillshaw.com); manatee photographs from USGS, reproduced with permission.

REPRODUCTIVE CYCLES

In this section we discuss the ovarian and testicular cycles of dugongs and manatees. Information on their mating, birthing and rearing behaviours is presented in Chapter 5.

The ovarian cycle

As in other mammalian ovaries, the cortex of the sirenian ovary contains most of the active cell types and tissues. The most important of these are the follicles with their oocytes, the corpora lutea and the corpora albicantia. As a follicle develops, there is a concomitant increase in the production and accumulation of fluid, leading to the development of a follicular cavity or antrum. An antral follicle appears similar to a blister on the surface of the ovary. A marked expansion of the follicle occurs before ovulation, and the enlarged follicle bulges from the ovarian surface. Ovulation is usually recognised by the presence of a stigma, a scar on the surface of the ovary resulting from damage to the surface when the oocyte is released. Ovulation scars are small in sirenians and difficult to confirm (Larkin 2000).

The corpus luteum (Figure 6.4) is the endocrine gland that normally develops from the cellular components of an ovarian follicle after ovulation. Regressing or regressed corpora lutea are referred to as corpora albicantia (irrespective of their colour, even though 'albicans' means 'white body'). An ovarian follicle that has not ovulated may undergo histological changes similar to those exhibited by a corpus luteum. Such a follicle is known as a luteinised follicle. In the Sirenia, it is generally very difficult to distinguish between a luteinised follicle and a corpus luteum by routine macroscopic or histological examination unless the ovulation scar is visible. Thus, structures classified as 'corpora lutea' have probably developed from both ovulated and unovulated follicles (Larkin 2000). Like elephants, dugongs and manatees produce a large and variable number of corpora lutea (e.g. a mean of 36.3 per ovary per pregnancy in the Florida manatee – Marmontel 1995). Sirenian corpora lutea are variable in size (e.g. ranging from 1–2 mm to 13–14 mm in the dugong – Marsh et al. 1984a). The function of these supernumerary corpora lutea is not well understood.

The simultaneous presence of a large number of corpora lutea (Figure 6.4), many apparently with stigmata, is paradoxical because sirenians typically bear only one calf at a time. Are sirenians monovular (ovulating from only one follicle at a time) as suggested by their litter size, or polyovular as suggested by their ovaries? This question has not been answered satisfactorily (see Larkin 2000 for the Florida manatee); nor has

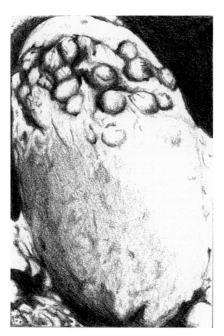

Figure 6.4. The dorsal surface of the right ovary of a female dugong with a foetus weighing 13.5 kg, showing the numerous vascular corpora lutea (endocrine glands) at the cranial pole of the ovary. The formalin-fixed ovary was 15 cm long. Drawn by Gareth Wild from a photograph in Marsh *et al.* (1984a), with permission from CSIRO Publishing, http://www.publish.csiroau/nid/90/paper/Z09840743.htm.

the question of whether a sirenian needs to accumulate corpora lutea from successive cycles to generate an adequate amount of progesterone required to maintain pregnancy. Larkin (2000) studied the hormonal cycles in captive female Florida manatees and concluded that they had an oestrous cycle of 28–42 days based on peaks in the hormone estradiol. She hypothesised that several estradiol peaks occur before a luteal phase is initiated in the ovaries. She speculated that female manatees communicate with males using chemical or behavioural cues across some distance during the interluteal phase, allowing males to find the female and determine some hierarchy of access or female choice as to who will father the calf. Such behaviours may be the function of the mating herds described in Chapter 5. Unfortunately, there are no comparable data available for other manatee species or dugongs. The hormonal basis of reproduction in dugongs is currently under study by Elizabeth Burgess (Burgess *et al.* 2009).

The testicular cycle

Immature male sirenians can be identified by their relatively small testes and testicular histology (narrow seminiferous tubules) (see Marsh *et al.* 1984b; Hernandez *et al.* 1995). As in other mammals, testicular size and histology change rapidly with maturity. Marsh *et al.* (1984b) and Hernandez *et al.* (1995) recognise and describe similar phases of testicular activity in dugongs and Florida manatees, respectively. However, it is not possible to use testicular size or histology alone to distinguish male sirenians approaching puberty from those whose testes are becoming active again after a period of quiescence. The reproductive condition of all pubertal and some adult sirenians is intermediate between complete testicular quiescence, with no spermatozoa stored in the epididymis and full spermatogenesis, with large numbers of epididymal spermatozoa. This situation indicates that mature male sirenians are not continuously in breeding condition, as discussed below.

SEXUAL MATURITY

Criteria of sexual maturity

The age at which animals in a population attain reproductive maturity is an important parameter in population biology. Knowledge of body size (usually body length) at sexual maturity allows an evaluation of the potential reproductive status of animals for which body size is the only available index of age (Perrin and Reilly 1984). These parameters have been estimated in various ways for species for which large data sets are available, such as some cetaceans (as summarised by Perrin and Reilly 1984). The data sets for dugongs and Florida manatees have been too small for such mathematical treatments, and most researchers have simply given the size or age ranges within which maturity is reached. There are no data on the age or length at sexual maturity of Antillean, Amazonian or West African manatees.

Assessment of sexual maturity in females

The most common criterion of sexual maturity in a female mammal is first ovulation, evidenced by the presence of at least one corpus luteum or corpus albicans in the ovaries (Perrin and Reilly 1984). This criterion assumes that corpora lutea/albicantia can be reliably distinguished from unerupted, large luteinised follicles, and that ovarian ovulation scars remain visible indefinitely. At least the first of these criteria is certainly not satisfied in sirenians, as explained above. Thus, it can be difficult to confirm from macroscopic ovarian examination whether a non-pregnant female sirenian which has not borne her first calf (nulliparous female) has ever ovulated.

Presumed corpora lutea or corpora albicantia have been observed in the ovaries of Florida manatees and dugongs which are not pregnant and have not yet given birth, as evidenced by the absence of placental scars in the uterus (Marsh et al. 1984a,c; Marmontel 1995; Kwan 2002). This result suggests that some dugongs and manatees may undergo several oestrous cycles before they first conceive. This finding is supported by behavioural observations of Florida manatees (Bengtson 1981; Rathbun et al. 1995). Some females are attended by cavorting herds of males (see Chapter 5) on several occasions, separated by inactive periods of a few weeks.

The distinction between age of sexual maturity (first ovulation) and age of first conception is probably of limited relevance to a sirenian population study. As summarised in Table 6.1, the usual criteria used to determine at necropsy whether or not a female sirenian is mature include at least one of the following: (1) the size, appearance and vascularisation of the ovaries and uterus; (2) presumed corpora lutea and/or corpora albicantia in the ovaries (Figure 6.4); (3) a foetus; (4) at least one placental scar in the uterus (Figure 6.5); and (5) colostrum or milk in the mammary glands. In addition, maturity may be inferred from body size, as explained below.

External assessment of the maturity of a female sirenian is used for mark–recapture studies (O'Shea and Langtimm 1995). Such assessment is difficult unless: (1) the animal is accompanied by a suckling calf; (2) her mammary glands are elongated and active (the presence of colostrum or milk is confirmed) or her abdomen is distended by a large foetus; or (3) her age and/or body length are known and outside the species-specific ranges within which females mature.

Alternative methods may be developed to assess the maturity of sirenians. In the 1970s indigenous hunters from Torres Strait claimed that they could tell the age and sex of a dugong from its wake, position in a group, surfacing and diving postures and the exhalation sounds made when they surface to breathe (Nietschmann and Nietschmann 1981); unfortunately, most modern hunters appear to have lost such skills (Kwan 2002). Researchers are developing tests to determine the reproductive status of live sirenians based on reproductive hormones in the faeces (Lanyon et al. 2005b; Tripp et al. 2008, 2009; Burgess et al. 2009) and blood serum (J. Reynolds et al., unpublished). The utility of these methods and other methods such as ultrasound is reduced by the need to capture the animal.

Table 6.1 *Characteristics of the ovaries, uteri and mammary glands of female sirenians in different stages of reproduction (modified from Table 6.2 in Boyd* et al. *(1999), which in turn was modified from Table 2 in Marmontel (1995) to apply to dugongs as well as Florida manatees). Corresponding data are not available for Amazonian, Antillean or West African manatees. Occasional pregnant and lactating Florida manatees (Marmontel 1995) and dugongs (Marsh 1989; Kwan 2002) have been recorded; see text for details.*

Stage	Ovaries	Uterus	Mammary gland
Immature, prepubescent	Smooth and flattened surface with numerous small follicles	Glands underdeveloped, lumen small, blood supply underdeveloped	Inactive
Immature, maturing	Several LGF[I]	Early proliferative endometrium, size intermediate between juvenile and adult	Inactive
Sexually mature, nulliparous	CL[I] and/or CA[I] present, may have several LGF	Adult-sized, no placental scar (dugong only)	Inactive
Parous	CA present, may have LGF or CL	At least one placental scar (dugong only)	Active or inactive
Ovulating	Luteal: LGF and CL	Secretory endometrium with or without placental scar, not pregnant	Inactive or active
Pregnant	CL present, may have CA	Embryo or foetus present, secretory endometrium	Usually inactive but may be active

[I] CA = corpora albicantia; CL = corpora lutea; LGF = large Graafian follicle.

Size at attainment of sexual maturity in females

Female Florida manatees and dugongs both reach sexual maturity over a range of sizes. In northern Australia, dugong females less than 2.0 m long have almost certainly not borne a calf (nulliparous), whereas those larger than 2.5 m are likely to have borne young (parous) (Marsh *et al.* 1984c; Kwan 2002; J. Grayson, unpublished). Nulliparous females with active ovaries have been recorded at lengths of 2.28–2.50 m (Marsh *et al.* 1984c; Kwan 2002). The smallest female with a placental scar was 2.27 m long (Kwan 2002). Reflecting their generally larger body size, Florida manatee females are longer than dugongs when they mature. The smallest mature female recorded by Marmontel (1995) was a 2.54-m-long three-year-old. The average length of immature three-year-old females was 2.55 m (n = 5;

Figure 6.5. The uterus of a female dugong showing one placental scar. H. Marsh photograph, reproduced with permission.

range = 2.42–2.72 m) and the average length of mature three-year-old females was 2.8 m (n = 7; range = 2.54–3.02 m).

Age at attainment of sexual maturity in females
The age of sexual maturity of female sirenians varies among individuals and species and possibly among and within populations. This result suggests that, as in many other species of large herbivorous mammals (e.g. Galliard et al. 2000; Eberhardt 2002), the fecundity of young female dugongs and manatees varies both with changes in their environment and the density of conspecifics (which tend to covary).

Wild Florida manatees apparently reach sexual maturity between the ages of 2.5 and six years. Marmontel (1995) reported that all 143 females five years old or older for which information on reproduction was available from salvaged carcasses were sexually mature. Probably as a result of the hormonal changes associated with puberty, the first year of sexual maturity in Florida manatees is typically associated with change in the width of the growth layer groups in the periotic bone. Study of a large sample of periotic bones by staff of the Florida Fish and Wildlife Research Institute's Marine

Table 6.2 *The ranges[1] of ages (years) at which female Florida manatees and dugongs bear their first calf. This table is modified from Table 6.3 in Boyd* et al. *(1999).*

Location	Dates of sample collection	Youngest parous female	Oldest nulliparous female	Type of study	References
Florida manatee: Florida					
Both coasts	1976–1991	4	5+[2]	Necropsy	Marmontel (1995)
Gulf coast	1978/1979– 1992/ 1993	3.5	7	Longitudinal	Rathbun et al. (1995)
Gulf coast	1993–1998	4	?	Longitudinal	Koelsch (2001)
Atlantic coast	1986–1991	4	7	Longitudinal	O'Shea and Hartley (1995); Reid et al. (1995)
Dugong: Australia–Papua New Guinea					
Townsville 19° 17′ S	1969–1981	10	9.5	Necropsy	Marsh et al. (1984c)
Mornington Island 16° 32′ S	1976–1977	14.5	17.5	Necropsy	Marsh et al. (1984c)
Daru 9° 04′ S	1978–1982	13	12[3]	Necropsy	Marsh (1986)
Mabuiag Island 9° 58′ S	1997–1999	6[4]	11	Necropsy	Kwan (2002)

[1] Ranges rather than means have been given here because of the small sample sizes.
[2] All female manatee carcasses aged five years or older were sexually mature.
[3] One 18-year-old had recently given birth to her first calf.
[4] One female aged eight years was pregnant with a 2.4 cm foetus and had two placental scars. The minimum age at which females first give birth should be about one year after the age of maturation if the gestation period lasts 12–14 months.

Mammal Pathobiology Laboratory suggests that sexual maturity occurs by five years of age in 86% of female manatees (Reep and Bonde 2006), even though some females attain sexual maturity by three years of age (Marmontel 1995), whereas some individuals do not give birth until they reach seven years old (Table 6.2). The youngest known-age Florida manatee to reproduce in captivity was a captured orphan that conceived when she was about 5.5 years old (Odell et al. 1995). These estimates indicate that Florida manatees reach sexual maturity at younger ages than those suggested by earlier studies based on much smaller samples (Hartman 1979; Odell *et al.* 1981). However, Marmontel (1995) reports that mature females, which have subsequently been aged as three years old, were collected in Florida as early as 1978, suggesting that the results reported in the earlier studies were likely due to small sample sizes rather than reductions in age at maturity over the intervening years.

Although less robust, the data for wild dugongs suggest that they generally mature several years later than wild Florida manatees (Table 6.2). All information is from carcass studies in which ages have been estimated from growth layer groups in the tusks. However, the dugong sample sizes are much smaller than those for Florida manatees, the age of maturity could not be detected in the growth layer groups in the tusk, and there are no longitudinal records for known individuals (Marsh 1980, 1986, 1995; Marsh *et al.* 1984a,c; Kwan 2002; Marsh and Kwan 2008; Grayson *et al.* 2008). The data suggest substantial individual variation in the age at the first pregnancy in dugongs, both between populations and within populations over time. Females from Torres Strait that were pregnant with their first calf or had borne only one calf ranged in age from 7 years to 14 years in 1997–1999 and from 13 years to 18 years in 1978–1982, suggesting that dugongs were reproducing at substantially younger ages in 1997–1999 than in 1979–1982. Indeed, the data suggest that the dugongs sampled in 1997–1999 were breeding for the first time several years earlier (under six years old compared to ten years old) than in any other dugong population studied (Table 6.2). These differences appear to be largely driven by the abundance and quality of seagrasses as forage, with delayed ages to maturity in years of low availability of a quality food supply (Marsh and Kwan 2008).

Assessment of sexual maturity in males

It is relatively easy to establish whether or not a female sirenian is sexually mature; assessing males is more problematic, although methods for measuring reproductive hormones in sirenians promise to overcome this problem (Burgess *et al.* 2009; J. Reynolds *et al.*, unpublished). Maturity can be inferred from testicular weight: in dugongs a single testis weight of 30 g or more is diagnostic of maturity (Marsh *et al.* 1984b). Histological examination of the testis and epididymis of necropsy specimens should be diagnostic of sexual maturity but, as explained above, without data on the animal's age or body length it may be difficult to distinguish the testicular histology of pubertal males (those approaching first spermiogenesis) from that of mature males with testes that are re-entering breeding condition (Marsh *et al.* 1984b; Hernandez *et al.* 1995).

External assessment of the reproductive maturity of a male sirenian is difficult unless his age or body length is known and outside the range within which males mature for that species, or he has erupted tusks (dugongs only – Marsh 1980). Participation in a mating herd is not diagnostic of sexual maturity (Chapter 5). Rathbun et al. (1995) recorded young of the year and yearlings in a mating herd of manatees. As with female sirenians, researchers are developing techniques to determine and monitor the reproductive status of males by measuring hormone concentrations in blood serum, saliva, lacrimal (tears), urine and faecal samples (Lanyon et al. 2005b; Larkin et al. 2005; de Souza et al. 2009, J. Reynolds et al., unpublished). Amaral et al. (2009) used experimental hormone challenge to monitor the time lag between androgen concentrations in the saliva and subsequent excretion in lacrimal secretions, faeces and urine of captive Amazonian manatees. Not surprisingly, they found that the time lag was 5–7 days longer in faeces than in serum, consistent with the long digesta times of sirenians (see Chapter 4).

As explained in Chapter 5, Florida manatees (and probably dugongs) are typically promiscuous, with multiple males mating with individual females during a single oestrus. This mating behaviour suggests that manatees may engage in sperm competition, wherein sperm from more than one male within the reproductive tract of a female compete with one another to fertilise the ovum. Species of mammals that are sperm competitors tend to have testes which are exceptionally large relative to their body size. Manatees have a tendency towards larger testes than typical for mammals of their size, but not exceptionally so (about twice the mass predicted by body size in mature males in breeding months – Reynolds et al. 2004). However, the body mass of sirenians may be inflated as explained in Chapter 4, making the relative mass of the testes 'artificially' low. Additionally, the large seminal vesicles which are a feature of the sirenian male reproductive tract might partially compensate for the size of the testes in sperm competition (Reynolds et al. 2004).

Size at attainment of sexual maturity in males

Like the female, male dugongs reach sexual maturity at a range of sizes. In northern Australia, dugong males less than 1.9 m long are almost certainly immature, whereas those larger than 2.5 m are likely to be mature (Marsh et al. 1984c; Kwan 2002). This size range at maturity is similar to that of females in the same population. Tusk eruption appears to succeed testicular competence in dugongs (Marsh et al. 1984c; Kwan 2002). Only 33% (8 out of 24) of the males classified as adult on the basis of their testicular histology by Kwan (2002) had erupted tusks. The youngest mature male with erupting tusks was 12 years old (Marsh *et al.* 1984c). Young, sexually mature males are usually smaller than older males. They tend to have smaller testes and their tusks may not have erupted. This combination may diminish their ability to compete with other males for mates (Marsh *et al.* 1984c).

As with females, male Florida manatees are longer than dugongs when they reach maturity. Hernandez *et al.* (1995) examined the testes of 67 Florida manatees. Animals with prepubescent or recrudescent testes ranged in body length from 2.37 m to 2.95 m. The most precocious male was a 2.37-m-long two-year-old individual that already had some sperm in his testes. The smallest male with fully spermatogenic testes was 2.52 m long, close to the minimum body length of mature females (2.54 m; see above).

Age at attainment of sexual maturity in males

As with females, the age at which male sirenians become sexually mature varies among individuals and species and perhaps between and within populations (Table 6.3). The age of sexual maturity of males tends to be similar to that of females in the same population. Male Florida manatees apparently mature at earlier ages than male dugongs.

PREGNANCY AND BIRTH

Length of gestation

Estimates of the gestation period of Florida manatees have not improved significantly since the observations of Hartman (1979), despite the births of tens of individuals in captivity (e.g. Odell *et al.* 1995). Estimates range from 12–14 months for captive Florida manatees (Cardeilhac *et al.* 1984; Qi Jingfen 1984; Odell *et al.* 1995). Rathbun *et al.* (1995) and Reid *et al.* (1995) obtained crude approximations of the gestation period from field observations of the time between individual females being observed in a mating herd and subsequently sighted with a small dependent calf. Most of these observations also suggested gestation periods within the 12–14-month range. Best (1982) assumed the length of gestation in the Amazonian manatee to be about 12–14 months, apparently based on information from the West Indian manatee. The capacity of ultrasound to detect early pregnancy in sirenians is complicated by their expansive and gas-filled intestines (M. Rodriguez, unpublished; cited in Tripp *et al.* 2008). The best estimate of the gestation period for the dugong (14.5 months, 95% CI 12.4–17.3 months) is close to the estimate for manatees. Kwan (2002)

Table 6.3 *The ranges[1] of ages (years) at which male Florida manatees and dugongs become sexually mature. This table is modified from Table 6.4 in Boyd* et al. *(1999).*

Location	Dates of sample collection	Youngest male with mature testes	Oldest immature or recrudescent testes	Oldest male with unerupted tusks[2]	References
Florida manatee					
All Florida	1975–1985	2	11	n/a	Hernandez et al. (1995)
Dugong: Australia/Papua New Guinea					
Townsville 19° 17′ S	1969–1981	9	6[3]	10.5	Marsh et al. (1984c)
Mornington Island 16° 32′ S	1976–1979	15	15.5	15.5	Marsh et al. (1984c)
Daru 9° 04′ S	1978–1982	11	16	18	Marsh (1986)
Mabuiag Island 9° 58′ S	1997–1999	7[4]	13	14	Kwan (2002)

[1] Ranges rather than means have been given here because of the small sample sizes.
[2] Dugongs only.
[3] No males examined aged between seven and nine years inclusive.
[4] One mature male was aged at 4 ± 0.6 years (95% CI) based on eye lens weight, which is less reliable than number of growth layer groups in the tusks of males with unerupted tusks (Kwan 2002).

obtained this estimate using data on the body lengths and dates of death of 26 foetuses from Daru in southern Papua New Guinea and 24 foetuses from Mabuiag Island in Torres Strait. However, as a result of the diffusely seasonal breeding pattern of the dugong (see 'Timing of reproductive activity in mature sirenians' below), the 95% confidence interval for this estimate is large. Also, Kwan (2002) used the method of Huggett and Widdas (1951) and Laws (1959), which assumes two different foetal growth rates at different stages of pregnancy, an assumption that has not been verified for sirenians, increasing the uncertainty of her estimates.

Litter size, sex ratio and size at birth
Sirenians usually bear a single young. Nonetheless, twins have been confirmed for both wild and captive Florida manatees on numerous occasions

(e.g. Marmontel 1995; Odell *et al.* 1995; O'Shea and Hartley 1995; Rathbun *et al.* 1995) and for the Antillean subspecies (Charnock-Wilson 1968). The records for the Florida manatee include twins of either and both sexes (Bonde 2009). The carcass studies of the Florida manatee (Marmontel 1995) suggest that the incidence of twinning is 4%. This estimate is higher than the percentages in observations of wild individuals from the Atlantic (1.79% – O'Shea and Hartley 1995) and Gulf (1.4% – Rathbun *et al.* 1995) coasts of Florida. The discrepancy could be explained by the estimates for wild manatees being lowered by the incidence of neonatal death of one or both twins, and the estimates for carcasses being inflated by an increase in female mortality attributable to twinning. The true incidence is likely to be somewhere between 1.4% and 4%; marginally higher than that in baleen whales (<1–2.2% – Kimura 1957).

Although there are also anecdotal reports of the occasional occurrence of twin foetuses in the dugong (Norris 1960; Jarman 1966; Thomas 1966; Bertram and Bertram 1968), only one foetus has been found in each of the 58 pregnant dugongs documented by scientists in recent years (Marsh 1995; Kwan 2002). Indigenous hunters report that twins are very rare (H. Marsh, unpublished). A litter size of one is reported for both the Amazonian manatee (Husar 1977; Best 1984; Timm *et al.* 1986) and the West African manatee (Beal 1939). Taken together, these results suggest that the mean litter size is very close to one for all species of modern sirenians.

Reliable estimates of the sex ratio at birth for the Florida manatee are not significantly different from 1:1 (O'Shea and Hartley 1995; Rathbun *et al.* 1995; Runge *et al.* 2004). The sex ratio of carcasses aged less than one year slightly favours males (Marmontel 1995), perhaps indicating a sex differ-ence in neonatal mortality rather than a difference in the sex ratio at birth.

The largest data set on sex ratios for the dugong is from Kwan (2002), who reported that the sex ratio of the 24 foetuses she examined was 2.7:1 in favour of males, significantly different from a 1:1 sex ratio. Her sample of animals less than five years old was also biased in favour of males, 1.9:1. This result contrasts with sex ratios in the dugongs caught in shark nets off Townsville between 1969 and 1981. Seven females and six males were estimated to be less than two years old, suggesting that the sex ratio of young dugongs is 1:1 (Marsh *et al.* 1984c). These sample sizes are too small to make robust conclusions about possible spatial and temporal variations in the sex ratio of dugongs at birth.

The size of sirenian neonates varies among and within species (Table 6.4). The variation probably mirrors specific and individual

Table 6.4 *Sizes of neonates and adults of various species/subspecies of manatees and the dugong. Adapted from Table 6.5 in Boyd et al. (1999).*

Species/subspecies	Neonatal body length (m) range (mean)[1]	Adult body length (m) mean	Neonatal body weight (kg) range	Adult body weight (kg) mean	References
Florida manatee	0.80–1.60 (1.22)	3	30–50	500	Odell (1982); O'Shea and Hartley (1995); Marmontel (1995); Rathbun et al. (1995)
Amazonian manatee	0.85–1.05	2.8[2]	10–15	480[2]	Best (1982)
West African manatee	1.00	3			Cadenat (1957)
Dugong[4]	1.00–1.30 (1.15)[4]	2.7	25–35	250–300	Marsh et al. (1984c); Marsh (1995)

[1] If available.
[2] Measurements of large individuals, not means.
[3] Not available but probably similar to Florida manatees.
[4] Kwan (2002) estimated the mean body length of 7 foetuses >100 cm long as 111.32 cm.

differences in adult sizes. We suggest three reasons for the variation within a species: (1) the difficulty of distinguishing true neonates from large aborted foetuses and young calves in studies of carcasses of small individuals found dead; (2) variation in the length of gestation; and (3) variation in neonatal size resulting from the size of the mother. None of the data are sufficient for the method of estimating size at birth recommended by Perrin and Reilly (1984), and information for both the Florida manatee and the dugong has been expressed as a mean birth length. This approach tends to overestimate the true size at birth (Perrin and Reilly 1984).

CALF DEPENDENCY

Like other large mammals, sirenian calves gain nutritional independence gradually as neonates shift from a milk diet to solid foods (Langer 2003). Thus, weaning is unlikely to occur abruptly and the time of lactation can be divided into two phases: (1) when only milk is digested; and (2) when milk is taken with solid food – the mixed-feeding or weaning period. Studies of other mammals indicate that the absolute length of the mixed-feeding period depends on the needs of the young animal. The longer this period, the more time is available for ontogenetic developments – for example, learning to obtain food, avoid predators and acquire immune competence (Langer 2003). The data for sirenians, below, do not differentiate between these two phases of calf dependency.

Field observations of the time during which calves accompany their mothers have been obtained for the Florida manatee based on long-term observations of individual females that are recognised at winter refugia on the basis of their scars or that give birth and wean young while being radio-tracked. Neither source of information is conducive to precise estimates. However, the trend is clear. Most calves (67% – O'Shea and Hartley 1995; 77% – Rathbun et al. 1995) are seen with their mothers during only one winter season; the remainder accompany their mothers for two winters. This pattern is independent of the sex of the calf, but may vary with the age of the mother; calves of young females seem more likely to be dependent for two winters than those of older females. Data from radio-tracked Florida manatees also indicate that some calves accompany their mothers for about 12 months, whereas other calves have been nursed for up to 24 months before weaning (Reid et al. 1995). Koelsch (2001) estimated that the calves she observed in the Sarasota area were dependent on their mothers for an average of 1.8 non-winter (March–November) seasons, with some spanning two years. Bonde (2009) presents some intriguing genetic evidence of wild Florida manatees suckling adopted calves (Figure 5.13). If adoption is wide-spread, the practice will confound estimates of life history parameters, including: (1) calf dependency based on the observations of wild individuals; and (2) estimation of annual reproduction using mark–recapture models.

Data on the duration of lactation in dugongs are sparse. Kwan (2002) used data from 47 lactating females collected from Mabuiag Island in 1997–1999 to calculate the length of lactation using the ratio of lactating to pregnant animals (confirmed and all possible pregnancies and gestation periods of 13, 14 and 15 months) and the method of Perrin and Reilly (1984). Kwan's estimates ranged from 13.6 (95% CI 9.9–18.7) to 17.6 (95% CI 12.7–21.5) months, suggesting that variability and duration of lactation in dugongs is similar to that in manatees.

There is some evidence to suggest that lactating female dugongs may be more likely to become pregnant than lactating Florida manatees. Four preg-nant lactating female dugongs have been necropsied and found with foetuses ranging in length from 17.2 cm to 76.6 cm (Marsh 1989; Kwan 2002), demonstrating that ovulation is not always suppressed in lactating female dugongs. The mammary glands of three dugongs with foetuses ranging from

61.2 cm to 85.4 cm had features characteristic of proliferating and lactating mammary glands, but it is uncertain whether they were nursing a previous calf or preparing for the new calf. Dugongs with large foetuses (>100 cm long) have also been recorded with lactating mammary glands (Kwan 2002). Even though the sample size of necropsied Florida manatees is much greater than that for dugongs, the only two confirmed cases of Florida manatee females that were simultaneously pregnant and lactating were carrying very large foetuses; Marmontel (1995) considered that the milk was probably for the new infants. Kendall *et al.* (2004) reported that it was very rare for Florida manatees to reproduce in two successive years, but recorded two apparent cases in the Atlantic coast population. Based on hormone measurements from four lactating females, Tripp *et al.* (2009) speculated that, at least during the early post-partum period, ovarian cyclicity may be inhibited by lactation in Florida manatees. If dugongs are more likely than Florida manatees to become pregnant while lactating, this capacity may reflect the need to capitalise on food availability in a fluctuating environment (see 'Timing of reproductive activity in mature sirenians' below).

CALVING INTERVALS

The calving interval of the Florida manatee has been estimated by direct observation of known wild individuals (O'Shea and Hartley 1995; Rathbun *et al.* 1995; Reid *et al.* 1995; Koelsch 2001) and in captivity (Odell *et al.* 1995). The estimates (Table 6.5) suggest that, on average, wild Florida manatees have a calf about every 2.5 years and that the calving interval is relatively stable over time.

Marsh *et al.* (1984c), Marmontel (1995), Marsh (1995), Kwan (2002) and Marsh and Kwan (2008) have estimated the inter-birth interval from carcasses of Florida manatees and dugongs. They estimated the inter-birth interval as the reciprocal of the annual pregnancy rate. The annual pregnancy rate is the percentage of mature females that are pregnant (including those pregnant and lactating) divided by the length of gestation in years (Perrin and Reilly 1984; Marsh *et al.* 1984c). Calculation of the annual pregnancy rate assumes that: (1) the length of the gestation period is known precisely and accurately; (2) all pregnancies are detected; (3) there are no biases due to seasonal birthing; and (4) the distribution of reproductive status in the sample is representative of the population. None of these assumptions is likely to be true for existing studies of sirenians. Researchers have attempted to address the first two assumptions by calculating calving intervals using three estimates of the gestation period (typically 12, 13 and 14

Table 6.5 *Inter-birth intervals of wild and captive Florida manatees based on observations of known individuals. See text for discussion of new statistical estimation procedures that refine some of these estimates. Adapted from Table 6.6 in Boyd et al. (1999).*

Location and dates of data collection	Number of females	Number of births	Inter-birth interval years mean ± SE (range)	References
Blue Spring 1978/ 1979–1992/1993	7	25	2.6 ± 0.17	O'Shea and Hartley (1995)
Crystal River 1976/ 1977–1990/1991	33	99	2.48 ± 0.08 (1–5)	Rathbun *et al.* (1995)
Atlantic coast 1978/ 1979–1991/1992	10	11	2.6 (2–4)	Reid *et al.* (1995)
Sarasota Bay 1993– 1997	18[1]		2.2 ± 0.19 (1–4)	Koelsch (2001)
Captivity			(1.2–8.6)	Odell *et al.* (1995)

[1] Censored data set limited to females observed in three consecutive seasons.

months, but Kwan (2002) used 13, 14 and 15 months) and by estimating the proportion pregnant based on both: (1) a conservative scenario using confirmed pregnancies only; and (2) an optimistic scenario based on all pregnancies suggested by various indicators of reproductive condition. Marmontel (1995) also calculated calving intervals based on an intermediate scenario. These data are summarised in Table 6.6, which indicates that the mean estimates of inter-birth interval for Florida manatees tend to be between 2.5 and five years. Although the minimum inter-birth interval for dugongs is not very different from that of manatees, the maximum estimates are longer, ranging up to 6.8 years (Table 6.6).

A maximum of nine placental scars has been counted in a dugong uterus. Assuming the scars persist, the calving interval can also be estimated by regressing the number of placental scars against age for parous dugongs. This calculation gave values in the same range as those in Table 6.6 (Marsh *et al.* 1984c). Placental scars do not persist in manatees and cannot be used to estimate the number of parous events (Marmontel 1995).

REPRODUCTIVE RATES

The studies of calving intervals for Florida manatees based on direct observation are not directly comparable with those obtained from carcass studies,

Table 6.6 *Estimates of inter-birth interval (± SE) of Florida manatees (Marmontel 1995) and dugongs (Marsh 1995; Marsh and Kwan 2008) based on the annual pregnancy rate of necropsy samples and several possible gestation periods.. Estimates are based on confirmed pregnancies and all possible pregnancies (bold). Adapted from Table 6.7 in Boyd et al. (1999).*

Location and dates of sample collection	Number of females	12 months	13 months	14 months	15 months
Florida manatee					
Florida	212	4.5 ± 1.4	4.9 ± 1.6	5.3 ± 1.9	n/a
1976–1991	**286**	**2.5 ± 0.7**	**2.7 ± 0.8**	**3.0 ± 1.0**	n/a
Dugong					
Numbulwar	86	3.1 ± 0.47	3.3 ± 0.55	3.6 ± 0.62	n/a
1965–1969	**86**	**2.7 ± 0.38**	**2.9 ± 0.43**	**3.1 ± 0.49**	n/a
Townsville	18	4.5 ± 1.98	4.9 ± 2.26	5.3 ± 2.5	n/a
1969–1990					
Daru	168	5.8 ± 0.98	6.3 ± 1.11	6.8 ± 1.25	n/a
1978–1982	**168**	**4.9 ± 0.76**	**5.4 ± 0.86**	**5.8 ± 0.97**	n/a
Mabuiag	73	n/a	2.84 ± 0.38	3.06 ± 0.43	3.28 ± 0.48
1997–1999	**73**	n/a	**2.37 ± 0.27**	**2.55 ± 0.39**	**2.73 ± 0.35**

because the sighting data are generally collected only in winter, when manatees seek refuge in warm water (see Chapters 5 and 7). (The exception is the year-round, long-term studies in Sarasota Bay, e.g. Koelsch 2001.) Thus, the resultant estimates of reproductive rate are a product of birth rate and the survival probability for the first few months of life. We know from the carcass recovery programme that mortality of Florida manatees can be high in the perinatal period (Chapter 7). Population models overcome this problem because the models begin at age-class six months (see 'Assessing population dynamics through modelling of life history parameters' below).

In addition, the observational data summarised above followed ad hoc sampling methods and did not address two potentially important problems: (1) possible differences in the sighting probabilities of females with and without first-year calves (conditional breeding probabilities); and (2) the potential for the observer to fail to detect the presence of a first-year calf whose mother was sighted, thus misclassifying the mother as non-breeding. Kendall *et al.* (2004) provided a new statistical estimation procedure to deal with these problems by further developing the multistate statistical model of Kendall *et al.* (2003).

This more sophisticated analysis, built in part on within-season re-sightings using Pollock's robust design (Pollock 1982), showed similar probabilities of sighting a calf with its mother (given that the mother was sighted) in two study areas (north-west Florida and the Atlantic coast, both at probabilities of 0.73 [0.06 SE]). This analysis was based on annual re-sighting efforts spanning 19 consecutive years at the two areas. Breeding probabilities that did not adjust for this misclassification would be under-estimated by 28%, providing ample justification for use of the more sophis-ticated model. Additionally, the method provides improved estimates of variance and uncertainty necessary for modelling population dynamics (see 'Assessing population dynamics through modelling of life history parame-ters' below). The adjusted conditional probabilities of females sighted with-out a calf in one year being with a first-year calf the next year were 0.43 (0.06 SE) for the Northwest subpopulation and 0.38 (0.05 SE) for the Atlantic coast subpopulation. Runge et al. (2004) used a similar technique to estimate the corresponding values for the upper St Johns River subpo-pulation (0.61, CI 0.51–0.71) and the south-western Florida subpopulation (0.60, CI 0.42–0.75). When these values are used in Runge et al.'s (2004) matrix population models, they result in ratios of first-year calves to females that have bred previously ranging from 0.32 for the Atlantic subpopulation to 0.42 for the upper St Johns subpopulation, suggesting mean inter-birth intervals for breeding adult females of 2.38–3.13 years.

Comparable data for reproductive rates are not available for any other sirenian; estimates of dugong reproductive rates have been based on calving intervals.

TIMING OF REPRODUCTIVE ACTIVITY IN MATURE SIRENIANS

Perhaps because sirenians are limited to the subtropics and tropics, some early researchers assumed that they bred year-round (e.g. see Hartman 1979). However, there is increasing evidence that the reproductive activity of Florida and Amazonian manatees and dugongs is diffusely seasonal rather than strictly synchronised (Figure 6.6). In addition, sirenians have the capacity to delay reproduction when their food supply is low (Marsh and Kwan 2008). In tropical areas, the abundance of resources can vary greatly between wet and dry seasons, particularly in the large inland river systems and lakes occupied by the three species of manatees (Arraut et al. 2010; see also Chapter 5). In Florida, seasonally cool temperatures and shorter photo-periods affect plant growth in winter and provide the principal natural

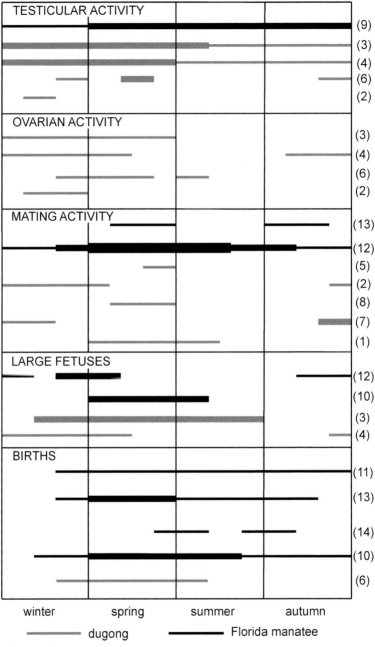

Figure 6.6.

constraints to seasonal forage production. Marine seagrasses in the range of the dugong can undergo periodic large-scale diebacks that can also result in reduced reproduction in dugongs (Marsh and Kwan 2008; Chapter 4).

Seasonality of reproduction in females

Data from necropsies indicate that female Florida manatees (Marmontel 1995) and dugongs (Marsh *et al.* 1984c; Marsh 1995; Kwan 2002) do not breed continuously. Graafian follicles and/or corpora lutea were absent from the ovaries of 33% of 27 mature female manatees and 56% of 25 mature female dugongs whose ovaries were examined histologically and that were neither pregnant nor lactating (Boyd *et al.* 1999). Although the reproductive activity of the females in a population is generally not highly coordinated, there is evidence of some seasonality. The ovaries of female dugongs in northern Australia and southern Papua New Guinea are more likely to contain large follicles or corpora lutea in the second half of the year (winter/spring) than in summer and autumn (Marsh *et al.* 1984c; Marsh 1995; Kwan 2002). Sterile cycles seem to be common (Marsh *et al.* 1984c) and may also occur in Florida manatees (Marmontel 1995; Rathbun *et al.* 1995).

Odell *et al.* (1995) noted that captive Florida manatees participate in mating behaviour throughout pregnancy, suggesting that females mate even when they are not fertile. This behaviour may be an artefact of captivity. However, mating herds have also been observed in Florida in winter, when at least some males are probably not spermatogenic (J. Reynolds, personal observation) Taken together, these observations make the temporal patterns observed in mating activity of wild sirenians difficult to interpret in the absence of information on the levels of

---

Caption for Figure 6.6.
Timing of the indices of breeding in Florida manatees and dugongs at the various locations where research has been conducted. The references on which this diagram is based follow. Dugong: (1) Shark Bay, Western Australia (Anderson 1997); (2) Mornington Island, Queensland (Marsh *et al.* (1984c); (3) Daru, Torres Strait (Marsh 1995); (4) Mabuiag Island, Torres Strait (Kwan 2002); (5) Princess Charlotte Bay, Queensland (H. Marsh, unpublished; (6) Townsville, Queensland (Marsh *et al.* 1984c); (7) Shoalwater Bay, Queensland (Anderson and Birtles 1978); (8) Moreton Bay, Queensland (Preen 1989a). Florida manatee: (9) Florida (Hernandez *et al.* 1995); (10) Florida (Marmontel 1995); (11) Florida (captives) (Odell *et al.* 1995); (12) Northwest Florida (Rathbun *et al.* 1995); (13) Blue Spring, Florida (O'Shea and Hartley 1995); (14) Atlantic coast, Florida (Reid *et al.* 1995). The months have been reversed for the northern and southern hemisphere populations to facilitate comparisons. The line thickness is an indication of the proportion of the population exhibiting the index. Drawn by Shane Blowes, based on a diagram in Boyd *et al.* (1999), reproduced with authors' permission.

reproductive hormones in female participants. Despite these methodological difficulties, observations suggest some seasonality of mating activity (Figure 6.6). In Florida, manatee mating is concentrated in spring and summer (Rathbun *et al.* 1995), a pattern similar to that observed in dugongs in Shark Bay (Western Australia) by Anderson (1997), southern Queensland by Preen (1989a) and northern Queensland and Torres Strait by H. Marsh (unpublished). However, exceptions to this pattern may occur. Anderson and Birtles (1978) and Adulyanukosol *et al.* (2007) observed presumed courtship/ mating behaviour in winter in central Queensland and Thailand, respectively. In the upper Orinoco River, Castelblanco-Martínez *et al.* (2009) report that mating of Antillean manatees occurs during the low-water period in restricted areas that retain deep water.

Not surprisingly, calving also appears to show diffuse seasonality rather than strict synchrony. Necropsy studies and observations of wild and captive Florida manatees indicate that very few calves are born in the winter months (Odell *et al.* 1995; O'Shea and Hartley 1995; Marmontel 1995; Rathbun *et al.* 1995; Reid *et al.* 1995). Although births have been recorded throughout the remainder of the year in Florida, calving peaks in spring and summer (Figure 6.6). The timing of births is similar for the dugong in Western Australia (Anderson 1998) and northern Australia/southern Papua New Guinea (Marsh *et al.* 1984c; Marsh 1995; Kwan 2002) (Figure 6.6). Anecdotal information suggests that the calving period may be longer at lower latitudes than at the southern limit of the dugong's range (Figure 6.6), but insufficient data are available to confirm this hypothesis. Even though the Amazonian manatee occurs in equatorial latitudes, calving is also diffusely seasonal. In Brazil, calves are born between December and July, especially during the period from February to May, when water levels are rising (Best 1982). In Ecuador, young are apparently born in January and June (Timm *et al.* 1986). Powell (1996) reports that African manatees mate during the rainy season between June and September, when water levels are rising. We know of no data on calving in this species.

The timing of births in the Florida manatee reduces the likelihood that calves will be exposed to potentially dangerous low winter temperatures at a time when food plant productivity may also be reduced (due to lower temperature and a shortened photoperiod). However, seasonal temperature changes cannot be the explanation for seasonal calving in sirenians living in the tropics. In most herbivorous mammals, birthing peaks coincide with high plant productivity. Timing calving to coincide with plant productivity helps the mother meet the high protein and energy demands of her off-spring during late pregnancy and early lactation. Both the Amazonian and

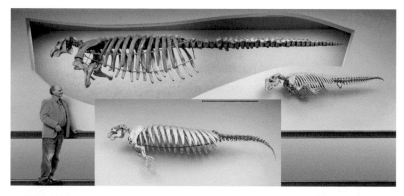

Figure 2.3

Figure 6.3a

Figure 6.3b

Figure 4.2a

Figure 4.2c

Figure 4.2b

Figure 4.2d

Figure 7.11

Figure 7.14

West African manatees have adopted this strategy, as explained above. Anderson (1998) also describes an aggregation of female dugongs with very young calves in summer in a nutrient-rich and sheltered environment (North Cove, Shark Bay in Western Australia) dominated by the seagrass *Halodule uninervis*. Anderson (1998) did not see young calves eating seagrass, but stomach contents analysis suggests that, like wild and captive Florida manatees (Hartman 1979; Odell 1982), wild dugongs (Marsh *et al.* 1982) start to eat plants a few weeks after birth. This strategy enables a calf born in spring or summer to supplement its milk intake with abundant new shoots that are low in fibre and high in soluble carbohydrates and nitrogen (see Chapter 4). The birth seasons reported for Amazonian manatees in Brazil (Best 1982) and for West African manatees (Powell 1996) also presumably result in mothers and calves having access to new plant growth.

Seasonality of reproduction in males
The reproductive status of mature male sirenians has been evaluated on the basis of histological examination of the testes and epididymides. This approach provides information on the reproductive status of a male at the time of his death, but gives few insights into his reproductive history. It is clear from these studies that not all of the mature males in a population of Florida manatees or dugongs produce spermatozoa continuously. For example, in dugongs more than half of 41 pubertal and mature males from north Queensland (Marsh *et al.* 1984c), 141 mature male dugongs from Daru (Torres Strait – Marsh 1995) and 42% of 24 mature males from Mabuiag Island (Torres Strait – Kwan 2002) were infertile at the time of sampling. Spermatogenic activity is not synchronised within a population. For example, the testes of a total of 17 mature male dugongs from Mornington Island in northern Australia were sampled in July–August in 1976, 1977 and 1979. The testes of five males were in active spermatogenesis, four were approaching spermatogenesis, four were intermediate and four were resting (Marsh *et al.* 1984b,c). However, it is difficult to confirm from such studies whether individual males have periods of sexual inactivity followed by periods of sexual recrudescence and activity, or if the reproductive activity of some males is permanently suppressed (Marsh *et al.* 1984b,c; Hernandez *et al.* 1995). It is likely that both of these explanations are true. Some dugong males had regressed testes (*sensu* Marsh *et al.* 1984b), suggesting that they had been sterile for long periods or even permanently. On the other hand, there is evidence of a seasonal pattern of gonadal activity that overlaps the equivalent period in females. Dugongs with active testes (fully spermatogenic or recrudescent *sensu* Marsh *et al.*

1984b) were a significantly higher proportion of the sample from Daru, in the Papua New Guinean waters of Torres Strait, between June and January than between February and May (Marsh 1995), a result supported by the sample collected at nearby Mabuiag nearly 20 years later (Kwan 2002).

Male manatees in Florida also show decreased spermatogenesis (Hernandez *et al.* 1995) and lower testicular mass (Reynolds *et al.* 2004) in winter, consistent with the lower frequency of observations of mating herds (Rathbun *et al.* 1995). Recent studies of seasonal fluctuations in testosterone concentrations and of anti-Mullerian hormone levels in the serum of free-ranging manatees (J. Reynolds *et al.*, unpublished) and in captive male Florida manatees are also consistent with the hypothesis that this subspecies is a diffusely seasonal breeder with peaks in spring and/or autumn (Larkin *et al.* 2005).

Seasonality of testicular activity is more difficult to explain than seasonality of ovarian activity and calving because there is no evidence that sperm production per se is energetically expensive. However, the reproductive activity of male sirenians must be demanding of time and energy. Male manatees seem to travel farther in search of females in oestrus during warm months (Bengtson 1981) and, if most females are not in oestrus in winter, selection would favour reduction in reproductive activity of males in winter to save energy. Geist (1974) points out that in the tropics the overall cost of social life for males with breeding seasons extending over several months should be higher than for those of temperate species where the mating season is shorter. He postulated that mechanisms may evolve to ration the mating activity of each male in species where the breeding season is prolonged over several months. The asynchronous, discontinuous breeding pattern of male sirenians would allow an individual male to recuperate from his reproductive activities and increase his life expectancy. This strategy would allow him to sire more offspring than if he were in rut continuously and died young (Marsh *et al.* 1984c).

Tendency to delay reproduction in adverse conditions

Like most other long-lived species, sirenians produce their offspring in a series of separate events, a pattern of reproduction known as iteroparity. As discussed above, neither mature males nor mature females are continuously in breeding condition. This situation cannot be attributed solely to the seasonality of reproduction. There is considerable individual variation in both the pre-reproductive periods and inter-birth intervals of Florida manatees and especially dugongs (see Tables 6.2, 6.3, 6.5, 6.6 and 6.7), suggesting that sirenians may postpone breeding under certain conditions.

Table 6.7 *Comparison of reproductive parameters of mature female dugongs sampled from two indigenous fisheries in Torres Strait: Daru 1976–1978 (Marsh 1995) and Mabuiag 1997–1998 (Marsh and Kwan 2008).*

Year	Number of females	Confirmed pregnancies	All possible pregnancies
Apparent pregnancy rate ± SE[1]			
1976–1977[2]	35	0	0
1978–1979	75	0.09 ± 0.03	0.11 ± 0.04
1979–1980	47	0.19 ± 0.06	0.26 ± 0.06
1980–1981	29	0.24 ± 0.08	0.28 ± 0.08
1981–1982	17	0.35 ± 0.12	0.35 ± 0.12
1998–1999	73	0.38 ± 0.06	0.46 ± 0.06
Annual pregnancy rate ± SE assuming 14-month gestation			
1978–1982	168	0.15 ± 0.03	0.17 ± 0.03
1997–1999	105	0.33 ± 0.05	0.40 ± 0.06
Inter-birth interval ± SE assuming 14-month gestation (see Table 6.6)			
1978–1982	168	6.8 ± 1.25	5.8 ± 0.97
1997–1999	105	3.06 ± 0.43	2.55 ± 0.39

[1] The apparent pregnancy rate is the proportion of mature females that were pregnant irrespective of gestation period.
[2] 1976–1977 data based on anecdotal information (Hudson 1986).

Studies of large terrestrial herbivore populations have shown that their population dynamics, particularly age at first breeding and breeding frequency, may be strongly influenced by uncontrolled factors in the environment (environmental stochasticity) and population density (both influences can be difficult to tease apart – Saether 1997; Galliard *et al.* 2000). Delaying breeding is a common response of long-lived iteroparous species to adverse environmental conditions and high population density (Galliard *et al.* 2000). Variation in the age at first reproduction is an important population regulatory mechanism in large mammals, including marine mammals (Fowler 1984) and terrestrial herbivores (Galliard *et al.* 2000) and is thus a useful index of changes in the reproductive potential of populations (DeMaster 1981). Age at first reproduction is ultimately determined by phylogenetic and life history traits such as body growth, development and longevity (Harvey and Zammuto 1985; Flowerdew 1987). Nonetheless, nutrition and thus body condition affect the onset of sexual maturity in several species of mammals (Flowerdew 1987; Saether 1997), and the fecundity of young females shows strong variation with changes in environmental conditions and population density (Galliard *et al.* 2000).

The evidence for Florida manatees delaying reproduction is equivocal. Necropsy data suggest that the annual pregnancy rate of Florida manatees significantly increased (almost doubled in some scenarios) between 1976 and 1991 (Marmontel 1995). The reason for such change is unknown but may be due, at least in part, to sampling issues in the early years of the Florida manatee carcass salvage programme. An increase in the annual pregnancy rate is expected in exploited populations as a density-dependent response and may be occurring in dugongs in Torres Strait, as discussed below. However, a density-dependent response is an unlikely explanation for Marmontel's observations, because the increases in numbers of Florida manatees counted in aggregation areas in winter suggest that the population was increasing (O'Shea and Ackerman 1995; Craig and Reynolds 2004), as does population modelling (see 'Assessing population dynamics through modelling of life history parameters' below). In contrast to Marmontel's results, Reynolds and Wilcox (1994) reported a decline in the percentage of calves in winter counts of manatees at some power plants during a ten-year period ending in winter 1991–1992, particularly at the Atlantic coast aggregation sites. Nonetheless, more recent percentages during exceptionally cold winters have been very high (J. Reynolds, unpublished). However, these percentages may be biased and an unreliable index of the situation. Kendall et al. (2004) used a sophisticated modelling approach to estimate the conditional breeding probability for manatees on both the north-west and Atlantic coasts of Florida. Their graphs of breeding probability in two-year intervals from 1982 to 2000 do not show major changes, and the best fitting model included constant probabilities for producing a calf.

The evidence for dugongs delaying reproduction is much stronger than for the Florida manatee. Marked temporal fluctuations have been documented in the pregnancy rate, age at first reproduction in both sexes, size at which sexual maturity is reached (see Tables 6.2 and 6.3; Marsh and Kwan 2008) and the incidence of reproductively active males (Marsh 1995, but see Kwan 2002). These fluctuations seem to track major changes in the status of the dugong's food supply, which is subject to episodic diebacks which are often associated with extreme climatic events (Johannes and MacFarlane 1991; Preen and Marsh 1995; Poiner and Peterken 1996; Marsh and Kwan 2008).

The strongest evidence for fluctuations in the dugong's reproductive rate comes from Torres Strait, a globally important habitat for the dugong (Marsh et al. 2002; see also Chapters 8 and 9). Information on dugong life history was acquired from specimens obtained from dugongs as they were butchered for food by indigenous hunters at two major dugong hunting

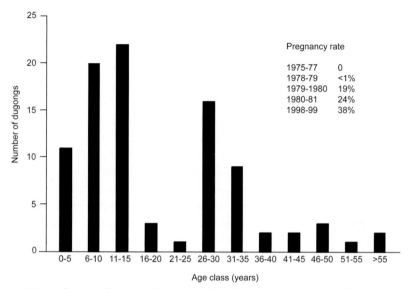

Figure 6.7. Age–frequency histogram of 92 female dugongs sampled from landings from the indigenous fishery at Mabuiag Island in 1997–1999. The females were aged on the basis of the growth layer groups in their tusks. The large changes in the apparent pregnancy rate with time in dugongs sampled from landings from the indigenous fishery is also provided. Redrawn by Shane Blowes from Marsh and Kwan (2008), with permission.

communities in Torres Strait: Daru in 1978–1982 (a time of seagrass die-back and recovery) and Mabuiag Island in 1997–1999 (when seagrasses were abundant) (Marsh and Kwan 2008). Dugongs sampled in 1997–1999 had their first calf at younger ages (minimum of six compared to ten years) (Table 6.2) and more frequently than those sampled some 20 years before (Tables 6.6, 6.7). Pregnancy rates increased monotonically during 1978–1982 (Table 6.7; Figure 6.7), coincident with seagrass recovery. The age distribution of the female dugongs collected in 1997–1999 at Mabuiag Island also suggested a low birth rate between 1973 and 1983 and/or a high level of mortality for calves born during that period (Figure 6.7).

These results suggest that the life history and reproductive rate of female dugongs are adversely affected by seagrass loss. However, the effect of seagrass loss cannot be separated from a possible density-dependent response to changes in population size in Torres Strait. There is some support for both explanations in the data from Torres Strait. The accounts of Bernard and Judith Nietschmann, who spent a year on Mabuiag Island studying dugongs, beginning in July 1976 (Nietschmann and Nietschmann

1981; Nietschmann 1984), and the oral history evidence recorded by Johannes and MacFarlane (1991) indicate that a high proportion of dugongs caught in Torres Strait during the 1970s were lethargic, with limited and poor-tasting fat. The Islanders attributed this unusually high proportion of 'wati dangal' to inadequate food availability (Johannes and MacFarlane 1991). This conclusion is supported by the oral history evidence of Islanders and the observations of Nietschmann (1984) that the stomach contents of the 'wati dangal' he collected contained larger amounts of brown and green algae than the largely seagrass-eating 'malu dangal' (deep-water dugongs which are fatter and considered good eating by Islanders). Evidence from other areas suggests that dugongs eat algae and more fibrous species of seagrass in greater quantities when seagrass is in short supply (Spain and Heinsohn 1973; Marsh *et al.* 1982; see Chapter 4) and that they are not well adapted to using algae or fibrous seagrasses as a food source (Marsh *et al.* 1982). The seagrass in Torres Strait allegedly recovered in the early 1980s, coincident with the increase in the dugong reproductive rate and in the proportion of reproductively active males (Marsh 1995), and was in good condition when Kwan (2002) collected her samples in the late 1990s (Long and Poiner 1997; Taranto *et al.* 1997).

Dugongs in poor condition are unlikely to breed. The Islanders claim that the females in the best condition are either pregnant or those seen mating, feeding or travelling with a male (Nietschmann and Nietschmann 1981). Marsh and Kwan (2008) showed that the mean fat depth differs significantly among female reproductive classes. Pregnant females were fatter than the other reproductive classes, providing empirical confirmation of Islander traditional knowledge.

The changes in the life history parameters of dugongs in Daru during 1978–1982 (Table 6.7; Figure 6.7) were also coincident with the collapse of the Daru dugong fishery, suggesting possible density-dependent responses to population decline. As discussed above, fecundity in dugongs from Daru increased monotonically from 1978 to 1982 (Table 6.7). The size of the catch went from 66 in 1978, peaked at 212 in 1979, then to 70 in 1981 and 18 in 1982 (Hudson 1986). The fishery was officially closed as a result of a ban on the sale of dugong meat in 1984. Thus, the change in fecundity observed across 1978–1982 could be a density-dependent response to the population being reduced by the fishery. Nonetheless, the significant increase in dugong fecundity in 1978–1979 and 1979–1980 is probably too rapid to be explained by a density-dependent response to population decline in the absence of environmental effects of food availability and quality. There are ongoing concerns that dugongs have been overharvested in the Torres Strait region since at least the 1980s (Hudson 1986; Marsh 1986; Johannes and MacFarlane 1991; Marsh *et al.* 1997; Chapter 7). Heinsohn *et al.* (2004) and Marsh *et al.* (2004) used two independent modelling techniques that suggest that the current levels of anthropogenic mortality of dugongs in Torres Strait are unsustainable (Chapter 8). We suspect that both seagrass dieback and the overharvest of dugongs have contributed to the changes.

The loss of more than 1000 km^2 of seagrass in Hervey Bay in southern Queensland in 1992, following two floods and a cyclone, demonstrated that when their food supply fails, individual dugongs variously exhibit one of two functional responses. They may emigrate from the affected area or remain, consuming any remaining seagrass and low-quality food such as algae, and postpone breeding. Twenty-one months after the extreme weather in early 1992, the regional dugong population was reduced to an estimated 500 ± SE 126 animals from an estimated 2206 ± SE 420 animals in 1988 (Preen and Marsh 1995). Although unprecedented numbers of dugong carcasses were found along 1500 km of coastline in 1992 and 1993, the dugong population of Hervey Bay recovered too fast for this population reduction to be caused by mortality in the absence of substantial immigration, reaching 2547 ± SE 410 in 2005 (Marsh *et al.* 2006). The dugongs that stayed in Hervey Bay delayed breeding and/or suffered high calf mortality. The proportion of the dugong population classified as calves during aerial surveys declined from 22% in 1988 to 2.2% in 1993 and 1.5% in 1994 (Preen and Marsh 1995; Marsh and Corkeron 1997), suggesting that the impacts of habitat loss on fecundity/calf survivorship may last several years. The percentage of calves then increased concomitant with the seagrass recovery, reaching 14.5% in 1999, 8.2% in 2001 and 7.2% in 2005 (Marsh and Lawler 2001; Grayson *et al.* 2008).

The data from Hervey Bay demonstrate that dugong calf counts are impacted by local weather events. El Niño episodes (which are associated with negative values of the Southern Oscillation Index and positive values of the NINO 3.4 sea surface temperature index) are usually accompanied by regional-scale responses, including sustained warming of the central and eastern tropical Pacific Ocean, a decrease in the strength of the Pacific trade winds, and a reduction in rainfall over the whole of eastern and northern Australia. The number of green turtles that breed each year in northern Australia increases two years after an El Niño episode; the time lag allows the female turtles to accumulate the fat required to breed (Limpus and Nichols 1988). This result is of interest to sirenian biologists because of the overlap in the diets of manatees and dugongs with that of green turtles (see

Chapter 4). Grayson *et al.* (2008) used the aerial survey records collected in Australia since the mid-1970s to quantify spatial and temporal patterns in percentage calf counts as a surrogate measure of dugong fecundity/calf survivorship. They used generalised linear models (GLMs) to investigate whether changes in percentage calf counts are associated with long-term weather indices. Large temporal and spatial differences were evident in the proportion of calves seen during aerial surveys. In regions that were comprehensively surveyed, the proportion of calves ranged from 0.2% in the northern Great Barrier Reef region in 1978 to 22% in Hervey Bay in 1988 and 1992. The modelling provided evidence that the calf counts were affected by the El Niño/La Niña cycle, lagged by two years in the northern Great Barrier Reef region, presumably as a result of: (1) the negative impact of increased turbidity on some of the coastal seagrass species eaten by dugongs (Preen *et al.* 1995; Longstaff and Dennison 1999); (2) the need for dugongs to be in good condition prior to and during pregnancy and lactation (Kwan 2002; Marsh and Kwan 2008); and (3) the life history of the dugong, with pregnancy lasting 12–14 months and lactation up to about 18 months or more (as discussed under 'Pregnancy and birth' above). The analysis suggests that similar relationships exist for other regions, but the data are not yet adequate to confirm them statistically (Type 2 error). The negative impact on dugong life history of the loss of coastal seagrass associated with exceptionally high rainfall and other extreme weather events is of major concern when considering the impact of climate change on dugongs (Chapter 7).

Best (1983) provides anecdotal evidence suggesting that Amazonian manatees may also be forced to delay reproduction during prolonged dry seasons when they may have to fast for up to almost seven months, leaving them in an emaciated condition similar to the 'wati dangal' described above. Best (1983) cautioned that large-scale deforestation in the Amazon may exacerbate the impacts of prolonged dry seasons on Amazonian manatees, a concern reinforced by Arraut *et al.* (2010) and discussed further in Chapter 7.

Evidence of reproductive senescence

As explained above, age determination studies indicate that both dugongs and manatees are very long-lived animals. Studies of terrestrial herbivores (Galliard *et al.* 2000) suggested that fecundity of old females might be expected to be lower and more variable than fecundity of prime-aged females, but there are no empirical data on reproduction at older ages in sirenians. Although there are reports of female manatees that have not given birth for long periods (O'Shea and Hartley 1995; Rathbun *et al.* 1995; Koelsch 2001) and old male dugongs whose testicular histology suggested prolonged aspermatogenesis (Marsh *et al.* 1984b), there is no evidence of a definite post-reproductive stage in the life cycle of female or male sirenians, similar to that recorded for some female cetacean species (Marsh and Kasuya 1986). Marmontel (1995) could find no evidence of age-related changes in the fecundity of Florida manatees older than four to five years, but warned that this might be an artefact of sample size. It may also reflect technical difficulties in estimating the ages of older manatees due to bone resorption. Although determining whether or not senescence in reproduction occurs in sirenians is of theoretical interest, the effect of this knowledge on our understanding of sirenian population dynamics is likely to be minor (see Eberhardt and O'Shea 1995; Runge *et al.* 2004) because so few sirenians live long enough to reach reproductive senescence, if it exists.

SURVIVAL

Estimation of survival in marine mammals is critical to understanding their population dynamics (Eberhardt and Siniff 1977). Adaptation of large mammals to a fully aquatic mode of existence appears to constrain reproduction to the fairly uniform litter size of one, long gestation periods, prolonged periods of calf dependency and the somewhat delayed sexual maturation characteristic of sirenians and cetaceans. This life history limits the reproductive potential of sirenians and cetaceans such that high survival of adults is required for population growth or stability (see 'Assessing population dynamics through modelling of life history parameters' below), as first pointed out by Eberhardt and Siniff (1977). Given the difficulties in estimating their population sizes and trends (see Chapter 8), application of survival rate estimates in a modelling context has proven to be the most practical means of inferring population status for the purpose of managing the Florida manatee population (Eberhardt and O'Shea 1995; O'Shea and Langtimm 1995; Langtimm *et al.* 1998, 2004; Runge *et al.* 2004, 2007a,b). Rigorous estimates of survival are not available for dugongs or for manatee populations other than in Florida.

Methods

Several methods are available for estimating survival in sirenians. All require great effort and long-term dedication for accruing the necessary data because of the lengthy life histories and complex annual cycles of these species. Survival in wildlife populations has typically been estimated using one of three approaches. One historical method involves the application of age-determination techniques to a sample of dead animals, and construction of a life table from which estimates of survival rates are possible (see Deevey 1947; Mech 1966; Spinage 1972; de la Mare 1986). This traditional approach has multiple drawbacks (Williams *et al.* 2002; Murray and Patterson 2006). Life table methods rely on certain key sampling assumptions that are difficult to validate and probably not often met. These assumptions (e.g. that population growth is stationary, that the age distribution of the population is stable and that sampling of the population is homogenous) will result in circular logic regarding population growth rates (Anderson *et al.* 1981; Williams *et al.* 2002). Most importantly, for purposes of statistical analyses and reliability, life table approaches are generally ad hoc in that they are not based on an underlying theoretically solid framework such as maximum likelihood theory (but see Udevitz and Ballachey 1998 for an exception). Without an underlying theoretical justification, life table analysis does not allow derivation of variance estimators or evaluation of other important sample properties required for models of population dynamics. One study of sirenians took the life table approach to estimating survival in Florida manatees while acknowledging some of these difficulties as caveats, in part because the analysis provided the only possible estimates for the entire population in Florida (Marmontel *et al.* 1997; see below). The resulting survival estimates were not dissimilar to those obtained using more statistically reliable approaches (see below), with population-wide adult survival at 0.91, first-year survival including neonates at 0.72 and second-year survival at 0.82.

A second technique for estimating survival involves tracking fates of individuals using radiotelemetry. Robust analytical procedures for estimating survival in wildlife using radiotelemetry are available (Williams *et al.* 2002; Murray and Patterson 2006). However, these methods are more efficient for smaller, less wide-ranging and more short-lived species than sirenians. A preliminary, simplified analysis of Florida manatee radio-tracking records (O'Shea and Langtimm 1995) suggested that major (and prohibitively costly and potentially dangerous), intensive capture, radio-tagging and tracking efforts would be required to arrive at estimates with precise confidence intervals, and that these attempts would be analytically complicated by the frequent loss of breakaway tags that is a necessary adjunct for manatee safety.

The third method available for estimating survival is based on the very firm statistical foundation of maximum likelihood theory, and involves the application of mark–recapture (sight–resight) analysis. Unlike calculation of simple return rates, this analytical method estimates survival based on adjusting for capture or sighting probability and relies on extensions of the general models first proposed by Cormack (1964), Jolly (1965) and Seber (1965). The method has been applied successfully to Florida manatees over several major study areas important to their management. The analysis uses scar patterns as marks and photographically documented re-sightings as recaptures. (In 1949 Moore (1956) first used scar patterns to identify individual Florida manatees for purposes unrelated to estimation of survival.) Statistical theory and available programmes and analytical methods have advanced rapidly in this area of wildlife population analysis. Survival can now be modelled as functions of various individual covariates and group characteristics, with the highest ranking models selected using Akaike's Information Criterion (Lebreton et al. 1992; White and Burnham 1999; White et al. 2001; Burnham and Anderson 2002). New model applications are being developed and applied to Florida manatees (Langtimm et al. 2004; Langtimm 2009), particularly to model non-random temporary emigration, which if not corrected can lead to spurious downward trends in survival at the ends of long time series.

Survival in Florida manatees

As explained in Chapters 5, 7 and 8, Florida manatees form transient aggregations in winter at sites where water is warm. These include natural, clear artesian springs such as Crystal River and Blue Spring, artificially heated industrial effluents such as those produced by power plants, and passive thermal basins. Each winter, biologists photographically document the presence of individual manatees at these sites, usually from above water at industrial plants but including underwater photographs taken while snorkelling in clear water (manatees are less tolerant of approaching scuba divers than they are of snorkellers). Photographs are dated and subsequently coded, sorted and matched using an automated computer system and following strict criteria for inclusion of observations as representing unique individuals (Beck and Reid 1995; Langtimm et al. 2004). Survival is computed and modelled in Program MARK (White and Burnham 1999).

Langtimm et al.'s (2004) analysis was based on records of 1560 individual manatees monitored over periods ranging from 8–22 years, ending in about 2001 at the four designated management areas: Northwest Florida (primarily Crystal River and Homosassa Springs); upper St Johns River (primarily Blue Spring); the Atlantic coast; and Southwest Florida. Records were based exclusively on adults, with the exception of 115 calves and subadults monitored intensively at Blue Spring (Langtimm *et al.* 2004). Further details of the analytical procedures and their statistical and bio-logical assumptions are available elsewhere (Langtimm *et al.* 1998, 2004; Langtimm 2009). Re-sighting probabilities of manatees in these studies were high, ranging from 0.49 to 0.93 at the four areas.

The models estimated the annual apparent survival of adults because permanent emigration cannot be distinguished from mortality. Annual apparent survival varied between 0.91 and 0.96 among three regions sampled over a ten-year period ending in 1999, and in south-western Florida over a six-year period ending in 2000 (Langtimm *et al.* 2004). Model selection procedures showed that factors influencing adult survival differed by area, but that there were no important differences between males and females. In the Northwest group (n = 342 adults) survival was constant over a 20-year period, with the exception of lower survival in three years marked by intense hurricanes or severe storms (Langtimm *et al.* 2004; see also Langtimm and Beck 2003; Langtimm *et al.* 2006). Survival probabilities in the upper St Johns River group varied by age-class, with lower estimates for first-year calves (0.81, CL 0.73–0.87), intermediate estimates for second-year calves (0.91, CL 0.83–0.96) and no evidence for differences in survival probabilities of three- to five-year-old subadults from those of adults (0.96, CL 0.93–0.99). Adult survival estimates were also constant over time in the upper St Johns River population. Initial estimates of adult survival for manatees from the Atlantic coast had a negative trend with time. The result for this region has since been identified as a statistical artefact of temporary emigration (Runge *et al.* 2007a). Models with constant survival probabilities over time ranked highest for the Southwest subpopulation, where adult survival estimates were lowest at 0.90 (CL 0.86–0.93).

The survival analyses that have been conducted for Florida manatees using the longitudinal re-sighting approach have thus far led to the following generalisations (O'Shea and Langtimm 1995; Langtimm *et al.* 1998, 2004). Apparent annual survival probabilities of adult Florida manatees were constant over time (stable adult survival is characteristic of most large mammalian herbivores – Galliard *et al.* 1998, 2000), except for years with intense storms in the north-west region of Florida. In these years survival was lower in this region, in part possibly due to mortality but also perhaps due to displacement of some individuals by storms or their after-effects. This emigration may not be permanent, however, with some evidence that certain individuals may eventually make their way back to the study area (Langtimm *et al.* 2004; Langtimm 2009).

At the only site where estimation of calf and subadult survival was possible (upper St Johns River), apparent survival was lowest in the first year of life and second lowest in the second year, but subsequent subadult survival was similar to the adult rate (Langtimm *et al.* 2004).

Langtimm *et al.*'s (2004) analysis suggested that survival probabilities differed by region and were highest in the two regions (north-west and upper St Johns River) characterised by intensive management for manatee conservation, more extensive areas of habitat protection and lower inciden-ces of manatees killed by boats (see also O'Shea and Langtimm 1995; Langtimm *et al.* 1998). Adult survival on the Atlantic coast and in south-west Florida was estimated to be lower, but with wider confidence intervals in the latter area. Although these analyses suggested that the temporal trend for survival was negative on the Atlantic coast (Langtimm *et al.* 2004), more recent analysis by Runge *et al.* (2007a) provides survival estimates for the Atlantic coast comparable to those from the Northwest and the upper St Johns River subpopulations.

ASSESSING POPULATION DYNAMICS THROUGH MODELLING OF LIFE HISTORY PARAMETERS

The ability to model the population dynamics of sirenians has progressed slowly over the past 25 years and has been far outpaced by advances in theory and application of modelling tools for wildlife management. Progress in modelling has also been very variable among the four species of living sirenians. There has been no progress in this area of research for manatees in any part of the world outside of Florida, but the science has steadily improved for Florida manatees given the improvements over the years in estimating their life history parameters. This result is important because the ability to assess population trends of marine mammals based on surveys is limited (Taylor *et al.* 2007; Chapter 8), making modelling of reproduction and survival a very valuable tool for assessing status for this subspecies. Modelling of Florida manatee populations based on parameters of reproduction and survival has an interesting history (see Supplementary Material Appendix 6.2) and has become increasingly sophisticated.

A sophisticated stage-based approach is now used for modelling life history data of Florida manatees (Runge *et al.* 2004, 2007a,b). The model (Figure 6.8) characterises the population of females in seven stages: first-year calves; second-year calves; third-year subadults; fourth-year suba-dults; pre-breeders; adults with calves; and breeders (mature females

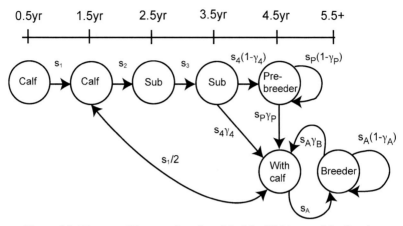

Figure 6.8. Diagram of the stage-based model of the life history of the female segment of the population of the Florida manatee, developed by Runge *et al.* (2004). The arrows govern transitions between the classes as a result of survival and reproductive events. In the matrix formulation, first-year calves are not tracked separately, but are inferred from the number of females with calves. Abbreviations: Sub (subadult); *s* (survival rate); (*γ*) breeding rate. s_1 is the probability a first-year calf survives to become a second-year calf; γ_P is the probability that an adult female that has not yet given birth to a calf breeds and successfully gives birth within the next year, given survival until that time. Prebreeders that survive either give birth to a calf (with probability γ_P) or remain as prebreeders. Females with a first-year calf that survive become breeders the next year (with probability = 1.0), regardless of whether they wean the calf after the first year. The model does not allow females to have calves two years in a row, reflecting the physiological limitations imposed by the length of pregnancy and early dependence of the calf. Breeders (without calves) that survive to the next year either give birth to a calf (with probability γ_B) or remain as breeders. A female with a first-year calf gives rise to a second-year calf (weaned or not weaned) in the next year with probability $s_{1/2}$, reflecting the probability of calf survival and an even primary sex ratio The litter size is assumed to be one calf. Redrawn by Shane Blowes from a diagram in Runge *et al.* (2004), reproduced with permission.

with second-year calves). Manatees make transitions among stages with rates based on nine parameters that are the survival probabilities and breeding probabilities of the respective stages. These probabilities are estimated based on re-sighting histories of distinctive individuals at winter aggregation sites grouped into the four recognised regional subpopulations (see 'Survival in Florida manatees' above and Chapter 8). The estimates of the nine parameters included confidence intervals that reflected sampling variation. Population growth rate is the primary model output, and distributions of the uncertainty associated with growth rates were determined through Monte Carlo simulation with 10 000 replications. The variance in growth rate estimates was further decomposed into contributions from the variances in the parameters. As explained below, elasticity and sensitivity analyses were then used to determine the relative importance of these life history parameters in influencing population growth rates. Elasticity analysis reveals the proportional changes in growth rate in response to proportional changes in breeding and survival probabilities, thus helping decide which of these parameters are most important to target in managing the population for conservation. The model can be considered to include environmental stochasticity implicitly because the reproduction and survival probability estimates encompassed a long time series of data when various environmental factors were operating on the populations (Runge *et al.* 2004). A key advantage of this modelling approach is that several critical components for population modelling were employed a priori and resulted in a full integration of the modelling, estimation methods and field work (Runge *et al.* 2004). Results were retrospective and describe population growth over the recent past (approximately ten-year periods), reflecting the status at the time of analysis (Runge *et al.* 2004). The model did not attempt to forecast future trends but could be modified to do so.

As will be discussed in Chapter 8, annual population growth rates as modelled by Runge *et al.* (2004) vary among the four areas in Florida and have been refined as methodology has improved (Runge *et al.* 2007a). In our opinion, the upper bounds to the growth rates for the upper St Johns River population (8.1% p.a.) probably reach a near biological maximum for Florida manatees (and possibly all extant sirenians) under modern environmental conditions. This maximum is not being achieved in the other areas of Florida, probably because manatees in these areas are experiencing more negative influences on their life history traits (Chapter 8), rather than experiencing a density-dependent decline in growth rates. Runge *et al.* (2004) performed an elasticity analysis (Figure 6.9) that strongly confirms the conclusions of previous modelling efforts (Packard 1985; Eberhardt and O'Shea 1995; Marmontel *et al.* 1997). In all regions studied, adult survival has the highest potential effect on growth rate, followed to a much lesser extent by subadult survival. All other life history parameters, including calf survival and adult breeding rate, have lesser effects on population growth rate.

The life history modelling approach for Florida manatees outlined above continues to be built upon for management purposes. A stochastic population projection model has been developed to serve the needs of managers interested in likely outcomes of various scenarios on life history parameters and population growth or extinction in separate regions of Florida over time

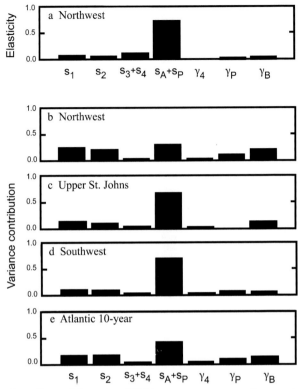

Figure 6.9. Effects of the various life history parameters on the growth rate of the Florida manatee population based on the model developed by Runge *et al.* (2004). In the graphs, the third and fourth bars refer to the combined effects of s_3 and s_4, and s_A and s_P, respectively. (a) Elasticity of growth rate to changes in the underlying life-history parameters, calculated from the mean matrix for the Northwest region of Florida. (b–e) Relative contributions to uncertainty in growth rate (λ) from uncertainty in the life history parameters of manatee management regions in the (b) Northwest, (c) upper St. Johns River, (d) Southwest, and (e) Atlantic Coast. For abbreviations, see Figure 6.8. Re-drawn by Shane Blowes from a diagram in Runge *et al.* (2004), reproduced with permission.

(Runge *et al.* 2007a,b). This model provides managers with a common tool for evaluating the effects of various threats to the population, such as the loss of warm-water refuges or occurrence of red tides (Runge *et al.* 2007a,b; Chapter 9). The continued refinement and increasing application of such modelling can be expected based on the longitudinal basic field research that generates reliable life history parameter estimates.

The background information available for dugongs is much more limited, but some modelling of their populations has also been conducted using some survival estimates of Florida manatees or African elephants (*Loxodonta africana*) as proxies (Marsh *et al.* 1984c; Marsh 1986, 1995; Heinsohn *et al.* 2004). Based on the lowest possible mortality schedule using African elephant survival data from life tables (Hanks and McIntosh 1973) and the most optimistic reproductive parameters available for dugongs at the time, Leslie matrix models suggested a maximum annual growth rate for a dugong population of 6.3%. Application of the reproductive parameters estimated from the large sample from the Daru population (Tables 6.2, 6.6, 6.7) at Torres Strait in 1978–1982 suggested a growth rate of only 2.4%. Subsequent analysis incorporating the lower age at reproduction and shorter calving intervals recently revealed by the anatomical study of dugongs at Torres Strait in 1997–1998 (Marsh and Kwan 2008), with proxy adult survival estimates from a growing population of Florida manatees, suggested a maximum rate of increase for dugong populations of about 5% (Marsh *et al.* 2004). The most serious shortcoming of all of these models and the population viability analysis of Heinsohn *et al.* (2004) is the lack of empirical data on dugong survival, and the maximum rates of increase calculated to date must be regarded as tentative. Most models have attempted to overcome this uncertainty by using a range of values for the salient life history parameters (e.g. Heinsohn *et al.* 2004; Marsh *et al.* 2004). Obtaining data on the survival of dugongs in the wild should be a priority for further research. The longitudinal study developed for dugongs in Moreton Bay, Australia, by Janet Lanyon and her co-workers (Lanyon *et al.* 2002, 2005b, 2006, 2009) will be particularly important in this regard.

LIFE HISTORY OF THE SIRENIA: COMPARATIVE OVERVIEW AND IMPLICATIONS FOR CONSERVATION

Compared with most species of mammals, including many of the large terrestrial herbivores (Galliard *et al.* 1998, 2000), dugongs and manatees are characterised by considerable longevity, long gestation, litter sizes of one, long intervals between births, sometimes prolonged periods until sexual maturity, and high and temporally stable adult survival. The suite of life history traits sirenians exhibit are marked by commonalities with the most long-lived of those species of large mammals that have been intensively studied. Eberhardt (2002) related life history modelling similar to that carried out for Florida manatees to independent assessments of trend analyses (population growth rates calculated through estimates of changes in population size) for several species and populations of long-lived large mammals. His synthesis provides a unifying comparative paradigm for interpreting sirenian life history and its implications for management. The paradigm also has been demonstrated in several other species of large mammals, including monk seals, Weddell seals, sea otters, killer whales, feral horses, elk, moose, caribou, red deer, grizzly bears and giant pandas (Eberhardt 2002).

Uniformly, adult survival is the most important determinant of population growth in all these species. Population growth is least sensitive to survival to breeding age, and reproductive rate is intermediate in importance. Under conditions where there is no harvesting or other human-caused mortality, adult survival in these species should be in the range of 0.95–0.99 per year, with low variability. The paradigm also concludes that where populations are not limited by resources, early-age survival up to breeding age approaches that of adults. These are conditions that modelling and survival estimation studies show are approached in Florida manatees in areas with optimal protection.

The paradigm also states that as populations of long-lived vertebrates increase towards resource limitation, early-age survival decreases first, which may be accompanied by slower growth and maturation of young, resulting in a delay in age of first reproduction. This sequence of population responses was first noted over 30 years ago (Eberhardt 1977, 2002). Resource limitation can occur when stochastic environmental factors (e.g. cyclones, episodic changes in turbidity) decrease food availability or quality, as has been shown for seagrasses in dugong habitats (Preen and Marsh 1995; Preen et al. 1995; Poiner and Peterken 1996). Resource limitations can also be expected when sirenian populations increase to high densities relative to their food supply. The recent findings regarding small sizes, lowered ages at first reproduction and shorter inter-birth intervals in Torres Strait dugongs provide a striking example of a shift in sirenian reproductive parameters in response to an increase in resource availability (Marsh and Kwan 2008). In the past, sirenian biologists have assumed that parameters of reproduction in dugongs were more conservative in potential for growth than in manatees (Marsh 1995; O'Shea and Ackerman 1995). However, it is likely that estimates of dugong reproductive parameters from earlier studies may have been based on samples from populations under greater resource limitation (Marsh and Kwan 2008). The plasticity in reproductive traits of dugongs revealed by the new analyses from a population under relaxed resource limitation showed estimates of minimum age at first breeding lowered from about 13 years to 7 years, an approximate doubling of annual pregnancy rates, and a lowering of inter-birth intervals from roughly six years to about 2.6–3.1 years (Tables 6.2, 6.7) (Marsh and Kwan 2008). These reproductive parameters for dugongs under excellent resource conditions are closer to those of Florida manatees than earlier estimates, indicating that life histories of these two sirenians may be more similar than previously assumed. It is unlikely that the vital parameters of African manatees, Amazonian manatees, dugongs in areas other than Australia, and West Indian manatees in areas other than Florida will be much different than those of Florida manatees and Australian dugongs. In the absence of comparable data for these other species and regions, the most prudent approach for managers will be to assume that similar parameters apply, and that their relative importance will also follow the Eberhardt (2002) paradigm for long-lived vertebrates.

Clearly, resource limitation has not been the case yet in Florida manatees, where both the high subadult survival and very early breeding ages predicted by the Eberhardt (2002) paradigm have been revealed in both carcass and observational studies. Indeed, reproductive parameters in most of the Florida study areas are probably constrained at phylogenetically determined anatomical and physiological limits. If conservation efforts are successful and density-dependent population regulation some day occurs, biologists and managers should expect changes in these parameters to become evident at some of the Florida study areas. Management might also use monitoring of life history parameters to ascertain if non-density-dependent resource limitation is taking place. For example, it has long been thought that seagrasses have declined throughout Florida, but that in some areas access to high-biomass, exotic, freshwater vegetation has provided an alternative food source for manatees (see Chapter 4). Availability of freshwater vegetation at the winter refuge in the north-west study area has shown a noticeable decline over the past several years (J. Reid, personal communication, 2009; USGS and USFWS, unpublished), while aerial counts of manatees and estimated growth rates from life history models have continued to increase or remain positive, presumably because of the continued availability of coastal seagrasses that are a consistently important component of the diet of Florida manatees outside this winter refuge (Chapter 4). If these two opposing trajectories intersect, breeding rates and early survival may be the first parameters to show declines. Continued monitoring will be necessary in order to ascertain and manage such effects.

The final phase of the Eberhardt (2002) paradigm based on the suite of life history parameters of large, long-lived vertebrates states that reproduction rates of females will decrease with resource limitation, followed by an ultimate response of reduced adult survival rates. This response appears to have been operative in the Torres Strait region when preferred seagrasses died back in the 1970s. Pregnancy was undetected in dugongs, and a noticeable gap in the age distribution of the population consistent with a lack of births during the dieback period was evident 20 years later (Figure 6.7) (Marsh *et al.* 2004; Marsh and Kwan 2008). Accrual of such long-term information about changes in life history traits can facilitate adaptive management of sirenian populations.

As discussed in Chapter 9, we believe that the life history parameters of sirenians dictate that the best management prescriptions will target actions that decrease human-caused mortality of adults, such as careful control of harvesting and restricting likelihoods of entanglement in nets, collisions with boats and crushing in flood gates (see Chapter 7). Such actions also may have a positive influence on other vital parameters: calf survival is dependent on survival of mothers up until weaning; and good habitat management that reduces human influences such as scouring of the substrate (Wright *et al.* 1995) and increased turbidity detrimental to plant growth may also increase habitat quality and the nutritional plane of breeding adults. In addition, management of sirenians should not neglect reproductive success and the survival of calves and subadults. In many large terrestrial mammalian herbivores, stochastic environmental conditions favouring these other life history attributes help populations increase (Galliard *et al.* 1998). However, it should be borne in mind that in many of these populations of large terrestrial herbivores, environmental variability is greater than in sirenians, adult survival somewhat lower, and most measures of fecundity higher (annual breeding pulses, shorter time to weaning, earlier female maturity and somewhat larger litters – Galliard *et al.* 2000).

We conclude that sirenian populations of at least several hundred individuals that are managed for high adult survival are likely to respond to stochastic environmental events (such as habitat loss caused by extreme storms, epizootics, poisonings or failure of multiple warm-water sources in Florida) by temporarily reducing recruitment or by emigration to less affected areas (e.g. Marsh *et al.* 2004). We are cautiously optimistic that if management acts to maintain high adult survival, the long-term population consequences of environmental stochasticity (within limits) may be reversible and less severe than some fear. Of course, this conclusion is limited to populations of at least several hundred individuals. Many sirenian populations are so small (Chapter 8) that demographic stochasticity is likely to have a major influence on their dynamics.

SUGGESTIONS FOR FUTURE RESEARCH

Runge *et al.* (2004) pointed out that a major research and monitoring goal for Florida manatees is the continued, long-term field assessment of quantitative life history traits, with increased efforts at reducing uncertainty in estimation of survival probability in all areas, closely integrated with future modelling and criteria for assessing population recovery. This approach will be enhanced by the development of new field and modelling techniques, especially techniques that reduce the uncertainty in parameter estimates and the comparative analysis of multiple data sets obtained using different techniques such as carcass analysis, photo-identification, genetics and aerial survey.

The measurement of vital rates will also be enhanced by the application of non-invasive techniques to identify individuals using their genetic signature and to determine their reproductive status, such as those based on measuring hormone levels in faeces or blood. Such studies should be extended to other populations of sirenians at several reference sites. Because of the magnitude and long-term nature of the research effort, such reference sites must be carefully selected on the basis of long-term research capacity and characteristics of the study population, such as its size, accessibility and the ease of identifying individual animals. As will be discussed in Chapter 8, mark–recapture techniques are labour-intensive, requiring multiple sampling occasions and a relatively high proportion of identifiable individuals; this means that the technique is logistically infeasible for most subpopulations of sirenians. At the least, the programme that has been developed over the past three decades for manatees in Florida must be maintained. The longitudinal study developed for dugongs in Moreton Bay, Australia, by Janet Lanyon and her co-workers (Lanyon *et al.* 2002, 2005b, 2006b, 2009) promises information of parallel significance to the Florida manatee programme. Such programmes will continue to provide insights into life history patterns of sirenians that will not only inform the management of the reference populations (see Runge *et al.* 2007a,b; Chapter 8), but also provide insights that will be applicable to management of manatees and dugongs in other parts of the world.

7

Threats

INTRODUCTION

The population biology of sirenians renders them particularly vulnerable to mortality as adults (Chapter 6). Threats to sirenians in various parts of the world have been documented extensively, especially those threats that cause death. In addition to ensuring the existence of adequate habitat, conservation of sirenian populations (Chapters 8 and 9) must include measures to reduce these threats and thereby facilitate the recovery of depleted populations. In this chapter we categorise and review the range of factors known to threaten sirenian populations, including those that are directly caused by humans and natural threats, and several that blur the line between these two categories: naturally occurring threats that are exacerbated by anthropogenic factors. Human-caused direct threats include hunting, incidental killing in fisheries, collisions with boats and ships and a variety of lesser known factors that can have cumulative impacts. Natural factors are limited to biotic threats such as predation (Chapter 5), severe climatic events, infectious diseases and macro-parasites. Diseases and parasites can operate in a density-dependent fashion, and their roles in population dynamics of marine mammals are not well understood. However, disease outbreaks can be devastating to small and isolated populations, and a background summary of what is known about diseases in sirenians is of importance for conservation planning. The dangers posed by diseases are not entirely natural: anthropogenic factors can stress populations, possibly affect immune system function and induce greater susceptibility to disease, and unwitting release or translocation of infected individuals can wreak havoc on immunologically naïve populations. Other naturally occurring threats that are exacerbated by anthropogenic factors include poisonings by harmful algal blooms and biotoxins, and cold stress near the margins of sirenian distributions. All these threats are operating against a backdrop of rising human population densities, future environmental degradation expected with global climate change, and an existing or potential loss of genetic diversity that may hamper the capability of sirenian populations to adapt over the long-term.

HUNTING

In their review of conservation and use of wildlife-based resources, Nasi *et al.* (2008) define hunting as 'the extraction of any wildlife, from the wild, by whatever means and for whatever purpose'. They point out that species of wildlife are extracted for food, trophies, medicines and other traditional uses, as well as pets. Individuals hunt wildlife primarily to eat or sell as food. A single manatee or dugong can provide a large quantity (100–200 kg or more) of succulent meat and offal, plus other products (Table 7.1), including fat or oil, hide, bone and tears (and tusks in the case of the dugong)

Figure 7.1. Tools used to capture West African manatees: (a) manatee trap from the Djilor Delta, Saloum, Senegal; (b) manatee net from Ntutu, Angola; (c) manatee hunter, Sebastian Domingos from Ntutu, Angola, with his harpoon; (d) box trap from Côte D'Ivoire, which is baited with cassava or mantioc root. The door at the end of the box is propped with a stick and weighted with a stone. When the manatee enters the box to eat the chum, it nudges the stick and the door shuts behind it. Photographs (a)–(c) by Lucy Keith; (d) by T. J. O'Shea; all reproduced with permission.

(Domning 1982b; Marsh *et al.* 2002; Rajamani *et al.* 2006). The coastal, estuarine and riverine habitats of sirenians (Chapters 4 and 5) make them accessible to humans with small canoes and boats, simple hunting platforms, traps, harpoons and nets (Box 7.1; Figure 7.1). Thus, it is not surprising that dugongs and manatees have been deliberately exploited for thousands of years. Indeed, there is also evidence that the hunting of Steller's sea cow by indigenous people over the last 20 000 years may have led to its disappearance around the North Pacific and contributed to the species' range reduction prior to its eighteenth-century discovery in, and eradication from, the Commander (Komandorskiye) Islands (Domning 1972, 1978a; Domning *et al.* 2007; Chapter 2).

Prehistoric and pre-European contact exploitation

Archaeological evidence confirms exploitation of dugongs and manatees by early humans, but at relatively few sites, a result which is not surprising given that much of the evidence must have been drowned by the Holocene sea-level rise (see Chapter 3). Nonetheless, these sites are geographically widespread, suggesting that many different human groups hunted sirenians in prehistoric times. For example, Sophie Méry and her co-workers (Méry *et al.* 2009) have described the remains of dugongs killed for food 6000 years ago on the small island of Akab in the United Arab Emirates. The dugong bones were precisely arranged, suggesting that dugong hunting was important to ancient rituals. This archaeological evidence is strikingly similar to sites in western Torres Strait, where excavations by Ian McNiven and his group have revealed that dugong hunting in the region dates back 4000 years, with major increases in intensity 2600 years ago (Crouch *et al.* 2007). At one site on Mabuiag Island, the remains of an estimated 10 000–11 000 dugongs caught between *c.* AD 1600 and *c.* AD 1900 have been documented (McNiven and Beddingfield 2008; Figure 7.2). (This region currently supports the largest dugong population in the world – Chapter 8).

Box 7.1 The contemporary importance of hunting as a threat to the species/subspecies of sirenians

	Current threat intensity[1]
Dugong	████
Amazonian manatee	████
Antillean manatee	████
Florida manatee	
West African manatee	████

[1] Shading intensity represents magnitude of threat.

Figure 7.2. Excavation of the Dabangai Dugong Mound on Mabuiag Island, Torres Strait. The mound is estimated to contain the remains of 10 000–11 000 dugongs. From McNiven and Bedingfield (2008). The scale bar is in 10 cm units.
Reproduced with permission of the publishers and the Chair of the Goemulgaw (Torres Strait Islanders Corporation) and Traditional Owner Terrence Wap.

McKillop (1985) documents evidence for prehistoric exploitation of West Indian manatees by Maya and other coastal peoples of the circum Caribbean area. Analysis of animal remains from an AD 400–700 midden at the Maya Island site of Moho Cay in Belize suggests that manatees were the focus of animal resource exploitation at this site (McKillop 1984). Other evidence of prehistoric exploitation of manatees has been reported from archaeological sites in the Antilles (Ray 1960; Wing and Reitz 1982), Honduras (Strong 1935) and Florida (Cumbaa 1980; see Lefebvre *et al.* 2001 and Laist and Reynolds 2005a for review).

Commercial exploitation
Commercial industries that exploited sirenians to supply both international and domestic markets developed in several countries following European settlement (e.g. Domning 1982b; Daley *et al.* 2008). Commercial use of meat and hides of Antillean manatees was recorded among seafarers and privateers extending to the 1600s (Lefebvre *et al.* 1989, 2001). Domning (1982b) reports that unknown quantities of meat from *Trichechus manatus* were shipped from Brazil to the Guineas and West Indies in the seventeenth century. For at least two centuries (1780–1974), Amazonian manatee hides

Figure 7.3. Commercial exploitation of dugongs in Queensland in the nineteenth century. From the collection of the State Library of Victoria, Australia; reproduced with permission.

were exported to southern Brazil and elsewhere for making glue and heavy-duty leather products such as machinery belts, hoses, gaskets and other items requiring heavy, durable leather (Domning 1982b). From 1954 to 1973, Amazonian manatees were heavily exploited for meat. Manatee meat preserved in its own fat, known as *mixira* (*michira*), was particularly valued.

Commercial fishing for dugongs by non-indigenous operators occurred intermittently at various places along the coast of Queensland, Australia, from 1847 to 1969 (Daley *et al.* 2008; Figure 7.3). The cottage industry was principally for dugong oil, which was sought for medicine (as a substitute for cod liver oil), cooking and cosmetics. The creation of this market for dugong oil is attributed to Dr William Hobbs, a Queensland Government Medical Officer, who exhibited samples at the 1855 Paris Exposition. Dugong hides, tusks, bones and meat were also sold; meat was sometimes given away as a by-product (Daley *et al.* 2008). Some Aboriginal missions in Queensland also earned income from dugong oil that was sold to the Queensland government for distribution to Aboriginal communities for medicinal use until the late 1960s (Daley *et al.* 2008).

Commercial industries for sirenian products generally ceased in the latter half of the twentieth century, when changed societal values led to banning of commercial exploitation and trade in most countries. By the time conservation laws were enacted, such exploitation was already headed for commercial extinction as the non-meat products were displaced by the increasing availability of plastics and pharmaceuticals (Domning 1982b). The use of sirenian body parts for purposes other than food is now largely limited to low-income countries.

Contemporary legal hunting

Although hunting is legally banned in most countries in the ranges of the dugong and the three species of manatees, there are important exceptions for the subsistence use of traditional people. For example, in Australia, the dugong hunting rights of 'Native Title Holders' (indigenous people who can prove continuous connections to their traditional land and sea country[1]) have been consistently upheld by the courts in a series of landmark decisions since 1993. Dugong hunting still has considerable cultural significance to coastal Aboriginal and Islander people across northern Australia (see Chapter 9), and dugong meat is considered by these groups to be the highest value traditional food (Marsh et al. 2004). In Torres Strait, the traditional fishery for dugongs is authorised by an international treaty, the Torres Strait Treaty, which was signed by Australia and Papua New Guinea in 1978. Subsistence dugong hunting is also allowed in the Pacific states of Papua New Guinea, Solomon Islands, Vanuatu and New Caledonia, where dugongs are still of great cultural and dietary significance (Marsh et al. 2002).

There are also some exceptions to the bans on hunting of Amazonian and West African manatees. For example, Brazilian law (No. 6905/98) allows subsistence hunting for those that can establish dire need (Lima et al. 2001 cited by Marmontel et al. 2012). In the Colombian Amazon, some indigenous communities have been granted some autonomy and allow hunting (Kendall and Orozco 2003 cited by Marmontel et al. 2012). The government of Nigeria permits a yearly killing quota of one or two manatees on application for cultural purposes during annual wrestling festivals (Obot 2002 cited by Awobamise 2008). In addition, some captures for symbolic local uses are permitted in Niger (Issa 2008). Nonetheless, concerns over indigenous hunting rights are apparently much less overt in developing countries than in developed countries such as Australia. This situation may change with increasing concern about the rate of loss of indigenous cultures (Chapter 9).

Poaching

Both dugongs and manatees are illegally poached throughout most of their ranges. Marsh (2008) estimates that dugongs are poached or harvested as food in 85–92% of their range (based on length of coastline). In most countries, dugongs are poached on an opportunistic basis by subsistence fishers because they represent a windfall of income from sale of meat and other products (Table 7.1).

The subsistence sale of meat and other products occurs virtually throughout the ranges of all three manatee species, with the exception of the United States (where illegal hunting was thought to be the major cause of declines in the first half of the twentieth century; e.g. Gunter 1942; Moore 1951b). Illegal hunting is still considered the main threat to manatees in the Amazon (Marmontel 2008). Only the Antillean subspecies of the West Indian manatee is still hunted (Deutsch *et al.* 2008). Poaching or hunting is still considered a major threat to Antillean manatees in Brazil, Colombia, Costa Rica, Cuba, Dominican Republic, French Guiana, Guatemala, Honduras, Mexico, Suriname, Trinidad and Tobago, and Venezuela (Deutsch *et al.* 2008), although it is apparently diminishing in some areas. O'Shea *et al.* (1988) and Correa-Viana *et al.* (1990) gained an impression of a general decrease in manatee hunting in Venezuela.

West African manatees are also illegally exploited in virtually all the countries in which they occur (Dodman *et al.* 2008). As outlined below, Moore *et al.* (2010) used interview surveys to collect information on incidental capture of marine wildlife in artisanal fisheries in three West African countries. Their interviewees volunteered information on the direct harvest of West African manatees using nets, traps and harpoons. In Sierra Leone, 12% of 693 respondents reported that they had collectively captured approximately 2100 manatees, with a single respondent claiming he had killed 500. In Nigeria, 648 interviewees reported a total of 180 manatee kills; 537 interviewees in Cameroon reported 290 kills.

Contemporary hunting methods

Dugongs mainly occur in open waters and are still primarily hunted from small boats; most are still hunted using a harpoon with a detachable head. Variants of traditional harpoons are now often made using modern components. For example, Australian indigenous peoples typically make the harpoon head from builders' nails (Marsh *et al.* 1981). The use of a traditional spear or '*wap*' is mandatory in the Australian waters of Torres Strait under the regulations of the Commonwealth *Torres Strait Fisheries Act 1984*. In Sabah, Malaysia, the fishers still enlist the help of indigenous medicine men ('*pawang*') who cast spells to locate dugongs (Marsh *et al.* 2002; Rajamani *et al.* 2006). In the late twentieth century in Palau and parts of South-east Asia, dugongs were hunted using dynamite (Brownell *et al.* 1981;

Table 7.1 *Products from dugong and manatee carcasses and their contemporary uses. The list is unlikely to be complete. Many uses are restricted to only some parts of the ranges of the various species*

Body Part	Use										
	Food	Folk medicine	Poison	Cooking/ lubrication	Lamp fuel	Aphrodisiac/ love potion	Amulets/ charms	Jewellery/ ornaments/ religious artefacts	Carvings	Leather products	Other
Meat and offal	D[1], AmM, AnM, WAM	D, WAM, AnM									
Genitals / mammary glands		WAM				D, WAM	D				
Lungs		WAM									
Oil/fat	D, WAM	D, AmM, AnM, WAM		D, AmM, AnM, WAM	WAM						D[2]
Tears						D	D				
Bile/bile ducts		D	WAM			D	D				D[3]
Blood		WAM									
Tusks		D				D	D		D		D[4]
Cheek teeth		D				D	D		D		D[5]

Body part				
Bones and/or skull	D, AmM, AnM, WAM	D, AmM, AnM	D, AnM	D[6], AmM[7], WAM[8]
Skin	D, AmM, WAM	D		D, AmM, AnM[9], WAM[9]
Hair	D			
External mucus	WAM			
Flippers		D		
Tail	WAM	D		

[1] D = Dugong; AmM = Amazonian manatee; AnM = Antillean manatee; WAM = West African manatee.
[2] Oil is mixed with salt to treat boat timbers (Hines *et al.* 2005b).
[3] Dugong tears can also be used in perfume (Marsh *et al.* 2002; Rajamani *et al.* 2006).
[4] Dugong tusks can be used to make pipes (Marsh *et al.* 2002; Rajamani *et al.* 2006).
[5] Teeth can be used as betel nut crushers (Hudson 1977).
[6] Dugong ribs are carved into spinning tops (Hines *et al.* 2008); dugong bones are made into hooks (Hudson 1977).
[7] Amazonian manatee scapulas are used to stir flour, bones are carved into utilitarian pieces, small pieces of rib are used as fish attractants (Marmontel 2008).
[8] Source of handles for walking sticks and spinning tops (Reeves *et al.* 1988).
[9] Leather is used to make whips.

Marsh *et al.* 1995). However, there is increasing overlap between directed hunting and incidental capture by fishermen (see 'Incidental capture in fishing gear' below).

The freshwater environments of manatees enable manatee hunters to use a much wider variety of technologies than those used by dugong hunters. West African manatees are exploited using an ingenious variety of methods, including harpoons, artificial feeding stations, elaborate systems of ambushing, nets, weirs, large box traps, drop traps using harpoons and snaglines with multiple hooks, hooks and corrals baited with cassava or mangrove fruits (Figure 7.1; Reeves *et al.* 1988; Powell 1996; Dodman *et al.* 2008; Powell and Kouadio 2008; Kouadio 2012). The methods make use of local knowledge of manatee biology, and manatees may be targeted when using movement corridors or when they are mating (Powell 1996). In Sierra Leone and Niger, West African manatees are intentionally hunted as pests because they damage rice crops (Dodman *et al.* 2008).

Amazonian manatee hunters also capitalise on seasonal changes in the availability of their prey. Fishermen close areas where Amazonian manatees are known to gather to mate in order to harpoon them (M. Leitao, personal communication to M. Marmontel, 2005, cited by Marmontel 2008). Some hunters hit the water with sticks to disorient the manatees and then harpoon them in channels and lakes closed with gill nets (Lima *et al.* 2001 cited by Marmontel 2008) or use captured calves to catch mothers (Marmontel *et al.* 2012). The hunters also take advantage of manatees isolated and entrapped by low water levels in extreme droughts (see Marmontel 2008). Traditional harpoons are still responsible for 55% to more than 90% of the take of Amazonian manatees (Marmontel *et al.* 2012); fishing traps with harpoons attached are used as well as harpoons in Peru (Reeves *et al.* 1996; Marmontel 2008) and on the Colombian–Peruvian border (Orozco 2001 in Marmontel 2008). Hunting records suggest that hunters use harpoons and nets almost equally when catching Antillean manatees in the Orinoco River in Colombia (Castelblanco-Martínez *et al.* 2009). Antillean manatee hunters also take advantage of the hydrological cycle (O'Shea *et al.* 1988; Castelblanco-Martínez *et al.* 2009); in some areas manatees are killed during the high-water season when hunting using a harpoon is easier because the signs of feeding in the vegetation allow the manatees to be tracked, and the manatees find it difficult to escape because the harpoon rope usually tangles in the vegetation. In other areas manatees are taken in isolated pools during the dry season, similar to the hunting of Amazonian manatees in the central and upper Amazon.

Contemporary uses

Numerous products are still obtained from manatees and dugongs and used in some parts of their ranges (Table 7.1). Almost all body parts are potentially used. Human consumption of meat and offal is the most wide-spread and enduring human use of all sirenians, but many body parts, particularly of dugongs and West African manatees, are used to make folk medicines (Marsh *et al.* 2002; Rajamani *et al.* 2006; Hines *et al.* 2005b, 2008) and the oil is widely used for cooking and sometimes even for lamp fuel. The bile ducts of West African manatees are used to make poison (Dodman *et al.* 2008) and dugong tears, genital organs, teeth, bones and flukes are used to make aphrodisiacs and love potions in some Asian countries (Marsh *et al.* 2002; Rajamani *et al.* 2006). The use of dugong and West African manatee body parts to make amulets and charms is widespread in Asia and West Africa, respectively (Marsh *et al.* 2002; Hines *et al.* 2005b, 2008; Rajamani *et al.* 2006; Dodman *et al.* 2008). In Mali and along the Benue River, parts of the penis of the West African manatee are used as a putative cure for impotency in men, and manatee skin is made into whips (Powell 1996; Dodman *et al.* 2008). In Sierra Leone, villagers consume all parts of the manatee carcass except for the heavy ribs (Reeves *et al.* 1988). The bones are used to make handles for walking sticks or spinning tops used in a local game called *cii*. Dugong ribs are also carved to make spinning tops in Cambodia (Hines *et al.* 2008); the atlas cervical vertebra has been used as a wristlet in Palau (Brownell *et al.* 1981; Marsh *et al.* 1995), fitted by breaking the bones of the recipient's hand. Similarly, oil from Antillean manatees is used as a liniment, skins are cooked and extracts used as medicine, and bones are pulverised and the powder used both externally and internally as cures for various ailments and conditions (Table 7.1). People along the Orinoco River and its delta in Venezuela keep certain bones of Antillean manatees as charms, particularly the ear bones (especially the tiny stapes) which are thought to be magical, attesting to the hearing ability and evasiveness of manatees that make them difficult to hunt (O'Shea *et al.* 1988; Correa-Viana and O'Shea 1992; Chapter 5). Bones of both manatees and dugongs are polished to resemble ivory and used to create objects of art (Figure 7.4).

Contemporary commercial values

Dugong and manatee meat is the aquatic equivalent of bushmeat in many countries. The value of a dugong or manatee is enhanced by the value of the various body parts. Rajamani *et al.* (2006) reported that a whole dugong sold

Figure 7.4. Antillean manatee bone polished to resemble ivory and carved into figurines by prisoners in Chetumal prison in Mexico in 1994. Benjamin Morales photograph, reproduced with permission.

for US$105 in rural Sabah, Malaysia. If the body parts are sold separately, the return can be much higher. Hines *et al.* (2008) were told that a pair of dugong tusks could be sold for US$100–150 and the meat for US$2 per kilogram in eastern Cambodia, where fishers had an average income of US$75 per month. The hunters of Amazonian manatees typically sell manatee products to neighbours and nearby communities, but the meat is sometimes sold in markets of local towns and there is some traffic between cities. Meat is sold untreated or as mixira or subproducts such as sausage. A 60 litre can of mixira sells for approximately US$168 (Lazzarini and Picanço, personal commication to M. Marmontel, 2005, cited by Marmontel 2008). Products from a large Amazonian manatee can provide some two months' wages for a subsistence fisher (Kendall 2001 in Marmontel 2008). West African manatee products are also valuable. In Senegal and Sierra Leone, manatee meat sells for US$2–3 per kilogram (Dodman *et al.* 2008). In the Côte d'Ivoire, a former manatee hunter reported that a slaughtered manatee could earn him US$350; he used to slaughter three manatees at the beginning of each school year to pay the secondary school fees for his five children (A. Kouadio, personal communication, in Dodman *et al.* 2008). According to sources reported by Idriss (in Dodman *et al.* 2008), manatee oil is highly valued as a traditional curative in Chad, and a litre of manatee oil can fetch US$300. A West African manatee typically yields 10–15 litres. Thus, sales of manatee oil can generate substantial income for impoverished people in Chad, one of the poorest countries in the world, ranking 175 out of 182 on the Human Poverty Index (UNDP 2009).

The sale of dugong meat and other dugong products is banned throughout Australia, including Torres Strait between Australia and Papua New Guinea; however, there are persistent reports of the meat being sold illegally and of informal bartering arrangements between indigenous communities living in remote areas and their diaspora living in cities and towns.

International trade

Trade in sirenian products is regulated or banned by the Convention for Trade in Endangered Species (CITES), as is explained in Chapter 8. At the time of writing (2010), the dugong, West Indian manatee, the Amazonian manatee and the West African manatee are all listed on CITES Appendices I or II. Despite these international conventions, there are contemporary reports of dugongs being traded between Sabah (East Malaysia) and the Philippines, allegedly involving Chinese–Malaysian merchants (Rajamani et al. 2006). Dugongs are sold across other national borders – for example, Cambodia–Vietnam (Hines et al. 2008). The Traffic Bulletin produced by CITES reported in 2009 that fishermen in Trang, Thailand, asked the government to control the illegal trade in dugongs because they claim that foreign fishermen hunt the animals and smuggle them out to Singapore (TRAFFIC 2009). The illegal hunt is reported to be carried out by fishermen from the Malaysian border who catch the dugongs using fish bombs (dynamite). Singapore is said to be the biggest market for the trade in dugongs for medicine and amulets. There are anecdotal reports of transnational sales of Amazonian manatee calves (Marmontel 2008; Marmontel et al. 2012). Powell (1996) reported that West African manatee meat and oil are illegally traded between Chad and Cameroon, and transborder incursions of poachers from Nigeria into Cameroon have been noted (Grigione 1996). As will be discussed in Chapter 8, wild-caught West African manatees from Guinea-Bissau are offered for sale on the internet for purchase by zoos.

INCIDENTAL CAPTURE IN FISHING GEAR

The unintended capture of marine mammals in both commercial and artisanal fisheries is considered to be a major and growing conservation issue (Read 2008; Moore et al. 2010). Species caught incidentally are often referred to as 'bycatch'. We have avoided this term because it may be inferred that the animals so-caught are discarded. In developing countries, most sirenians caught incidentally are retained for food.

All species of sirenians are subject to possible incidental capture in fishing gear throughout their ranges, and this is one of the largest and most widespread threats to manatees and dugongs (Box 7.2). The incidental take of sirenians typically results in utilisation and sale of the meat and by-products, helping to perpetuate the market. Dugongs are caught incidentally in purse seine nets, bay nets, trap nets, beach seine nets, shark nets, trawl nets, fish traps, bag nets, longlines and fish weirs (Marsh *et al.* 2002) and are killed as a result of cyanide and dynamite fishing (fish bombing) (e.g. Marsh *et al.* 2002; Rajamani *et al.* 2006). West African manatees are caught in fishing trawls, stationary funnel nets placed across the inlets of major rivers to catch shrimp and in fishing weirs made of sticks (Powell and Kouadio 2008). The 2006 status review of the Florida manatee concluded that over the previous decade there had typically been 10–15 rescues per year to disentangle manatees, mostly from crab trap lines (Haubold *et al.* 2006). The injuries resulting from entanglement can hamper the ability to survive and reproduce, and can lead to infections and death (Beck and Barros 1991).

The introduction of nylon nets during the 1960s led to a global expansion in the commercial use of gill nets. The new materials were cheaper and easier to handle, lasted longer and required less maintenance than natural fibres, and were almost invisible in water. Gill nets have been provided as aid to many developing countries. Outside Florida, where inshore gill netting has been banned since mid-1995 under Amendment 3 to the Florida State Constitution, capture in commercial gill nets is the most significant source of incidental capture of sirenians. In addition, in many areas gill nets are increasingly used to deliberately capture dugongs and manatees (Figure 7.5). As explained above, in most developing countries, dugongs and manatees are worth much more dead than alive, because the sale of their products represents several months' income to a

Figure 7.5. Dugongs drowned in nets off Cairns, Queensland, in 2010. The nets are believed to have been set by poachers. Australian Navy photograph, reproduced with permission.

low-income artisanal fisher and enforcement is non-existent or ineffective. Thus the incentive for an artisanal fisher to kill a dugong or manatee caught in his fishing gear and to sell the products is considerable. This problem is exacerbated when the target product of gill netting is also extremely valuable. For example, as discussed further in Chapter 9, Mozambique is one of the poorest countries in the world (Human Development Report 2009). In the Bazaruto National Park, gill netters target shark fin, which can be worth up to US$200 per kilogram; dugong meat is worth US$2 per kilogram in the local market (information provided to H. Marsh and J. Reynolds at the Dugong Workshop, Maputo, Mozambique, May 2009). The link between the capture of sirenians in nets and the illegal, unreported and unregulated shark-fin trade is unlikely to be limited to Mozambique, given the widespread concern about dugongs being caught in shark nets (Marsh *et al.* 2002) and the Asian focus of the shark-fin trade (Lack and Sant 2008).

Marsh (2008) estimates that dugongs are caught in drift and bottom-set gill nets in 87–99% of their range (based on length of coastline). Gill netting is also considered an increasingly significant threat to Amazonian manatees. Gill nets are deliberately used to catch adult manatees (see 'Hunting' above), a practice that has resulted in a concomitant rise in incidental calf mortality, which is now regarded as a major threat to Amazonian manatees in all range countries (Marmontel 2008). Most entangled young manatees drown; even if they survive, they are usually kept alive for later sale when they have increased in size and have more meat. Entanglement in fishing gear and the resultant stranding of live orphaned calves are also considered major threats to the Antillean manatee (Deutsch *et al.* 2008). Castelblanco-Martínez *et al.* (2009) suggest that incidental drowning in fishing nets causes almost half the mortality and wounding of Antillean manatees in the Orinoco River, Colombia, and deaths in artisanal fisheries occur virtually everywhere throughout their range (Lefebvre *et al.* 2001). Incidental killing of manatees occurs in many areas in West and Central Africa (Dodman *et al.* 2008; Powell and Kouadio 2008), often in nets used for catching sharks.

There are few data on the incidental capture of marine mammals in small artisanal fisheries (Read 2008). With artisanal fishers comprising seven-eighths of the world's fishers (Pauly 2006), this knowledge gap presents a major challenge to threatened species conservation and sustainable fisheries initiatives. The most extensive and robust data set of relevance is for dugongs caught in shark nets set for bather protection by the Queensland government at popular swimming beaches along the urban coast between Cairns (16.5° S) and the Gold Coast (28° S). Eight hundred and thirty-seven dugongs were caught between 1962 and 1992 (Anonymous in Marsh *et al.* 2005) and the catch per unit effort in 1999 was only 3.1% of that in 1962 (Figure 7.6).

In an attempt to quantify the rate of incidental capture of marine wildlife in artisanal fisheries in developing countries, Moore *et al.* (2010) collected data from interviews with a total of more than 6100 fishermen in seven developing countries: Tanzania, Comoros, Malaysia (Sabah) in the range of the dugong; Jamaica (Antillean manatee); Sierra Leone, Nigeria, Cameroon (West African manatee). The incidental catch of sirenians occurred almost exclusively in gill nets, but was also reported for hook-and-line gear in the Comoros and Cameroon. The incidental catch of dugongs (in Tanzania, Comoros and Sabah) and Antillean manatees (Jamaica) was rare, with many fishermen indicating that the capture of a sirenian was a once-in-a-lifetime event. This result is likely to reflect the low densities of sirenians in most of these range states. (Jaaman *et al.*'s (2009) estimates for Sabah are much higher: 479 (95% CI 434–528) per annum). However, the proportion of fishers reporting an incidental catch of West African manatees was generally higher; up to 18% of fishers in Nigeria (Moore *et al.* 2010). In most countries, the incidental catch of sirenians was described as a localised phenomenon (e.g. associated with specific lagoons and river mouths), making estimating total takes impossible with the survey technique used. Nonetheless, the fact that the take tended to be localised increases the potential of using spatial risk assessment techniques to reduce incidental

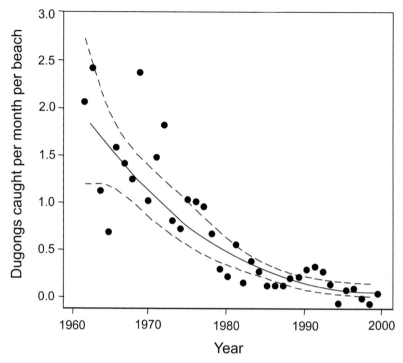

Figure 7.6. Profile of the annual estimated mean numbers of dugongs caught in shark nets set for bather protection in Queensland for the period 1962–1999 showing a strong overall decline in the number of dugongs caught per month per beach. The confidence bands have 95% pointwise coverage. The estimated rate of decline averages 8.7% per year. From Marsh *et al.* (2005), reproduced with permission.

catch (Chapter 9), if the food security incentives to keep and sell the take can be addressed.

INTERACTIONS WITH VESSELS

Accidental killing and wounding of sirenians in collisions with boats was first noted in Florida during the 1940s (O'Shea 1988). The threat grew considerably with time as boat motors increased in size and power, the number of vessels proliferated and new hull designs allowed high-speed-vessels to cruise through shallow waters. Although the scope of the problem may be best known and most well studied in Florida, the issue is more widespread than generally appreciated, and it is likely that the threat to other sirenian populations will grow as long as human activities and habitat encroachment continue to expand (Box 7.3). For example, deaths of

	Current threat intensity[1]
Dugong	
Amazonian manatee	
Antillean manatee	
Florida manatee	■
West African manatee	

[1] Shading intensity represents magnitude of threat.

sirenians from collisions with boats have been reported from Asia (Hines *et al.* 2012a), Africa (Marsh *et al.* 2002), Australia (Hill *et al.* 1997; Hodgson and Marsh 2007; Eros *et al.* 2007), the Caribbean (Mignucci-Giannoni *et al.* 2000) and Central and South America (O'Shea *et al.* 1988; Castelblanco-Martinez *et al.* 2009; see also Deutsch *et al.* 2008).

Increases in mortality and injury due to boat strikes could have serious impacts on the small and patchy populations of manatees and dugongs found in many countries (Chapter 8). Dugongs living along the urban Queensland coast in Australia also appear to be at increasing risk of collision, as the number of registered boats has increased at an average of 4.45% per year over the last five years to almost 230 000 registered recreational marine vessels in Queensland on 30 June 2009 (Blackman and Jones 2010) Although watercraft-related injury and mortality of dugongs in Queensland is much less than for manatees in Florida, the deaths of 26 dugongs were attributed to collisions with boats between 2000 and 2008; 12 of these were in Moreton Bay near Brisbane (StrandNet 2010; Figure 7.7). Here, we primarily consider threats from boat collisions as studied in Florida, where the problem is most acute and best documented.

Impacts of watercraft on Florida manatees

Three decades ago, Hartman (1979) noted that collisions with watercraft represented one of the greatest conservation threats to Florida manatees. Indeed, between 1979 and 2004, despite the focus on this threat, 1253 manatee deaths in Florida were documented as a result of the impact of a boat collision, being cut by the propeller or both; in many cases the impact alone caused death without wounding by propeller blades. This statistic represents about one-quarter of the total manatee mortality (5033) documented in Florida over 1979–2004 (Rommel *et al.* 2007). These figures *dramatically under-represent* the true impact that watercraft can have, directly

Figure 7.7. Dugong cow and calf killed by collision with a ferry in Moreton Bay, Queensland. Rachel Groom photograph, reproduced with permission.

or indirectly, on sirenians. Many of the carcasses recovered in Florida died from causes that were not determined. Runge *et al.* (2007b) calculated that if these deaths occurred at the same proportion as in the known-cause categories, then 50% of the manatee deaths on the Atlantic coast of Florida from 1986 to 2000 would have been caused by collisions with boats. Although this percentage may be an overestimate, the adverse impact of watercraft on Florida manatees is certainly considerable. In addition to death, indirect impacts of watercraft considered below include: serious injuries which can affect reproduction and health of individuals over the long-term; increased water turbidity that can affect seagrass productivity; scarring of seagrass beds, which also impairs productivity; disturbance of manatees, resulting in displacement from foraging habitat; and possible separation of females and dependent calves (Reynolds 1999). It is important to note that sirenians are not the only resource impacted: effects on seagrass meadows can also detrimentally affect fisheries' productivity and ecosystem health. Although the issue is often portrayed by members of the media as 'manatees vs boats', the reality is far more complex.

The presence of an excellent and comprehensive carcass salvage programme in Florida (e.g. Lightsey *et al.* 2006; Rommel *et al.* 2007) has permitted analyses that shed considerable light on the extent of mortality, the circumstances and locations associated with collisions and deaths, and

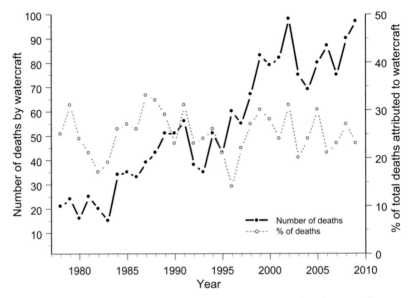

Figure 7.8. Number of Florida manatees killed by watercraft each year and as a proportion of total manatee deaths. Graph drawn by Rhondda Jones. Reproduced with permission.

possible mitigation. Pathologic descriptions of the extensive traumatic lesions inflicted by collisions can be found in Bossart (2001), Bossart *et al.* (2004), Lightsey *et al.* (2006) and Rommel *et al.* (2007). Figure 7.8 summarises the annual levels of watercraft-related deaths and total mortality of Florida manatees (Marine Mammal Commission 2009; Florida Fish and Wildlife Conservation Commission 2010c,d). Lower numbers from the early to mid-1970s likely reflect to some extent the lack of public profile of the carcass salvage programme at that time. The overall increase in watercraft-related mortality until 1995 was 9.3% per year (Ackerman *et al.* 1995). Remarkably, the proportion of total deaths due to collisions with boats has showed low variation over a 32-year time span, averaging 25% annually (95% CI 23.4–26.7%, CV 18%), with lower proportions occurring primarily in years where unusual mortality from sources such as cold stress or red tide took place (e.g. 1982 and 1996 – see related sections below). However, shifts in the predominance of boat-related mortality have occurred over the years in broad geographic regions as coastal development in those areas increased (O'Shea *et al.* 1985; Ackerman *et al.* 1995). Boats continue to increase in number, power and ability to speed through shallow depths (Wright *et al.* 1995). Thus a constant proportional fraction of mortality despite conservation efforts and regulation could ultimately be analogous to the impact of increasingly sophisticated catch technologies in driving collapses of fisheries (Eberhardt and O'Shea 1995). Even though the Florida manatee population has grown substantially during the past 30 years (Chapter 8), this constant proportional loss suggests a 'technology creep' that does not bode well for the future. Furthermore, most manatees killed by boats in Florida are adults (O'Shea *et al.* 1985; Ackerman *et al.* 1995), and life history modelling shows that adult survival is the most important parameter in maintaining positive population growth rates (Chapter 6). Indeed, models comparing the impacts of various mortality sources show that reduction in deaths from boats will have a greater positive impact on future Florida manatee populations than reduction in any other mortality factor (Runge *et al.* 2007a,b; Chapter 9).

Several authors (e.g. Beck *et al.* 1982; Wright *et al.* 1995) have analysed propeller-inflicted wounds on manatees to attempt to understand which sizes of boats posed the greatest risk. Most recently, Rommel *et al.* (2007) conducted forensic analyses of such wounds on dead manatees and con- cluded that a disproportionate number of the lethal vessel interactions involve large boats more than 12.2 m long. This size class of boat represents only 2% of the more than one million registered boats in Florida, but accounted for 58 of 115 (or 50.4%) of the cases assessed by Rommel *et al.* (2007). Most collisions are not reported by boat operators (Lightsey *et al.* 2006), who allegedly are often unaware of the nature of the object they have struck. Calleson and Frohlich (2007) reviewed the circumstances surround- ing the 21 collisions between 1978 and 2006 where the responsible boats were known. Except for two incidents in which tug boats were involved, *all* cases involved boats travelling at speeds ranging from 24 km/hour to 64 km/hour (15–40 miles/hour).

Most analyses of the effects of boat collisions on manatees have concentrated on the wounds and scars left by propellers. Nonetheless, more than half of the total watercraft-related mortality has been due to impact (Wright *et al.* 1995; Lightsey *et al.* 2006), which can produce massive internal injuries without wounds by propellers. Additionally, despite their density and heaviness, manatee bones are more brittle and have lower fracture toughness than bones of other mammals; they are, thus, especially susceptible to breakage from impacts of boats and sub- sequent pathology (Yan *et al.* 2006a,b; Clifton *et al.* 2008a,b). Mitigation of the problem must involve more than just shrouding the propeller blades with guards; the solutions must focus on avoidance of high-speed- impacts altogether.

O'Shea *et al.* (2001) documented that many Florida manatees survive multiple collisions, sometimes with grotesque wounding and maiming. Calleson and Frohlich (2007) stated that two manatees necropsied by the Florida Fish and Wildlife Conservation Commission had scars indicative of more than 50 encounters with boats. Thus, sublethal injuries to manatees are common, although their extent has not been documented. Furthermore, the effects of sublethal injuries are unknown at the population level, although severe injuries of mature female manatees may affect their reproduction (O'Shea *et al.* 2001). The latter authors also note that comprehension of the effects of boats on Florida manatees should move beyond tallies of total deaths or projecting impacts on numbers at the population level: the level of wounding, pain and suffering of individual manatees due to boat strikes is inhumane, and in Florida intentional inflicting of similar injuries on domestic animals would result in criminal charges.

Management actions to reduce impacts of boats on Florida manatees

Conservation actions to date have been based on results of the mortality analyses for Florida manatees, particularly when such analyses reveal areas where collisions with boats are concentrated or areas that must be avoided because of manatee aggregations. However, these actions have been controversial as they involve restrictions on activities about which people have strong feelings, such as their perceived right to travel by boat over publicly owned waters at unregulated speeds wherever they wish. Speed-limited zones and restrictions on boat access have been challenged in court, as have restrictions on dock construction and locations of boat ramps that increase boater access to manatee habitat. Despite this controversy, the total area in which boat speeds are regulated for manatee protection in Florida is a small fraction of the many thousands of kilometres of available waterways (Florida Fish and Wildlife Conservation Commission 2007). In contrast to the special interest groups that object most vocally, the general boating public favours speed zones for manatee protection (see reviews in O'Shea 1995; Aipanjiguly *et al.* 2003).

Glaser and Reynolds (2003), Laist and Shaw (2006) and Calleson and Frohlich (2007) are among the most recent authors who have argued that slower vessel speeds constitute an important component of reducing the risk of vessel strikes for manatees. Their arguments include the speeds at which documented strikes have occurred, as well as the logical statements that slower speeds allow greater response time for boaters and manatees, and reduced severity of injuries in the event a strike does occur. By comparison,

Laist *et al.* (2001) assessed watercraft collisions with great whales and concluded that most known incidents occur at vessel speeds greater than 13 knots (24 km/h); vessel speeds of less than 10 knots (18.5 km/h) rarely cause collisions and subsequent serious injury or mortality.

Although the slowing of watercraft seems logical and easy to support as a conservation action, scientific challenges have occurred. Gerstein *et al.* (1999) and Gerstein (2002) interpreted results of their intensive audiogram studies to claim that manatees have poor hearing ability for the frequencies of sounds produced by slow-moving boats, and that therefore manatees may hear and be better able to avoid faster boats (see Chapter 5 for additional information on acoustic perception in sirenians). These authors also have advocated the mandatory use of acoustic warning devices on all Florida boats with the intent of modifying the behaviour of the manatees. On the other hand, Nowacek *et al.* (2004) and Miksis-Olds *et al.* (2007b) demonstrated that manatees hear boats well, can detect and respond even to slow-moving vessels at distances of at least 25 m, and that slowing of boats provides time (as noted above by Calleson and Frohlich (2007) and others) for manatees to perceive and avoid oncoming boats. Mann *et al.* (2007) suggested that manatees have the ability to localise the direction of underwater sounds; a result experimentally verified by Colbert *et al.* (2009) and supported by the anatomical and biochemical work of Ames *et al.* (2002). Nowacek *et al.* (2004) noted an interesting response by manatees to approaching watercraft: manatees move towards deeper water, which could sometimes place them directly into the path of an oncoming boat if water depth is insufficient for the animals to dive below the boat.

Despite conflicting interpretations of results of auditory and avoidance research, conservation agencies continue to attempt to regulate behaviour of the boaters, rather than that of the manatees per se. The proposed application of underwater alarm devices to boats has caused concerns unrelated to manatees. Anthropogenic underwater sound levels are increasing globally at an average rate of 7% annually or more (Hildebrand 2005), with several possible negative impacts on marine organisms including manatees (Tyack 2008). For example, manatees use sound for communication (Chapter 5) and elevated ambient noise may reduce the range over which they can communicate. Consequently, there is reluctance among many managers to consider potential solutions that mandate the introduction of a new source of underwater sound pollution on one million boats. Such intervention may create unforeseen problems, and there is a real possibility that manatees would simply habituate to the noise and remain susceptible to collisions.

> **Box 7.4 Florida Manatee Recovery Plan: principal actions designed to mitigate vessel strike**
>
> - Minimise causes of manatee disturbance, harassment, injury and mortality.
> - Determine and monitor status of the manatee population.
> - Protect, identify, evaluate and monitor manatee habitats.
> - Facilitate recovery through public awareness and education.

The effectiveness of boat speed regulations as a conservation tool has not been fully explored. Gorzelany (2004) examined boater compliance with boat speed regulatory zones and documented substantial non-compliance (25–50%), but Laist and Shaw (2006) have suggested that the zones have the potential to be beneficial for conservation if they are well designed and enforced. Citing mortality data from two locations (the Barge Canal and Sykes Creek) in Brevard County, Florida, where deaths of manatees had been notably high (O'Shea *et al.* 1985; Ackerman *et al.* 1995), Laist and Shaw (2006) documented that watercraft-related deaths 'decreased sharply' after enforcement of new boat speed rules began in 2002. They attributed the decline in mortality to speed reductions that, as described above, provide manatees and boaters with more time to avoid a collision.

In the most recent version (as well as all previous versions) of the Florida Manatee Recovery Plan, the US Fish and Wildlife Service (2001) recognised that solving the issue of watercraft-related injury and mortality was complex. The four principal components of the plan that were designed to address the issue are summarised in Box 7.4.

As stated in Chapter 8, the issue in Florida is not really a matter of having the scientific data to support appropriate conservation decisions. Rather, it essentially involves two factors: (1) finding an appropriate balance between human activities, economics and conservation of wildlife and habitat; and (2) having the will to make the tough decisions to achieve that balance. Proper management of watercraft speed and access will also have positive impacts on seagrass habitats, fisheries and other human interests (people also are routinely killed and injured in accidents with fast boats). Thus, solutions will need to be broadly inclusive of multiple stakeholders.

Indirect effects of vessel traffic on sirenians
Aside from deaths and serious injury of sirenians, vessel interactions can have more subtle but still serious impacts. Disturbance by watercraft

displaces manatees (Provancha and Provancha 1988; Buckingham *et al.* 1999; Miksis-Olds *et al.* 2007a) from important habitats, including foraging areas or warm-water refugia. Such effects have been claimed for sirenians in Australia (Johannes and Macfarlane 1991; Kwan 2002) and manatees in the wider Caribbean (Reynolds *et al.* 1995; Smethurst and Nietschmann 1999). In contrast, Jiménez (2005) developed predictive models to explain the distribution of West Indian manatees in watercourses in north-eastern Costa Rica and southern Nicaragua. He detected no significant effect of vessel traffic on manatee presence. In addition, despite the fact that local informants repeatedly claimed that boat traffic was one of the causes of the decline of West Indian manatees from these areas, Jiménez found frequent signs of manatee feeding in the areas of greatest boat traffic; note, however, that manatees in that area may be feeding primarily at night (Reynolds *et al.* 1995), when boats operate less frequently.

Dugongs may temporarily change location coincident with the approach of vessels. Anderson (1982) described the movement of a herd when a boat approached to within 150 m, and Preen (1992) suggested that dugongs may detect watercraft as far away as 1 km. In contrast to mortality, which is an established problem (see 'Interactions with vessels' above), Hodgson and Marsh (2007) found limited evidence of disturbance to dugongs by vessel traffic in Moreton Bay, near Brisbane, Australia, presumably because the density of boats is still relatively low. Nonetheless, levels of boating in Moreton Bay are already estimated to reduce dugong feeding time by 0.8–6.0% (Hodgson and Marsh 2007); the problem will be exacerbated by anticipated increases in human population size and boating traffic. Dugongs can be expected to ingest less food at preferred feeding sites and to expend more and ingest less energy as they seek quieter locations. As discussed in Chapter 6, inter-calf intervals and onset of sexual maturation in dugongs are profoundly affected by forage availability, so limiting access to high-quality habitats could decrease the rate of increase of the population.

Vessel traffic can also have direct impacts on the seagrasses on which sirenians depend for food. Seagrass scarring, turbidity (which reduces light penetration and, hence, productivity) and bottom-shear stress are among the documented, adverse effects of vessel traffic on seagrasses in Florida (Reynolds 1999). Factors which affect the availability of forage for sirenians can have a range of effects at the individual and population levels. Miksis-Olds *et al.* (2007a) found that in seagrass beds of equivalent species composition and density, manatees selected those that had lower ambient noise levels at frequencies below 1 kHz; use of any seagrass beds was negatively correlated with high boat presence in the early morning hours.

COLD STRESS

With the notable exception of the Steller's sea cow lineage (Chapters 2 and 3), the distribution of most sirenians has been limited to tropical and subtropical waters throughout their evolutionary history (Chapter 3). Captive manatees held in water below 16 °C become lethargic and anorectic (Campbell and Irvine 1981; Irvine 1983) and individuals may remain at warm-water sources without feeding for at least a week (Bengtson 1981).

Near the high-latitude limits of their distribution, sirenians (notably manatees in Florida and dugongs in the Arabian/Persian Gulf) make seasonal movements to warm-water sources (Chapter 5), where they may aggregate (Figure 7.9). The warm water can be in various forms: (1) permanent seeps and temporary warm water trapped by bathymetry and salinity stratification (e.g. Stith *et al.* 2010 for dredged canals and Florida manatees; Moore 1953 for Florida manatees at small seeps in the Everglades; Preen 2004 for large

Figure 7.9. Florida manatee aggregation at the warm-water discharge at the Riviera power plant in Florida, which began operations in the 1960s. The plant is currently not operating, as Florida Power & Light Company is modernising it, which will dramatically enhance efficiency of power production while reducing air pollution. During construction of the new plant, warm water is being produced by large electric water heaters. John Reynolds and Florida Power & Light Company photograph, reproduced with permission.

dugong herds in the Arabian/Persian Gulf in winter); (2) major warm spring outflows (e.g. Hartman 1979 for manatees at Crystal River and Blue Spring); (3) industrial heated outflows in Florida (Reynolds and Wilcox 1994; Figure 7.9; see below); and (4) warm currents in Eastern Australia at the southern limits to the range of the dugong (Preen 1992).

The high use of warm-water refugia in Florida is best known, and illustrates a critical dependence on warm-water sources (Figure 7.9; Box 7.5), contrary to Hartman's (1979, p. 23) speculation that manatees' sensitivity to cold was exaggerated, and that they may be less reliant on warm-water sources for survival but use these places more for their 'salubrious sensation'. Indeed, the influence of natural selection in limiting the distribution of sirenians to regions with warm water or access to warm-water refugia has been dramatically illustrated by manatee deaths in winter in Florida. It has long been known that considerable numbers of dead Florida manatees are found after extended or deep winter cold snaps (e.g. Bangs 1895; Cahn 1940). Hartman (1979) and Campbell and Irvine (1981) speculated that cold weather might induce a susceptibility to pneumonia in manatees but improved diagnostics now indicate a more complex aetiology, with pneumonia not playing a major role (Bossart et al. 2002a). Subadult manatees rather than dependent calves or mature adults seem to be disproportionately, but not exclusively, impacted by exposure to cold water (O'Shea et al. 1985; Ackerman et al. 1995; Bossart et al. 2004).

Box 7.5 The contemporary importance of cold stress as a threat to the species/subspecies of sirenians

	Current threat intensity[1]
Dugong	
Amazonian manatee	
Antillean manatee	
Florida manatee	
West African manatee	

[1] Shading intensity represents magnitude of threat.

Cold stress syndrome in manatees

Pathologists now recognise a 'cold stress syndrome' in Florida manatees (Bossart et al. 2002a). The cold stress syndrome typically involves emaciation, fat store depletion, serous atrophy of fat, lymphoid depletion, epidermal hyperplasia, pustular dermatitis, enterocolitis and myocardial degeneration (Bossart et al. 2002a). Death is due to a complex syndrome of processes involving metabolism, nutrition and immune suppression, along with secondary opportunistic and idiopathic diseases (Bossart *et al.* 2002a). Inflammation and infection by bacteria and fungi cause cutaneous lesions on some manatees, primarily on the head, fluke and flippers (Bossart 2001; Bossart *et al.* 2002a). Gross lesions of cachexia and gastrointestinal tract conditions are the most consistent findings (Bossart *et al.* 2002a). Lymphoid tissue depletion, lesions in the heart, liver, pancreas and other tissues, and opportunistic lung infections may also occur (Bossart *et al.* 2002a). The stress of cold and malnutrition likely impacts the immune system (evidenced by lymphoid depletion), allowing a cascade of effects that can account for the skin and organ pathology and opportunistic infections. Immune suppression due to cold stress in manatees is also found in vitro (Walsh *et al.* 2005). By some criteria, cold stress can be considered anthropogenic, particularly when manatees become habituated to industrial warm-water effluents that become insufficient in deeper or longer winter cold periods (Bossart *et al.* 2004).

Loss of warm-water refugia in Florida

In January 2010, several events highlighted the inadequacy of warm-water refugia for manatees in Florida and increased focus on questions salient to the future of the Florida manatee (Box 7.6). A cold front of extremely unusual duration and intensity drove manatees to seek warm water in unprecedented numbers: 5067 manatees were counted in an annual 'synoptic survey', shattering the previous high count by approximately 1200 manatees (well over 30%!; see Chapter 8). However, the weather pattern also led to the deaths through April 2010 of at least 236 manatees from cold stress (Figure 7.10), with another 65 unrecovered carcasses and 147 carcasses for which cause of death was undetermined (Florida Fish and Wildlife Conservation Commission, unpublished), underscoring the

---

Box 7.6 Questions relevant to the adequacy of warm-water refugia for manatees in Florida highlighted by the unusually cold winter weather in 2010

(1) Was the cold weather an anomaly or a harbinger of the future in the face of extreme weather caused by global climate change?
(2) Will existing warm-water resources persist?
(3) Will mitigation options be stepped up to provide extra protection during such extreme events?
(4) Do the scientific and management communities have the resources to deal with such events?
(5) Is it time to invest in solutions?

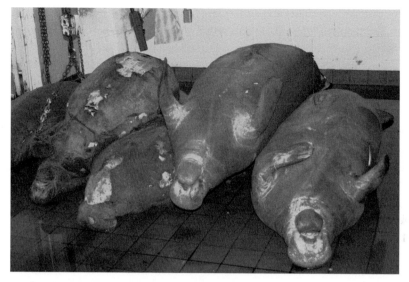

Figure 7.10. Massive red tides and winter cold stress can be devastating to the health and survival of Florida manatees. Deaths can occur at a rate of tens to hundreds of manatees in a few weeks. Such mortality events can overwhelm scientists at the Marine Mammal Pathobiology Laboratory (MMPL), where necropsies of recovered carcasses occur. This photograph depicts accumulated carcasses (acquired during a cold-related die-off in early 2010) awaiting processing at the MMPL. This event killed at least 236 manatees through April 2010; even more manatees likely died but were not recovered for examination or were too decomposed to ascertain cause of death. Reproduced with permission of the Florida Fish and Wildlife Conservation Commission.

inherent vulnerability of Florida manatees as they live on the fringe of the species' range.

The news about the very high manatee counts and mortality was tempered in mid-January by news articles (e.g. Klas 2010a,b; Patel 2010) that indicated that a requested customer rate increase by Florida Power & Light Company had been essentially rejected by the Florida Public Service Commission. On the face of it, that decision would seem to have little to do with manatees, but the reality is quite different.

Since at least 1997 (Rose 1997), it has been clear that the looming loss of industrial warm-water discharges and the reduction in flow at Florida's thermal springs, where manatees sought shelter during cold winter weather, were arguably the single greatest threats to manatees in Florida. Many of the industrial plants were ageing and no longer cost-effective to operate, and the aquifer that fed the springs had been tapped too heavily by housing developments and agriculture (Taylor 2006). What would happen to the manatees when the warm water disappeared, or was so reduced in flow volume that it could not provide adequate protection?

Rose and others warned of an impending disaster unless mitigation options were developed quickly. Reynolds (2000) provided recommendations involving a non-industry-dependent network of smaller warm-water refugia. These recommendations were extremely well received by the Manatee Recovery Team, and Laist and Reynolds (2005a,b) offered further recommendations (including enhancing availability to manatees of natural springs) which also received considerable verbal support. Nonetheless, the responsible agencies (Florida Fish and Wildlife Conservation Commission and US Fish and Wildlife Service) focused their efforts and funds mostly on monitoring, modelling and workshops, and little has been done to develop effective mitigation options.

This lack of response is not due to lack of appreciation of the consequences of loss of warm-water resources for manatees in Florida. C. Deutsch and J. Reynolds (unpublished) note that the elimination of several key industrial thermal refuges could be a catastrophe for Florida manatees, causing immediate mortality and a long-term reduction in carrying capacity. Models clarifying the future population status of manatees have simulated scenarios including loss of carrying capacity associated with loss of warm-water refuges (Florida Fish and Wildlife Conservation Commission 2002; Haubold et al. 2006; Runge et al. 2007b). Even the power companies with the best environmental track records in the state (Florida Power & Light Company and Reliant Energy), advocated for and invested in certain possibilities as mitigation options. Nonetheless, there have been few solutions. In the face of imminent closures of key plants for economic reasons, the sirenian conservation community lacked any beneficial tested options.

Then, in 2008, Florida Power & Light Company announced that it planned to 'modernise' two of its oldest and least efficient plants at Cape Canaveral and Riviera Beach. These plants are especially important for manatees (with one-day counts frequently exceeding 400 manatees at each, and counts approaching 1000 at the Canaveral plant in winter 2010; Reynolds 2010). The location of the Cape Canaveral plant in east-central Florida, where water temperatures may drop below 4 °C, makes this power plant perhaps the single most important in existence for manatees. The modernisation process was expected to take several years, during which Florida Power & Light Company would provide warm water using a series of large water heaters, but after that the plants would be expected to function efficiently and cost-effectively for several decades. The slow-moving scientific and management communities were handed a gigantic reprieve.

Box 7.7   Answers to questions raised in Box 7.6 with regard to the adequacy of warm-water refugia for manatees in Florida

(1) Was the cold weather an anomaly or a harbinger of the future in the face of extreme weather caused by global climate change? POSSIBLY.
(2) Will existing warm-water resources persist? UNLIKELY.
(3) Will mitigation options be stepped up to provide extra protection during such extreme events? NOTHING CURRENTLY EXISTS OTHER THAN THE GIGANTIC HOT WATER HEATERS.
(4) Do the scientific and management communities have the resources to deal with such events? NO.
(5) Is it time to invest in solutions? YES, A GREATER FOCUS ON SOLUTIONS IS NEEDED URGENTLY.

Working closely with those communities, Florida Power & Light Company negotiated in 2009 to proceed as requested. Then, as noted, in January 2010, the Florida Public Service Commission rejected the request for a rate increase by the company that would in part support the changes. As a result, Florida Power & Light Company advised that it might not proceed with the modernising of the two old power plants. Instead, both plants might be mothballed.

However, on 30 April 2010, Florida Power & Light Company announced that it had decided again to move forward with the plans to repower the two plants. Company President Armando Olivera indicated that the decision was not easy, but was made 'to deliver the best value, service and reliability over the long term ... to ... our customers' (http://www.fpl.com/news/2010/043010.shtml).

For the present, the reprieve in solving the problem seems to be reinstated. In Box 7.7, we return to the critical questions in Box 7.6 with answers that indicate that it is time to focus on solutions to this problem.

Reynolds et al. (2005) urged proactive, anticipatory management instead of reactive, crisis-driven responses. The above example shows that, even when it is clear what the conservation challenges and solutions are, people tend to not respond in ways that prioritise implementation of solutions. Rather, additional research and monitoring is often seen as easier than taking action to resolve an obvious impending crisis. Such issues must be addressed if humans are to succeed in conserving sirenians and other species into the future. In addition, scientists and managers must recognise and take steps to make the connections between seemingly unrelated economic and political decisions and conservation imperatives (see Chapter 9).

Box 7.8 The contemporary importance of disease as a threat to the species/subspecies of sirenians

	Current threat intensity[1]
Dugong	
Amazonian manatee	Unknown
Antillean manatee	Unknown
Florida manatee	
West African manatee	Unknown

[1] Shading intensity represents magnitude of threat.

In summary, the threat associated with the loss of warm-water refugia in Florida is actually two-fold: (1) the real threat to Florida manatees associated with losing habitat and dying during cold weather events; and (2) the threat imposed by our inability to solve conservation problems, even when the issues and potential mitigation options are identified proactively.

INFECTIOUS DISEASES AND MACROPARASITES

Die-offs due to disease

Numerous infectious disease agents and parasites have been reported in sirenians. These organisms could play important roles in sirenian ecology and conservation (Box 7.8), particularly when manatees and dugongs are under stress or when infections may act synergistically or opportunistically in conjunction with other threats. However, to date there have been no reports of major mortality events in sirenians caused by infectious diseases (unlike the case with harmful algal blooms, see 'Harmful algal blooms and biotoxins' below). Although this situation may reflect the low sirenian densities in some regions, we find it surprising that disease outbreaks have not been recorded in Florida manatees aggregating in close contact (sometimes engaging in coprophagy) in unnaturally large aggregations at warm-water sources in winter, when they are also under greatest nutritional stress. This situation contrasts, for example, with the dramatic and widespread epizootics and die-offs seen in some pinnipeds and cetaceans caused by infections with morbilliviruses or other viruses (Miller et al. 2001). Morbillivirus epizootics have never been fully documented in sirenians, although they are suspected of having occurred in dugong die-offs in the Arabian/Persian Gulf (Preen et al. 2012). In the case of Florida manatees, infectious disease is uncommon and some veterinary biomedical experts have suggested that this may be due to their 'remarkably efficient and responsive immune system' (Bossart et al. 2004, p. 434).

Viral diseases

Morbillivirus-neutralising antibodies have been detected in Florida manatees and provide evidence of past exposure to these agents, but do not give evidence of clinical disease (Duignan *et al.* 1995). Morbilliviruses are paramyxoviruses, similar to agents of rinderpest, distemper and measles, diseases that can occur with high rates of transmission and mortality in a variety of other mammals, and effects of which can include immune suppression (Kennedy 1998). Sera were analysed from 148 Florida manatees, 12 Antillean manatees from Guyana and 12 captive Amazonian manatees from Brazil (Duignan *et al.* 1995). Morbillivirus-neutralising antibodies were found only in the Florida manatees, but at a low seroprevalence (detected in 4%). Highest titres in the manatees were against porpoise and dolphin morbilliviruses (neutralising tests for seven different morbilliviruses were used); however, the manatees showed no evidence of morbillivirus disease. As titres can persist for years, these findings suggest that the manatees had exposure to these viruses at some unknown time in the past, possibly through contact with infected small cetaceans. The manatees may have served as dead-end hosts that had only subclinical infections and no further transmission spreading to the population, yet it cannot be ruled out that they could be subject to effects of morbilliviruses in a future epizootic (Duignan *et al.* 1995).

Florida manatees are known to host a unique papilloma virus, the first genetically characterised and isolated virus known from the Sirenia (Bossart *et al.* 2002b; Rector *et al.* 2004; Woodruff *et al.* 2005). This virus causes cutaneous lesions that are concentrated in contact areas such as the head and muzzle, and found more diffusely in association with areas of the skin marked by scratch lines or trauma (Rector *et al.* 2004). The potential for malignancy is not known and sites of infection have thus far been cutaneous only (different papilloma viruses can cause disease and malignancies in internal organs of other species), with transmission likely by body contact among manatees (Bossart *et al.* 2002b; Woodruff *et al.* 2005;). This virus occurs in both captive and wild Florida manatees (Bossart *et al.* 2002b; Woodruff *et al.* 2005). Latent infections may be typical, with lesions expressed as a result of trauma or impaired immunity (Bossart *et al.* 2002b; Rector *et al.* 2004; Woodruff *et al.* 2005; Bossart 2011).

There are records of one other virus, a possible viral disease of unknown aetiology, and serological evidence of exposure to several other viruses in West Indian manatees. A new herpes virus (trichechid herpes virus 1, TrHV1) was identified from Florida manatees based on single-gene DNA analysis of skin swabs and blood (buffy coat) samples (Wellehan *et al.* 2008). The virus was not isolated in culture. Viral DNA was found in about half of the 47 manatees sampled, although negative cases may simply have had latent infections without shedding of the herpes virus at the time of sampling. Positive cases were found more often in samples from manatees with active skin lesions (74%) than those without skin lesions (25%), suggesting both subclinical and clinical effects. Related herpes viruses can have more serious effects, including death, in elephants (Richman *et al.* 1999, 2000). However, TrHV1 can be expected to be widespread among healthy Florida manatees without indications of harm unless individuals become immunosuppressed, in which case removal of the cause for immunosuppression may be the best course of action to curtail disease (Wellehan *et al.* 2008). In a case suspected to be related to a viral disease, neoplastic lymphoma was diagnosed as the cause of death in a captive seven-year-old *T. manatus manatus* from Guyana held at the Odense Zoo in Denmark (Hammer *et al.* 2005). Diagnostic efforts included gross pathology, histology and immunohistochemistry. Up to six organs were affected, and it was noted that this disease would not have been diagnosed based on gross necropsy alone. The cause of this malignancy was unknown, but was speculated to have been an oncogenic virus (Hammer *et al.* 2005). Serologic evidence of exposure to other viral diseases has been reported in Florida manatees, but with no evidence for disease. These include San Miguel sea lion viruses type 1 (a calicivirus), eastern, western and Venezuelan equine encephalitis viruses (togaviruses transmitted by arthropod vectors) and pseudorabies (reviewed in Bossart 2001). Although the number of viral diseases found in sirenians to date is low, such diseases are likely to increase as virologists continue to apply molecular tools not previously available.

Bacterial diseases
Among bacterial diseases, mycobacterial infections have caused deaths of at least two species of manatees in captivity (the genus *Mycobacterium* includes bacteria that cause tuberculosis and leprosy in humans). Systemic infection with *M. marinum* resulted in the death of one captive Amazonian manatee (Morales *et al.* 1985), and a second Amazonian manatee died as a result of infection with *M. chelonei* (Boever *et al.* 1976); two captive West Indian manatees were infected by three species of *Mycobacterium* (*M. fortuitum, M. kansasi* and *M. marinum* – Sato *et al.* 2003). In all cases multiple internal organs were affected (particularly the lungs) and skin lesions were evident in early stages.

A wide variety of other disease-causing bacteria has been reported from sirenians, but not necessarily implicated in their deaths. Forrester (1992) listed 48 species of bacteria isolated from Florida manatees, including various species of *Aeromonas, Bacillus, Bacteroides, Citrobacter, Corynebacterium, Edwardsiella, Enterobacter, Escherichia, Eubacterium, Fusobacterium, Klebsiella, Lactobacillus, Pasteurella, Plesiomonas, Proteus, Pseudomonas, Salmonella, Serratia, Staphylococcus, Streptococcus* and *Vibrio*. Additional bacteria also have been identified in other sirenians (e.g. *Clostridium sporogenes* in Australian dugongs; *Salmonella panama, Providencia rettgeri* and *Stenotrophomonas maltophila* in Antillean manatees in Brazil; and *Flavobacterium* and *Xanthomonas* in Antillean manatees in Colombia (Montoya-Ospina *et al.* 2001; Vergara-Parente *et al.* 2003a,b; Eros *et al.* 2007)). This list will undoubtedly grow. The agent of the abortion-causing bacterial disease brucellosis has not been reported in sirenians, although antibodies to *Brucella* have been reported in Florida manatees (reviewed in Bossart 2001); antibodies to *Leptospira* have also been reported (Geraci and Lounsbury 2005).

Some of the bacteria cultured from sirenians can be adventitious or opportunistic and cause illness and death, particularly in animals that are debilitated through wounding or other stressors. Examples of fatal bacterial infections among wild sirenians include: (1) death due to infection of the umbilical area and peritonitis in a wild Florida manatee calf (Walsh *et al.* 1987); (2) bacterial meningoencephalitis in a subadult Florida manatee (Buergelt *et al.* 1984); (3) bacterial colitis, typhlitis, pneumonia and pustular dermatitis in a Florida manatee calf (Forrester 1992); (4) other bacterial infections in Antillean manatee calves from Colombia (Montoya-Ospina *et al.* 2001); (5) peritonitis and septicaemia in a wild subadult male dugong from Australia (Eros *et al.* 2007); (6) salmonellosis in an Antillean manatee calf (Vergara-Parente *et al.* 2003b); and (7) likely bacterial pneumonia in wild dugongs and Florida manatees (Beusse *et al.* 1981a; Campbell and Ladds 1981). Several bacteria have also been isolated in association with pustular dermatitis in wild and captive manatees (reviewed in Forrester 1992). Deaths of sirenians in captivity from bacterial infections include: (1) a dugong calf from enteritis (likely due to *Salmonella*); (2) a presumed adult dugong from bacterial pneumonia (Elliott *et al.* 1981; Campbell and Ladds 1981); and (3) orphaned Florida manatee calves with a variety of bacterial infections (Cardeilhac *et al.* 1981; Bossart *et al.* 2001). Chronic abscessation in the absence of trauma is also a common finding in Florida manatees, both shallow in the skin and dermis, and deep within skeletal muscle, often involving infection by *Staphylococcus aureus* (Bossart 2001).

Bacterial infections are fairly common sequelae to wounding by propellers or other physical forces, best known in Florida manatees, and can lead to septicaemia, peritonitis, pleuritis, osteomyelitis and other infections, depending on the site of wounding (Beusse et al. 1981a,b; Forrester 1992; Bossart 2001). Having noted this, we reiterate that manatees recover from incredible wounds (e.g. O'Shea et al. 2001), even though they frequent sewage outfalls and other locations where exposure to pathogens must be astounding. Osteomyelitis from infections with various opportunistic bacteria associated with wounds, thought to be due to harpooning during capture, have caused death in a captive dugong and debilitation but recovery following treatment in a captive Amazonian manatee (Frye and Herald 1969); osteomyelitis from boat wounds and other trauma (floodgates, entanglement) is also commonly seen in Florida manatees (Bossart 2001). Chronic cold stress can lead to a variety of opportunistic bacterial infections from some of the bacteria listed above, resulting in bronchopneumonia, enterocolitis and dermatitis, sometimes with secondary fungal infections (Bossart 2001).

Mycotic (fungal) infections

Mycotic infections have also been reported in sirenians, often in association with pustular dermatitis, including Florida manatees (Forrester 1992; Bossart 2001), Amazonian manatees (Mok and Best 1979) and captive West Indian manatees (Tabuchi et al. 1974). Fungi have included members of the genera *Cephalosporium*, *Cercospora* and *Mucor* (some identifications tentative).

Protozoal diseases

Infections by protozoans resulting in death have been documented in sirenians. Florida manatees have died as a likely result of neuropathology resulting from infection with the protozoan *Toxoplasma gondii* (Buergelt and Bonde 1983). This organism is often transmitted to other mammals through faeces of domestic cats, likely reflecting the use of highly developed coastal areas by Florida manatees. Oocysts can persist in salt water (Lindsay et al. 2003). *Toxoplasma gondii* infection (and resulting myocarditis) was responsible for the death of an Antillean manatee held semi-captive in the canal system in Georgetown, Guyana (Bossart 2001; Dubey et al. 2003), and was also found in a wild dugong from Australia (Eros et al. 2007). Other protozoans known from sirenians include three species of the coccidian *Eimeria*, one species from Amazonian manatees and two from Florida manatees, all with oocysts found at a high prevalence (31–87%) in faeces

(Lainson *et al.* 1983; Upton *et al.* 1989). It is unknown if these are pathogenic in manatees. Infection with the protozoan *Cryptosporidium parvum* has been reported in a wild dugong (Hill *et al.* 1997; Morgan *et al.* 2000). This protozoan was the human genotype that can cause severe diarrhoea outbreaks in people. Morgan *et al.* (2000) speculated that the dugong could have been infected through contaminated sewage, because oocysts of this protozoan can also persist in sea water. A *Cryptosporidium* has also been found in an Antillean manatee (Borges *et al.* 2009).

Macroparasites and commensals

A wide variety of endoparasitic helminths (worms and flukes) is known from sirenians. Steller (see Chapter 2) was the first to record macroparasites in sirenians: likely roundworms found in his dissections of *Hydrodamalis gigas* in the 1700s (Blair 1981). The known helminth parasites of extant sirenians include at least 26 species that specialise in different organ systems (see Supplementary Material Appendix 7.1). Beck and Forrester (1988) reviewed these parasites, but over the past two decades there have been further advances in understanding the systematics and occurrence of these groups, resulting in some major changes to nomenclature, as well as new records (see summary in Supplementary Material Appendix 7.1).

Verminous pneumonia with nasal flukes caused the death of a dugong in Australia. Ruptures of trematode-caused abscesses also resulted in death (from subsequent bacterial peritonitis) in another Australian dugong (Eros *et al.* 2007). Verminous pneumonia due to heavy infection with nasal flukes and a case of haemorrhagic enteritis from infection with over a million intestinal flukes have been reported in Florida manatees (Buergelt *et al.* 1984; Forrester 1992), and helminth parasite loads may have contributed to death from cardiac failure in an Antillean manatee (Moore *et al.* 2008). In most cases, however, parasitic infections are not severe and, in the case of Florida manatees, most individuals are not infected with the full complement of helminth species (Beck and Forrester 1988). Clear patterns in the intensity of infection of these macroparasites in sirenians by season, region, sex and age of host are not apparent, although a lower prevalence in calves and juveniles and regional differences in prevalence have been reported for helminths of Florida manatees and dugongs (Beck and Forrester 1988; Blair and Hudson 1992). The life cycles of sirenian macroparasites are unknown, although a variety of plausible scenarios has been proposed for some species (Beck and Forrester 1988; Blair and Hudson 1992). It is likely that additional species of helminths will be discovered in sirenians, particularly in poorly studied populations.

Figure 7.11. Dugong with remoras in Palau. Mandy Etpison photograph, reproduced with permission.

Ectoparasitic fauna found on sirenians are generally thought to be commensal, and include large crustaceans (copepods and cirriped barnacles), microcrustaceans, protozoans, rotifers, ostracods, amphipods, isopods, larval dipterans, gastropods, small nematodes and leeches (Hartman 1979; Beck and Forrester 1988; Forrester 1992; Williams *et al.* 2003; Bledsoe *et al.* 2006; Suarez-Morales *et al.* 2010). Tanaid crustaceans (*Hexapleomera robusta*) form tube patches on the skin of Antillean manatees (Morales-Vela *et al.* 2008). The skin of Florida manatees is also a substrate for biofouling algae (the cyanobacteria *Lyngbya* and *Oscillatoria*, the red algae *Polysiphonia*), including several species of diatoms (Bledsoe *et al.* 2006). Some cyanobacteria secrete toxins, and the biotoxin debromoaplysiatoxin from *Lyngbya* has been detected in algal growths on Florida manatees, which may perhaps account for some cases of dermatitis (Harr *et al.* 2008a). Dermatosis, perhaps of autoimmune origin, has also been reported in captive Antillean manatees (Leistra *et al.* 2003).

Commensal vertebrates associated with sirenians include sharksuckers (or remoras) (*Echeneis naucrates* and *E. neucratoides*), commonly observed on West Indian manatees and dugongs (Figure 7.11), and known to feed on sirenian faecal material (Williams *et al.* 2003). Fish of several species feed on shedding skin and commensal invertebrates of Florida manatees and cause occasional irritation (Hartman 1979). This habit has recently been documented for the invasive armoured catfish (*Pterygoplichthys* sp.), sometimes with many individual fish attached or feeding on a single manatee to the extent that physical discomfort and disturbance occurs to manatees using warm-water springs in cold weather (Nico et al. 2009). Little blue herons (*Egretta caerulea*) have been observed feeding on fish or invertebrates at openings made in floating vegetation while Florida manatees fed at the surface (Scott and Powell 1982).

Other pathological conditions

A variety of other conditions has been observed in a low percentage of sirenians. Formation of fatal kidney stones and renal pathology of unknown aetiology have been reported in Florida manatees and Antillean manatees from Cuba (Keller et al. 2008). Congenital malformations of flippers have been observed in Florida manatees, including 'cleft hands' (Watson and Bonde 1986), and some young Florida manatees show endocardiosis and valvular fibrosis, which seem to regress with age (Buergelt et al. 1990).

Application of modern diagnostic tools for health assessments

Until recently, sampling of live sirenians for health assessments has not been extensive, and most medical attention has been placed on rehabilitation of sick or injured individuals in captivity or diagnoses through postmortem necropsy of carcasses. It is increasingly recognised that health assessments can be highly informative of the overall condition of sirenian populations and their environments (Bonde et al. 2004; Bossart 2006, 2011; Chapter 8), although this approach has only been applied in the United States and Australia. Likely much more will be learned with greater attention towards application of more sophisticated diagnostic tools and methods to sirenians sampled in the field. Biomedical approaches used in sirenian health studies have been expanded in recent years to include tools such as field electrocardiography on Antillean and Florida manatees (Siegal-Willott et al. 2006), vital-sign monitoring in dugongs (Lanyon et al. 2010) and more intensive biochemical, haematological (Bossart et al. 2001; Harr et al. 2006, 2008b; Harvey et al. 2007, 2009; Tripp et al. 2010), immunologic (e.g. Bossart et al. 1998; Walsh et al. 2005; Wetzel et al. 2010), cytologic (Varela et al. 2007), reproductive (J. Reynolds and D. Wetzel, unpublished), molecular genetic (Rector et al. 2004; Wellehan et al. 2008), genotoxic (D. Wetzel and J. Reynolds, unpublished) and image (J. Lanyon, personal communication, 2009) sampling. Such techniques may lead to improved understanding of the roles of chronic stress and other anthropogenic factors as potential threats to sirenian populations.

CHEMICAL CONTAMINANTS

Persistent organic contaminants

The potential impacts of persistent organic contaminants on marine mammals have long been a concern (for history, see O'Shea and Tanabe 2003). There have been hundreds of investigations into the occurrence of residues of these chemicals in tissues of marine mammals, and several in-depth studies have pointed out associations between some of these contaminants and disruptions in various biological processes in cetaceans and pinnipeds (for reviews, see Vos *et al.* 2003; O'Hara and O'Shea 2005). Much of the concern stems from findings of unusually elevated concentrations of certain organochlorine compounds in blubber of piscivorous or carnivorous mammals high in marine food chains. The impacts of organochlorine compounds have been of less concern in sirenians because they are exposed to much lesser concentrations of these chemicals due to their lower positions in food webs (Box 7.9), although their tendency to ingest fine-grained sediments to which organic contaminants can adsorb may create an important avenue of exposure. Even during the 1970s and early 1980s, when wet weight concentrations of chemicals like DDE (dichlorodiphenyldichloroethylene) or PCBs (polychlorinated biphenyls) were found at 1000–2000 mg kg wet weight in blubber of some other marine mammals (reviewed in O'Shea 1999), the amounts in blubber of Florida manatees were to the order of 1 mg kg (O'Shea *et al.* 1984; O'Shea 2003). Florida manatees and dugongs sampled in the 1990s had even lower concentrations of DDE and PCBs in blubber, typically less than the chemical analytical limits of detection employed (Ames and Van Vleet 1996; Vetter *et al.* 2001; Haynes *et al.* 2005). Unpublished studies on manatees sampled in Florida and Chetumal Bay, Mexico, referenced by Wetzel *et al.*

Box 7.9 The contemporary importance of chemical contaminants as a threat to the species/subspecies of sirenians

	Current threat intensity[1]
Dugong	
Amazonian manatee	Unknown
Antillean manatee	Unknown
Florida manatee	
West African manatee	Unknown

[1] Shading intensity represents magnitude of threat.

(2012), showed continued low levels of organochlorine pesticides in Florida manatees, but also detected compounds not reported in past studies (endosulfan and lindane). Concentrations of total PCBs in Antillean manatees from Chetumal Bay sampled in 2006–2007 (0.38–5.7 ppm lipid weight, presumed blubber) were very similar to those in Florida manatees sampled about 25 years earlier (O'Shea et al. 1984; Wetzel et al. 2012), when PCBs had just previously been banned and were more predominant in the US biota. Analysis of the more recent samples from manatees in Florida and Chetumal Bay for polycyclic aromatic hydrocarbons (PAHs, often related to petroleum pollution) were negative (Wetzel et al. 2012). Similarly, flame retardants (polybrominated diphenyl ethers, PBDEs) were detected at much lower concentrations (low ng/g) in an Australian dugong than in other marine mammals from around the world (Hermanussen et al. 2008), and minor amounts of brominated compounds (perhaps of natural origin) have been found in blubber of dugongs from north-eastern Queensland (Vetter et al. 2001). In contrast to what has been found for other organic contaminants, dugongs sampled in Queensland had higher concentrations of the polychlorinated dibenzofurans (PCDFs) and polychlorinated dibenzodioxins (PCDDs) than typically found in other marine mammals (Haynes et al. 1999). Descriptions of other contaminants of modern concern, their biological activities and methods for sampling and analysis can be found in Wetzel et al. (2012).

Although sampling of sirenian tissues to date has not revealed the alarming concentrations of organic contaminants that were found in some populations of other marine mammals, the unique sirenian character of herbivory could mean greater exposure to different kinds of chemical contaminants, with potentially different consequences. Thus, continued surveillance and monitoring of tissues is called for, particularly for emerging environmental contaminants. Dugongs and some populations of manatees feed extensively on bottom-growing vegetation (Chapter 4), where they may be subject to, or even ingest, toxic elements and synthetic compounds that adsorb to or accumulate in sediments and plants. For example, the source of the PCDDs reported from tissues of Queensland dugongs was contaminated sediments and associated seagrass (McLachlan et al. 2001). The nearshore, estuarine and riverine habitats of sirenians place them close to terrestrial sources of pollution from run-off, wind deposition and waste disposal. Florida manatees and dugong carcasses stranded in more urbanised areas tend to show greater concentrations of metals and some organochlorine compounds in tissues than concentrations found in carcasses from areas with less human impact (O'Shea et al. 1984; Haynes et al. 2005).

Organotin compounds are another example. Used in various industrial applications and as antifouling agents on hulls of ships, they are common pollutants in nearshore marine waters and occur in sediments and some biota (Kan-Atireklap *et al.* 1997; Midorikawa *et al.* 2004; Jones *et al.* 2005). Butyltins in tissues of dugongs exceed concentrations reported in cetaceans or pinnipeds (Tanabe 1999; Harino *et al.* 2007), although specific effects, if any, in dugongs have not been determined.

Toxic elements, herbicides and oil

Several studies have examined tissues of dugongs and manatees for other metals and toxic elements (reviewed in Haynes and Johnson 2000; O'Shea 2003). Haynes and Johnson (2000) report that some studies (e.g. Denton *et al.* 1980) have detected unusually high concentrations of iron and zinc in liver tissues and high concentrations of cadmium in kidney tissues of dugongs in Australia, including animals from remote areas. These concentrations are considered unlikely to reflect anthropogenic impacts, but levels of some metals may have health implications for human consumers (Gladstone 1996). Nonetheless, alarming concentrations or evidence of toxicity from metal poisoning generally have not been reported, with the possible exception of copper in Florida manatees (see below). Recent analyses of 19 metals and trace elements in skin and blood of Florida manatees also found concentrations typical of or less than other marine mammals (Stavros *et al.* 2008).

Contaminants can indirectly impact sirenians through their effects on aquatic plants. Many metals can be taken up into plant tissues from sediments and water, and herbicides can both intentionally and unintentionally impact sirenian habitats. Use of the herbicides diuron, simazine and atrazine in terrestrial agriculture in Australia has resulted in their transport to nearshore sediments in seagrass habitat important to dugongs (Haynes *et al.* 2000a; McMahon *et al.* 2005). Diuron was detected in seagrasses (but atrazine, chlorpyrifos and organochlorines were not; Haynes *et al.* 2000a). In follow-up experiments, diuron was shown to suppress photosynthesis in seagrasses, including dugong food plants (Haynes *et al.* 2000b). Bengtson Nash *et al.* (2005) also showed that the herbicide diuron affected seagrass growth rates. Nothing is known about the ingestion of these herbicides by dugongs or manatees, although it is likely they are readily metabolised. Copper herbicides were of concern in specific areas of Florida where manatees foraged extensively in winter (O'Shea *et al.* 1984), and organic herbicides continue to be used against 'nuisance' plants in Florida and other areas of the world inhabited by sirenians.

Although PAHs and other metabolites have not yet been reported in sirenian tissues (we are aware of only a single study that carried out such analyses – Wetzel *et al.* 2012), oil pollution can be both an acute and a chronic threat to sirenians and their habitats. About 150 dugongs were estimated to be killed in the Arabian/Persian Gulf during the Nowruz oil spill in 1983–1984 (over one million barrels of oil flowed from seven wells damaged during the Iran–Iraq war; Preen *et al.* 2012). Oil spills have taken place on seagrass meadows used by dugongs and manatees (e.g. Mignucci-Giannoni 1999; Preen *et al.* 2012). Dramatic effects of oil spills on seagrasses, however, may not persist for long periods (e.g. Kenworthy *et al.* 1993). Indirect effects of oil extraction can be more chronic and long lasting, and can include increased turbidity that decreases light penetration, greater ship traffic and mechanical disruption.

Box 7.10 The contemporary importance of algal blooms and biotoxins as threats to the species/subspecies of sirenians

	Current threat intensity[1]
Dugong	
Amazonian manatee	Unknown
Antillean manatee	Unknown
Florida manatee	
West African manatee	Unknown

[1] Shading intensity represents magnitude of threat.

HARMFUL ALGAL BLOOMS AND BIOTOXINS

Harmful algal blooms and marine mammals

There is unequivocal evidence that biotoxins from harmful algal blooms cause mortality and morbidity of marine mammals, including sirenians (Box 7.10). Globally there has been an increase in the occurrence of harmful algal blooms in nearshore environments, and there is a scientific consensus among specialists in this field that these blooms are on the increase due to human activities (Heisler *et al.* 2008). In particular, eutrophication from increased run-off of nutrients originating in terrestrial and freshwater systems has played a major role in the increase in the frequency, extent and duration of many of these blooms (Heisler *et al.* 2008). Management of nutrient loads in run-off is key to preventing and minimising the occurrence and effects of many algal blooms that are detrimental to other marine life. Human activities also can play a role in causing these blooms by changing relationships among various nutrients in complex ways, transporting algal cells or cysts, or otherwise altering food webs and disrupting the ecological balance of the natural invertebrate grazers of algae (Heisler *et al.* 2008). Harmful algal blooms are of major concern worldwide because of their economic impacts on fisheries, wildlife, recreation and human health.

Many species of microalgae can form harmful blooms that poison vertebrates, and different species of algae each release unique biotoxins. The effects of these biotoxins differ based on their biochemical structure and pharmacological modes of action. The biotoxins most well known for causing poisonings of marine mammals (including endangered species), other wildlife and humans are all potent neurotoxins that can be categorised as saxitoxins, ciguatoxins, domoic acid and brevetoxins (Van Dolah *et al.* 2003). Saxitoxins are produced by some dinoflagellates, cause paralytic shellfish poisoning in humans and have been implicated in the deaths of humpback whales (*Megaptera novaeangliae*) in New England; the whales were exposed through feeding on fish that concentrated the saxitoxin through the food chain (Geraci *et al.* 1989). Dinoflagellate saxitoxins were also likely responsible for mortality of about one-third of the total remaining population of Mediterranean monk seals (*Monachus monachus*) in 1997 (Hernandez *et al.* 1998; Reyero *et al.* 1999).

Ciguatoxins are polyether neurotoxins of the dinoflagellate (*Gambierdiscus toxicus*) that grows in association with filamentous algae of coral reefs. These toxins bioaccumulate in reef fish, can cause serious acute and chronic pathology in humans, and are suspected to be responsible for mortality of Hawaiian monk seals (*Monachus schauinslandi*; see summary by Van Dolah *et al.* 2003).

On the Pacific coast of North America, blooms of the diatom *Pseudo-nitzschia australis* release the neurotoxic agent domoic acid, first recognised for its toxicological effects in humans (including permanent damage to short-term memory – hence the term amnesic shellfish poisoning) based on exposure through eating filter-feeding mussels that concentrate the toxin (Van Dolah *et al.* 2003). Many deaths and illnesses in California sea lions (*Zalophus californianus*) over the years are now recognised to have been a result of domoic acid exposure, and there is evidence that similar harmful exposures have also occurred in threatened California sea otters (*Enhydra lutris*) and grey whales (*Eschrichtius robustus*; Van Dolah *et al.* 2003). Scholin *et al.* (2000) pinpointed domoic acid bioconcentration in fish as the route of exposure in a large die-off of California sea lions in 1998. This event centred around an irruption of these diatoms coincident with terrestrial run-off (confirmed by satellite imagery), increases in domoic acid in sardines and anchovies, and deaths or neurological impairment in sea lions associated with unique lesions in the brain and domoic acid in tissues (Scholin *et al.* 2000).

Brevetoxins from dinoflagellates in the Gulf of Mexico have caused massive fish kills, deaths of bottlenose dolphins (*Tursiops truncatus*) through bioconcentration in fish prey, and respiratory distress and neuro-toxic shellfish poisoning in humans (Kirkpatrick *et al.* 2004; Fleming *et al.* 2005); they are also definitively known to impact Florida manatees (Bossart *et al.* 1998; Flewelling *et al.* 2005; see '*Karenia brevis* red tides and Florida manatees' below).

Karenia brevis red tides and Florida manatees

The only well-documented mortality of sirenians due to harmful algal blooms involves poisonings of Florida manatees by brevetoxins produced by the diatom *Karenia brevis* (O'Shea *et al.* 1991; Bossart *et al.* 1998; Landsberg *et al.* 2009). This organism produces the Florida red tide blooms in the Gulf of Mexico, where such blooms have been recorded historically since 1648 (Magaña *et al.* 2003). Although the occurrence of red tides thus cannot be directly related to anthropogenic eutrophication, there is evidence that Florida red tides are increasing in frequency and extent (Brand and Compton 2007). The vernacular term 'red tide' is also applied to blooms of other species of microalgae in other parts of the world, toxins of which may have very different modes of action and impacts on marine mammals. However, none is as frequent or long-lasting as the Florida red tides (Steidinger 2009). The brevetoxins from *Karenia brevis* are complex neuro-toxins that include up to ten polycyclic ether compounds. In addition to acting on nerve channels subsequent to oral ingestion, they can also be inhaled from aerosols produced by lysis of diatom cells through physical disruption (e.g. wave action) at the air–water interface (Van Dolah *et al.* 2003; Steidinger 2009).

Deaths of a total of about 300 Florida manatees due to brevetoxicosis were documented in 1996, 2002, 2003 and 2005 (Landsberg *et al.* 2009). High concentrations of brevetoxins occurred in the tissues of affected manatees (Flewelling *et al.* 2005). The quantification of brevetoxins in stomach contents and in seagrasses confirmed poisoning through inges-tion, with most of the brevetoxin occurring in epiphytes on seagrasses rather than in blade or rhizome tissue (Flewelling *et al.* 2005). Although most manatee deaths tend to occur coincident with blooms, lag effects can exist between the time of the red tide event and continuing manatee mortality, with deaths occurring in areas with low concentrations of brevetoxin and low *Karenia brevis* cell counts in water samples, yet higher levels in seagrasses (Landsberg *et al.* 2009). In most of the major die-offs of manatees due to brevetoxicosis, the events took place in south-western Florida when red tide blooms came inshore during late winter and spring months and persisted for weeks or even months (O'Shea *et al.* 1991; Landsberg *et al.* 2009). Coastal salinities in lagoons and river mouths were higher than usual (often >24 ppt; associated with reduced freshwater inputs).

In addition to toxicity through ingestion noted by Flewelling *et al.* (2005), many manatees that were killed by brevetoxin during a large die-off in 1996 showed effects of inhalation toxicity. Gross and histopathological lesions were found throughout the respiratory tract as well as in the liver, kidney and brain (Bossart *et al.* 1998, 2002c). These tissues (including nasopharyngeal passages) were all brevetoxin-positive, as revealed by immunohistochemical staining with polyclonal antibodies to brevetoxin. Subsequent testing also revealed that the archived tissues collected from manatees that died during a 1982 die-off before this technique was available were similarly brevetoxin-positive (O'Shea *et al.* 1991; Bossart *et al.* 1998). The analyses of these tissues suggested that inhalation was an important route of exposure to brevetoxins in 1996, and that a toxic shock reaction may have been an acute terminal event in some of the cases; haemosiderosis and a chronic haemolytic anaemia were also involved in a more chronic pathogenesis of brevetoxin exposure, suggesting the involvement of at least two molecular and cellular processes of brevetoxin poisoning (Bossart *et al.* 1998, 2002c; Bossart 2011). Inhalation of brevetoxins can have an immunosuppressive effect in rats (Benson *et al.* 2005); manatees exposed to these toxins also show evidence of immunosuppression (Walsh *et al.* 2005, 2007). Manatees can recover from the neurotoxic effects of brevetoxins, as evidenced by the recovery in captivity of neurologically distressed manatees rescued during red tide events (O'Shea *et al.* 1991).

Retrospective analysis of manatee liver samples has shown that lethal or sublethal exposure to brevetoxin elicits a significant lipidomic change in fatty acid composition (Wetzel *et al.* 2010). In some cases, manatees that died due to watercraft collisions manifested the distinctive hepatic fatty acid profile, raising questions regarding vulnerability of brevetoxin-exposed manatees.

In the last decade, sophisticated research has aided understanding the ecology of *Karenia brevis* red tide blooms at multiple scales from trans-oceanic atmospheric circulation patterns to cellular processes (Walsh *et al.* 2006). Blooms of *Karenia brevis* initiate from a 'seed stock' of cysts or cells of this organism on the mid-continental shelf off western Florida at depths of 12–37 m (Steidinger 2009). Satellite imagery shows that many blooms do not move inshore and would be of no consequence to manatees. When blooms occur, the dinoflagellates undergo slow population growth (doubling every 3–5 days) at salinities from 31–37 ppt. Nutrient concentrations are generally low offshore and can limit bloom development. Fish kills occur once a bloom reaches about 10^5 cells per litre, at which point the blooms are detectable by satellite on the basis of chlorophyll a concentrations; at 10^6 cells per litre blooms are discernible to the human eye, a phenomenon referred to as a 'maintenance stage'. At this point the *Karenia brevis* blooms (extending to depths of 50 m) can be moved inshore by winds and currents to places where manatees are more abundant and susceptible to exposure (Steidinger 2009).

Close to the Florida coast, blooms of *Karenia brevis* are initiated by conditions of nutrient availability. Remarkably, prime conditions are sometimes triggered by climatic conditions in another hemisphere, and can begin with dust storms near the Sahara Desert in Africa (Figure 7.12). The winds lift iron-rich sand particles from drought-stricken sub-Saharan soils and carry them aloft across the Atlantic Ocean to the Gulf of Mexico (Figure 7.13), where they are deposited by rainfall (Prospero and Lamb 2003). Iron can be a limiting factor for the cyanobacterium *Trichodesmium*, a nitrogen fixer, in the Gulf of Mexico. The deposition of iron-rich African dust promotes the irruption of *Trichodesmium* into its own bloom (Figure 7.12). This cyanobacterium takes advantage of mineral phosphorous available in mid-shelf geological deposits in the Gulf, and abundantly fixes nitrogen so that it is biologically available upon death of its cells (Lenes *et al.* 2008). At that point, nitrogen is no longer a limiting factor for *Karenia brevis* and a Florida red tide bloom ensues (Walsh *et al.* 2006; Lenes *et al.* 2008; Steidinger 2009). The amount of transported dust can depend on the level of drought conditions in Africa (Prospero and Lamb 2003). Thus terrestrial climatic conditions on a different continent on the far side of the Atlantic Ocean can influence the death of manatees in western Florida (Figure 7.12). The overall model of red tide development is based on a complex ten-step hypothesis that also involves other factors such as light tolerance of competing species, upwelling of bottom currents and fish kills as additional sources of nutrients (Walsh *et al.* 2006).

Methods for prevention and control of Florida red tides have been discussed and subject to experimentation, but have not been implemented or routinely applied (Sengco 2009; Steidinger 2009). The global connections revealed by the findings regarding red tides in the Gulf of Mexico may

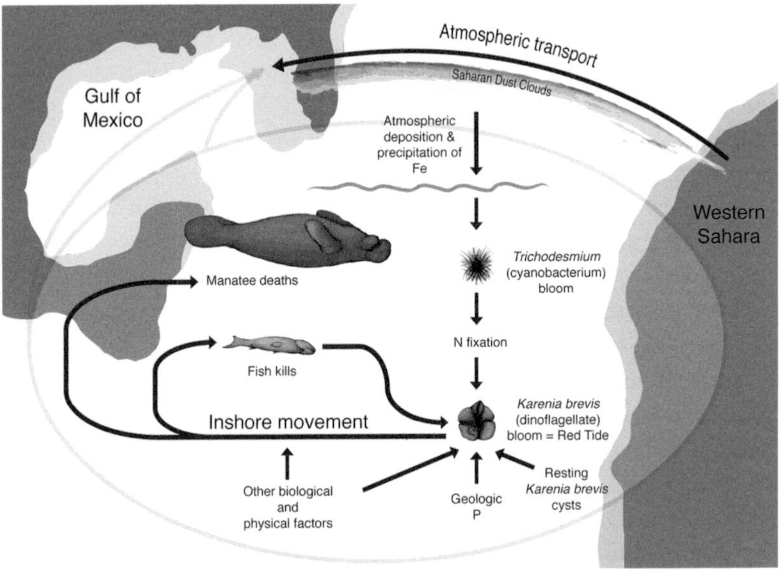

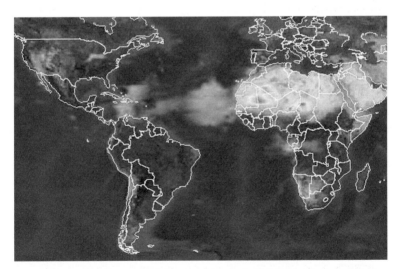

Figure 7.12. Schematic diagram of events that can trigger *Karenia brevis* red tide blooms in the Gulf of Mexico, beginning with atmospheric deposition of iron-rich dusts carried aloft from dust storms in the Sahara Desert. Drawn by Gareth Wild, reproduced with permission.

Figure 7.13. Satellite image of dust clouds carried aloft from Africa towards the Caribbean and Gulf of Mexico, where their deposition of iron-rich dust can lead to a *Karenia brevis* red tide bloom. Photo courtesy of Jay Herman, US National Aeronautics and Space Administration, Goddard Space Flight Center, Laboratory for Atmospheres, reproduced with permission.

extend to other waters: drought-exacerbated deserts may also feed wind-deposited nutrients that facilitate blooms of similar dinoflagellates near the Atlantic coasts of South America and West Africa, and the coast of south-eastern Asia (Walsh *et al.* 2006). This tale of complexity may be a foreboding example of as yet unknown consequences to sirenians of future changes in global meteorological phenomena.

Dinoflagellates and dugongs

Dinoflagellates of the genus *Procentrum* that occur as epiphytes on seagrasses in dugong habitat may contain significant amounts of the algal toxin okadaic acid (Takahashi *et al.* 2008). Dinoflagellates and okadaic acid were found on all four species of seagrasses analysed from Moreton Bay,

Box 7.11 Contemporary assessment of the likely importance of climate change as a threat to the species/subspecies of sirenians

	Predicted threat intensity[1]
Dugong	
Amazonian manatee	
Antillean manatee	Unknown but of considerable concern
Florida manatee	
West African manatee	

[1] Shading intensity represents magnitude of threat.

Queensland, raising concern because of the tumour-promoting potential of this biotoxin. However, at the level of sensitivity employed, okadaic acid was not detected in tissues of 41 dugongs stranded in Moreton Bay, and no cancerous tumours have been reported in dugongs from this area (Takahashi *et al.* 2008). Seagrass beds in parts of coastal Queensland and Western Australia are also vulnerable to the blooms of toxic *Lyngbya majuscula* fuelled by agriculturally derived nutrients (Albert *et al.* 2005; Bishop 2010), with consequent degradation of these areas as dugong habitat.

GLOBAL CLIMATE CHANGE

The impacts of changes in global climate on sirenian populations have not been well explored and are open to conjecture (Box 7.11). However, impacts will certainly occur and in the aggregate they are unlikely to be positive. The precise factors that will be involved are not easy to predict, and effects could be through indirect interactions between meteorological and biotic factors involving terrestrial, freshwater and marine ecosystems. At present, the nature of these interactions may not even be recognised. For example, the finding that manatee-killing blooms of *Karenia brevis* red tides in the Gulf of Mexico can be triggered by drought in sub-Saharan Africa (as reviewed under '*Karenia brevis* red tides and Florida manatees' above) was not even a hypothesis under consideration during manatee die-offs a little more than a decade ago. If climate change results in more extensive continental droughts, then the scenario of increased red tide blooms described for the Gulf of Mexico is likely to be repeated across the globe through impacts of desertification, dust transport and eutrophication on coastal areas inhabited by West African manatees, West Indian manatees and dugongs (Prospero and Lamb 2003; Jickells *et al.* 2005; Walsh *et al.* 2006).

The impacts of climate change on terrestrial ecosystems may be felt more directly by manatees than by dugongs or other coastal and pelagic marine mammals because manatee populations extend deep into continental landmasses through large river systems. For example, the vast Amazon rainforests appear to be suffering increased droughts induced by climate change, and a major Amazon drought in 2005 was driven by elevated sea surface temperatures in the tropical North Atlantic (Marengo *et al.* 2008; Phillips *et al.* 2009). As has been discussed (Chapters 5 and 6), Amazonian manatees are adapted to surviving annual droughts, but the limits to their ability to fast and survive are uncertain, and even more extensive droughts may lead to manatee mortality. Increases in such events may also change these tropical forests from sinks for sequestering atmospheric carbon to sources of carbon that will accelerate global change (Phillips *et al.* 2008).

Loss of tropical rainforest over the Amazon watershed is a major possibility (e.g. Betts *et al.* 2008; Cox *et al.* 2008; Harris *et al.* 2008) that will undoubtedly cause great alterations to the hydrological cycle and the ability of the Amazon River system to support manatees. Such changes are also likely elsewhere in the Americas in watersheds and rivers occupied by West Indian manatees. Impacts on rivers occupied by manatees in South America are expected to be complex and will interact with ongoing deforestation and other land-use practices by humans (Coe *et al.* 2009). Other impacts of climate change on terrestrial ecosystems may include stresses on food sources for indigenous communities that respond in turn by increasing subsistence killing of sirenians. This pattern may also occur in the marine environment, as explained below.

Alterations to seasonal flooding regimes and river flows due to climate change expected in parts of Africa also are likely to impact habitats used by West African manatees (Conway *et al.* 2009; Goulden *et al.* 2009; Paeth *et al.* 2009; Figure 7.14). Climate change is only one of many factors which

Figure 7.14. West African manatee rescued after being trapped behind an agricultural dam in Senegal after the water dried up. Tomas Diagne and CBD-Habitat photograph, reproduced with permission.

are expected to negatively impact tropical lowland rivers around the world during the immediate future (Malmqvist and Rundle 2002; Palmer *et al.* 2008, 2009; Chapter 6). Decreased discharge rates are expected in some large rivers within the ranges of all three species of manatees, especially those already impacted by dams (e.g. Milly *et al.* 2005; Palmer *et al.* 2008), and have already taken place in Central and Western Africa (Dai *et al.* 2009).

There are many complex factors with dangerous consequences that are likely to operate on the world's marine ecosystems as global climate change progresses (Hoegh-Guldberg and Bruno 2010). A simplified view of the effects of human-induced climate change on sea level and water temperature is that they possibly may be beneficial to sirenians. After all, climate changes that provided warmer intervals during the Pleistocene resulted in a northward expansion of manatees in North America (Chapter 3). Similarly, increased water temperature is likely to allow a southward extension of the range of dugongs in Australia (Lawler *et al.* 2007). Nonetheless, such expansion is likely to be limited by the narrowing of the Australian continental shelf at higher latitudes and the resultant limited availability of conditions suitable for the growth of large seagrass meadows.

We believe that possible beneficial effects of climate change on sirenians will be overshadowed by the potential for harm. An increase in sea level alone could in theory provide more inshore habitats for food plants and expanded access to them, but corresponding increases in depths over existing deep-water seagrasses could reduce light penetration and eliminate such resources (Short and Neckles 1999). The landward extent of sea-level expansion could in many areas be limited by buttresses put in place to protect existing developed shorelines, and expanding waters may simply penetrate over what may be unsuitable substrates for sirenian habitats. Similarly, the present thermal constraints on the distribution of modern sirenians may be somewhat relaxed, allowing manatees and dugongs to expand their ranges. However, warmer water temperatures may also favour shifts in global ocean currents (Wilkinson and Souter 2008; Florida Oceans and Coastal Council 2009; Hoegh-Guldberg and Bruno 2010) that could influence sirenian movement and dispersal patterns as well as the distribution of their food resources. Warmer temperatures may have a strong negative impact on seagrasses in shallow intertidal zones (Waycott et al. 2007).

Global climate change is also likely to result in increased intensity and perhaps frequency of severe storms (Webster et al. 2005; Florida Oceans and Coastal Council 2009; Bender et al. 2010); mortality of dugongs due to storm events has been well documented (Heinsohn and Spain 1974), including strandings (Figure 7.15) in receding high waters (Marsh 1989). Florida manatees also have lower survival rates during years with intense storms (Langtimm and Beck 2003). In the latter case, the mechanisms that cause mortality are unknown, but likely vary with timing, intensity and duration of storms (Langtimm et al. 2006). Intense tropical storms and rainfall events can also negatively disrupt seagrass communities (Short and Neckles 1999), with seagrass diebacks linked to suppression of reproduction and increased mortality and emigration of dugongs (Preen and Marsh 1995; Gales et al. 2004; Grayson et al. 2008; Marsh and Kwan 2008; Chapter 6). Indirect effects of increased intensity and frequency of storms will include increased run-off of anthropogenic contaminants, light-blocking silt, and generally decreasing water quality in nearshore areas and estuaries (Florida Oceans and Coastal Council 2009). The increased run-off of sediments and nutrients from terrestrial ecosystems likely to occur with global climate change will cause loss of seagrass (Orth et al. 2006). Seagrasses are highly dependent on water clarity, and run-off and other factors that increase turbidity will negatively impact important sirenian food sources (Short and Neckles 1999; Orth et al. 2006).

Even without many other negative anthropogenic effects, the initial impacts of global change could superficially be of benefit to seagrass growth because of

Figure 7.15. Rescuing one of the dugongs stranded by a storm associated with Cyclone Kathy in northern Australia, 1984, up to 9 km from the coast over an area of 250 km^2 and behind stands of mangroves up to 3 m high (Marsh 1989). Helene Marsh is the person on the left. C. Limpus photograph, reproduced with permission.

increased water temperature and availability of carbon dioxide (CO_2) for photo-synthesis (Short and Neckles 1999; Orth *et al.* 2006). However, increased CO_2 also results in increased ocean acidity, which could reach a threshold at which it becomes detrimental to seagrass survival (Florida Oceans and Coastal Council 2009). Increased storm frequency and intensity can cause physical disruption of seagrass meadows and substrates, as well as increased turbidity. High mortality and emigration of dugongs were observed following the occurrence of two floods and a cyclone within three weeks which impacted seagrasses in Hervey Bay, Australia, in 1992 (Preen and Marsh 1995). Overall uncertainty in the degree and scale of effects of climate change on seagrass resources has been expressed for the Great Barrier Reef region of Australia (Waycott *et al.* 2007); however, there are worrisome signs of seagrass in the Gulf of Carpentaria suffering irreparable photosynthetic stress or 'burning' as a result of periods with temperatures 10 °C above the seasonal average, especially if these high temperatures occur during periods of day-time low tides (Campbell *et al.* 2006).

Some scientists (e.g. Schiedek *et al.* 2007; Noyes *et al.* 2009) anticipate that climate change in tropical environments may increase both the bio-availability of contaminants (due to perturbation of sediments to which organic contaminants are adsorbed and to heavy rainfall during increasingly intense tropical storms) and the toxicity of contaminants in warmer waters. In their review, Noyes *et al.* (2009) recommended (1) identifying taxa that are especially vulnerable to climate–pollutant interactions (which could include the long-lived sirenians); and (2) predicting 'tipping points' that might trigger or enhance synergistic interactions.

In addition, climate change is projected to increase pressure on the world's fisheries. Twenty-one of the 33 countries where the impacts of climate change on fisheries are projected to have the greatest national economic impacts (Allison *et al.* 2009) are in the ranges of dugongs and manatees. The potential national economic impacts of climate change on fisheries could not be assessed in a further 21 countries in the range of sirenians; many of these are classified as least developed countries or small island states or both, and are thus potentially especially vulnerable to climate change. A negative correlation between the catches of subsistence fishers and 'wild meat' has already been established for West Africa (Brashares *et al.* 2004; Rowcliffe *et al.* 2005) and is likely to apply in other regions of the developing world. These projected changes in human population and climate are significant to the future of manatees and dugongs because the population growth rate of all sirenians is highly sensitive to changes in adult survival rate (Chapter 6), and most local populations of sirenians cannot withstand the human-induced mortality of even a few animals per year.

Although diseases have not yet had a known major impact on sirenian populations (see 'Infectious diseases and macroparasites' above), it is expected that warmer waters will favour an increased proliferation of disease organisms (Harvell *et al.* 2002, 2009), putting sirenian populations at greater risk. Warmer temperatures and increased run-off and eutrophication associated with climate change effects on terrestrial ecosystems may not only favour increased occurrence of harmful algal blooms (Van Dolah 2000; Paerl and Huisman 2008), but could also allow changes to seagrass communities that result in shifts in dominance to marine macroalgae such as *Lyngbya* and *Caulerpa* that are not important sirenian foods (Chapter 4). Such changes in dominance have been noted in contemporary manatee freshwater foraging areas, and increased homogenisation of floras and proliferation of non-native plants is yet another expected consequence of global change (Stachowicz *et al.* 2002). Warming sea surface temperatures have caused episodic diebacks of seagrasses in nearshore marine habitats (Wilkinson and Sauter 2008; Florida Oceans and Coastal Council 2009), and estuaries occupied by sirenians will have altered salinities and changes in circulation patterns due to alterations in rainfall regimes and freshwater inputs (Easterling *et al.* 2000; Alber 2002; Lau and Wu 2007; Florida Oceans and Coastal Council 2009).

It is possible that steps taken to ameliorate climate change could unintentionally impact sirenians: efforts to control emissions that contribute to climate change could result in loss of older electricity-generating facilities in coastal Florida that currently serve as winter refugia for Florida manatees (Laist and Reynolds 2005a,b). Irrespective of the consequences of climate change, such changes will take place over decades, whereas even optimistic scenarios for control of factors causing climate change suggest that reversibility may take centuries, during which time the growth and ecosystem demands of human populations are expected to inexorably increase.

Given the uncertainty of how climate change may affect sirenians and their habitats in the future, perhaps the best precautionary actions for today are to control other factors already known to adversely impact dugongs and manatees (Lawler *et al.* 2007), while carefully considering the implications of new predictions. Advances in data and refinements in modelling should reduce uncertainty about courses of action needed to benefit sirenian conservation in the face of climate change.

Box 7.12 The contemporary importance of miscellaneous threats to the species/subspecies of sirenians

	Current threat intensity[1]
Dugong	
Amazonian manatee	
Antillean manatee	
Florida manatee	
West African manatee	

[1] Shading intensity represents magnitude of threat.

MISCELLANEOUS THREATS AND SOURCES OF MORTALITY

Other anthropogenic sources or potential sources of sirenian mortality have been identified (Box 7.12). Dugongs in Australia drown in nets set along beaches to protect bathers from sharks (see 'Incidental capture in fishing gear' above). In Florida, crushing and drowning in locks and gates on canals and other water control structures (Odell and Reynolds 1979) accounted for about 4–5% of total deaths of manatees between 1976 and 1992 (Ackerman *et al.* 1995), but may have declined in recent years with installation of

Box 7.13 A contemporary assessment of the loss of genetic diversity as a threat to the species/subspecies of sirenians

	Current threat intensity[1]
Dugong	
Amazonian manatee	
Antillean manatee	
Florida manatee	
West African manatee	

[1] Shading intensity represents magnitude of threat.

pressure sensors and acoustic arrays at some structures (Runge *et al.* 2007b). Deaths of West African manatees in turbines and intake structures of hydroelectric dams have been reported (Powell 1996). Mortality due to internal injuries from ingesting rope, fishing line, hooks or other foreign objects have been reported in dugongs and manatees (Forrester *et al.* 1975; Buergelt *et al.* 1984; Beck and Barros 1991; Eros *et al.* 2007); entanglement in marine debris also can cause infection, loss of flippers and death in Florida manatees (Beck and Barros 1991). Manatees are occasionally injured or killed in Florida by vandals shooting or throwing impaling objects. Large-scale modifications of habitat due to constructions of dams and reservoirs in areas occupied by the three species of manatees will likely have serious negative effects. Other natural sources of mortality include accidents from impaling on barbs of rays in both manatees and dugongs (e.g. Eros *et al.* 2007), and predation (Chapter 5). Small Florida manatees classified as perinatal or dependent calves are recovered dead each year (Schwarz 2008). Specific causes of death in many of these calves are largely unknown but may be related in part to some of the factors described elsewhere in this chapter that impact their mothers or cause separation from them.

LOSS OF GENETIC DIVERSITY

Loss of genetic diversity is a widespread conservation issue for many species of wildlife that persist in small or isolated populations or that have experienced population bottlenecks in the past. Loss of genetic diversity can render populations susceptible to a variety of threats, increase the prevalence of deleterious genetic conditions and decrease the evolutionary potential for adaptation to changing environments. Molecular genetic studies are beginning to shed light on the level of genetic diversity that currently exists in populations of manatees and dugongs (Box 7.13). The more comprehensive sampling to date has been based on variation in the mito-chondrial DNA (mtDNA) control region. Microsatellite DNA markers applicable to manatees (Pause *et al.* 2007; Tringali *et al.* 2008), dugongs (Broderick *et al.* 2007) and both groups (Garcia-Rodriguez *et al.* 2000; Hunter *et al.* 2010a) are also beginning to reveal important patterns in genetic diversity for some populations (see below). Additional findings regarding genetic diversity in sirenians can be expected with advances in molecular technology.

The Amazonian manatee appears to exhibit good genetic diversity based on mtDNA control region haplotypes (Table 7.2; Cantanhede *et al.* 2005; Vianna *et al.* 2006). Little genetic structuring was observed among localities, and the calculated effective female population size was 3–5 times as great as in any one of four genetically defined population clusters of West Indian manatees (Garcia-Rodriguez *et al.* 1998; Cantanhede *et al.* 2005; Vianna *et al.* 2006; Chapter 3). Overall, Garcia-Rodriguez *et al.* (1998) considered that haplotype diversity and nucleotide diversity across all populations of West Indian manatees were not low (Table 7.2). However, unlike Amazonian manatees, populations of West Indian manatees were strongly structured with most of the variation partitioned among locations; there was much lower diversity within populations, including zero diversity in samples from Florida, Mexico and coastal Brazil (Garcia-Rodriguez *et al.* 1998; see also Vianna *et al.* 2006). Low to no mtDNA diversity in the Florida subspecies suggests either a major recent bottleneck, or the complete loss of manatees from Florida in relatively recent geologic time (the last glaciation) followed by subsequent recolonisation by limited numbers from the West Indies (Garcia-Rodriguez *et al.* 1998; Chapter 3). This low diversity is cause for conservation concern (Garcia-Rodriguez *et al.* 1998; Bonde 2009). Documented hybridisation between West Indian manatees and Amazonian manatees in north-eastern South America may also pose conservation problems, especially for the small coastal population of *T. manatus* in north-eastern Brazil (Cantanhede *et al.* 2005; Vianna *et al.* 2006).

Limited genetic analysis has been performed on samples from West African manatees (although samples from more than 200 individuals from nine countries have recently been collected: L. Keith, personal communication, 2010), but six individuals from four locations (Guinea-Bissau, Ghana, Niger and Chad) showed high haplotype diversity (Table 7.2); West African manatees also appear to have greater population structuring than Amazonian manatees (Vianna *et al.* 2006). Genetic diversity based on mtDNA was also assessed in dugongs over a wide area (but concentrated in Australian waters). Haplotype diversity was higher than in any species of manatee, with

Table 7.2 *General summary of results of mitochondrial DNA analyses of the four species of sirenians pertinent to genetic diversity*

Taxon	Sample	N_H	H	π	S	Comments	Source
D. dugon	188 dugongs across the range; emphasis on Australian waters; 411 base pairs	56	0.95	0.03	60	Geographic structuring of lineages	Blair et al., unpublished
T. inunguis	Up to 92 manatees along the Amazon River system from Japurá to Pará; 410 base pairs	31	0.88	0.005	34	No geographic structuring	Vianna et al. (2006); Cantanhede et al. (2005)
T. senegalensis	6 manatees from 4 locations; 410 base pairs	5	0.93	0.02	15	Geographic structuring of lineages	Vianna et al. (2006)
T. manatus	Up to 224 manatees from throughout the range; 410 base pairs	20	0.86	0.04	45	Geographic structuring of lineages	Garcia-Rodriguez (1998); Vianna et al. (2006)
T. manatus manatus	Belize, 113 manatees; 410 base pairs	3	0.53	0.03	28	Fine-scale geographic structuring	Hunter et al. (2010b)
T. manatus manatus	Puerto Rico population; 114 manatees; 410 base pairs	4	0.49	0.001	3	Fine-scale geographic structuring	Hunter et al. (in press)
T. m. latirostris	Up to 28 manatees from throughout Florida; 410 base pairs	1	0.00	0.00	0	No geographic structuring	Garcia-Rodriguez (1998); Vianna et al. (2006)

N_H = number of haplotypes; H = haplotype diversity; π = nucleotide diversity; S = number of polymorphic sites.

comparable nucleotide diversity (Table 7.2; D. Blair, personal communication, 2010). Structuring was evident at a large geographic scale, with two geographic lineages recognised around Australia that were also distinct from the few samples from outside Australia ($n = 11$), which likely represent additional lineages (D. Blair, personal communication, 2010).

Application of microsatellite markers to questions of genetic diversity in sirenians is increasing, with measures available for West Indian manatees and dugongs that begin to provide a basis for comparison with other mammals (Table 7.3). Calculated heterozygosities and numbers of alleles at polymorphic loci in West Indian manatees are generally lower than in dugongs (sampled at one location; Broderick et al. 2007). Several analyses of microsatellites in the Florida manatee confirm and underscore the low genetic diversity in this subspecies also revealed by the mtDNA analyses (Table 7.3). However, a very recent and more extensive sampling in Florida (362 manatees) suggests a higher diversity, more similar to the dugong (Pause 2007). The population of Antillean manatees in Belize has low levels of diversity, and shows evidence for slight inbreeding and a recent bottleneck, concordant with a history of intense exploitation (Hunter et al. 2010b). Heterozygosity and numbers of alleles at polymorphic loci in both dugongs and manatees were lower than or comparable to averages for other species of mammals from populations considered disturbed or demographically challenged, but definitely lower than averages for populations considered undisturbed or 'healthy' (Table 7.3). Species-level genetic diversity is not as depauperate in sirenians as in some species of mammals for which it may be a major conservation concern (examples are listed in Schultz et al. 2009), but regional populations such as those of manatees in Belize, Florida and Puerto Rico show substantial reductions in genetic diversity that are comparable to the low values of some well-known imperilled mammalian populations (Hunter et al. 2010a,b, in press).

INCREASING HUMAN POPULATIONS AND HABITAT LOSS

Most of the identified threats to sirenians globally are either directly or indirectly influenced by humans, ranging from deaths caused by human-associated protozoans to trauma from boat or ship propellers. As human populations continue to increase (Figure 7.16), so too will the pressures on sirenians and their habitats (see Chapter 5 for additional information on habitat needs) as people seek more of the basic necessities of food, clean

Table 7.3 *General summary of results of microsatellite DNA studies of sirenians pertinent to genetic diversity*

Sample	S	A	He	Source
Dugongs (n = 98, southern Queensland; 37 polymorphic dugong–manatee cross-species microsatellites)	–	4.9 ± 0.37 (range 2–10)	0.52 ± 0.03	Hunter *et al.* (2010a)
Dugongs (n = 50, Moreton Bay, Queensland; 45 loci amplified, 26 selected for summary analysis)	32	4.7 ± 0.44 (range 2–10)	0.51 ± 0.04	Broderick *et al.* (2007)
Antillean manatees, Belize population (n = 118; 16 loci)	16	3.1 ± 0.28 (range 2–5)	0.45 ± 0.04	Hunter *et al.* (2010b)
Antillean manatees, Puerto Rico population (n = 114; 18 candidate loci)	15	3.9 ± 0.36 (range 2–6)	0.45 ± 0.03	Hunter *et al.* (in press)
Florida manatees (n = 50; 14 candidate loci)	8	2.9 ± 0.49 (range 2–6)	0.46 ± 0.04	Garcia-Rodriguez *et al.* (2000)
Florida manatees (n = 116; 39 candidate loci)	18	2.6 ± 0.20 (range 2–4)	0.35 ± 0.04	Tringali *et al.* (2008)
Florida manatees (n = 91; 35 polymorphic dugong–manatee cross-species microsatellites)	–	3.4 ± 0.26 (range 2–8)	0.43 ± 0.03	Hunter *et al.* (2010a)
Florida manatees (n = unspecified)	10	4.2 ± 0.59 (range 2–7)	0.51 ± 0.04	Pause *et al.* (2007)
Florida manatees (n = 362; 18 polymorphic loci)	18	5.3	0.48	Pause (2007)
Mammals (49 species in 'undisturbed populations')	–	8.2 ± 0.69	0.65 ± 0.03	DiBattista (2008)
Mammals (34–35 species in 'hunted/harvested populations')	–	6.6 ± 0.55	0.60 ± 0.02	DiBattista (2008)
Mammals (38–39 species in 'fragmented' populations)	–	6.2 ± 0.54	0.59 ± 0.03	DiBattista (2008)
Placental mammals (74 species in 'healthy' populations)	–	–	0.68 ± 0.01	Garner *et al.* (2005)
Placental mammals (28 species in 'demographically challenged' populations)	–	–	0.50 ± 0.03	Garner *et al.* (2005)

A = mean number of alleles per locus at polymorphic sites; He = mean expected heterozygosity (observed heterozygosities and expected heterozygosities were in close agreement); S = number of polymorphic loci; NA = data not available or not applicable. Values are reported as means ± standard error.

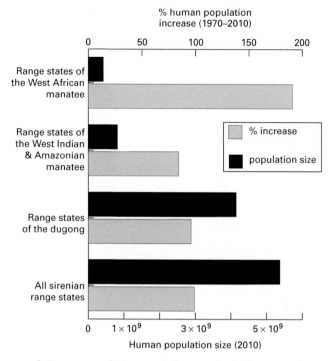

Figure 7.16. Human population growth in countries in the range of sirenians, 1970–2010. Data from the US Census Bureau (2009). Drawn by Shane Blowes.

water and shelter from their surrounding environments (Box 7.14). Intentional and incidental deaths in fisheries can be expected to increase in destitute areas of the developing countries that encompass much of the range of sirenians in Africa, Asia, South America, Oceania and the Caribbean (Chapter 9).

Habitat loss is already considered to be a major threat to all species and subspecies of sirenians (Deutsch *et al.* 2008; Marmontel 2008; Marsh 2008; Powell and Kouadio 2008). For example, habitat loss due to human settlement is considered a threat to dugongs in 82–85% of their huge range (Marsh 2008). In the more advanced nations, pressures for more coastal recreational activities such as boating will continue to occur, and globally rising expectations for improved lifestyles beyond basic necessities will foster greater environmental stresses in countries that move up the scale of development. The synthesis of new chemicals by industry occurs at an accelerating rate and will likely produce compounds with unintended impacts on sirenians and their environments. The increased

Box 7.14 The contemporary
importance of increasing human
populations and habitat loss as threats to
the species/subspecies of sirenians

	Current threat intensity[1]
Dugong	███
Amazonian manatee	███
Antillean manatee	███
Florida manatee	███
West African manatee	███

[1] Shading intensity represents magnitude of threat.

global shipping industry will result in marine ecosystem degradation in many ways (impacts of noise, as one example; Tyack 2008). Additional people will require clearing of more land for food, wood products and minerals with resultant loss of sirenian habitat, particularly for riverine populations. There will be more construction of dams on tropical rivers for electricity, irrigation and water supplies. Changing habitat features and the growth of coastal urban centres may fragment sirenian populations by blocking traditional long-distance migration routes (see Chapters 5 and 8). The number of isolated areas that allow sirenians to persist today by virtue of a low human presence will shrink in size and number, making populations more vulnerable and accelerating loss of genetic diversity. Wars will also take at least an indirect toll on regional populations of sirenians (e.g. Preen 2004). Seagrass habitats have shrunk globally, with a worldwide loss of nearly 30% in extent since the late 1800s, ranking them among the world's most endangered ecosystems (Waycott et al. 2009). Freshwater ecosystems are also considered to be endangered (Dudgeon et al. 2006), and climate change threatens to imperil freshwater species such as manatees. Increasing human use of fresh water will drive engineering responses that will further imperil freshwater biota (Strayer and Dudgeon 2010).

In the future, climate change will not only directly change aspects of the marine environment, but it will affect terrestrial habitats through altered water quality and run-off, which can impact marine plants and enhance harmful algal blooms. All of the threats noted in this chapter will be compounded by likely synergistic interrelationships among the numerous negative factors associated with higher human densities, and new threats are likely to emerge that at this time cannot be predicted. The authors of this book began their professional research on the Sirenia in the 1970s. From 1970 to 2010 the total human population in countries inhabited by sirenians (Chapter 8) has doubled, increasing from 2.7 billion to 5.4 billion (US Census Bureau 2009; Figure 7.16). Although the Sirenia have a rich history of evolutionary responses to long-term changes in the Earth's history (Chapter 3), such changes have not occurred at the extremely condensed time scale of recent human population expansion. Keeping sirenians from extirpation worldwide and allowing their populations to persist as functional components of ecosystems in the face of human population growth will be a daunting challenge for future generations of biologists, conservationists and responsible citizens.

CONCLUSIONS

The modern Sirenia are subjected to numerous threats that directly reduce their populations and degrade their habitats. The various threats impact the four species of sirenians differently (see Boxes 7.1–7.3, 7.5, 7.8–7.14). However, most threats fall into three categories: threats experienced in the developing world; threats more prevalent in developed regions; and threats that may be more universal. Use of meat and other by-products in poor countries, often with a long traditional basis, results in direct poaching by hunters and opportunistic killing when captured in nets set for fish. This situation is exacerbated by food insecurity and the drive for lucrative products such as shark fins. In more developed countries, deaths typically result from collisions with boats, nets set to protect recreational beachgoers from sharks, physical trauma from a variety of other mechanical factors imposed on their environments and dependency on artificial warm-water sources. Throughout the world, the pollution of coastal and riverine environments with chemical contaminants is pervasive and growing, as is the incidence of harmful algal blooms. Infectious diseases are poorly known in both manatees and dugongs, but recent research shows that some sirenians are exposed to diseases that can result in epizootics in other species. It is probably only a matter of time before sirenian populations experience infectious disease outbreaks.

Some sirenian populations are fragmented and isolated, as revealed in analyses of genetic diversity, and likely more susceptible to threats. However, overall genetic diversity is not as depauperate as in some other mammals of major conservation concern.

Contemporary conservation must work in this context. Clearly, different approaches to conservation will be required in the developed world than in poor countries (Chapters 8 and 9). Nonetheless, the habitat needs and life history characteristics (Chapters 5 and 6) of the Sirenia are fundamental to conservation efforts across the ranges of sirenians. Multiple threats operate simultaneously, but targeting those that affect the more critical phases of the life history will have the greatest impact on population dynamics (Runge *et al.* 2007b; Chapter 9), as will prioritising actions that protect the most important habitats (Chapters 5, 8 and 9). Clearly, much is known about present-day threats to the Sirenia and it will be expedient to accelerate their reduction now, because the future promises even greater challenges. Increasing human populations will provide additional stressors, including a rapidly changing climate with uncertain but likely harmful ramifications for sirenian populations throughout the world.

NOTES

1 Around much of the coast of Australia, Aborigines and Torres Strait Islanders own (according to their own law) both the land (country) and the surrounding waters (sea country). This association between traditional people and tracts of land and sea with which they can establish unbroken association is recognised by the Australian legal system through a body of law associated with Native Title. Native Title is the recognition by Australian law that some indigenous people have rights and interests to their land and sea country that come from their traditional laws and customs. Native Title rights and interests may include rights to harvest living resources, including dugongs.

8

Conservation status

Criteria, methods and an assessment of the extant species of Sirenia

Understanding the conservation status of species is very important because it is an indicator of the likelihood of their continuing to exist. In this chapter we first review the criteria used for designating the conservation status of sirenians. We then provide an overview of the methods used for estimating abundance and trends in population sizes. The remainder of the chapter provides an updated summary of findings and our assessment of the status of the extant sirenians. This assessment includes reviews of their status under various international agreements, population sizes and trends, major threats and pertinent conservation impediments and actions.

CRITERIA FOR DESIGNATING CONSERVATION STATUS

The IUCN Red List of Threatened Species (Red List) produced by the Species Survival Commission (SSC) of the International Union for Conservation of Nature (IUCN) is the best-known global conservation status listing and ranking system (IUCN 2009). In 2005 the World Conservation Congress passed a resolution mandating the use of the Red List for national legislation, international conventions, conservation planning and scientific research (IUCN 2005).

The IUCN has assessed sirenians since the Red List was initiated in the 1960s. The first assessments were qualitative and relied on expert opinion (Rodrigues *et al.* 2006). In 1965 all five species of Recent sirenians were classified as 'Status inadequately known – survey required or data sought' (Scott 1965). In 1986 Steller's sea cow was classified as Extinct (IUCN 1986); the four extant species have been classified as Vulnerable since 1982 (Thornback and Jenkins 1982; IUCN 2009).

The nature of the assessments changed in the early 1990s with the implementation of data-driven and objective criteria for estimating extinction probability (Mace and Lande 1991; Mace *et al.* 1992). The process was refined further in 2001 (IUCN 2001). The Red List currently classifies species into nine categories: Extinct (EX), Extinct in the Wild (EW), Critically Endangered (CR), Endangered (EN), Vulnerable (VU), Near Threatened (NT), Data Deficient (DD), Least Concern (LC) and Not Evaluated (NE). Species classified as Critically Endangered, Endangered and Vulnerable are referred to as Threatened (see IUCN 2009).

These categories are assigned by one or more of five criteria or decision rules denoted as A–E (IUCN 2009). The most robust is Criterion E, which is based on a quantitative analysis of extinction probability, such as a population viability analysis. The drawback of this criterion is that it requires data such as absolute population size, trends in abundance, life history traits, natural variability, trends in habitat loss and estimates of parameter uncertainty (Beissinger and Westphail 1998). Such data are generally not available for sirenians, and Criterion E has not been used for their IUCN Red List assessments. (Nonetheless, Haubold *et al.* (2006) used this approach for Florida manatees for the Florida Fish and Wildlife Conservation Commission.) Criteria A–D draw on warning signals that are highly correlated with extinction probability: population reduction (Criterion A); a small distribution area in combination with fragmentation, decline or extreme fluctuations in population size (Criterion B); small population size in combination with population decline (Criterion C); and extremely small population (Criterion D). The numerical thresholds associated with each criterion are different for each of the threat categories. For example, an extremely small and declining population (Criterion C) must contain fewer than 250 mature individuals to be classified as Critically Endangered; the corresponding threshold for Endangered is fewer than 2500 mature individuals (IUCN 2009).

The Red List criteria were developed to assess the probability of extinction of a species at a global scale. However, most conservation policy is developed and implemented within geopolitical borders, such as nation states, or provinces and states within nations. As discussed in Chapter 5, dugongs and manatees are mobile and do not recognise geopolitical borders. Thus there are several scale mismatches (*sensu* Cumming *et al.* 2006) in using the conservation status of sirenians from the Red List. These include spatial scale mismatches for the global scale of IUCN listings, the geopolitical scales of conservation policy and the ecological and genetic scales at which sirenian populations operate. In addition, the Red List assessments use time frames of one to three generations or 100 years, whichever is the longer (i.e. ~20–75 years for sirenians; see 'Estimating generation length' below). This condition means that there is a temporal scale mismatch between the time frames used in the IUCN Red List and the political time frame; the latter is generally a few years at most. In addition, the time frame used for IUCN assessment is so long that there are very few reference points for quantifying past changes in abundance. These mismatches also fail to establish a sense of urgency for action to prevent future changes over similarly long time scales (Godfrey and Godley 2008).

The IUCN Marine Turtle Specialist Group has long debated the relevance and usefulness of using the global scale Red List categories and criteria for globally distributed, long-lived species (e.g. Mrosovsky 1997; Broderick *et al.* 2006). Godfrey and Godley (2008) suggested that the Marine Turtle Specialist Group desist from using current Red List criteria to 'generate implausible global assessments of extinction risk and instead concentrate their efforts on developing more realistic and credible criteria, perhaps for application at the regional level'. Some of their concerns are very relevant to sirenians, particularly the dugong, with an Extent of Occurrence spanning 128 000 km of coastline across at least 38 and up to 44 countries and territories. Nonetheless, we reject the notion of abandoning the IUCN Red Listing process for assessing the conservation status of sirenians, especially given the World Conservation Congress' 2005 resolution mandating the use of the Red List for national legislation, international conventions, conservation planning and scientific research (IUCN 2005).

The IUCN (2009) lists the definitions associated with their categories and criteria. Box 8.1 lists the definitions that are used in the IUCN assessments of the extant sirenians (which are all based on Criteria A and C) and our updates of these assessments.

METHODS FOR GENERATING THE DATA FOR A STATUS ASSESSMENT

In this section we discuss the methods used to acquire the data required to determine the status of sirenians.

Estimating abundance

Estimating the absolute abundance of dugongs and manatees as required by the IUCN Red Listing process (see Box 8.1) is very difficult. Sirenians mostly occur in turbid waters, surface cryptically and only for short periods, and can spend long but variable periods on the bottom (Chapter 5; Figure 8.1). For example, when Florida manatees aggregate at warm-water sites they may rest on the bottom for up to 20 minutes (Edwards *et al.* 2007). In addition, the probability of detecting sirenians varies greatly with environmental conditions both among and within sites.

....

Box 8.1.   IUCN definitions relevant to the assessment of the conservation status of extant sirenians (from IUCN 2001)

Area of occupancy (Criterion A): Area of occupancy is defined as the area within its 'extent of occurrence' (see below) which is occupied by a taxon, excluding cases of vagrancy. The measure reflects the fact that a taxon will not usually occur throughout the area of its extent of occurrence, which may contain unsuitable or unoccupied habitats.

Continuing decline (Criterion C): A continuing decline is a recent, current or projected future decline (which may be smooth, irregular or sporadic) which is liable to continue unless remedial actions are taken.

Extent of occurrence (Criterion A): Extent of occurrence is defined as the area contained within the shortest continuous imaginary boundary which can be drawn to encompass all the known, inferred or projected sites of present occurrence of a taxon, excluding cases of vagrancy. This measure may exclude discontinuities or disjunctions within the overall distributions of taxa (e.g. large areas of obviously unsuitable habitat).

Generation length (Criteria A and C): Generation length is the average age of the parents of the current cohort (i.e. the average age of the parents of the newborn individuals in the population).

Mature individuals (Criteria A and C): The number of mature individuals is the number of individuals known, estimated or inferred to be capable of reproduction.

Reduction (Criterion A): A reduction is a decline in the number of mature individuals of at least the amount (%) stated under the criterion over the time period (years) specified, although the decline need not be continuing.

Population and population size (Criteria A and C): Population is defined as the total number of individuals of the taxon.

Subpopulations (Criterion C): Subpopulations are defined as geographically or otherwise distinct groups in the population between which there is little demographic or genetic exchange (typically one successful migrant individual or gamete per year or less).

....

The shape of the habitats of many sirenians is another challenge. Manatees often reside in narrow, irregularly shaped, linear bodies of water (rivers, creeks, lakes, canals). As outlined in Chapter 5, the coastal populations of manatees are restricted to inshore areas. Dugongs occupy similar habitats in areas where the continental shelf is narrow. These

---

Caption for Figure 8.1
Senegal. The elusive behaviour of the animals is probably exacerbated by hunting pressure. The presence of manatee(s) can be detected as: (a) the outline of the animal below the water and slight surface disturbance; (b) three sets of nostrils at the surface; (c) the back of a surfacing individual; (d) a trail of bubbles in the water. Lucy Keith and Tomas Diagne photographs, reproduced with permission.

Figure 8.1. The challenge estimating the abundance of sirenians in turbid waters is illustrated by these photographs of West African manatees in Angola, Gabon and

habitats are difficult or impossible to survey effectively using standard methods such as line transects because of highly clumped distributions which make it difficult to estimate density (Calambokidis and Barlow 2004).

Sirenians make individual movements over variable spatial scales (Chapter 5), making it difficult to separate changes in abundance from distributional shifts. These difficulties have been compounded by researchers using different techniques for subpopulations of the same species or for the same subpopulation over time, making temporal and spatial comparisons mostly unreliable and inappropriate.

Because of these difficulties, Deutsch *et al.* (2008) estimated population sizes using orders of magnitude (tens, hundreds, thousands) for each of the 20 or more countries where populations of Antillean manatees are likely to exist. We have largely followed their precedent in our assessment of the conservation status of manatees and dugongs. We also outline the techniques used to estimate the abundance of sirenians below, to assist readers in interpreting the primary literature and in planning future assessments. We stress that there is no best technique for estimating the abundance of sirenians. The challenge is to identify the most appropriate technique for the subpopulation of interest, bearing in mind the financial resources available. In choosing a technique it is vital to make all the associated assumptions explicit.

Aerial survey techniques

Aerial surveys are a cost-effective method for estimating the abundance of wildlife because they cover large areas quickly. Such surveys have been the main method used to estimate the size of the Florida manatee population and dugong populations in Australia for many years (see Heinsohn *et al.* 1976; Ligon 1976; Irvine and Campbell 1978 for accounts of early surveys). Aerial surveys have been used to estimate abundance in approximately 20 countries for dugongs, and approximately 16 countries for West Indian manatees, plus some coastal populations of West African manatees. The riverine habitats of Amazonian manatees and most populations of West African manatees are unsuitable for aerial surveys. In addition, cost, logistical constraints and concerns about the cost-effectiveness of aerial surveys in turbid waters have precluded aerial surveys in many developing countries. There are also safety risks inherent in aerial surveys that should be paramount in planning. People have died while conducting aerial surveys for sirenians.

Aerial surveys that intensively cover well-defined areas of Florida manatee habitat, such as winter aggregation sites (see Chapters 5–7 and below), have been useful in obtaining minimum counts as crude estimates of subpopulation size (Shane 1984; Packard *et al.* 1985; Reynolds and Wilcox 1994; Ackerman 1995; Garrott *et al.* 1995). A minimum estimate of the entire Florida manatee population has been obtained from range-wide synoptic surveys which have been conducted nearly every winter since 1991. The surveys coincide with periods of cold weather when manatees aggregate at a limited number of warm-water sites (Haubold *et al.* 2006; Chapter 5). Count estimates of dugong abundance have also been obtained from spatially and temporally limited aerial surveys, mostly conducted parallel to the shoreline (for references, see Marsh *et al.* 2002 and Reynolds *et al.* 2012). Like the surveys of Florida manatees, these surveys provide minimum counts only.

Such surveys have generally aimed to develop indices that reflect spatial patterns of abundance or temporal trends in population size. Unfortunately, such methods do not yield robust indices or accurate population estimates for several reasons: (1) only part of the potential habitat is sampled, so some animals are not present to be counted (absence bias *sensu* Lefebvre *et al.* 1995); (2) results cannot be extrapolated to areas not covered in the survey because the area is not sampled in a random or stratified-random design (sampling bias); and (3) the minimum counts are not adjusted for imperfect detection by the observers. The probability of detecting the sirenians that are present in a survey area has two components: (1) the probability of an animal being near enough to the surface to be seen by an observer (availability bias *sensu* Marsh and Sinclair 1989b); and (2) the probability of an animal being detected and counted by an observer, given that the animal is available to be seen (perception bias *sensu* Marsh and Sinclair 1989b). Many variables may prevent the observer from seeing a dugong or manatee which is available, including weather conditions such as glare and cloud cover, sea state, water turbidity, observer inexperience or fatigue (Marsh and Sinclair 1989a,b; Ackerman 1995; Lefebvre *et al.* 1995; Wright *et al.* 2002; Pollock *et al.* 2006; Edwards *et al.* 2007; Reynolds *et al.* 2012). Therefore, aerial surveys are generally thought to underestimate numbers because it is assumed that observers will not see every animal (Caughley 1977; Eberhardt 1982; Pollock and Kendall 1987; Marsh and Sinclair 1989a,b, Lefebvre *et al.* 1995; Marsh *et al* 2002). However, surveys in areas where a high proportion of animals is available can overestimate numbers because of double counting resulting from movement of animals within the study area during the survey. It is crucial to consider this possibility when designing a survey. Sirenian researchers have attempted to avoid the problems of imperfect detection by using standard count protocols and assuming that the detection probability is constant over time and place, so the counts can be used as robust population indices. However, as is true for many wildlife surveys (Anderson 2001), this assumption is not valid for sirenians (Marsh and Sinclair 1989b; Pollock *et al.* 2006; Edwards *et al.* 2007) because the biases listed above can vary among and within surveys. For example, in a few minutes within a single transect during dugong surveys, the water clarity can vary from clear with the bottom visible and all animals potentially available, to turbid so that only animals at the surface are visible.

Marsh and Sinclair (1989b), Pollock *et al.* (2006) and Edwards *et al.* (2007) have developed techniques to estimate the different components of the detection probability for dugongs and manatees. Pollock *et al.* (2006) found line transect techniques, one of the more usual variants of distance sampling (Buckland *et al.* 2001), unsuitable for dugongs in extensive surveys in turbid waters in tropical Australia because of the difficulty in assigning animals to distance categories. Consequently, most surveys have used strip transects over very large areas (tens of thousands of square kilometres). The researchers standardised the availability and perception bias, using the techniques of Marsh and Sinclair (1989b) or improvements by Pollock *et al.* (2006). Extensive aerial surveys using these techniques have resulted in a comprehensive knowledge of dugong distribution and relative abundance in the coastal waters of most (but not all) of the dugong's range in Australia (Marsh *et al.* 2002, 2004; 2006, 2007a, 2008; Holley *et al.* 2006 and below), the Arabian region (see Preen *et al.* 2012 for an overview) and New Caledonia (Garrigue *et al.* 2008). In contrast, Cockcroft *et al.* (2008) and Provancha and Stolen (2008) successfully used line transect techniques to survey a much smaller population of dugongs in relatively clear water in Mozambique. Strip transects were also used by Miller *et al.* (1998) to estimate the population of Florida manatees using the Banana River estuary, an important warm season habitat.

The Pollock *et al.* (2006) method attempts to estimate absolute abundance. Both population viability analysis and potential biological removal modelling based on estimates of the abundance of the Torres Strait dugong population (Heinsohn *et al.* 2004; Marsh *et al.* 2004) suggested that the population was seriously overharvested. The failure of five large-scale aerial surveys spanning 20 years to detect any evidence of decline in the Torres Strait dugong populations despite an estimated unsustainable harvest level and relatively precise population estimates suggests that the surveys continue to underestimate population size; this probably occurs largely because the availability correction factor is underestimated (Marsh *et al.* 2007a), likely due to the failure to correct for changes in the diving pattern of dugongs with bathymetry and time of day (R. Hagihara, personal communication, 2010). Nonetheless, the Pollock *et al.* (2006) method should improve the homogeneity of the detection bias because it is based on fewer assumptions.

Accurate and precise indices of population size that account for heterogeneity in detection probability are needed for both dugongs and manatees to assess temporal population trends (discussed under 'Estimating population trends' below). In addition, accurate estimates of population size can help in determining the impacts of various sources of mortality on the population, including human-related mortality, using various modelling techniques such as population viability analysis (Beissinger and Westphal 1998) and potential biological removal (Wade 1998).

Mark–recapture techniques

Mark–recapture techniques are much more labour-intensive than aerial surveys. They require at least two sampling occasions and a relatively high proportion of identifiable individuals, rendering the approach logistically infeasible for most subpopulations of more than 1000 sirenians. The assumptions are rigorous and include that: (1) animals have permanent identifiable marks which are recorded correctly; (2) sampling events are independent; (3) animals are independent; and (4) the mark rate is quantified (Williams et al. 2002). The assumption of equal capture probability of all individuals in a group is very difficult to meet because of heterogeneity in behaviour. Mark–recapture probability models generate likelihood functions for catchability for given encounter histories. Estimators have been developed to account for as much heterogeneity as possible, given that the assumptions of the proposed model and estimator are not violated. Methods exist for open or closed populations: a closed population remains effectively unchanged during the study; an open population changes through births, deaths, immigration or emigration. If the assumptions can be met, mark–recapture techniques have many advantages over count methods, especially for small subpopulations occupying defined areas where the risk of double-counting during surveys is considerable.

Mark–recapture (sight–resight) methods based on the photo-identification of naturally marked individuals consistently provide greater precision than counts or line-transect surveys for small populations (< 500) of coastal and riverine dolphins (e.g. Sutaria 2009). Sirenians are less amenable to mark–recapture techniques than coastal dolphins because individuals are not usually identifiable by observers on the surface. Beck and Clark (2012) summarise methods for identifying individual sirenians, including: (1) marks and tags applied at capture, such as passive integrated transponder (PIT) tags, radio tags and freeze brands; (2) photographic identification of unique features such as scars; and (3) genetic analysis of tissues. As they point out, the most appropriate technique depends on species and locality;

not all techniques lend themselves to estimating population size. For example, as explained in Chapters 6 and 7, a high proportion of Florida manatees are individually identifiable from propeller scars, allowing mark–recapture theory and photographic methods to estimate life history parameters. However, the records of scarred Florida manatees are not appropriate for determination of abundance because of violation of underlying assumptions of mark–recapture models for population size estimation. In contrast, Lanyon and her co-workers use mark–recapture methodology to estimate the size of the dugong population of Moreton Bay, Queensland, where large herds predictably occur in shallow water. Dugongs are captured and tagged using turtle tags, PIT tags, photographs and micro-satellite primers in order to meet Assumption 1 above (Lanyon *et al.* 2002).

Providing it is acceptable to local stakeholders, capture and marking of sirenians for population estimation may have promise for small subpopulations in defined areas. Determining the proportion of the population that is unmarked is usually very difficult. The technique needs to be carefully evaluated if the target population is Critically Endangered because of possible risks associated with sampling live animals (even taking small skin samples for genetics may not be risk-free if the animals must be chased).

As pointed out by Calambokidis and Barlow (2004), population size estimates from a mark–recapture study are not usually comparable with estimates from transect surveys, even when the detection probability challenges associated with the surveys are quantified. Mark recapture techniques estimate the size of the population using the study region over a period that is typically months or years. In contrast, transect survey estimates are snapshots of the subpopulation using the area at the time of the survey, typically a few days at most for an aerial survey. The two estimates will not be the same unless the population or subpopulation being studied is closed both geographically and demographically during the sampling period, assumptions unlikely to be met in studies of sirenians (see 'Long-distance movement patterns and migrations' in Chapter 5).

Estimating population trends

The IUCN criteria (IUCN 2001; Box 8.1) recognise that trend in population size is a significant component of estimating conservation status. Population trends are very difficult to estimate for sirenians because of: (1) the difficulties outlined above in obtaining precise estimates of absolute abundance or robust and precise indices of relative abundance; and (2) changes in population size are typically slow. The life history constraints of sirenians mean that their populations are unlikely to grow at more than about 8% per year, usually less (Chapter 6). There are no similar constraints on population decline, but in the absence of catastrophes most declines are likely to be gradual.

Statistical power measures the probability of detecting a significant change in a population if it is actually increasing or declining. Assuming that biases associated with the population estimate are constant or can be estimated, the power of a time series of surveys to detect trends depends on: the length of the monitoring period; the rate of change in the population; the frequency of surveys; and the precision (coefficient of variation or CV) of the population estimate. The last is typically negatively correlated with the size of the population (Gerrodette 1987). Thus, small populations become almost impossible to census for trend as animals become increasingly rare, although improved statistical methods are being developed in an attempt to improve this situation (Dixon and Pechmann 2005; Gerrodette et al. 2011). This situation applies to most local populations of sirenians. Barlow and Reeves (2001) suggest a useful rule of thumb for cetaceans: at least ten annual surveys with high precision (CV < 20%) are required to yield a high probability (> 80%) of detecting a 50% change in total population size. The same may be true for sirenians, with the added qualification that detection probability is usually more variable for coastal and riverine sirenians than for oceanic cetaceans. Thus, a population can be reduced in size by 50% or more before strong statistical evidence of a decline becomes available, despite frequent replicate surveys of high precision. It is also difficult to confirm increases in abundance, making it difficult to confirm whether management interventions are effective (Chapter 9).

Population experts have consistently cautioned against using most of the quantitative data on the sizes of sirenian subpopulations for trend analyses because of the varying detection probability and the confounding effects of movements between survey regions. In addition, count data (as in synoptic winter surveys for manatees) typically do not include an estimate of precision (uncertainty), a deficiency that increases the challenge of modelling trends and optimising a monitoring programme. There are complicated methods for dealing with such problems when a long time series based on a standard protocol is available. Counts at Florida power plants are the prime, and perhaps only, example of such a time series analysis. Craig et al. (1997) and Craig and Reynolds (2004) estimated size and trends of the manatee population along the east coast of Florida by analysing repeated winter counts at power plants surveyed for over 20 years by using a Bayesian hierarchical model that allowed for uncertain detection that varied by region and by water temperature.

Using case studies, Taylor *et al.* (2007) demonstrate that the ability to detect declines in marine mammal stocks with current monitoring programmes is generally poor, even when the decline is considerable. They concluded that it would be impossible to detect even precipitous declines in most marine mammal populations with present levels of investment, survey technology and design. Improvement of performance in detecting declines depends on increasing survey extent and frequency, and developing different methods to detect decline. Such improvements would require a substantial increase in funding, unlikely to be available in the developing countries that comprise most of the ranges of sirenians (see below, especially Tables 8.1, 8.6 8.9). In addition, in many countries the numbers of dugongs and manatees are now apparently so low that it may be impossible to detect trends in abundance. Thus, management intervention should not require the trigger of statistical evidence of reduction in abundance; indeed such a requirement can be a red herring that unduly delays conservation action.

In the absence of more robust information, most of the inferences about past trends in sirenian populations have come from expert opinion and local knowledge of fishers and hunters (Deutsch *et al.* 2008; Marmontel 2008; Marsh 2008; Powell and Kouadio 2008). However, there are two problems with this approach. First, there is the problem of 'shifting baselines' *sensu* Jackson *et al.* (2001); informants' memories cover relatively short timespans and do not incorporate knowledge of historical abundances. Second, anecdotal information may be unreliable if the spatial distribution of the population changes in unpredictable ways. For example, Nietschmann and Nietschmann (1984) inferred that dugongs were so abundant in Torres Strait in the mid-1970s that they were overgrazing seagrass beds close to the major hunting communities. It now appears likely that dugong distribution had changed due to seagrass dieback (Marsh and Kwan 2008). The habitat changes associated with climate change (Chapter 7) are likely to increase the temporal variability in the spatial distribution and population density of all sirenians, making it even more problematic to rely on anecdotal information, and increasing the need for developing new methods for monitoring population trends or new approaches to prioritising and assessing conservation actions, such as risk assessment (Chapter 9).

Survival estimates coupled with breeding rates calculated from longitudinal studies of marked animals provide a robust alternative measure of the status (increasing, stable or decreasing population growth rates) of the various subpopulations of Florida manatees (Chapter 6). High priority should be given to similar studies for suitable reference populations of other sirenians when feasible.

Estimating the percentage of mature individuals

The number of mature individuals in the population is another important component of a conservation assessment (Box 8.1). The percentage of mature Florida manatees was estimated to be 70% (95% CI 0.67–0.73) utilising age-dependent population modelling (Haubold et al. 2006) and 46% from carcass recovery data (Hernandez et al. 1995; Marmontel 1995; Marmontel et al. 1997; Deutsch et al. 2008). The latter figure is probably biased low because of over-representation of calves in the sample. The only potentially unbiased estimate for dugongs is 54%, based on age estimates (Marsh 1980) of the very small sample of 39 females drowned in shark nets deployed for bather protection in Queensland. Kwan's (2002) sample of hunted dugongs is certainly biased against young (small) animals. However, the size distribution of the sample of 234 hunted dugongs from Numbulwar in Australia's Northern Territory reported by Bertram and Bertram (1973) appears less biased, and suggests that 60% of dugongs were of adult size.

Different subpopulations of sirenians are quite likely to have very different age structures and thus different conservation prospects, as discussed under 'Assessing the status of individual sirenians' below. For example, if hunters target large (i.e. mature) animals, the percentage of mature individuals could be substantially lower than in unharvested populations. In view of this uncertainty, we have followed Deutsch et al. (2008) in assuming that the percentage of mature individuals for the various subpopulations is likely to lie somewhere between 45% and 70%.

Estimating generation length

Generation length is an important component of a status assessment because it determines the time frame over which the assessment is made, as explained above. Generation length, the average age of parents of the current cohort (i.e. newborn individuals in the population), is greater in sirenians than the age at first breeding (Chapter 6), and less than the age of the oldest breeder because sirenians breed more than once. Where generation length varies under threat, the more natural (i.e. pre-disturbance) generation length should be used in conservation assessments. Generation length has been estimated as approximately 22–25 years for the dugong and approximately 20 years for the Florida manatee. The dugong estimate is based on the average age of the adults in two relatively small samples: (1) females drowned in the Townsville (Queensland) shark nets (Marsh 1980); and (2) animals caught by hunters at Mabuiag Island in Torres Strait (Kwan 2002; Marsh and Kwan 2008). The estimate for the Florida manatee is based on population modelling and should be more robust than that for the dugong (Deutsch *et al.* 2008). Generation length is likely to be approximately similar for the other sirenians. Thus, using the IUCN thresholds, the periods for consideration of evidence of past or future population decline is between 20 and 75 years, depending on the species of sirenian and IUCN criterion under consideration (IUCN 2009).

Assessing the status of individual sirenians

The approaches described above provide insights into the numbers of individuals in a population or species of concern, but they do not address the *quality* of those individuals. For example, the number of mature individuals in a population is intended to provide information regarding reproductive potential, but the use of biomarkers has demonstrated (J. Reynolds *et al.*, unpublished) that many large animals that would be included in models as 'mature' in fact have low fertility. Thus, a comprehensive approach to understanding the 'status' of a population or species should include assessments of both numbers and quality (e.g. fertility potential, immune health, etc.) of the individuals comprising the group in question. Such assessments are starting to be done for Florida manatees and Moreton Bay dugongs as 'health assessments' become a routine component of animal captures (see also Chapter 7). Useful baselines are emerging from studies of adrenocortical function (Tripp *et al.* 2010) and reproductive hormone levels in faeces (e.g. Larkin *et al.* 2005; Chapter 6).

STATUS ASSESSMENTS OF THE EXTANT SIRENIANS

Obtaining the data required for robust evaluations of the status of sirenians is clearly very difficult. This situation is unlikely to change, especially for species that occur mainly in developing countries. In the remainder of this chapter, we critically evaluate, sometimes for the first time, the status of populations and subpopulations of dugongs and manatees at spatial scales that we regard as appropriate for coordinated conservation management given our present state of knowledge. Gärdenfors (2001) and IUCN (2003) developed guidelines for using the IUCN Red List approach at regional and national scales as a two-step procedure: (1) evaluate the target population against the Red List criteria as if it were isolated; (2) assess the likelihood of other populations of the species influencing extinction risk of the target population and modify the assessment accordingly. We have used this approach for our assessment of the conservation status of the dugong and present regional assessments below. We follow IUCN (2009) and assess the manatee species and subspecies at global scales because their ranges are not as large or fragmented as that of the dugong. Our assessments update the most recent assessment of the status of the extant sirenians, which was conducted by the IUCN–SSC Sirenia Specialist Group from 2006 to 2008 (IUCN 2009).

International conservation conventions

The extant Sirenia are specifically covered under two international conservation conventions: the Convention on International Trade in Endangered Species of Wild Fauna and Flora (CITES) and the Convention on Migratory Species of Wild Animals (also known as CMS or the Bonn Convention). CITES controls international trade in specimens of selected species (CITES 2010). All imports, exports, re-exports and introductions of species covered by CITES must be authorised through a licensing system. Each party to the convention must designate one or more management authorities in charge of licensing, and one or more scientific authorities to advise on the effects of trade on the status of the species. The species covered by CITES are listed in three appendices, according to the degree of protection needed. Appendix I lists species threatened with extinction. Trade in specimens from these species is permitted only in exceptional circumstances. Appendices II and III provide a lower level of protection.

The CMS is an intergovernmental treaty, concluded under the aegis of the United Nations Environment Programme, concerned with the conservation of species and habitats of wildlife that migrate across international borders (CMS 2010). Additionally, the Convention on Wetlands (the Ramsar Convention) protects important sirenian habitats. The Ramsar Convention aims to foster the conservation and wise use of all wetlands through local, regional and national actions, and international cooperation, as a contribution towards global sustainable development (Ramsar 2010). The flagship activity of the convention is its List of Wetlands of International Importance (the 'Ramsar List'). Many of the sites on the list are important habitats for manatees or dugongs. The listing of these sites under the Convention has increased their international profile and in some cases provided an impetus for conservation action.

Capacity for implementation

The capacity to implement conservation initiatives varies across the range states of sirenians. Leverington et al. (2008) evaluated the capacity of different countries to manage protected areas. Management effectiveness varied with a country's degree of socioeconomic development. Not unexpectedly, the capacity scores are much higher in those countries with very high, high and medium Human Development Index (HDI) ratings than for countries with low HDI ratings. In our assessments of conservation status, we use this index as a measure of the capacity of a country to implement strategies to conserve dugongs and manatees, acknowledging that there are other indices that we could have used for the same purpose (e.g. the Human Poverty Index; Human Development Report 2009).

DUGONG

The IUCN (2009) lists the dugong as Vulnerable at a global scale. The dugong is also listed in Appendix I of CITES and under the CMS. All of the dugong's 38 confirmed range states are parties to CITES, except Timor-Leste and Bahrain; about half the range states are also parties to the CMS (Table 8.1). A Memorandum of Understanding (MoU) has been developed to protect the dugong under the auspices of the CMS, and increasing numbers of range states are agreeing to sign this document (Table 8.1).

A crude estimate of the dugong's extent of occurrence (see Box 8.1 for definition) is 860 000 km² (see Figure 8.2). This estimate is based on potential habitat (waters < 10 m deep in its known range). This spans approximately 128 000 km of coastline[1] across at least 38, and up to 44, countries and territories (Table 8.1). Although the dugong still occurs at the extremes of its range, Marsh (2008) concluded that the species is declining or extinct in at least one-third of its range, of unknown status in about half its range, and possibly stable in the remainder – mainly the remote coasts of parts of tropical Australia. Husar (1978) and IUCN (2009) list the dugong as extinct in the waters of several islands, including: the Maldives, the Lakshadweep Islands, Mascarene Islands of Mauritius and Rodrigues, and Taiwan (Husar 1978). Although the species may only have been vagrant at some of these islands (e.g. Taiwan; Hirasaka 1932), there is historical evidence of substantial dugong populations off Mauritius and Rodrigues Islands; these populations were harvested in the eighteenth century (Cheke 1987). Cheke quotes historical accounts of a dugong fishery on Rodrigues in the 1730s.

The dugong's confirmed range states include a mix of developing and developed countries and territories (Figure 8.2; Table 8.1), including eight countries and territories with a very high HDI, three countries with a high HDI, 11 countries with a medium HDI and ten countries with a low HDI. Five countries/territories with a very high HDI (Australia, Bahrain, New Caledonia, Qatar and the United Arab Emirates) support substantial populations of dugongs (thousands), suggesting that the capacity to implement effective conservation initiatives is likely to be the best of any sirenian, with the possible exception of the Florida manatee. The global population of the

Table 8.1 *Definite and possible range states of the dugong showing their Human Development Index (UNDP 2010) and whether or not they are parties to some of the international conventions important to the conservation of sirenians. As of October 2010, all the confirmed dugong range states are signatory to the Convention on International Trade in Endangered Species, except Timor-Leste and Bahrain.*

Human Development Index	#	Range states including territories[1] showing parties to the Convention on Migratory Species and Dugong Memorandum of Understanding as of October 2010
Confirmed range states and territories		
Very High	8	Australia[2,3] (including Cocos Keeling); Bahrain[3], Brunei Darussalam; France (Mayotte; New Caledonia)[2,3]; Japan (Ryukyus); Qatar; Singapore; United Arab Emirates[3]
High	3	Iran[2]; Malaysia; Saudi Arabia[2]
Medium	11	Cambodia; China; Egypt[2]; India[2,3] (including Andaman Island, Laccadive Island, Nicobar Island); Indonesia; Philippines[2,3]; Solomon Islands[3]; Sri Lanka[2]; Thailand; Timor-Leste; Viet Nam
Low	10	Comoros (Union of)[3]; Djibouti[2]; Kenya[2,3]; Madagascar[2,3]; Mozambique[2]; Myanmar[3]; Papua New Guinea[3]; Sudan; Tanzania (United Republic of)[2,3]; Yemen (Socotra)[2,3]
N/A	5	Eritrea[2,3]; Palau[2,3]; Seychelles[2,3]; Somalia[2]; Vanuatu[3]
Total	37	
Possible range states and territories		
Very High	1	Israel[2]
High	2	Jordan[2]; Kuwait
Low	1	Bangladesh[2]
N/A	2	Iraq; Oman
Total	6	
TOTAL	43	Confirmed and possible ranges states and territories

[1] Updated from IUCN (2009).
[2] Party to Convention on Migratory Species.
[3] Party to Dugong Memorandum of Understanding.

dugong is also much greater than for any other extant sirenian, totalling in the tens of thousands in northern Australia alone.

Because of the dugong's huge and fragmented range and the geographic variation in its status, we have made eight regional assessments of its status (below). We have also made subregional assessments in two very extensive regions: (1) East and South-east Asia: major archipelagoes; and (2) Australia. In the absence of robust genetic information, the regions have been chosen based on the apparent fragmentation of the range and geopolitical boundaries and thus function as 'designatable units' *sensu* Green (2005).

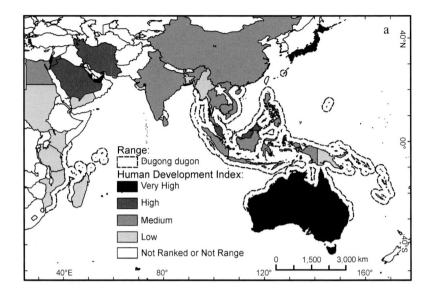

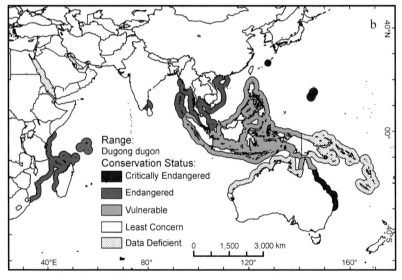

Figure 8.2. The global range and conservation status of *Dugong dugon* as assessed in this chapter. (a) The Human Development Indices (UNDP 2010) of the countries and territories in the dugong's range; (b) the conservation status of the regions as assessed in the text. Drawn by Adella Edwards, reproduced with permission.

East Africa

The known range of the East African dugong population extends south from at least southern Somalia through Mozambique, and includes several offshore islands such as Madagascar and parts of the Comoros Archipelago (see Figure 8.3 for key habitats). Occasionally vagrants are recorded in South African waters (V. Cockroft, personal communication, 2010). This

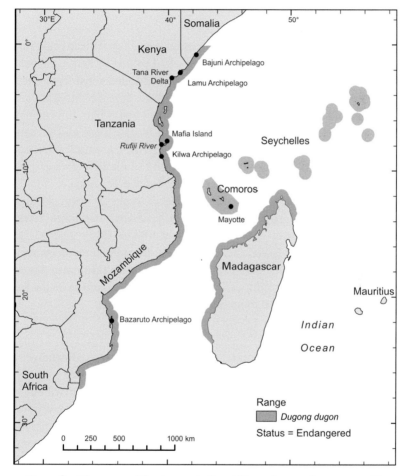

Figure 8.3. The range of the dugong in East Africa, where we have assessed its conservation status as Endangered. Key habitats include the Bajuni Archipelago in Somalia; the Tana River delta and Lamu Archipelago in Kenya; the Rufiji River, Mafia Island and Kilwa Archipelago in Tanzania; Mayotte in Comoros; and the Bazaruto Archipelago in Mozambique. Northern Madagascar is also likely to be important, but to date survey effort has been limited. Drawn by Adella Edwards, reproduced with permission.

region (including Somalia) comprises some 11 000 km of coastline (~8.5% of the dugong's global extent of occurrence based on length of coastline and 4% of the dugong's potential habitat < 10 m deep). There are no data on the connectivity of the East African dugong population with that in the Red Sea. Connectivity is likely to be low: the known Somalian population occurs about 100 km north of Kenya, and 1600 km from the entrance to the Red Sea (Figure 8.2).

Population size and trends

Estimates of the size of the dugong population in the East African region are largely anecdotal, apart from uncorrected aerial counts in Kenya, Mayotte and parts of Madagascar, and much more sophisticated aerial survey estimates from the Greater Bazaruto Archipelago area in Mozambique (Cockcroft *et al.* 2008; Provancha and Stolen 2008). This information suggests dugong numbers in the hundreds in Mozambique and perhaps north-western Madagascar (where J. Kiszka, personal communication, saw six animals in 2010 but none in west-central and south-west Madagascar), and in the tens in Kenya, Mayotte, the Seychelles and Tanzania.

Anecdotal information on population trends suggests major declines since the 1960s and 1970s in Kenya and Tanzania, and more recent declines in Madagascar, Mayotte and Mozambique (Muir and Kiszka 2012). No information is available for Somalia. Declines seem to coincide with the introduction of monofilament nylon gill nets (Muir and Kiszka 2012). In most of these countries, dugong numbers are likely to be too low to confirm trends in a time frame useful for management.

Threats

The most commonly cited contemporary threat is incidental capture in fishing gear, especially gill nets. Moore *et al.* (2010) used interview surveys to conduct a rapid assessment of the incidental capture of marine wildlife in parts of Tanzania and the Comoros. They found that dugong bycatch was rare, presumably reflecting the species' low density. Much of the gill-netting in parts of East Africa targets sharks; fine mesh nets are also used in some areas and also catch dugongs, especially in tidal channels (V. Cockroft, personal communication, 2010). Shark fins are one of the world's most valuable fish products because they are the main ingredient in shark-fin soup, a prestige dish in China. The value of shark fin increases the incentive for illegal gill-netting in East Africa (Attwell in press; Chapter 7). When dugongs are caught incidentally by fishers they are usually killed and sold for meat (see also Chapters 7 and 9). Dugongs are now so rare that direct hunting is no longer considered a problem. Human settlement on coasts, agricultural pollution and destructive fishing are all processes that can damage dugong habitats (see details for Tanzania in Ochieng and Erftemeijer 2003; Mozambique in Bandeira and Gell 2003; and Madagascar in Parent and Poonian 2009). Halpern *et al.* (2008) provide a global context for the severity of human impacts on dugong habitats in East Africa. They conducted a spatial analysis of the cumulative anthropogenic impacts on the world's oceans at a scale of one square kilometre, and list the status of the coastal seas of this region as mainly medium and medium to high impact, indicating that threats to dugong habitats are significant.

Conservation actions

Dugongs are protected by national legislation in most countries in the region (WWF Eastern African Marine Ecoregion 2004). Protected area initiatives aim to protect dugongs in Kenya, Mozambique, Seychelles and Tanzania. Effective enforcement of regulations is a problem because of limited personnel and resources.

Assessment

We conclude that the East African dugong population is likely to be isolated from other dugong populations. We consider this population to be Endangered (ENC1 *sensu* IUCN) on the basis of: (1) a population size less than 2500 mature individuals; and (2) an estimated continuing decline of at least 20% within two generations (~44–50 years) without effective conservation actions, as a result of current and projected future anthropogenic threats. There is a high likelihood of this decline continuing because of high poverty, especially in Mozambique (Table 8.1), which apparently supports the highest numbers of dugongs in the region.

Red Sea and Gulf of Aden

The Red Sea is a long, deep, narrow, semi-enclosed sea. Much of the coast has a narrow fringing reef (Preen *et al.* 2012). This entire region, including the adjoining Gulf of Aden, encompasses some 7000 km of coastline (~6% of the dugong's global extent of occurrence; ~4% of the potential habitat < 10 m deep). Dugongs in the Red Sea are isolated. They are some 1600 km from known dugong habitats in southern Somalia and about the same distance from known dugong habitats in the Arabian/Persian Gulf.

Aerial surveys conducted in 1986 by Preen (1989b) indicated that dugongs occur in three core areas along the Saudi Arabian coast of the Red Sea (Figure 8.4). Dugongs have been reported in the coastal waters of

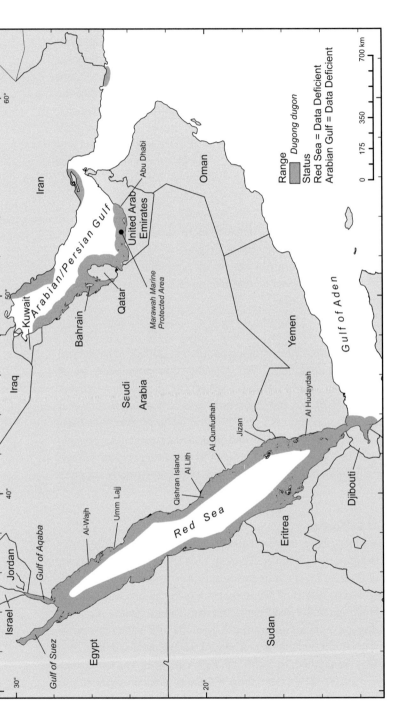

Figure 8.4. The range of the dugong in the Red Sea, Gulf of Aden and Arabian/Persian Gulf. We have assessed its conservation status as Data Deficient throughout this region. Key habitats along the west coast of Saudi Arabia are centred around Sharm Munaibira, which is south of Al-Wajh; around Qishran Island and Al Lith; and extend along the coast from just north of Jizan to Al Hudaydah in Yemen. In the Arabian/Persian Gulf, the Marawah Marine Protected Area is a key conservation initiative in the United Arab Emirates, and should make a significant contribution to dugong and habitat conservation in the southern Arabian/Persian Gulf. Drawn by Adella Edwards, reproduced with permission.

Egypt, Sudan, Djibouti (Marsh *et al.* 2002) and Eritrea (Marsh *et al.* 2002; Mahmud 2010; Figure 8.4). Dugongs have not been recorded in the Gulf of Aden, although Phillips (2003) reports isolated seagrass beds in this region. Preen worked on the Saudi Arabian and Yemen coasts of the Red Sea, and thus little information is available for its African coastline. There have been no follow-up surveys of dugongs along the Arabian coast since 1987.

Population size and trends
Preen (1989b) estimated the size of the dugong population of the Red Sea coast of Saudi Arabia to be 1820 ± 380 (SE) in 1986, on the basis of aerial surveys using the method of Marsh and Sinclair (1989b). Dugong numbers in Yemen were thought to be in the low hundreds (200) based on fishers' comments, gill net mortality, habitat and bathymetry; the African coast of the Red Sea could potentially support a dugong population similar in size to that along the Arabian coast (Preen 1989b). If these estimates are correct, the Red Sea supported several thousand dugongs in 1986. There is no trend information.

Threats
The threats reflect the importance of the Red Sea for artisanal fishing, coastal development and as a globally important shipping route (Gladstone *et al.* 1999; Gerges 2002). Gladstone *et al.* (1999) and Preen *et al.* (2012) report that incidental catches of dugongs in gill nets are widespread and considered a moderate to severe problem. The Red Sea is a major sea route for oil tankers, and oil refinery capacity, loading capacity and exports are increasing. Extensive construction and habitat alteration have occurred along the coastline, especially in Egypt and Saudi Arabia (Preen *et al.* 2012). Serious environmental threats to seagrass in the region include untreated sewage disposal, coastal dredging and reclamation, inshore commercial trawling (including illegal trawling from foreign vessels) and agricultural pollution, especially from shrimp farms (Gladstone *et al.* 1999). Boat traffic is believed to cause disturbance and mortality to dugongs. Halpern *et al.* (2008) consider the anthropogenic impacts on the Red Sea as medium and medium to high, and impacts on the Gulf of Aden as medium to high.

Conservation actions
The dugong is protected by legislation in most countries in the Red Sea region. Most of the marine protected areas in the region with the potential to protect dugongs are 'paper parks', where laws and regulations exist but are not implemented. A Strategic Action Programme for the Red Sea and Gulf of Aden (Gerges 2002) was developed by countries of this region and is coordinated by PERSGA (the Regional Organisation for the Conservation of

the Environment of the Red Sea and Gulf of Aden), with funding provided by the Global Environment Facility and implementation support from the UN Development Programme, UN Environment Programme and the World Bank. The Strategic Action Programme aims to develop a regional framework for protection of the environment and sustainable development of coastal and marine resources. The proposed framework includes increasing public understanding of the threats to the environment, introducing and strengthening environmental legislation and enforcement and improving information systems about the health of the marine and coastal environment.

Assessment
Our regional assessment of the Red Sea and Gulf of Aden dugong population is that it is Data Deficient (see Figure 8.4).

Arabian/Persian Gulf

The Arabian/Persian Gulf (Figure 8.4) is a shallow (average 35 m) semi-enclosed sea about 1000 km long and 200–300 km wide (Phillips 2003), supporting vast areas of seagrass in waters less than 15 m deep on its western and southern shores. Only three species of seagrasses (*Halodule uninervis*, *Halophila ovalis* and *Halophila stipulacea*) occur in the region because of harsh natural conditions: inshore sea temperatures range seasonally from 10 °C to 39 °C, offshore temperatures from 19 °C to 33 °C, and salinities from 38 psu to 70 psu (Phillips 2003). The connectivity (if any) between the dugong populations in the Arabian/Persian Gulf with those in the Red Sea (> 2000 km away) and the Gulf of Kachchh in India is unknown.

Dugongs mainly occur along the southern and western coastal waters of the Gulf (Preen *et al.* 2012; Figure 8.4) over a coastline of some 2000 km (< 2% of the dugong's global range based on both length of coastline and potential habitat < 10 m deep). The core area of dugong habitat is from the central coast of Saudi Arabia to east of Abu Dhabi in the United Arab Emirates (Figure 8.4) (Preen *et al.* 2012). Small numbers of dugongs occur east of this region to the Omani border (Figure 8.4). The bathymetry and latitude along the southern Iranian coasts suggest that some of this region is potential dugong habitat. The dugong's presence has been recently confirmed in Iran, but there are few details (Braulik *et al.* 2010). Cold water (Chapter 5) prevents dugongs from using the seagrass areas of northern Saudi Arabia and Kuwait in winter.

Population size and trends
In the summer of 1986 Preen (1989b) conducted an aerial survey of virtually all the dugong habitat in the Arabian/Persian Gulf using the

strip transect technique of Marsh and Sinclair (1989b), yielding a regional population estimate of approximately 6000 dugongs. Between 1999 and 2006, at least six strip transect aerial surveys were conducted over key parts of the Arabian/Persian Gulf. The resultant estimates confirm that the Arabian/Persian Gulf supports several thousand dugongs (Preen *et al.* 2012), the largest known population outside Australian waters.

Comparison of the various aerial surveys conducted for dugongs in the Arabian/Persian Gulf does not provide robust trend information. Surveys covered different parts of the dugong's range in the Arabian/Persian Gulf, and in some cases used different analytical techniques. Nonetheless, there is no evidence that numbers have declined in the United Arab Emirates or Bahrain, but this assessment must be treated with caution.

Threats

The most serious chronic threats to dugongs in the Arabian/Persian Gulf are incidental and deliberate capture in mesh nets, and habitat loss (Preen *et al.* 2012). Considerable areas of seagrass have been dredged or reclaimed along the Gulf coastline and development is continuing rapidly. Trawling is common in dugong habitat and is likely to cause disturbance to seagrass beds as well as mortality of dugongs. In addition, a massive bridge/cause-way being planned to link Bahrain and Qatar will pass along the edge of one of the premier dugong habitats and may affect current flows, salinity gradients and turbidity (Preen *et al.* 2012).

The Arabian region is a globally important centre of oil production. Oil extraction, treatment and transfer are common activities undertaken throughout much of the dugong habitat in the Arabian/Persian Gulf. Although dugongs and seagrasses can be relatively resilient to chronic, low-level oil pollution (Preen *et al.* 2012), the risk of a catastrophic spill is more serious. At least three die-offs of dugongs have occurred in the western Arabian/Persian Gulf. One was associated with the Nowruz oil spill. The other two die-offs may have been the result of an epizootic, possibly a morbillivirus (Preen *et al.* 2012; Chapter 7).

Given the harsh environment of the Arabian/Persian Gulf, its habitats are particularly vulnerable in a warming world. Halpern *et al.* (2008) rate the threats to the seas in the region as variable, ranging from low to very high impact.

Conservation actions

The dugong is protected by national legislation in Bahrain and the United Arab Emirates (Preen *et al.* 2012). The Marawah Marine Protected Area was declared in 2001, covering 4255 km^2 of core dugong habitat in the United

Arab Emirates (Figure 8.4). This marine protected area is professionally managed and should make a significant contribution to the conservation of dugongs and their habitats in the southern Gulf (Preen *et al.* 2012). Attitudes toward dugongs are improving in the region under the leadership of the United Arab Emirates, which hosts the Dugong MoU under the CMS.

Assessment

We conclude that the information to assess the dugong's risk of extinction in this region is inadequate and have classified its status in the Arabian/ Persian Gulf as Data Deficient (Figure 8.4). The HDI is very high for several countries in the region (Figure 8.2; Table 8.1); conservation initiatives for dugongs should be affordable. There is increased interest in obtaining the knowledge base required to underpin such initiatives, encouraging confidence that they will be informed by science.

Indian subcontinent and Andaman and Nicobar Islands

The dugong's contemporary extent of occurrence on the Indian subcontinent is limited to the Gulf of Kachchh in the state of Gujarat and the coastal waters of Tamil Nadu and Sri Lanka between Colombo and Jaffna. Small numbers of dugongs also occur in the coastal waters of the Andaman and Nicobar Islands (Das and Dey 1999; Hines *et al.* 2012a). Collectively, these regions encompass some 3000 km of coastline (< 3% of the dugong's global extent of occurrence based on both the length of coastline and potential habitat < 10 m deep) (Figure 8.5).

The dugong (sub)populations of the Indian region are fragmented and isolated. The Iranian border is more than 1000 km to the west of the Gulf of Kachchh, and the Gulf of Mannar lies about 3100 km to the west and south. Dugongs are not known from Pakistan. There are no records from the east coast of India or the Sundarban region of India and Bangladesh (in total some 4000 km of coast). It is unknown if dugongs still occur in other parts of Bangladesh, but there are old records from the Chittagong coast (O'Malley 1908 in Jones 1981). Dugongs are apparently extinct in the Maldives and the Lakshadweep Islands (Husar 1978), but may have only occurred there as vagrants.

Population size and trends

Anecdotal information suggests that dugong numbers are now very low in the Indian region. Information on trends is also anecdotal, but strongly suggests that the range in the Indian region has contracted over the last 100 years. Frazier and Mundkur (1991) report that dugongs once occurred along

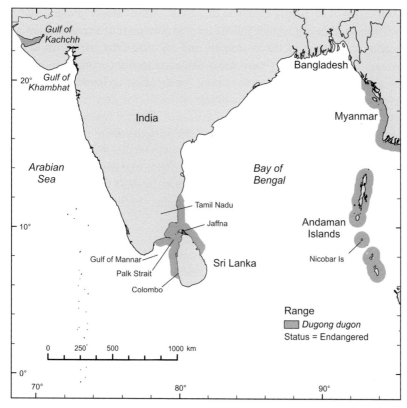

Figure 8.5. Range of the dugong (*Dugong dugon*) in the Indian subcontinent and the Andaman and Nicobar Islands. We have assessed the dugong as Endangered in this region. Note the isolation of the Gulf of Kachchh subpopulation from other populations in the region. Part of the range of the dugong in Myanmar is also shown. Drawn by Adella Edwards, reproduced with permission.

the west coast of India, where seagrasses are patchily distributed (Jagtap *et al.* 2003). There have been no records of dugongs in this region since the early twentieth century outside the Gulf of Kachchh (where dugong feeding trails were photographed in 2009, K. Tatu, personal communication, 2010). Reports suggest a major decline since the 1950s in the Gulf of Mannar–Palk Strait region, and continuing decline in the Andaman and Nicobar Islands (Das and Dey 1999). The population appears too low to confirm trends in a useful time frame.

Threats

The most commonly cited contemporary threats are illegal hunting and incidental capture in fishing gear, particularly gill and shark nets

(Ilangakoon *et al.* 2008; Hines *et al.* 2012a). When dugongs are caught incidentally, they are usually killed and sold (Chapter 7). Some traditional hunting continues in the Andaman and Nicobar Islands (Hines *et al.* 2012a). Seagrass habitats were badly damaged in the Nicobar Islands by the 2004 tsunami. Destructive fishing (push nets), cyclones and coastal development damage dugong habitats. In India, the construction of the Sethu Samudran shipping canal off the coast of Tamil Nadu will not only cause habitat destruction by bisecting the seagrass beds in the Gulf of Mannar, but will also increase the risk of ship strikes (Ilangakoon *et al.* 2008). Halpern *et al.* (2008) list the status of the coastal seas of this region as mainly medium and medium to high impact, indicating that the anthropogenic impacts on the dugong's seagrass habitats are relatively high.

Conservation actions
The dugong is protected by national legislation in both India and Sri Lanka. Protected area initiatives for dugongs exist in the Gulf of Kachchh and in small parts of their known range in both Indian and Sri Lankan waters in the Gulf of Mannar–Palk Strait region. However, enforcement is not effective (especially in Sri Lanka) because of limited personnel and resources. The most pressing need is for incentives to promote alternative livelihoods for fishers using gill nets and destructive fishing gear.

Assessment
We consider that the Indian dugong population is isolated and have classified it as Endangered (ENC1 *sensu* IUCN) on the basis of: (1) estimated size less than 2500 mature individuals; and (2) an estimated continuing decline of at least 20% within two generations (~44–50 years) without effective conservation actions, as a result of current and projected future anthropogenic threats. There is a strong likelihood of this situation continuing because of the high level of rural poverty in the region. (Table 8.1; Figure 8.2).

Continental East and South-east Asia, including the coastal islands
Dugongs are believed to occur in small fragmented populations in the inshore waters of all coastal countries from Myanmar east and north to the southern coast of mainland China, south of Hong Kong (Figure 8.6). The entire region encompasses more than 11 000 km of coastline (~9% of the dugong's global extent of occurrence based on length of coastline; 13% of the potential habitat < 10 m deep). The level of connectivity among the fragmented dugong populations in this region and between the continental and archipelagic regions of South-east Asia is unknown.

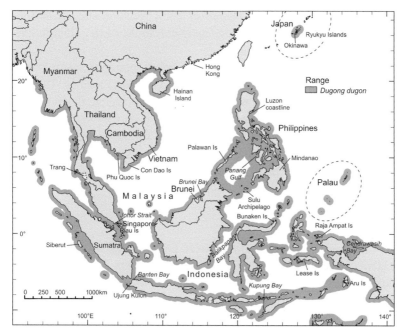

Figure 8.6. The range and conservation status of the dugong (*Dugong dugon*) in South-east Asia, as assessed in this chapter. We have assessed the conservation status of the dugong in Continental South-east Asia and adjacent coastal islands as Endangered; the conservation status for the major archipelagoes in the region as Vulnerable, with the exception of Japan and Palau where the dugong is considered Critically Endangered. Key habitats include: Thailand: Trang Province; Vietnam: Phu Quoc Island and Con Dao Islands; Malaysia/Singapore: Johor Strait and islands of the south-east coast of Peninsula Malaysia; Malaysia/Brunei Darussalam: Brunei Bay; Indonesia: Ujung Kulon and Banten Bay (Java), Riau Islands, Balikpapan Bay (East Kalimatan); Siberut (West Sumatra); Lease Islands (Malaku Province); Aru Islands; Bunaken Island (North Sulawesi); Cendrawasih Bay and Raja Ampat (West Papua); Philippines: Palawan Island, Panang Gulf, Luzon coastline (Quezon-Isabela-Aurora area), coast of Mindanao and the Sulu Archipelago. Part of the range of the dugong in northern Australia is also shown. Drawn by Adella Edwards, reproduced with permission.

Population size and trends

Estimates of dugong numbers in the region are largely anecdotal, but suggest that most populations are patchy at a local scale and very small (tens of dugongs – Marsh *et al.* 2002; Hines *et al.* 2005a,b, 2008, 2012a; Tun and Ilangakoon 2007, 2008). The exception is Trang Province on the Andaman coast of Thailand, which supports that country's largest documented area of seagrass (Supanwanid and Lewmanomont 2003).

Estimates, obtained after the December 2004 tsunami, suggest that the region supports more than 100 dugongs (Adulyanukosol and Thongsukdee 2005; Hines *et al.* 2012a).

Information on trends is based predominantly on interview surveys and suggests that dugongs are declining, except in Trang Province, Thailand (Hines *et al.* 2012a). At local scales throughout this region, dugong populations are mostly too small to confirm trends, even with dedicated monitoring programmes. The only monitoring of which we are aware is in Trang Province, where aerial and interview surveys have been conducted intermittently since 1991 (Marsh *et al.* 2002; Adulyanukosol and Thongsukdee 2005; Hines *et al.* 2005a,b, 2012a).

Threats

The most serious contemporary threat for dugongs of this region is incidental capture in fishing gear (gill and mesh nets, fish traps and fish weirs) or through dynamite and cyanide fishing (Marsh *et al.* 2002; Perrin *et al.* 2005; Hines *et al.* 2012a). Dugongs caught incidentally are unlikely to be released alive because of the high value of their body parts: their meat and tusks represent a windfall for poor fishers (Chapter 7). Dugong habitat loss and damage is widespread due to coastal development, agricultural expansion (especially shrimp farms) and destructive fishing such as push netting (Marsh *et al.* 2002; Hines *et al.* 2012a). According to Halpern *et al.* (2008), the cumulative human impacts on the coastal seas of this region range from low to very high, suggesting a variable status of dugong seagrass habitats which are under threat from illegal fisheries, fishing practices, unmanaged development, reclamation and land-based pollution, particularly from mining (Supanwanid and Lewmanomont 2003; Bujang and Zakaria 2003).

Conservation actions

The dugong is protected by legislation in all countries in this region (Marsh *et al.* 2002; Hines *et al.* 2012a) and by some marine protected areas. The Hepu National Reserve was established to protect dugongs in China (Hines *et al.* 2012a). Important dugong areas in Trang Province in Thailand are protected by Had Chao Mai Marine National Park and Talibong Island non-hunting area, also designated as a Ramsar wetland site. Effective enforcement of conservation rules is a problem throughout most of the region because of poverty, lack of resources and personnel. The most pressing need is for alternative sustainable livelihoods that address poverty and provide incentives for conservation.

Assessment

We have classified dugongs in this region as Endangered (ENC2a(i) *sensu* IUCN) because the available data suggest that: (1) the population size is less than 2500 mature individuals; (2) the decline in the number of mature individuals throughout the region will almost certainly continue as a result of high levels of rural poverty (Table 8.1; Figure 8.2); and (3) no subpopulation contains more than 250 mature dugongs. We acknowledge that a more evidentiary assessment might classify dugongs as Data Deficient in this region.

East and South-east Asia: major archipelagoes

Our distinction between the 'archipelagic' and 'continental' populations of dugongs in East and South-east Asia is arbitrary and influenced by geo-political as well as natural boundaries. We justify this distinction on the basis of the size of the East and South-east Asian region and the vulnerability of fragmented archipelagic populations.

Dugongs are believed to occur in small fragmented populations in the inshore waters of all island groups considered here (Figure 8.6). The entire region encompasses more than 50 000 km of coastline (40% of the dugong's global extent of occurrence based on coastline length; ~50% of its potential habitat < 10 m deep). Connectivity between the various island populations of dugongs with those in Australia and continental South-east Asia is unknown, but dugongs have been tracked moving more than 100 km in the Moluccas (de Iongh *et al.* 2009a,b). The Palau population is very isolated: the closest dugongs are 800 km to the south in Papua Barat and 850 km to the west in the Philippines.

Population size and trends

Estimates of population size in the region are largely anecdotal and based on interviews and/or limited aerial surveys of Palau, Sabah (Malaysia) and parts of Indonesia, especially the Rajah Empat Islands and the Lease Islands (Marsh *et al.* 2002; Perrin *et al.* 2005; Rajamani *et al.* 2006; Shirakihara *et al.* 2007; Rajamani 2008; de Iongh *et al.* 2009a,b; Hines *et al.* 2012a; Ikeda and Mukai 2012). All the evidence suggests that local population sizes are small (tens of dugongs) and patchy, especially in areas where the continental shelf is narrow. Information on trends is anecdotal. The distribution and abundance of dugongs are now believed to be vastly reduced throughout the region and so low that monitoring will be of limited value. The only monitoring of which we are aware is in Palau, where there have been intermittent aerial surveys since the 1970s (Marsh *et al.* 1995, 2002), consistently counting low numbers of dugongs (tens), almost certainly an underestimate. There is a strong likelihood of continuing decline because of the high poverty in some countries in the region, especially in rural coastal areas.

Threats

Incidental capture in fishing gear is the most commonly cited contemporary threat (Marsh *et al.* 2002; Perrin *et al.* 2005; Rajamani *et al.* 2006; Shirakihara *et al.* 2007; Rajamani (2008); de Iongh *et al.* 2009a,b; Jaaman *et al.* 2009; Hines *et al.* 2012a – but see Moore *et al.* 2010). Rajamani *et al.* (2006) and Rajamani (2008) consider that dugongs caught incidentally are unlikely to be released alive because of the high value of their body parts. The deliberate harpooning of dugongs for local use is reported from the Aru Islands and mortality of dugongs from vessel strike has also been reported both in Balikpapan Bay and in Ambon (de Iongh *et al.* 2009a,b). Coastal development and destructive fishing damage dugong habitats in many areas (Bujang and Zakaria 2003; Kuriandewa *et al.* 2003). In Okinawa, the proposed construction of an offshore landing facility for a military base has been extremely controversial, in part because of the projected loss of critical dugong habitat (Shirakihara *et al.* 2007; Chapter 1). Halpern *et al.* (2008) list the status of the coastal seas of this region as mainly medium to high impact, but there are regions of very high impact around Malaysia and off the coast of China, and regions of very low impact such as coastal waters south of Borneo, suggesting variable status of seagrass habitats.

Conservation actions

Dugongs are protected by national legislation throughout this region (Marsh *et al.* 2002). In addition, Indonesia (de Iongh *et al.* 2009b) has developed action plans for dugong conservation. Indonesia, the Philippines, Malaysia, Papua New Guinea, the Solomon Islands and Timor-Leste all endorsed the Coral Triangle Initiative on Coral Reefs, Fisheries and Food Security. This initiative is a multilateral partnership to safeguard the rich marine resources of the Indo-Pacific region (including threatened species such as dugongs) and may provide a vehicle for coordinated dugong conservation efforts throughout the region (Coral Triangle Initiative 2009).

Effective enforcement of management regulations is a problem, because of lack of resources and personnel. The most pressing need is for incentives to promote alternative livelihoods for fishers using gill nets and destructive fishing methods.

Table 8.2 *Assessment of the dugong subpopulation in various parts of the archipelagos of South-east Asia. These subregions have been tentatively selected on the basis of their isolation from each other, and are not based on genetic or demographic criteria.*

Subregion	Tentative IUCN category	Criteria
Indonesia, East Malaysia, Brunei and Timor-Leste	Vulnerable	VUA2bcd: an observed, estimated, inferred or suspected population size reduction of ≥30% over the last ten years or three generations, whichever is the longer, where the reduction or its causes may not have ceased OR may not be understood OR may not be reversible, based on: (b) an index of abundance appropriate to the taxon; (c) a decline in area of occupancy, extent of occurrence and/or quality of habitat; (d) actual or potential levels of exploitation.
Philippines	Vulnerable	VUA2bcd: as for Indonesia (above)
Japan	Critically endangered	CRC1: population size estimated to number fewer than 250 mature individuals and an estimated continuing decline of at least 20% within two generations (~44–50 years)
Palau	Critically endangered	CRC1: as for Japan.

Assessment

We consider the small isolated dugong populations in Japan and Palau to be Critically Endangered, and the (sub)populations in the remainder of the region Vulnerable (Table 8.2). We acknowledge that a more evidentiary assessment might classify the dugong populations of Indonesia, East Malaysia, Brunei, Timor-Leste and the Philippines as Data Deficient. Our assessment assumes limited movement among the various parts of the region and between eastern Indonesia and Australia. The latter conclusion is supported by limited genetic evidence (D. Blair *et al.*, unpublished).

Australia

Dugongs occur from Shark Bay in Western Australia (25° S) across the northern coastline of the continent to Moreton Bay in Queensland (27° S) (Marsh *et al.* 2002; Figure 8.7). Archaeological analyses and contemporary records indicate stranded dugongs as far south as ~36.5° S on the east coast, with occasional sightings south to 32–33.5° S in summer (Allen *et al.*

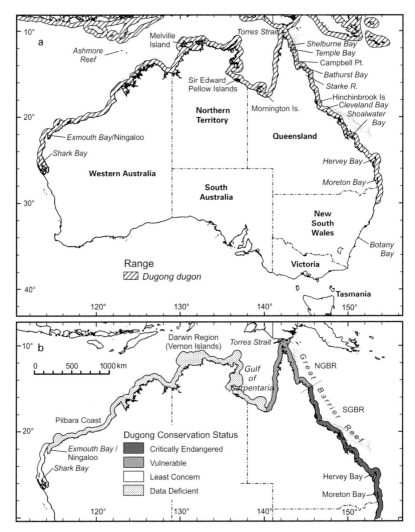

Figure 8.7. (a) Range and key habitats and (b) conservation status of the dugong in Australia as assessed in this chapter. Key habitats include: Shark Bay, Exmouth Bay and Ningaloo (Western Australia); Melville Island and Sir Edward Pellew Islands (Northern Territory); Mornington Island, Torres Strait, Shelburne Bay, Temple Bay, Campbell Point, Bathurst Bay, Starke River, Hinchinbrook Island, Cleveland Bay, Shoalwater Bay, Hervey Bay, and Moreton Bay (Queensland). Part of the range of the dugong in Papua New Guinea and Indonesia is also shown. Drawn by Adella Edwards, reproduced with permission.

2004). The winter range, which extends to 25–27° S on the east coast, encompasses some 24 000 km of coastline (19% of the dugong's global extent of occurrence based on coastline; 16% of the potential habitat < 10 m deep).

Connectivity between Australian populations and those in South-east Asia is unknown (see above). Dugongs genetically similar to both Australian and South-east Asian dugongs (which are genetically distinct) have been sampled at Ashmore Reef on the edge of the north-west Australian continental shelf, suggesting some intermingling of different stocks (D. Blair *et al.*, unpublished). A dugong (probably a vagrant) has recently been recorded at Cocos (Keeling) Islands (Hobbs *et al.* 2007).

Population size and trends
Dugongs in Australian waters exhibit high levels of genetic diversity and population structure at large spatial scales (hundreds of kilometres; see Chapters 3 and 7). Estimates of population size (Table 8.3) are based on quantitative aerial surveys that correct for the sampling fraction and various visibility biases (Marsh and Sinclair 1989b; Pollock *et al.* 2006). Thus the population estimates are not comparable with estimates from most other areas in the dugong's range (except Bazaruto Bay in East Africa, the Arabian region and New Caledonia). These surveys indicate that the dugong is the most abundant marine mammal in the coastal waters of northern Australia, with estimates from the > 120 000 km² area surveyed since 2005 totalling almost 70 000 dugongs (Table 8.3). Estimates are unavailable or outdated for large regions of Australia, including the Western Australian coast north of Exmouth Gulf, most of the Northern Territory coast outside of the Gulf of Carpentaria, much of the Dugong Sanctuary in Torres Strait, Ashmore Reef and offshore territories such as Cocos (Keeling) Islands.

Population estimates based on aerial surveys (Table 8.3) are almost certainly underestimates. For example, the 2005 estimate for Moreton Bay reported by Marsh *et al.* (2006) is 421 ± 60 (SE) dugongs. Lanyon has marked approximately 650 dugongs in Moreton Bay since 2001 for her mark–recapture study (Lanyon *et al.* 2010). As pointed out above, survey and mark–recapture studies are not strictly comparable unless the population is closed (not the case in Moreton Bay). Nonetheless, the discrepancy between the aerial survey results and the estimate of the minimum number known to be alive (Lanyon *et al.* 2010) reinforce that the surveys underestimate dugong numbers, despite the attempts to correct for visibility biases.

The catch per unit effort data collected by the Queensland Shark Control Program indicate that the dugong population on the urban coast of Queensland declined precipitously between the 1960s and early 1980s (Marsh *et al.* 2005;

Table 8.3 *Summary of the distribution and relative abundance of the dugong in Australia obtained from aerial surveys using two transect techniques which use different methods to correct for availability bias. The estimates developed using the Pollock et al. (2006) method are likely to be the more robust. Note: aerial surveys are likely to significantly underestimate dugong populations and have a low power to detect trends unless trends are very large.*

Bioregion	Region	Jurisdiction	Area (km²)	Date of last survey	Reference	Population estimate ± SE (Marsh and Sinclair 1989b method)	Population estimate ± SE (Pollock et al. 2006 method)	Status of population suggested by surveys
North-western	Shark Bay	WA[1]	13 000	2007	Hodgson et al. (2008)	14 022 ± 1230	9347 ± 1204	Stable
	Exmouth Gulf and Ningaloo	WA and Commonwealth	3180	1999 (July)	Hodgson et al. (2008)	1411 ± 561	704 ± 354	Stable
	Pilbara coast	WA and Commonwealth	N/A	2000 (Apr)	Prince et al. (2001)	2046 ± 376	–	Unknown
Northern	Northern coast of the Northern Territory	NT[1]	N/A	1983	Bayliss (1986)	13 800 ± 2683	–	Unknown
	Darwin region (Vernon Islands)	NT	1750	2002 (Mar., Aug.)	Whiting (2008)	956 ± 80	–	Unknown
	Gulf of Carpentaria	NT, Qld[1] and Commonwealth	35 592	2007	Marsh et al. (2008)	–	12 438 ± 1951	Stable

Torres Strait	Torres Strait	Qld, Commonwealth and PNG	30 560	2006 (Mar.–Apr.)	Marsh et al. (2007a)	19 583 ± 995	14 767 ± 2292	Fluctuating
Eastern	Cape York coast[2]	Qld and Commonwealth	25 440	2006 (Mar.–Apr.)	Marsh et al. (2007a)	8239 ± 992	8812 ± 1769	Stable
	Southern Great Barrier Reef	Qld and Commonwealth	4652	2005	Marsh et al. (2006)	2580 ± 271	2059 ± 413	Stable
	Hervey Bay	Qld	4936	2005	Marsh et al. (2006)	2547 ± 410	2077 ± 543	Fluctuating
	Moreton Bay	Qld	1627	2005	Marsh et al. (2006)	454 ± 41	421 ± 60	Fluctuating

[1] NT=Northern Territory; PNG=Papua New Guinea; Qld=Queensland; WA=Western Australia
[2] The boundary between the Cape York and urban coasts (southern Great Barrier Reef, Hervey Bay, Moreton Bay) of Queensland is ~15° 30′ S.

Table 8.4 *Threats to the dugong subpopulation in various parts of the Australian region and the responses to these threats. These subregions have been selected based on the Australian marine bioregions and patterns of human use and are not based on genetic or demographic criteria.*

Threat	Region impacted	Response
Causes of anthropogenic loss of seagrass		
Coastal development	Urban coast of Queensland (south of 15° 30′ S) and major ports in the remainder of the range (e.g. Weipa, Karumba, McArthur River, Groote Island, Gove, Darwin, Port Headland)	Some attempts to limit or offset impacts through state and commonwealth Environmental Impact Assessment Process.
Agricultural land use and/or mining	Of most concern on urban coast of Queensland (south of 15° 30′ S)	Addressed by Reef Water Quality Protection Plan (2009) along urban coast of Great Barrier Reef and South-east Qld Healthy Waterways Strategy (2007–2012). Limited response in other areas.
Trawling in seagrass areas		Largely prohibited in Great Barrier Reef region, Gulf of Carpentaria and Western Australia; seagrass beds not adequately mapped in most of the remainder of the range.
Fishing practices that catch dugongs as bycatch		
Gill and mesh nets including shark nets	Problem throughout much of the dugongs' range in Australia, especially in Queensland	Shark nets for bather protection replaced with drum lines in most locations on urban coast of Queensland (Marsh *et al.* 2005). Risk of dugongs being killed in commercial gill nets largely managed by zoning in Great Barrier Region (Dobbs *et al.* 2008; Grech *et al.* 2008) and near mouth of McArthur River in the Northern Territory (Marsh *et al.* 2002). In Western Australia, impact has been reduced by shift from demersal gill netting to wetlining (Department of Environment and Conservation 2007).

Ghost fishing	Of most concern in the Gulf of Carpentaria	People from (indigenous) communities all around the Gulf of Carpentaria work together to get rid of marine debris in their sea country through the Gulf of Carpentaria Ghost Net Program funded by the Australian government
Trawl	Rare occurrence	Turtle Excluder Devices are mandatory in Australian trawl fisheries and have been found effective at excluding other species of marine wildlife.

Other causes of dugong mortality

Hunting and poaching	Occurs throughout most of the dugongs' range	Hunting is a Native Title right for Traditional Owners and Article 22 fishery under the Torres Strait Treaty between Australia and Papua New Guinea (see Chapter 7). Some communities have adopted a voluntary moratorium on the urban coast of Queensland, but poaching still occurs. Indigenous communities in northern Australia are being funded by the Australian government to develop and implement community-based hunting management plans to manage their dugong harvest as part of a more comprehensive programme of sea country management (e.g. see Havemann *et al.* 2005 and NAILSMA 2010; Chapter 9).
Vessel strike and disturbance	Close to major cities (especially Brisbane) on the urban coast of Queensland (south of 15° 30′ S) and large ports in the remainder of the range (e.g. Weipa, Karumba, McArthur River, Groote Island, Gove, Darwin, Port Headland) and associated with tourist operations in Shark Bay and Ningaloo in Western Australia.	Addressed by speed limits (e.g. Moreton Bay) and voluntary vessel lanes (e.g. Hinchinbrook Channel) in some areas on the urban coast of Queensland. Dugong tourism is regulated by license in Queensland and Western Australia.

Figure 7.6). In contrast, aerial surveys since the mid-1980s suggest that populations are now stable in Shark Bay, the Exmouth/Ningaloo Reef region of Western Australia, the Gulf of Carpentaria, the northern Great Barrier Reef and the southern Great Barrier Reef (Marsh et al. 2006, 2007a, 2008; Hodgson et al. 2008). However, as discussed above, the power of these surveys to detect declines is weak unless the declines are very large.

The surveys suggest that dugong numbers in Moreton Bay, Hervey Bay and Torres Strait fluctuate over time (Marsh et al. 2006, 2007a; Table 8.3). The fluctuations in estimates for Hervey Bay and Moreton Bay are attributable to dugongs moving between the two bays or from shallow to deeper water within bays, especially after 1000 km² of seagrass were lost from Hervey Bay following two floods and a cyclone in 1992 (Preen and Marsh 1995; Marsh et al. 2006; Chapter 6). Movements of dugongs from shallow to deeper water within the survey region and between that region and the unsurveyed areas to the west may contribute to the population fluctuations observed in Torres Strait (Marsh et al. 2004, 2007a). The fluctuations may also reflect over-harvest as indicated by modelling using both potential biological removal and population viability analysis (Heinsohn et al. 2004; Marsh et al. 2004).

Threats

Throughout much of northern Australia, the greatest source of dugong mortality is legal indigenous hunting (Chapter 7). In contrast, threats to dugongs on the urban coast of Queensland are similar to those in most other parts of their range. Dugongs are also killed by illegal poaching and incidental capture in nets (Figure 7.5). However, the sale of dugong meat is illegal and the imperative to sell incidental catch is less than in countries where food security is a problem.

The remote tropical waters of much of northern Australia are subject to very low levels of human impact (Halpern et al. 2008) and threats to dugong habitats are low. Grech (2009) and Grech et al. (2011a) used expert opinion to evaluate the relative impact of hazards to seagrass habitats in the Great Barrier Reef region, the southern two-thirds of which is more urbanised than the remainder of the dugong's Australian range. The following threats to habitat were identified: agricultural, urban and industrial run-off; urban and port infrastructure development; dredging; shipping accidents; trawling; recreational and commercial boat damage; and commercial fishing other than trawling (see Table 8.4 for details).

Conservation actions

As a developed country, Australia has been able to implement significant measures to protect dugongs. Conservation is occurring at national,

Table 8.5 *Assessment of the dugong subpopulations in various parts of Australia. These subregions have been selected based on the Australian marine bioregions and patterns of human use and are not based on genetic or demographic criteria. The population estimates on which these assessments are based are listed in Table 8.3.*

Subregion	Tentative IUCN category	Criteria	Key references
Urban coast of Queensland: Queensland–New South Wales border (28° S) to Cooktown (15° 30′ S)	Critically Endangered	A2b. Reduction in population size based on estimated size reduction of ≥80% over the last three generations (66–75 years) where the cause of reduction may not have ceased.	Marsh *et al.* (2005, 2006)
Eastern bioregion: Cape York coast of Great Barrier Reef World Heritage Area (10° 41′ S to 15° 30′ S)	Vulnerable	A2d. Reduction in population size based on suspected population size reduction of ≥30% over the last three generations (66–75 years) where the cause of reduction may not have ceased.	Heinsohn *et al.* (2004); Marsh *et al.* (2004, 2007a)
Torres Strait bioregion: Torres Strait between Cape York and Western Province of Papua New Guinea	Vulnerable	A2d. Reduction in population size based on suspected population size reduction of ≥30% over the last three generations (66–75 years) where the cause of reduction may not have ceased.	Heinsohn *et al.* (2004); Marsh *et al.* (2004, 2007a)
Northern bioregion: Cape York (10° 41′ S 142° 30′) west to Northern Territory–Western Australia border (129° E)	Data Deficient	Some of the region has never been surveyed. Data for dugongs and the threats are not mapped or quantified.	Saalfeld and Marsh (2004)
North-west bioregion: Northern Territory–Western Australian border (129° E) south-west to Muldron Islands off North-west Cape (21° 47′ S 114° 10′ E)	Data Deficient	Some of the region has never been surveyed. Data for dugongs and the threats are not mapped or quantified. Increasing exploration for offshore oil and gas in coastal waters.	
North-west bioregion: North-west Cape–Shark Bay (~22–28° S)	Least Concern	This region supports an apparently stable population of approximately 10 000 dugongs protected by its World Heritage status and the Shark Bay and Ningaloo Marine Parks. Anthropogenic threats are low.	Hodgson *et al.* (2008)

state/territory and local levels. The responses to the various threats are summarised in Table 8.4.

Assessment

We subdivided this region for our assessment because of the size of the dugong's range and the spatial variability of impacts (Table 8.5; Figure 8.7b): (1) urban coast of Queensland: Critically Endangered; (2) northern Great Barrier Reef and Torres Strait: Vulnerable; (3) the northern tip of Cape York west to North-West Cape in Western Australia: Data Deficient; (4) North-West Cape to Shark Bay in Western Australia: Least Concern. We acknowledge that a more evidentiary assessment might classify the northern Great Barrier Reef and Torres Strait as Data Deficient.

Western Pacific islands

Dugongs are widely distributed in the tropical and subtropical island waters of the Western Pacific region, mostly in scattered subpopulations (Marsh *et al.* 2002; Garrigue *et al.* 2008; Kinch 2008; Bass 2009, 2010; Figure 8.8). Vanuatu is the eastern limit of the range. This entire region encompasses more than 11 000 km of coastline (9% of the dugong's global extent of occurrence; 6% of the potential habitat < 10 m deep). Connectivity between the Western Pacific island populations and populations in Australia and South-east Asia is unknown, although Australia and Papua New Guinea certainly share the Torres Strait dugong population.

Population size and trends

Estimates of dugong numbers in the region are anecdotal, apart from New Caledonia where a quantitative aerial survey in 2003 using the methodology outlined in Pollock *et al.* (2006) resulted in an estimated population of about 2000 dugongs (Garrigue *et al.* 2008). Thus the regional population is likely to be in the thousands. There is no reliable information on trends, although a majority of the fishers interviewed by Kinch (2008) in the autonomous region of Bougainville (part of Papua New Guinea) claimed that the numbers were increasing, a conclusion supported by Yen (2006) and Bass (2009), who interviewed fishers in the Samarai region of Milne Bay and on the islands of Bougainville and Manus (all in Papua New Guinea), respectively.

Threats

Dugongs have high cultural value and legal traditional hunting is widespread and probably the main source of dugong mortality in this region. In some regions there has been a technology switch from harpoons to gill nets. For

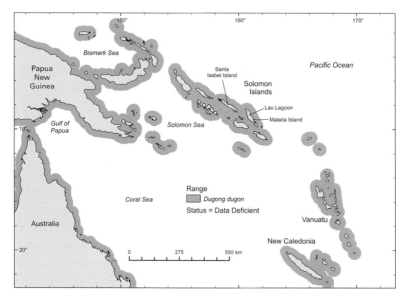

Figure 8.8. Range and key habitats of *Dugong dugon* in the western Pacific, where we have assessed its conservation status as Data Deficient. Key habitats include the Papua New Guinea waters of Torres Strait; the southern coast of Santa Isabel Island; Lau Lagoon on Malaita in the Solomon Islands; and the coastline of New Caledonia – particularly the centre and southern parts of the west coast. Part of the range of the dugong in northern Australia is also shown. Drawn by Adella Edwards, reproduced with permission.

example, in Bougainville, Kinch (2008) reports that consumed dugongs were all captured in gill nets. The dugong harvest of the Papua New Guinea villagers along the northern coast of Torres Strait is significant; many animals are now caught using nets (H. Marsh, unpublished information from a 2009 workshop in Daru, Papua New Guinea). Other concerns include 'swimming with dugongs' tourism in Vanuatu (Marsh *et al.* 2002). Halpern *et al.* (2008) assess the status of the coastal seas of this region as low impact in the Gulf of Papua and New Caledonia, medium to high off northern Papua New Guinea, and high in the waters off Vanuatu, presumably because of the offshore fishing impacts. Coles *et al.* (2003b) provide information on the threats to seagrasses of this region, including coastal reclamation, tourist development, port developments and small boat marinas.

Conservation actions

The dugong is protected by legislation in New Caledonia, Papua New Guinea, Solomon Islands and Vanuatu, although the capacity to implement protection is limited. A regional Action Plan for Dugongs has been developed by the South Pacific Regional Environmental Programme (SPREP) (Gillespie 2005). The Australian and Papua New Guinea governments are working bilaterally to attempt to address the issue of the dugong harvest by the Papua New Guinea villages along the northern coast of Torres Strait.

Assessment
We consider the Western Pacific Islands dugong population to be Data Deficient.

AMAZONIAN MANATEE

The Amazonian manatee is listed as Vulnerable at a global scale by IUCN and on Appendix I of CITES. All four range states are signatories to CITES and the Ramsar Convention on Wetlands. Peru and Ecuador are also parties to the CMS (Table 8.6). All range states are developing countries with a high HDI (Table 8.6).

Trichechus inunguis occurs in the freshwater systems in the vast (7 000 000 km²) Amazon Basin from headwaters in Ecuador, Peru and Colombia to the Atlantic estuary in Brazil (Figure 8.9), where it is sympatric and apparently hybridises with the Antillean manatee (Garcia-Rodriguez *et al.* 1998; Vianna *et al.* 2006; Chapters 3 and 7). We undertook a single assessment for this species as genetic studies based on mitochondrial DNA provide little evidence of population structure (Cantanhede *et al.* 2005) and because most of the threats are similar throughout the region. Our assessment is mainly based on information from Marmontel (2008) and Marmontel *et al.* (2012).

We could not reliably estimate the Amazonian manatee's extent of occurrence because of the apparent lack of a recent scientific estimate of the area covered by the water of the Amazon River and its tributaries, which varies greatly both within and among years. Using 94-year records from *in situ* gauge stations, Sippel *et al.* (1998) estimated that the mean flooded area of the Amazon River main stem alone was 46 800 km², of which the open water surfaces of river channels comprised about 20 700 km². Hess *et al.* (2003) mapped the wetland area for the central Amazon region, using satellite imagery 'ground-truthed' by high-resolution aerial digital videography. The wetlands were 96% inundated at high water and 26% inundated at low water. Flooded forest constituted nearly 70% of the entire wetland area at high water, although there were large regional variations in the proportions of wetland habitats. Frappart *et al.* (2005) used remote sensing to estimate that the total area subject to flooding in the basin of the Negro River, which comprises only 12% of the entire Amazon Basin, ranges from approximately

Table 8.6 *Range states of the extant species and subspecies of manatees showing their Human Development Index (UNDP 2010) and whether or not they are parties to two international conventions, important to the conservation of sirenians. As of October 2010, all states were parties to the Convention of International Trade in Endangered Species, with the exception of Angola.*

Human Development Index	#	Range States including Territories[1] showing parties to the Convention on Migratory Species and the 2008 Memorandum of Understanding under the Convention on Migratory Species to conserve the small cetaceans and manatees of West Africa and Macronesia[3] as of October 2010
Amazonian manatee		
Confirmed range states and territories		
High	4	Brazil; Colombia; Ecuador[2]; Peru[2]
Total	4	
Antillean manatee		
Very high	2	France (French Guiana)[2]; United States (Puerto Rico)
High	9	Belize; Brazil; Colombia; Costa Rica[2]; Jamaica; Mexico; Panama[2]; Trinidad and Tobago; Venezuela
Medium	6	Dominican Republic; Guatemala; Guyana; Honduras[2]; Nicaragua; Suriname
Low	1	Haiti
N/A	1	Cuba[2]
Total	19	
Florida manatee		
Confirmed range states and territories		
Very high	1	United States
Total	1	
Possible range states and territories		
High	1	Bahamas[1]
Total	1	
TOTAL	2	Confirmed and possible range states and territories
West African manatee		
Confirmed range states and territories		
Medium	3	Congo[2,3]; Equatorial Guinea[3]; Gabon[2]
Low	18	Angola[2,3]; Benin[2,3]; Cameroon[2]; Chad[2,3]; Congo (Democratic Republic of)[2]; Côte D'Ivoire[2,3]; Gambia[2]; Ghana[2,3]; Guinea[2,3]; Guinea-Bissau[2,3]; Liberia[2,3]; Mali[2,3]; Mauritania[2,3]; Niger[2,3]; Nigeria[2]; Senegal[2,3]; Sierra Leone; Togo[2,3]
Total	21	

[1] Updated from IUCN (2009), which lists the Bahamas as a range state for the Antillean manatee. Recent information suggests that manatees in the Bahamas may be Florida manatees. At least one Florida manatee has also been seen in Cuban waters.
[2] Party to Convention on Migratory Species.
[3] Party to Convention on Migratory Species MoU to conserve the small cetaceans and manatees of West Africa and Macronesia.

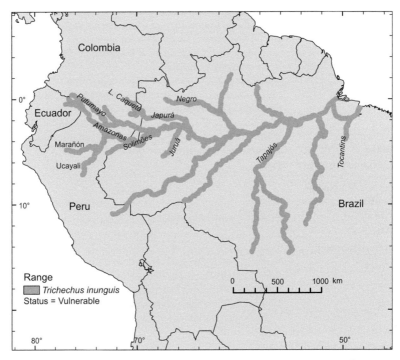

Figure 8.9. Range of the Amazonian manatee (*Trichechus inunguis*). We have assessed its conservation status as Vulnerable. Key habitats include the Brazilian Amazon, particularly in the Japurá, Juruá, Madeira, Negro, Nhamundá, Purús, Uatumã, Solimões, Tapajós Tocantins, Trombetas and Xingú river systems; the extreme north-east of Ecuador; rivers in the Yanayacu catchment area in Peru and the Amazonas rivers near the Colombian border; the Amazon, Putumayo and lower Caquetá rivers in Colombia. Note the range of the West Indian (Antillean) manatee is not shown on this figure (see Figure 8.10 for details). The ranges of the two species overlap at the mouth of the Amazon River. All range states have a high Human Development Index (UNDP 2010). Not all these locations can be shown at the scale of this map. Drawn by Adella Edwards, reproduced with permission.

36 000 km² at low water in the dry season to about 153 000 km² at high water in the wet season. The extent of flooded forest throughout the entire drainage may be as high as 300 000 km² (Marmontel 2008). Collectively, this information suggests that the extent of occurrence of the Amazonian manatee is both vast and seasonally highly variable (but at least in the dry season probably much less than the 860 000 km² of the dugong).

The area of occupancy (see Box 8.1) of Amazonian manatees is believed to have been greatly reduced by human activities, particularly hunting. Nonetheless, Amazonian manatees are still widely distributed in the lowland

(<300 m above sea level) forested areas of the Amazon Basin. Marmontel *et al.* (2012) consider that the most important populations are in the following areas (Figure 8.9): (1) the Brazilian Amazon, especially in the Solimões, Negro, Japurá, Juruá, Tocantins and Tapajós river systems; (2) the extreme north-east of Ecuador (Denkinger 2010); (3) the Samiria, Pacaya and Yanayacu-Pucate rivers in Peru, as well as in the Javari and Amazonas rivers near the Colombian border; and (4) the Amazon, Putumayo and lower Caquetá rivers in Colombia. Although Amazonian manatees inhabit lakes, rivers and channels of white, black and clear water, Marmontel *et al.* (2012) report that highly productive white water is their preferred habitat, particularly 1–4 m-deep, quiet murky waters with easy access to aquatic vegetation. Manatees also occupy areas up to 10 m deep in Lake Mamirauá (E. Arraut, personal communication, 2010).

Habitat use is seasonally variable. In the central Amazon and higher tributaries, manatees make seasonal migrations using *várzea* lakes and flooded forest during rising and high water, when aquatic plant productivity increases, and move to waters that remain deeper during the lowering and low-water seasons (Arraut *et al.* 2010; Chapters 4 and 5).

Population size and trends

There are no reliable population estimates. Amazonian manatees have proved extremely difficult to census and are not suitable for mark–recapture studies because of their secretive nature and turbid habitat. Estimates of numbers at local scales from feeding patches, sightings and interview surveys have provided limited information. Genetic diversity is relatively high (Garcia-Rodriguez *et al.* 1998; Cantanhede *et al.* 2005; Vianna *et al.* 2006; Chapter 7), suggesting that the population size must have been substantial in the recent past or that bottlenecks were not long-lasting. The estimated effective population sizes are high for the Amazonian manatee compared with those for the West Indian manatee (Cantanhede *et al.* 2005).

The 2008 IUCN assessment (Marmontel 2008) determined that the species was declining, despite some genetic evidence of possible recent population increases (Cantanhede *et al.* 2005; Vianna *et al.* 2006). The IUCN assessment was inferred from largely anecdotal evidence suggesting unsustainable exploitation.

Threats

The main threat to Amazonian manatees is illegal hunting for meat and other products (Chapter 7). Extreme droughts make manatees easier to catch. They become isolated and trapped in small water bodies, and the opportunity to capture migrating manatees passing through narrow channels increases (Arraut *et al.* 2010).

Although the use of traditional harpoons remains the most widespread harvesting technique (Marmontel 2008), the adoption of new technologies is increasing hunting efficiency. Fishing traps with harpoons attached have been recorded for Peru and on the Colombian–Peruvian border (Reeves *et al.* 1996; Orozco 2001). The use of nets is growing (Marmontel *et al.* 2012). Marmontel (2008) reports the increasing use of dedicated manatee nets and netting incidents in which a large number of manatees (perhaps 60) have been caught together. The increased use of gill nets has caused greater incidental mortality of calves and their mothers (Marmontel 2008). Surviving calves are usually sold as pets or kept alive for later sale, because they have little meat for immediate consumption (Marmontel 2008). There are reports of transnational calf sales despite the range states being signatories of CITES (Marmontel 2008; Marmontel *et al.* 2012).

Expanding human activities are causing the pollution, loss, alteration and fragmentation of Amazonian manatee habitats (Marmontel 2008). The manatee's food supply is threatened by contamination by mercury and oil as a result of mining. The impact of deforestation on manatee habitats is complex and difficult to evaluate (E. Arraut, personal communication, 2010). Although macrophytes are extremely abundant in deforested *várzea* areas (such as in the Lago Grande de Curuai region, Pará), there is concern that the removal of forest may change species composition, making the community less valuable as manatee habitat. Construction of hydroelectric dams threatens to isolate populations, reducing genetic variability (Rosas 1994). Although human population density is generally low outside urban centres, human population growth is accelerating habitat loss through growth of towns, deforestation and agriculture (Marmontel 2008). Increased boat use and road access threaten to open up remote areas to hunting.

Climate change is also a threat (see below; Chapter 7). Droughts are predicted to increase in the Amazon under all the scenarios developed by the Intergovernmental Panel on Climate Change (Malhi *et al.* 2008), causing reduction in manatee habitat, especially during the low-water period when they are most accessible to hunters (Arraut *et al.* 2010).

Conservation actions

The Amazonian manatee is legally protected in all of its range states. Colombia has developed a national manatee management plan and the species is listed in the Brazilian Federal Law for the Protection of Endangered Fauna and the Brazilian Conservation Action Plan for Aquatic Mammals. Management plans also exist for two protected areas (Pacaya Samiria in Peru and Mamiraua Reserve in Brazil), and two communities in Colombia (Puerto Narino and

Mocagua) have informal local management agreements (Marmontel 2008). Amazonian manatees occur in two protected areas in Ecuador (Denkinger 2010), two in Colombia, four in Peru and 23 in Brazil (Marmontel 2008). However, the protection is generally not well enforced, although there are exceptions such as at Mamirauá and Amanã. (E. Arraut, personal communication, 2010).

Captive rescue and rehabilitation programmes are conducted by several facilities in Brazil (Marmontel 2008). Most research to date has been in Brazil, with studies of captive manatees since the 1970s and long-term research on movement patterns since 1993 (Arraut et al. 2010; Marmontel et al. 2012). The Omacha Foundation started work in Colombia in 1998, and the Durrell Foundation at the University of Kent maintains a long-term monitoring programme in the Samiria River. Research efforts in Ecuador have been less intensive (Marmontel et al. 2012; but see Denkinger 2010). Educational programmes are underway to raise public awareness about the importance of manatee conservation. In the last decade, conservation efforts have increasingly involved working with local people, especially in Colombia (Marmontel et al. 2012)

There are only four range states, all at a relatively similar stage of development (Figure 8.9; Table 8.6). Thus coordinated action should be possible under the aegis of the CMS. The active involvement of local people, particularly fishermen and hunters, will be crucial. It may be effective to formally harness the cultural taboos outlined by Marmontel et al. (2012) as a basis for conservation initiatives in at least some areas (Chapter 9). As Arraut et al. (2010) have pointed out, local people are necessary stakeholders in conservation of manatees and their environments, and conservation must be seen to provide them with tangible benefits (Chapter 9).

Assessment

We consider that the IUCN's (2009) assessment of the Amazonian manatee as Vulnerable (VUA3cd *sensu* IUCN) is valid based on a projected population decline of at least 30% within the next three generations (~60 years), due primarily to ongoing levels of hunting, increasing incidental calf mortality, climate change and likely increased habitat loss and degradation. As highlighted by Meybeck (2003) in his global analysis of the future of river systems: 'In a scenario of global warming and modified climate variability, increased population and economic growth for the next 100 years, water demand and flood control demand will rise (Falkenmark 1997a and b; 1998).' The Amazonian manatee's habitat is thus almost certain to be degraded further.

WEST INDIAN MANATEE

The West Indian manatee is considered by the IUCN to be Vulnerable (VUC1 *sensu* IUCN) due to the small estimated number of mature individuals (< 10 000), and assumptions about a likely decline in population size of more than 10% over three generations (~60 years) due to threats associated with habitat loss and anthropogenic influences. The species is listed in Appendix I of CITES. Within the wider Caribbean, the species is protected under the Specially Protected Areas and Wildlife (SPAW) Protocol of the Cartegena Convention. The Ramsar Convention on Wetlands and the Convention on Biological Diversity protect West Indian manatee habitat. Not all countries with manatees have ratified these three conventions.

Country-specific legislation exists within each of the 21 range states to protect the species and, in some cases, its habitat (except Haiti, which has ratified certain international agreements that protect wildlife, including manatees; UNEP 2010; Figure 8.10). Each subspecies was considered Endangered by the IUCN (2009) in its most recent assessment because the respective estimated populations of mature individuals were below the threshold value of 2500 mature individuals.

Florida manatee

The Florida manatee occupies coastal and riverine waters and inland lakes, primarily of the south-eastern United States (Deutsch and Reynolds 2012, Figure 8.10). The extent of occurrence in Florida is ~19 500 km² (Haubold *et al.* 2006); including other US states used seasonally would result in a much larger area. Nonetheless, in winter months, the subspecies is essentially confined to peninsular Florida due to its need for warm water (Chapter 5). In warm seasons along the Atlantic coast, manatees are common in waters of Florida and Georgia, and relatively common as far north as the Carolinas, with occasional vagrants reaching as far as New England. In the Gulf of Mexico in the non-winter months, manatees are common in Florida, including the Florida Keys (as far as the Dry Tortugas) and are also reported from waters of Alabama, Mississippi, Louisiana and Texas. Manatees from Florida have been reported with increasing frequency from the Bahama Islands; 25 sightings were recorded between 1995 and 1999. At least one manatee originally photo-identified in Florida was observed in Cuba (Alvarez-Alemán *et al.* 2010).

In winter, Florida manatees aggregate in warm-water refugia to avoid cold stress (Chapters 5 and 7). The refugia include natural springs, such as those at Crystal River, Florida, and power plant discharges, but even deep

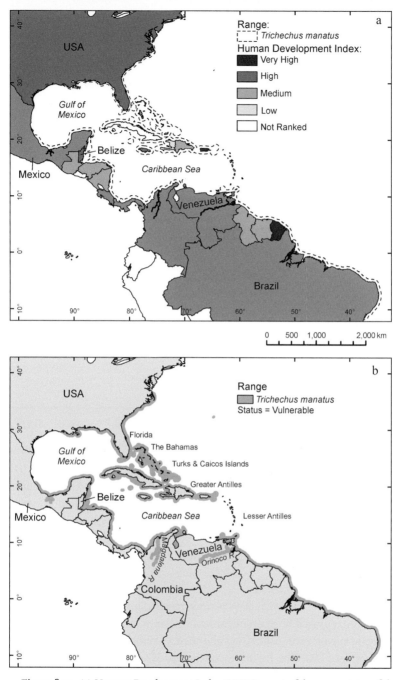

Figure 8.10. (a) Human Development Index (UNDP 2010) of the range states of the West Indian manatee (*Trichechus manatus*). (b) Range of the species, which we have assessed as Vulnerable. We have assessed the Florida manatee as Vulnerable; the Antillean manatee as Endangered. Key habitats include: Florida (US); the Mexican states of Veracruz, Tabasco, Campeche, Chiapas and Quintana Roo (especially Chetumal Bay); coastal and inshore waters of Belize and Nicaragua; the Greater Antilles (i.e. Cuba, Hispanola and Puerto Rico); Orinoco River (Venezuela); Magdalena River (Colombia); and discontinuous locations in Brazil, between the states of Amapá and Alagoas. Not all these locations can be shown at the scale of these maps. Drawn by Adella Edwards, reproduced with permission.

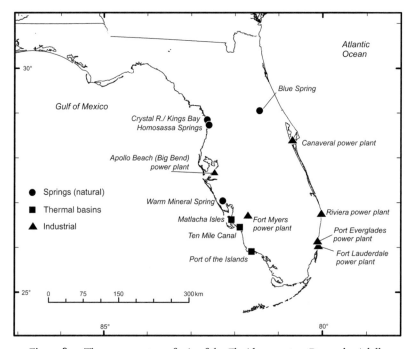

Figure 8.11. The warm-water refugia of the Florida manatee. Drawn by Adella Edwards, reproduced with permission.

boat basins or canals can serve as temporary warm-water refugia, especially during moderately intense cold spells (Figure 8.11). During the non-winter months, manatees disperse from locations of warm-water refugia. Areas with notably high manatee use outside the winter season are the Banana River and Indian River Lagoon (eastern Florida) and coastal waters of Charlotte and Lee counties (south-western Florida) (Figure 8.11).

Population size and trends
The number of Florida manatees is uncertain, despite considerable research. The main approach to assessing at least the minimum population size is a 'synoptic survey' (see 'Estimating abundance' above). These surveys are not accurate censuses. The highest count under ideal conditions totalled 5067 manatees in January 2010 (2779 manatees on Florida's east coast and 2288 on the west coast – Florida Fish and Wildlife Conservation Commission 2010a).

There are conflicting assessments of population trends. The 2009 stock assessment indicated that 87 manatees is the annual estimated *average* human-caused mortality. In 2009 that number was 112 (Florida Fish and Wildlife Conservation Commission 2010b). For the entire manatee population of Florida, the potential biological removal level *sensu* Wade (1998) of 16 manatees (adjusted for 2010 count) is exceeded by an average of about 70 animals per year (Federal Register 2009). This assessment suggests that the population is declining, an assessment at odds with assessments of each of the four somewhat distinctive regional 'subpopulations' or 'management units' recognised in the Florida Manatee Recovery Plan (USFWS 2001): the Atlantic, upper St Johns River, Northwest and Southwest subpopulations. The regional assessments use two other techniques: (1) population growth rates using a stage-based model developed by Runge *et al.* (2004) that integrates information from mark–recapture (photo-identification) estimates of survival and reproduction (Kendall *et al.* 2004; Langtimm *et al.* 2004; Chapters 6 and 7); and (2) for the Atlantic subpopulation, Bayesian analysis of winter aerial survey data (Craig and Reynolds 2004).

The Atlantic subpopulation is the largest manatee subpopulation. Most of the 2779 manatees counted in eastern Florida during synoptic surveys in January 2010 were from the Atlantic subpopulation. Based on aerial survey data and Bayesian models, Craig and Reynolds (2004) estimated an adult manatee population size of 1607 (95% Bayesian CI: 1353–1972) manatees using power plants on the Atlantic coast in 2001–2002, and suggested that the subpopulation increased at a rate of 5–7% per year between 1982 and 1989, levelled off (0–4% growth per year) between 1990 and 1993, and then increased at 4–6% per year between 1993 and 2001. Growth rates based on mark–recapture data and the stage-based model were more modest, suggesting essentially no growth (or decline) for the subpopulation (Runge *et al.* 2004). Recent re-analyses (Runge *et al.* 2007a,b) of the photo-identification data indicate that the results of the two approaches to trend analysis are now more similar.

The upper St Johns River subpopulation is the smallest, numbering slightly over 300 manatees (highest count was 317 in February 2010; W. Hartley and H. Edwards, personal communication, 2010). Photo-identification at Blue Spring (the primary winter aggregation site) over a period of nearly four decades demonstrates strong population growth, estimated (as of 2004 publications) to be 6.2% (95% CI: +3.7–8.1%; Runge *et al.* 2004).

The Northwest subpopulation is exceptionally well studied around winter, warm-water refugia. Counts totalling over 400 manatees regularly occur at Crystal River during cold weather. The subpopulation has been estimated (as of 2004 publications) to be growing at a rate of 3.7% per year (95% CI: +1.6–5.6%; Runge *et al.* 2004) based on photo-identification studies since the late 1960s.

The Southwest subpopulation is the second largest subpopulation. Runge *et al.* (2004) estimated that this subpopulation declined at an estimated rate of 1.1% per year (95% CI: −5.4−+2.4%). The wide confidence interval (compared to other subpopulations) reflects greater uncertainty of survival and reproductive rate estimates. Recent unpublished re-analyses (S. Barton, personal communication, 2009) indicate that the Southwest subpopulation appears likely to be increasing.

Threats

Florida manatees exist in close proximity to large and growing human populations. As discussed in detail in Chapter 7, the primary threats are watercraft collisions; loss of habitats, especially warm-water habitats; climate change; red tides; pathogens and contaminants. Many of these threats are human-induced and can therefore be mitigated, in principle. Deutsch and Reynolds (2012) note that on an annual basis, 52–88% of all adult manatee deaths are directly attributable to human activities (Figure 8.12; see also Chapter 7). Because the rate of manatee population change is especially sensitive to adult survival (Chapter 6), the high proportion of mortality that is human-induced is significant.

The Florida Fish and Wildlife Conservation Commission (2010b) has a remarkably capable system for acquisition and necropsy of dead manatees. This programme is costly but productive, and serves as a model for other marine mammal necropsy programmes. In 2010, the total mortality was

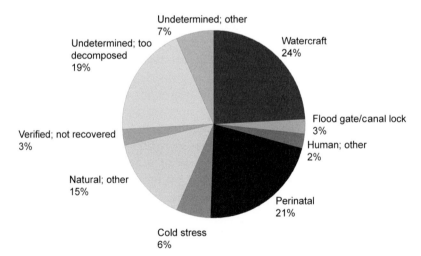

Figure 8.12. Causes of mortality of the Florida manatee, *Trichechus manatus latirostris*. Perinatal is a size category that can be further subdivided by causes. Drawn by Shane Blowes from data supplied by the Florida Fish and Wildlife Conservation Commission in August 2010. Reproduced with permission.

the highest on record (767). The highest number of deaths due to collisions with watercraft (97) occurred in 2009 (Figure 8.12). These data do not include serious injuries and indirect effects of human activities. Thus the true impact of human-related stressors on manatees is underestimated. In addition, the cause of death remains unknown for a large proportion of carcasses. Thus the proportion of manatee deaths resulting from human impacts may be even higher than the official statistics indicate (see also Chapter 7).

Conservation actions

Concerted conservation actions on behalf of Florida manatees began in earnest in the latter part of the 1970s with the start of on-the-ground implementation of provisions of the US Marine Mammal Protection Act of 1972, the US Endangered Species Act of 1973 and, equally importantly, the state of Florida's Manatee Sanctuary Act of 1978. Non-government agencies such as the Save the Manatee Club also began to increase their influence at this time. Many of the conservation actions discussed below had their roots in this framework. Since the 1970s, many millions of dollars have been spent on research, management and education. Overall, the effect of these initiatives has been very positive, as evidenced by generally increasing population growth rates based on aerial survey and life history data, and a change in the baseline minimum population size acquired from synoptic serveys from about 1000 manatees in the late 1970s to over 5000 in 2010.

Reynolds (1995, 1999) describes the many stakeholders (federal, state and private) that contribute to conservation efforts for Florida manatees (Box 8.2). Despite this huge investment, the Florida subspecies continues to be threatened by several factors, many of which are inadequately understood or controlled (Chapter 7). Reynolds *et al.* (2009) noted that many marine mammals (including Florida manatees) are sufficiently well studied to inform appropriate conservation decisions, if the will exists to prioritise conservation over other values. The primary conservation action needed for Florida manatees is to place a greater emphasis on solving problems, while maintaining the long-term research that has yielded the important insights outlined in Chapters 4–7. It will be extremely challenging to garner the political will to confront the conservation (and animal welfare and cruelty) needs of Florida manatees. The human population of Florida has undergone a more than six-fold increase over the last 60 years, and this growth will continue. Many people come to Florida for a lifestyle that is incompatible with the habitat needs of manatees (Chapters 4–5).

Box 8.2. Leading organisations involved in research, conservation and management of the Florida manatee (not including academic institutions and most independent laboratories)

Federal: US Fish and Wildlife Service, US Geological Survey, US Marine Mammal Commission.

State: Florida Fish and Wildlife Conservation Commission, Georgia Department of Natural Resources.

Private: NGOs (non-governmental organisations) including but not limited to the Save the Manatee Club, Mote Marine Laboratory, Sea2Shore Alliance, the Nature Conservancy, and the Ocean Conservancy.

Box 8.3. Critical habitat as defined under Section 3(5)(A) of the US *Endangered Species Act* of 1973

(1) The specific areas within the geographical area occupied by the species, at the time it is listed in accordance with the Act, on which are found those physical or biological features: (a) essential to the conservation of the species and (b) which may require special management considerations or protection; and
(2) Specific areas outside the geographical area occupied by the species at the time it is listed, upon a determination that such areas are essential for the conservation of the species.

Under the US Endangered Species Act of 1972, provisions exist to designate and protect 'critical habitat' for listed species (Box 8.3). Critical habitat has been designated for Florida manatees since 1976. The areas designated exist throughout Florida, and they are often specific at a local scale and impractical to include here (e.g. canals, areas of coastline between named bridges or other landmarks). In late 2009 the US Fish and Wildlife Service initiated a review to consider whether critical habitat designations for Florida manatees warranted revision. The review was prompted by a formal petition (dated December 2008) from several conservation groups, based on several factors: (1) improved scientific understanding of manatee biology, ecology and habitat needs and use patterns; (2) a long time interval having passed since critical habitat was originally designated for the sub-species; and (3) changes in how the term 'critical habitat' is used. The list of essential features recommended by the petitioners for the regions that encompass each of the management units or subpopulations of Florida manatees includes warm-water sources and refugia (natural springs, passive thermal basins and power plant thermal discharges); locations of seagrasses and freshwater vegetation; travel corridors; shelter (for calving and from disturbances); fresh water; and other habitat features (water depth, water quality and salinity).

Conservation actions implemented in an attempt to reduce watercraft collisions include: development of Manatee Protection Plans in 'key counties' (see below); creation of regulatory zones (e.g. sanctuaries and refuges) that limit boat speed and access, and in a few cases even restrict access by people under any circumstances; improved enforcement of regulations; and consideration of technological improvements that may have potential to reduce the collisions with boats, or crushing or drowning in water-control structures.

Habitat acquisition programmes are supported by federal, state, county and private entities. Although manatees and their needs represent only a subset of the criteria that might compel purchase and protection of key habitats, in some cases land acquisition motivated by other needs may still carry benefits for manatees. Not only are habitats being purchased and set aside, but several agencies and private groups are involved in permitting of activities that could affect manatees or their habitats (e.g. construction of marinas; dredge and fill activities; repowering of existing power plants).

Manatee Protection Plans exist in at least two forms. At the county level, they mandate that manatees and their needs be considered and factored into planning and decisions about development, habitat use and other factors. In addition, electric power companies are required to develop site-specific Manatee Protection Plans that outline requirements by each plant to ensure that manatees in winter will have adequate access to warm water.

Two additional categories of activities have the potential to affect conservation of manatees and their habitats. The first involves the rescue, rehabilitation and release of injured or diseased manatees. Since 1973, more than 180 manatees have been treated and returned to the wild (N. Adimey, personal communication, 2010), but many may not survive long enough to contribute reproductively to the population, so the net conservation gains are equivocal. However, rescue, rehabilitation and release contributes to another category of conservation-related work: public education and awareness. Education and awareness incorporates personnel from an enormous range of stakeholder groups. Not only can people encounter and passively observe wild manatees at selected sites, but educational opportunities abound in books, films, pamphlets, boater guides/decals, courses and websites. Some sources of information are more credible and reliable than others, but it is certainly true that such efforts 'encourage informed public participation in regulatory and other management decision-making processes and provide constructive avenues for private funding of . . . recovery . . . research and land acquisition efforts' (Deutsch *et al.* 2008).

Assessment

The January 2010 synoptic count indicates that there are almost certainly more than 2500 mature Florida manatees (70% of ~5000 is 3500). Thus the population exceeds the IUCN population size threshold for Endangered (Criterion C). Accordingly, we assess the status of the Florida manatee as Vulnerable (VUC1 *sensu* IUCN 2009) due to an estimated population size of fewer than 10 000 mature individuals and a projected continuing population decline of at least 10% over the next two generations (~40 years) due to threats associated with loss of warm-water habitat in winter and increasing watercraft traffic in Florida waters.

Antillean manatee

The Antillean subspecies has an extensive range in the wider Caribbean region. Politically (see Marine Mammal Commission 2009), that region consists of 13 island nations, 12 continental nations and 13 territories variously under the jurisdictions of France, the Netherlands, the United Kingdom or the United States (but note changes in political status as of 10 October 2010 of the various Caribbean islands associated with the Kingdom of the Netherlands). Of these 38 political entities, two (the United States and the Bahamas) contain the Florida subspecies of West Indian manatee; and at least 19 contain resident or vagrant Antillean manatees (Deutsch *et al.* 2008; UNEP 2010; Table 8.6).

The Antillean manatee ranges to Alagoas in Brazil, far south of the mouth of the Amazon River (Lefebvre *et al.* 2001). It can hybridise with *T. inunguis* near the mouth of the Amazon (Garcia-Rodriguez *et al.* 1998; Vianna *et al.* 2006). To the north of Brazil, Antillean manatees are found discontinuously along the Caribbean coast of South and Central America, as far north as the Rio Soto La Marina in north-eastern Mexico. Antillean manatees are also found in Trinidad and Tobago off the north-eastern coast of Venezuela. Each of the Greater Antilles (i.e. Cuba, Jamaica, Hispaniola and Puerto Rico) is home to small numbers of manatees. Manatees are no longer found in the Lesser Antilles (the islands extending from the Virgin

Islands to Grenada). The distribution of the subspecies is fragmented and populations are thought to be very low in most countries (UNEP 2010; Lefebvre *et al.* 2001; Tables 8.7, 8.9). Deutsch *et al.* (2008) include the Bahamas as part of the range of *T. m. manatus*, but at least one manatee reported from the Bahamas was originally sighted and identified in Florida waters. It may be more appropriate to include the Bahamas in the range of the Florida manatee (Table 8.6).

Population size and trends
The population size is unknown. Deutsch *et al.* (2008) suggested (based more on anecdotal than scientific evidence) a likely total population size of 2800–5600. UNEP (2010) cited information provided by C. Self-Sullivan and A. Mignucci-Giannoni (unpublished), who considered mostly anecdotal information to suggest the population 'estimates' summarised in Table 8.7. Many of these estimates have limited validity. We believe there to be more manatees than reported in some of the above countries. Nonetheless, Belize and Mexico are the only countries that provide contiguous habitat for Antillean manatees and have been described (O'Shea and Salisbury 1991) as 'the last stronghold' for the subspecies in the Caribbean Sea.

Table 8.7 *The estimated size and status of the Antillean manatee population in its range states based on anecdotal information from C. Self- Sullivan and A. Mignucci-Giannoni (unpublished). The only state where the population is thought to be increasing is the Bahamas, with around ten manatees likely to be Florida manatees (Deutsch et al. 2008).*

Estimated population size for each range state	Declining	Likely population status		
		Unknown/ probably declining	Stable/ probably declining	Stable
< 100	Costa Rica, Dominican Republic, Guyana, Suriname, Trinidad and Tobago, Venezuela	Cuba,		
	Guatemala, Haiti, Jamaica, Panama	French Guiana, Honduras		
> 100 and < 500	Nicaragua	Colombia	Brazil	Puerto Rico[1]
> 500		Mexico	Belize	

[1] Known deaths in Puerto Rico exceed sustainable take estimated using Potential Biological Removal technique (Wade 1998). See text.

386 | Conservation status

In the most recent stock assessment report by the US Fish and Wildlife Service (http://www.fws.gov/caribbean/es), the Puerto Rico population of Antillean manatees was estimated to number at least 72, with a maximum rate of reproduction of 0.04, and able to sustain a potential biological removal of zero animals. The average human-caused mortality per year averages two manatees, which clearly exceeds the potential biological removal.

Population trends are also based mainly on anecdotal information. The assessment in Table 8.7 is based on C. Self-Sullivan and A. Mignucci-Giannoni (unpublished) unless indicated otherwise.

Threats

The primary documented threats are anthropogenic and include poaching, habitat loss, chemical contamination, entanglement in nets and, increasingly in some locations, collisions with watercraft (Chapter 7). In some areas, such as parts of Belize and Brazil, ecotourism may be a burgeoning threat. Effects of climate change are a growing long-term concern.

Manatees were hunted to the point of extirpation on Guadeloupe (French Antilles) many decades ago (Chapter 9). Poaching is still common throughout the region. For example, in Nicaragua, poachers take up to nine manatees annually, and Bonde and Potter (1995) documented 35 poached carcasses near Port Honduras, Belize, in 1995. Manatees continue to be hunted for their meat, oil and bones (UNEP 2010). The meat is highly prized and in some markets in Central America can bring an extraordinary price, especially for developing countries: up to US$100 per pound (>US$200 per kilogram; UNEP 2010; Chapter 7). The bones are used in some locations (e.g. Mexico and Belize; Morales-Vela and Olivera-Gomez 1992; Auil 1998) by craftsmen to make jewellery and sculptures (Chapter 7).

As discussed in Chapter 7, incidental capture in fishing nets is an increasing problem. For example, Castelblanco-Martinez *et al.* (2009) compiled 90 reports of manatees killed in the Orinoco River in Colombia, 1980–2008; incidental drowning in fishing nets was responsible for 43% of deaths; hunting with harpoons 39%. Moore *et al.* (2010) reported low numbers of manatees caught in gill nets in Jamaica, presumably a consequence of the low manatee population size.

Levels and effects of contaminants are not well studied in Antillean manatees, but fears of effects of contaminants on wildlife and people of the Caribbean run high (UNEP 2010). Manatees around Chetumal Bay, Mexico, had concentrations of organochlorine pesticides and PCBs in their tissues comparable to those in Florida manatees 30 years ago, shortly after the use of such chemicals was first discontinued (Wetzel *et al.* 2012). In Panama, the deaths of 'several' manatees in Bocas del Toro near extensive banana plantations have been suggested to be due to exposure to pesticides (I. Anino, personal communication, 2008), although K. Ruiz (personal communication, 2009) is uncertain of the extent to which pesticide exposure is lethal to manatees in Panama. This issue has long been a concern in this region (Mou Sue et al. 1990), but no definitive data exist. Oil and gas exploration and pollution with polycyclic aromatic hydrocarbons (PAHs) is obvious in many parts of the Caribbean. Effects on Caribbean wildlife have been documented (e.g. Siung-Chang 1997; Noreña-Barroso et al. 2004), but have not been examined in manatees. In some areas, mining activities introduce mercury into local ecosystems.

Strandings of living, orphaned, dependent manatee calves are frequent in north-eastern Brazil (Parente et al. 2004 cited by Deutsch et al. 2008).

Conservation actions

The Antillean manatee is protected regionally under the SPAW Protocol and by laws specific to almost every country or territory where it exists. Several Caribbean countries have started to create protected areas, including ones specifically designed for protection of manatees (UNEP 2010). Unfortunately, enforcement remains a critical issue, especially where the value of a dead manatee for meat and other products is perceived to outweigh the value of a living one (Chapter 7). UNEP (2010) included an extensive discussion of short- and long-term recommendations for the improvement of both knowledge and conservation of manatees in the wider Caribbean. The capacity to implement conservation initiatives is variable, although most range states have a very high, high or medium HDI (Table 8.6).

Assessment

The Antillean manatee is considered Endangered (ENCI) by the IUCN (Self-Sullivan and Mignucci-Giannoni 2008) because of: (1) the small number (< 2500) of mature individuals estimated to exist; (2) the likelihood of a population decline of at least 20% over the next two generations (~40 years); (3) the paucity of effective conservation actions throughout its range; and (4) effects of current and projected future anthropogenic threats (loss and degradation of habitat, hunting, incidental mortality in fisheries, pollution and disturbance). We concur with this assessment, noting that it may have to be revised if the population estimate is found to be too low, which we consider possible.

WEST AFRICAN MANATEE

The West African manatee is listed as Vulnerable by the IUCN (2009). It was listed in CITES Appendix II as of June 2010. All range states except

Angola are parties to the CITES Convention. Most are also parties to the CMS and the Ramsar Convention on Wetlands. An increasing number are signatories to the MoU to conserve the small cetaceans and manatees of West Africa and Macronesia under the CMS (Table 8.6).

West African manatees occur in wetlands, rivers and coastal and inland ecosystems in West Africa from the Senegal River at the Senegal–Mauritanian border to the Longa River in Angola (Figure 8.13). The region is very poor, a significant impediment to effective conservation. Eighteen of the 21 range states have a low HDI; the remaining states have a medium HDI (Table 8.6; Figure 8.13).

Genetic diversity appears high and is suggestive of more population structure than is evident for *T. inunguis* (Vianna *et al.* 2006; Chapter 3). Additional genetic samples have been collected, but the results were not available at the time of writing. We decided not to assess the status of this species at a subpopulation level, as few data are available on any aspect of its ecology. West African manatees are habitat generalists that live in a diversity of habitats from the Sahel to equatorial rainforests (Chapter 5). They can be found up to 75 km offshore among the shallow coastal flats and mangrove creeks of the Bijagos Archipelago of Guinea-Bissau (Powell 1996; Silva and Araújo 2001). In some other parts of the range, manatees are found in estuaries, lagoons or larger rivers in the dry season, and move into smaller rivers, flood plains and flooded forests during the rainy season. They occur in most major rivers within their range until cataracts or shallow water prevents upstream progress. During the dry season, they seek refuge in permanent lakes that communicate with the rivers during high water but may be cut off when river waters subside in some areas such as the Cote D'Ivoire (Powell and Kouadio 2008; L. Keith, personal communication, 2010). Some populations are naturally isolated (e.g. in Chad and in the

---

Caption for Figure 8.13
(a) Human Development Indices (UNDP 2010) of the range states of the West African manatee (*Trichechus senegalensis*). (b) The range of the species, which we have assessed as Vulnerable. We consider that the long-term prospects for this species are the worst of any sirenian because of the widespread poverty throughout much of its range. Key habitats include: the Senegal and Casamance Rivers and Saloum Delta, Senegal; Bijagos Archipelago, Guinea-Bissau; Fresco, Nioumozou, Tadio Lagoons Complex, Côte d'Ivoire; Lake Volta, Ghana; Lakes Lere and Trene, Chad; the inland Niger Delta, Mali; Oueme River Valley, Benin; the Niger, Benue and Cross Rivers and Badagry Creek (Lagos), Nigeria; Lake Ossa and the Sanaga River, Cameroon; N'dogo and N'gowe Lagoons and the Ogooue River, Gabon; the lower Congo River, the Cuanza and Longa Rivers, Angola. Not all these locations can be shown at the scale of this map. Drawn by Adella Edwards, reproduced with permission.

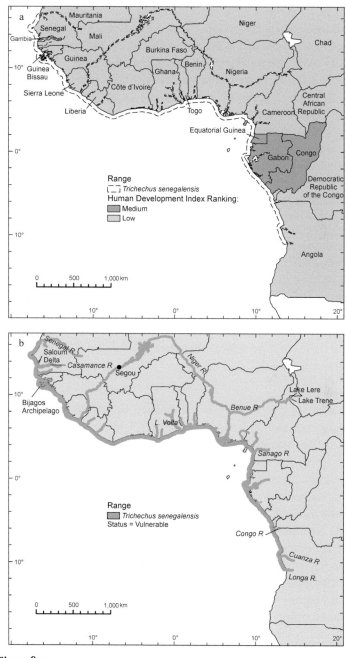

Figure 8.13.

upper reaches of the Niger River in the inland delta of Mali as far as Segou, over 2000 km from the ocean; Figure 8.13). Other populations are now permanently isolated due to dams (e.g. the Senegal River – Diama Dam – and Lake Volta – Ghana – populations). Although the overall population is reportedly reduced, range contraction has not been confirmed, apart from the species' contemporary absence from the Chad Basin in Chad (Salkind 1998 in Dodman *et al.* 2008). Information is insufficient to estimate the species' extent of occurrence. Dodman *et al.* (2008) state that: 'In practical terms, the manatee has a restricted distribution in Mauritania, Togo, Equatorial Guinea and Chad, whilst in most other range states it is fairly widespread in suitable wetland habitats, especially along the coast and in the lowland reaches of the main rivers.'

Population size and trends

The problems of estimating population size in inland waters (see Figure 8.1) are similar to those outlined for the Amazonian manatee. In addition, range states often lack the resources required for aerial surveys of coastal waters or consider it inappropriate to use funds for such surveys when waters are turbid. Thus, there is no information about the population size and trends of West African manatees.

Threats

The threats are typical of those encountered by sirenians in low-income developing countries. Hunting is a major problem across the range (Reeves *et al.* 1988; Dodman *et al.* 2008; Kouadio 2012; Chapter 7). Specialist hunters are widespread and have been recently reported from many areas: the Bijagós Archipelago in Guinea-Bissau, the Sine Saloum Delta of Senegal, the Ogouue River in Gabon, the Congo River in Angola, and throughout Cameroon, Lake Volta (Ghana), Togo, Benin and Nigeria (Dodman *et al.* 2008; L. Keith, personal communication, 2010). In other areas, fishers supplement their income by hunting (Chapter 7), and incidental capture in fishing nets was the most frequently reported threat in surveys carried out in preparation for the UNEP/CMS Action Plan (Dodman *et al.* 2008; see also Silva and Araújo 2001). Incidental capture was reported to account for 72% of 209 deaths reported in Guinea-Bissau from 1990 to 1998 (Silva and Araújo 2001). Incidental capture occurs in many areas around West Africa using: (1) nets for catching sharks; (2) fishing trawls; (3) stationary funnel nets across the inlets of major rivers to catch shrimp; and (4) fishing weirs made of sticks. Manatees are also occasionally killed in the turbines and intake of the hydroelectric generators of dams (Powell 1996). Collisions with large vessels in some rivers and lagoons may be a burgeoning threat (Powell and Kouadio 2008).

International trade in live West African manatees is apparently not well regulated. In Nigeria, manatees are taken for exhibit in local zoos. Toba Aquarium in Japan also has a West African manatee exhibit. Wild-caught manatees from Guinea-Bissau were offered for sale on the internet in 2010 (River Zoo Farm 2010).

Loss of manatee habitat is widespread due to urban development, large-scale conversion of wetlands to other land uses, damming of rivers, cutting of mangroves for firewood and destruction of wetlands for agricultural and other coastal development (Dodman *et al.* 2008; Powell and Kouadio 2008). In some areas manatees are treated as pests to agricultural production of rice (Reeves *et al.* 1988; Silva and Araújo 2001). Habitat loss is likely to increase as the human population is projected to grow exceptionally fast in West Africa in comparison to every other region in the world (United Nations 2004).

Table 8.8 *A summary of research and conservation actions for the West African manatee by country, based on Dodman* et al. *(2008), Powell and Kouadio (2008) and Kouadio (2012).*

Initiative	Range states
National laws[1]	Angola, Benin, Cameroon, Chad, Congo, Cote d'Ivoire, Democratic Republic of Congo, Gabon, Gambia, Ghana, Guinea-Bissau, Liberia, Mali, Niger, Nigeria, Senegal, Sierra Leone
Included in relevant National Conservation Strategy or equivalent	Congo, Côte d'Ivoire, Ghana, Gabon, Guinea, Sierra Leone
Some manatee habitat within protected areas	Angola, Cameroon, Congo, Democratic Republic of Congo, Côte d'Ivoire, Equatorial Guinea, Gambia, Guinea, Guinea-Bissau, Mauritania, Niger, Nigeria, Senegal, Togo
Rescue efforts for individual manatees	Mali, Senegal
Research and/or population monitoring at some sites	Angola, Benin, Cameroon, Chad, Côte d'Ivoire, Gabon, Gambia, Ghana, Guinea-Bissau, Guinea, Mali, Niger, Senegal, Togo
Education/sensitisation	Chad, Côte d'Ivoire, Ghana, Guineau-Bissau, Togo
Other conservation initiatives	Benin, Chad, Congo, Niger

[1] Adequate enforcement is rare.

Climate change is projected to increase pressure on the world's fisheries (Allison *et al.* 2009), with resultant increased food insecurity. A negative correlation between the catches of subsistence fishers and 'wild meat' has already been established for West Africa (Brashares *et al.* 2004; Rowcliffe *et al.* 2005) and thus the impact of climate change on fisheries is likely to increase the pressure to hunt West African manatees (Chapter 7).

Conservation actions

The West African manatee is protected by national laws in all range states. However, enforcement and control of hunting appears negligible, although efforts are being increased to protect manatees from local sale in some regions. Dodman *et al.* (2008), Powell and Kouadio (2008) and Kouadio (2012) list the research and conservation actions by country, as summarised in Table 8.8. This list may not be complete. Significantly, 16 West African countries and three NGOs had by 2010 signed an MoU under the CMS to conserve the small cetaceans and manatees of West Africa and Macronesia with an associated Manatee Action Plan (UNEP and CMS 2008). The initiative seeks to provide a legal and institutional framework to facilitate and coordinate local initiatives. In addition, IUCN, Wetlands International, World Wildlife Fund and Fondation Internationale du Banc d'Arguin (FIBA) are supporting a conservation project for the West African manatee within the region from Mauritania through Sierra Leone (Dodman *et al.* 2008).

Assessment

We agree with the IUCN's 2009 assessment of Vulnerable (A3cd; C1) because of the high probability that a 30% or greater reduction in population size will result within three generations (~60 years). The level of threats, particularly hunting and incidental catch, will continue to increase throughout the range, resulting in near extirpation in some regions. We expect this situation to be exacerbated by human population increase and the resultant habitat destruction from activities such as urban development, mangrove harvesting and silting of rivers and dams. Threats will be amplified by the high poverty that is expected to increase as a result of climate change. Nonetheless, the status of the species is almost certainly variable across its wide range; when more information becomes available, regional assessments will be required the better to inform conservation policy. Meanwhile, we consider that the West African manatee is the extant sirenian species at greatest risk of extinction because of the high levels of human poverty in many parts of its range.

CONCLUSIONS

This overview reveals differences between the conservation status and prospects of the three species of manatees and the dugong in developed and developing countries. The populations at greatest risk are those that are small (< 100) and/or in countries with a low HDI; those that are most secure are those numbered in the thousands and/or in countries with a high or very high HDI (Table 8.9).

The most secure sirenian subpopulation is the dugong in Shark Bay, Western Australia, a 14 000 km² bay that is more than 70% as large as the extent of occurrence of the Florida manatee in Florida (19 500 km²). Although the dugong is under threat in most of the rest of its range in Australia, there is no evidence for a conservation crisis, except perhaps on the urban coast of Queensland, where significant management interventions are now being implemented. It is very encouraging that Florida manatee numbers have increased since the 1970s, despite the huge human population increase in Florida. The situation in the more than 70 developing countries that constitute most of the ranges of the extant sirenians is much more concerning and challenging. Legal protection, although nearly ubiquitous, is seldom implemented effectively. The infrastructure required to support increasing human populations in coastal regions is destroying sirenian habitats; in most areas dugongs and manatees are perceived as being worth more dead than alive, a situation likely to be exacerbated by climate change.

The IUCN has repeatedly emphasised the importance of not automatically linking conservation action to the inclusion of a species in any particular category of the IUCN Red List, stressing that management interventions need to be applied after a careful analysis of the processes driving the threats and the measures needed to counteract them (see also Possingham et al. 2002). Resources for wildlife management are inevitably assigned on the basis of factors in addition to extinction probability (Marsh et al. 2007b). As charismatic megafauna of high social and cultural value, manatees and dugongs have attracted significant management attention in most range states, although the resultant interventions have not always been effective. Opportunities to develop more effective approaches are discussed in Chapter 9.

NOTE

1. Coastline length estimates vary with the method and data sets used. All the estimates quoted here used a standard GIS technique, the same projected coordinate system (Cylindrical Equal Area World) from the same global bathymetry shape file and excluded small islands. The estimates are more appropriately used for comparisons rather than as absolute estimates.

Table 8.9 *A comparison of the estimated capacity of range states 'for which the data are available to effectively conserve sirenians. The approach assumes that the capacity to implement effective conservation is a function of sirenian population size (larger is easier) and the country capacity for which the Human Development Index (UNDP 2010) is a proxy. Sirenian population sizes are order of magnitude 'guesstimates' based on the best available information. The countries where the challenge will be greatest are shaded in dark grey, those with an intermediate challenge in light grey, while the range states judged to have the best prospects of conserving sirenians are unshaded. The range of population estimates differs between species.*

Population estimate	Human Development Index			
	Low	Medium	High	Very High
Dugong[1]				
10^4–10^5				
10^3–10^4	Pacific: Papua New Guinea	Archipelagic South-east Asia: Indonesia, Philippines; Pacific: Solomon Islands	Arabian/Persian Gulf: Saudi Arabia	Australia Arabian/Persian Gulf: Bahrain, Qatar, UAE; Pacific: New Caledonia
10^2–10^3	East Africa: Madagascar, Mozambique; Arabian/Persian Gulf: Yemen; Continental South-east Asia: Myanmar	Indian region: India; Archipelagic and Continental South-east Asia: Thailand, Timor-Leste;	Archipelagic and Continental South-east Asia: Malaysia	
10–10^2	East Africa: Comoros, Kenya, Tanzania; Red Sea: Djibouti, Sudan;	Red Sea: Egypt; Indian region: Sri Lanka; Continental South-east Asia: Cambodia, China, Vietnam	Arabian/Persian Gulf: Iran	East Africa: Mayotte; Archipelagic South-east Asia: Brunei, Japan, Singapore

Amazonian manatee

Population estimate	Range states
$10^3–10^4$	
$10^2–10^3$	Brazil, Colombia, Ecuador, Peru

West Indian manatee (Florida and Antillean subspecies)

Population estimate	Range states
$10^3–10^4$	Florida manatee: USA
$10^2–10^3$	Antillean: Nicaragua; Antillean: Belize, Brazil, Colombia, Mexico, Puerto Rico; Antillean: Costa Rica, Jamaica, Panama, Trinidad and Tobago, Venezuela; Antillean: French Guiana
$10–10^2$	Antllean: Haiti; Antillean: Dominican Republic, Guatemala, Guyana, Honduras, Suriname

West African manatee

Population estimate	Range states
$10^3–10^4$	
$10^2–10^3$	Angola, Benin, Cameroon, Democratic Republic of Congo, Côte d'Ivoire, Gambia, Ghana, Guinea, Guinea-Bissau, Liberia, Mali, Niger, Nigeria, Senegal, Sierra Leone; Congo, Gabon
$10–10^2$	Chad, Mauritania, Togo; Equatorial Guinea

---

1 Human Development Index not available for Cuba, Eritrea, Oman, Palau, Seychelles, Somalia or Vanuatu (UNDP 2010).

2 Bahamas also has a high Human Development Index. There is some debate about the subspecies identity of its very small West Indian manatee population (see Table 8.6) and so it has not been included in this table

9

Conservation opportunities

As discussed in Chapters 3–6, manatees and dugongs are long-lived, slow breeding, aquatic herbivores with a low metabolic rate. This life history means that their populations are sensitive to both mortality, especially adult mortality, and habitat loss. Dugongs and manatees are subject to a range of threats (Chapter 7), the relative importance of which varies with location and particularly whether the range state is a developed or developing country. The capacity of a country to implement effective conservation actions also varies with its developmental stage; in general, developing countries are likely to have a lower capacity to implement effective conservation than developed countries (Leverington *et al.* 2008; Table 8.9).

Our review of sirenian research (Chapters 3–6) is largely based on Florida manatees and Australian dugongs, because those populations are the ones that have provided the bulk of our knowledge. Ironically, these are among the sirenian populations most likely to persist (Table 8.9). This situation is not a coincidence, as extensive scientific information can inform effective conservation decisions. However, as noted in Chapter 1 and by Reynolds *et al.* (2009), lack of science is *not* the primary factor that is preventing the conservation of sirenians and other marine mammals. Indeed, developing tropical nations should not wait for science at the pace and sophistication of the developed world before implementing conservation actions. Rather, the science learned elsewhere in broad outline will be for the most part transferable to conservation of *Trichechus senegalensis*, *T. inunguis*, *T. m. manatus* and most populations of dugongs, as assumed in the sirenian conservation primer by Hines *et al.* (2012b).

As the twenty-first century progresses, there is growing appreciation of the urgent need for conservation actions at the local, regional, national and international levels. In response to increasing worldwide biodiversity loss, the United Nations Environment Programme (UNEP) started working in the late 1980s/early 1990s to develop an international treaty to address the issue. The result was the creation of the Convention on Biological Diversity (CBD), its intent being to enable the conservation of biological diversity, the sustainable use of its components and the fair and equitable sharing of benefits from using genetic resources (United Nations 1993: www.cbd.int/convention/text). Although more properly considered species conservation rather than biodiversity conservation, the conservation of sirenians is congruent with the goals of this convention.

The year 2010 was declared the International Year of Biodiversity, and the parties to the CBD made a commitment 'to achieve by 2010 a significant reduction of the current rate of biodiversity loss at the global, regional and national level as a contribution to poverty alleviation and to the benefit of all life on Earth'[1]. It is generally agreed that we have missed the CBD's target of reducing the rate of biodiversity loss by 2010. Indeed, Butchart *et al.* (2010) found no evidence for a significant reduction in the rate of decline of biodiversity loss, confirming that world leaders have failed to deliver commitments made in 2002 to reduce the global rate of biodiversity loss by 2010.

Sirenians are important components of tropical and subtropical coastal and freshwater ecosystems. In the years following the creation of the CBD, some groups, such as the Pew Charitable Trusts (see Pew Oceans Commission 2003), focused on marine environments and resources, and noted that society has done an inadequate job with regard to research on and management of the seas and living resources contained therein. This report concluded (p. V) that the 'oceans are in crisis and the stakes could not be higher ... without reform, our daily actions will increasingly jeopardise a valuable natural resource.' In 2009 the Joint Ocean Commission Initiative of the US government cited a 'lack of rational management strategy and a substantially weakened ocean science enterprise' as causes of clear declines in the health, productivity and sustainability of marine environments and marine resources. The status of the coastal zone is of greatest concern (Halpern *et al.* 2008) and affects dugongs and all manatee species other than the Amazonian manatee.

The other major sirenian habitat, fresh water, makes up only 0.01% of the world's water and about 8% of the world's surface, but supports almost 6% of described species (Dudgeon *et al.* 2006). Freshwater biodiversity is the overriding conservation priority for the International Decade for Action – 'Water for Life' – 2005–2015. Nonetheless, freshwater habitats are experiencing declines in biodiversity far greater than those in most terrestrial systems. If trends in human demand for fresh water remain unaltered, and species losses continue at present rates, the opportunity to conserve much of the remaining freshwater biodiversity will disappear before the 'Water for Life' decade ends (Dudgeon *et al.* 2006). Vörösmarty *et al.* (2010) concluded that massive investment in water technology coupled with a lack of precautionary investment is jeopardising freshwater biodiversity, with habitats associated with 65% of continental freshwater discharge classified as moderately to highly threatened. This assessment, if accurate, will have major implications for all three species of manatees.

Recent publications (Marsh *et al.* 2003; Marine Mammal Commission 2007; Reynolds *et al.* 2009) also underscore the untenable status of many stocks and species of marine mammals and recognise the reality that, in the past 60 years, nearly 3% of the world's marine mammal species have become extinct due to human activities. Chapter 8 highlights that some sirenian populations are in jeopardy of imminent extirpation in many parts of the world, with Marsh (2008) suggesting that dugong populations are either in decline or extirpated in a very concerning one-third of the species' range. The future of the West African manatee is of particular concern: all of its 21 range states are developing countries; 18 have a low Human Development Index (HDI; UNDP 2010), suggesting a limited capacity for delivering effective conservation (Chapter 8). Thus the 'forgotten sirenian' seems at high risk of disappearing *as a species* unless fast and effective changes occur (Table 8.9), despite its current global conservation status of 'Vulnerable' being similar to that of the other extant sirenians.

There are many published exhortations to change the relationship humans have with natural ecosystems in order to avoid such catastrophes. We agree with the sentiments expressed by Swaisgood and Sheppard (2011) that the pervading culture of hopelessness among conservation biologists is likely to have a negative influence on our ability to mobilise conservation action among the general public. Thus instead of pursuing a discussion of a possibly gloomy future of coastal and freshwater ecosystems devoid of sirenians, we offer the current chapter with the hope that it will facilitate engagement with possible solutions to guide a generation more attuned to and, we hope, more successful in conservation efforts than our own.

PROBLEM RECOGNITION AND DEFINITION

As is the case with any problem to be solved, recognition and definition are requisite and sometimes difficult first steps. However, the real work involves developing and implementing strategies to eliminate past problems and to facilitate meaningful and sustained improvement for the future. Poverty represents an enormous challenge to the conservation of natural resources in most sirenian range states, especially in view of the recent human population increases documented in Chapter 7 and predicted to occur in the next 30–40 years. The growth in the number of people will not be uniform: the locations expected to experience most of the more than two billion person projected increase are developing countries, where poverty already exerts excessive pressure on many species and resources (United Nations 2004), including manatees and dugongs.

The environmental Kuznets curve (EKC) hypothesis (Stern *et al.* 1996) argues that environmental performance and per capita wealth follow an inverted U-shaped relationship among countries. The Kuznets curve predicts that beyond a certain threshold, wealthier societies can reduce environmental degradation via cleaner technologies and higher demand for sustainable behaviour from their citizenry. The evidence for the EKC hypothesis with respect to biodiversity is equivocal and controversial; some analyses suggest that the number of threatened birds and mammals increases initially with economic growth, but then declines after a threshold (Naidoo and Adamowicz 2001; Hoffman 2004). However, others suggest that increasing economic development leads to higher species endangerment (Naidoo and Adamowicz 2001; Clausen and York 2008) and general levels of species threat (Czech *et al.* 2000). When the costs of economic activity are borne by the poor, as is the case for artisanal fishers and hunters of sirenians in developing countries, the incentives to correct threats to biodiversity are likely to be weak (Arrow *et al.* 1996).

It is vital that local and regional social, political and economic issues are confronted and addressed if conservation is to have a chance of success (Marsh *et al.* 2003). The explicit recognition by the CBD of the relationship between conservation and poverty alleviation (above) underscores the complexity and difficulty of preserving the world's species, including sirenians. Meffe *et al.* (1999) were explicit that 'solutions will require the development and integration of policy based on expertise in biology ... economics, law, political science, human behaviour, adaptive management, statistical uncertainty, sociology, philosophy, ethics and property rights'.

Lavigne *et al.* (2006) provided a blueprint for reinventing wildlife conservation. We agree with their general approach and have added some ideas of our own (Box 9.1; see also Chapters 7 and 8). These general ideas provide a useful framework, but do not give specific guidance.

The challenges for and the inertia against conservation may seem enormous, but our world has been fortunate to have individuals who have tenaciously worked to ensure that conservation has a chance to succeed. One of the leaders of the effort, Aldo Leopold, wrote in his classic work in 1949, *A Sand County Almanac and Sketches From Here and There*:

Box 9.1.   A blueprint for sirenian conservation, modified from Lavigne *et al.* (2006)

(1) Establish clear goals.
(2) Recognise that conservation is value based.
(3) Adopt a geocentric conservation ethic.
(4) Recognise the central role of values in the formation of public policy.
(5) Clarify the role of science in public policy formation.
(6) Establish fundamental principles for twenty-first-century conservation.
(7) Identify real conservation problems and work towards finding real solutions.
(8) Put old myths to rest.
(9) Clarify and address issues of scale (temporal and spatial).
(10) Establish long-term funding programmes.
(11) Develop a coherent ideology.
(12) Proactively frame the debate.
(13) Provide inspired and inspiring leadership.
(14) Build infrastructure, including but not limited to setting up think tanks, developing grassroots initiatives, forming novel alliances and coalitions and actively marketing the ideology.
(15) Be proactive, not reactive.
(16) Be creative rather than tied to traditional approaches.
(17) Build interdisciplinary teams.
(18) Err on the side of too much, rather than too little, communication.

We abuse land because we regard it as a commodity belonging to us. When we see land as a community to which we belong, we may begin to use it with love and respect.

Thus Leopold, like Lavigne *et al.* (2006), links conservation with values. The issue before us in the twenty-first century is to encourage culture-specific values that help to change human behaviours in ways that dramatically increase the likelihood that human activities are compatible with species and ecosystem conservation.

As a closing note, we refer to the various 'conservation actions' noted in different sections of Chapter 8. These existing and needed actions include tools and processes for conserving sirenians that we consider in the remainder of this chapter. We have divided these tools and processes into two non-exclusive categories: (1) regulatory; and (2) enabling. We attempt to describe practical tools and processes that will establish an elevated value for conservation and promote a reversal of past destruction.

REGULATORY TOOLS AND PROCESSES
FOR CONSERVING SIRENIANS

Legal protection

As noted in Chapter 8, almost all sirenian populations are legally protected in their range states as well as by several international treaties and agreements. It seems likely (e.g. UNEP 2010) that many people who harvest or incidentally take a dugong or manatee are aware that killing a sirenian is against the law. Thus, the primary issue is not a matter of creating appropriately protective legislation; nor in many cases is it a matter of making people aware of the laws. Rather, the challenge is to ensure that consequences of other needs do not outweigh consequences of ignoring or breaking the law. For sirenian range states in developing countries, making conservation laws work as intended is a socioeconomic issue, further complicated at times by weak governance and high levels of corruption (Laurance 2004). In the relatively few developed range states, the primary impediment is a lack of political will to enforce laws.

In summary, what is needed most is *not* only legal protection or commendable intentions on the part of law makers. What is needed are mechanisms (dependent on cultures, economics, etc.) to allow the laws to work as intended.

Enforcement

In virtually all sirenian range states, there is a need for adequate funding and staffing to ensure that existing laws are enforced. This problem is not restricted to prohibitions on hunting in developing range states. The United States is an extremely wealthy country – the density of people in Florida is relatively high and the extent of occurrence of Florida manatee habitat relatively small compared with, for example, the extent of Amazonian or West African manatee habitat or dugong habitat in Australia (Chapter 8). Nonetheless, the area of Florida manatee habitat, coupled with a large and rapidly growing human population that uses that habitat for various activities, makes enforcing laws designed to protect habitat and slow boat speeds extremely difficult (US Fish and Wildlife Service 2001). In addition, statutory and regulatory vagueness (e.g. what constitutes 'harassment' of sirenians) accentuates the problems with enforcement of seemingly good conservation laws.

Situational crime prevention (Clarke 1997) analyses the circumstances giving rise to particular types of crime to reduce the opportunity for those crimes to occur, focusing on the settings for crime rather than on those committing criminal acts. This approach seeks to make criminal action less attractive to offenders. Situational crime prevention identifies opportunity-reduction measures that have been developed for particular crimes by making them more difficult and risky, or less rewarding and excusable. Many of the enabling tools and processes for conserving sirenians outlined later in this chapter are consistent with this approach. The challenge is to optimise these tools for specific sirenian populations and to rationalise control measures to be in line with local capacity without surrendering key conservation outcomes.

Aquatic protected areas

The IUCN (1994) declares that protected areas are 'especially dedicated to the protection and maintenance of biological diversity, and of natural and associated cultural resources, and managed through legal or other effective means'. Marine protected areas have proliferated worldwide, although the extent to which they are designed to protect resources varies from strict reserves to areas for management and sustainability of particular resources (IUCN 1994). A growing body of literature (e.g. Salm *et al.* 2000; Hooker and Gerber 2004; Fernandes *et al.* 2005) documents the establishment, successes and shortcomings of marine protected areas. Because manatees occur in fresh water as well as marine habitats, we use the term 'aquatic protected areas'.

Marsh and Morales-Vela (2012) provided an overview of protected areas as a tool for sirenian conservation. They outlined the potential importance of protected areas for sirenian conservation; they also noted that for a variety of reasons (e.g. lack of explicit goals, enforcement, funding or assessment) marine (or aquatic) protected areas often fail to accomplish what their creators intended. Marsh *et al.* (2002) and UNEP (2010) documented that dozens of protected areas exist specifically to protect dugongs and manatees (see also Chapter 8), but noted that many are 'paper parks' that exist only as documents, and functionally contribute little to conservation. Even protected areas that have functioned well in the past, such as the Chetumal Bay Manatee Protected Area in Quintana Roo, Mexico, can become ineffective when budgets for enforcement are reduced or management goals shift (Marsh and Morales-Vela 2012). However, marine protected areas can be successful at conserving sirenians. The Great Barrier Reef World Heritage Area, for example, contains a network of ecosystem-scale marine protected areas and utilises other management approaches that protect a high proportion of habitats that receive high use by dugongs (Grech *et al.* 2008).

Box 9.2.   Desirable attributes of aquatic protected areas for sirenians as identified by the participants at the 2009 workshop at the International Marine Conservation Conference (Washington, DC)

- Ensure community involvement that incorporates local knowledge.
- Develop management planning that reflects the regional legal framework and includes goals that are specific to sirenians.
- Encourage legal frameworks and the political will to make them work.
- Develop strong education and awareness programmes.
- Create protected areas that are sufficiently large to: (a) include a high percentage of the sirenian population throughout the year; and (b) protect ecological processes.
- Ensure long-term funding to implement management plans.
- Develop co-management involving government agencies, non-government organisations, local communities and scientists.
- Ensure effective enforcement.
- Build capacity, including succession planning, for all co-management partners.
- Maintain active research programmes to inform management.
- Develop alternative livelihoods for community members affected by the implementation of the protected area plan.

Participants in a 2009 workshop at the International Marine Conservation Conference (Washington, DC) discussed attributes of protected areas that would help to ensure their success for conserving sirenians. The primary recommendations of that group included many of the points that are raised elsewhere in this chapter with regard to stakeholder involvement and communication, as summarised in Box 9.2 and in Marsh and Morales-Vela (2012).

Thus, the creation of an effective protected area, as opposed to a 'paper park', requires considerable (a) information with regard to the species or ecosystems being conserved; (b) communication and feedback among stakeholders; (c) knowledge of and attention to mitigation of threats; and (d) commitments of funding for research, management and enforcement. Reserve-design software is available (e.g. Ball and Possingham 2000; Possingham et al. 2000) to allow multiple data sets, objectives and social costs to be assessed to provide several alternative reserve designs. However, reserve-design software does not reduce the need for effective face-to-face communication between managers and stakeholders (see Nursey-Bray et al. 2010) or for funding for management, enforcement and research.

Once a plan for a protected area is implemented, it is important to evaluate whether that process promotes achievement of the goals of the plan. Pomeroy *et al.* (2004) developed a practical guidebook for evaluating effectiveness based on carefully selected social and ecological indicators. Managers of protected areas should endeavour to practise 'adaptive management' (see below), an iterative process in which new initiatives are attempted, their success evaluated and refinements in practices implemented based on results of the evaluation (Pomeroy *et al.* 2004).

Creation of effective aquatic protected areas can be extremely useful for conservation of sirenians in both developed and less developed countries. The challenges and 'ingredients' of management plans will naturally vary as a function of stakeholder needs and perceptions, but the end result can help achieve species conservation goals and provide alternative livelihoods for some community members as enforcement rangers, guides and conservation educators.

Working with local communities to enlist their support for effective measures to conserve sirenians, and by extension other species as well, is essential. Setting aside areas where hunting and/or fishing currently do not occur because of problems of access may be one of the tools that will be effective (Nasi *et al.* 2008), provided that such areas can be enforced. Naturally, if there is interest in creating such off-limits areas, the local people should be heavily involved in discussions and negotiations; without their involvement and support, the efforts will likely fail.

ENABLING TOOLS AND PROCESSES
FOR CONSERVING SIRENIANS

Education and awareness

It has become a cliché that conservation must involve education. Unfortunately, managers often ignore the important step of assessing the effectiveness of education programmes to ensure that they truly facilitate achieving conservation goals.

In designing an education programme, it is vital that processes and materials be developed to reach a particular audience with a particular message or set of messages. Children respond to different materials and messages than adults; adults whose literacy is limited require different forms of communication than adults who read well; and poor, subsistence users in coastal communities in developing countries have vastly different perspectives than bureaucrats in large, prosperous cities.

Box 9.3.   Vietnamese dugong hunter turned educator

'Before working as an ordinary fisherman, (Nguyen Van) Khanh, now 46, was known as the 'sea monster' on Phu Quoc Island because he and his father, who died several years ago, had caught and slaughtered hundreds of dugongs . . .. In Viet Nam, the mammals only live in the sea off Kien Giang Province's Phu Quoc Island and Ba Ria–Vung Tau Province's Con Dao Island.

'Khanh began joining his father on long-day trips out to sea when he was only eight years old. When he grew up, he became the captain of a ship of dugong hunters.

'Aware of his past mistakes, Khanh gave up his hunt for dugongs in recent years. He has also traveled to other areas in the region to persuade fishermen to stop catching dugongs . . . .

'Khanh once caught a 20 kg baby dugong. It cried and struggled in the net while tears fell from the mother's eyes. A few days later he returned to the same spot and caught the mother, as expected, because she was still searching for her baby.

"When I saw the mother dugong lying on the ship, her body looked like a woman with breasts full of milk, I was hurt and haunted", Khanh says. "I decided to give up." . . .

'When Khanh learns that someone has caught a dugong, he immediately tries to persuade them to release the creature back to sea. "Sometimes the people agree but sometimes they scold me and drive me away," he says. Khanh and several other fishermen have joined a group of volunteers established by the local authorities to propagandise information and urge fishermen to stop catching dugongs and other rare animals.

"We meet every week to talk about our plans and distribute leaflets to help people understand why the animal should be protected," Khanh says.

Extract from 'Fisherman fights to protect rare mammal' by Monh Thu. *Viet Nam News* 12 April 2010: http://www.vietnamnews.vnagency.com.vn/ Sunday/Features/198603/Fisherman-fights-to-protect-rare-mammal.html, downloaded 13 June 2010.

The awareness that education and communication must be tailored to an audience is an important step. Even more important is finding the proper individuals to develop education and awareness materials. In developing countries, the best people to develop culturally appropriate materials and activities are the teachers, students and key stakeholders of the local communities, such as former hunters (see Box 9.3). Traditionally structured Western approaches to education simply do not work in other settings (and may not even be optimal in the settings for which they were designed).

Box 9.4. The range of educational tools that have been found to work well in manatee habitats such as Puerto Nariño, Colombia and Mamiraua, Brazil (from Aragones *et al.* 2012b)

(1) Posters, videos, flyers and booklets;
(2) presenting songs, plays, puppet shows, dances and story telling to promote a conservation message at local festivals;
(3) local handcrafts (wood carvings, paintings, jewellery) or foods (Sirenia-shaped cookies or chocolates);
(4) statues, murals or other objects dedicated to local sirenian species.

Aragones *et al.* (2012b) have noted a range of sirenian educational tools that have been found to work well in locations such as Puerto Nariño, Colombia and Mamiraua, Brazil. These tools extend the range typically used in high-income countries (Box 9.4).

The development, assessment and improvement of effective education and awareness programmes is a community-wide activity, involving continuous communication, feedback and adjustment. Such programmes provide long-term benefits because they have the potential to affect local perceptions, values and behaviours in fundamental ways. However, they also require a long-term commitment, not just an occasional site visit; for this reason, if no other, the careful development and nurturing of education programmes is ideally suited to the missions and long-term funding capacities of non-governmental environmental organisations. It is especially important to garner long-term funding and to consider succession planning. Too often, successful programmes collapse when their champion leaves. For example, arguably the most effective and innovative dugong conservation effort in the late 1970s and early 1980s was the Dugong Conservation, Management and Public Education Program in the Western Province of Papua New Guinea (Hudson 1981). The programme collapsed when international funding ceased and its champion, Brydget Hudson, left Papua New Guinea, indicating the need to gain the commitment of local policy makers as well as international non-governmental organisations.

Community partnerships
Without the support of local communities, conservation of any resource is unlikely to succeed. This assertion is exemplified by specific case studies of declining sirenian populations for which community involvement was

> **Box 9.5.   Factors to consider in developing community-based conservation programmes**
>
> - Respect for and integration of local knowledge of species and habitats;
> - open communication;
> - identifying stakeholder interests, especially those that may conflict with goals of the programme or project;
> - developing education programmes specifically for particular community audiences; and regular feedback to and interaction with the community.
> - the availability of long-term funding.

deficient, as well as in situations for which conservation prospects improved in the wake of appropriate team-building (Aragones *et al.* 2012b).

Community partnerships for sirenian conservation are becoming more prevalent. Efforts have been initiated for West Indian manatees (e.g. Chetumal, Mexico), Amazonian manatees in Brazil and Colombia, and dugongs in a number of locations (e.g. Phuket, Thailand; Myanmar; parts of Australia, such as Torres Strait Islands; see 'Reinforcement of cultural protocols' below). Although such efforts are relatively new and have not been described in primary publications to any great extent, efforts by Ilangakoon and Tun (2007) and Hines *et al.* (2005b) for the dugong, and Kendall and Orozco (2003) for the Amazonian manatee, are instructive.

Aragones *et al.* (2012b) note that a successful programme to integrate local communities in sirenian conservation activities needs to be tailored to the perceptions, needs and culture of the local people. Thus, every programme will be, in some critical ways, unique. Nonetheless, there are general categories of factors to consider in developing community-based conservation programmes, as summarised in Box 9.5.

One of the more successful of the emerging community-based sirenian conservation programmes, which embodies the general approaches noted above, has been developed by Sarita Kendall and her colleagues in the Colombian Amazon. The Manatee Conservation Program originated in 1998. Manatees were traditionally hunted for food and other products in the area but, using interview surveys, the group was able to identify and gain support from the hunters themselves – a key step.

The Manatee Conservation Program encouraged people to discuss their perceptions of manatees, including local myths about the origin of the species. Education programmes were developed to help inform the community of the existence of laws to protect manatees, as well as reasons why the laws made sense.

A manatee calf that was captured accidentally in a fishing net was rehabilitated by the community and the Manatee Conservation Program. The latter activity promoted a change in local perceptions of manatees as simply a source of meat. When the animal (named Airuwe) reached three years old, the Manatee Conservation Program hosted a community party to celebrate the event. Activities such as baking manatee-shaped biscuits, dancing, painting and plays included the entire community. When Airuwe was finally released, local fishers took turns following him using VHF telemetry. Some of the local fishers who helped with Airuwe became manatee observers, assisting with the collection of scientific data by the biologists involved with the programme.

Ultimately, Kendall and her colleagues organised community workshops that encouraged open communication among scientists, hunters and fishers about manatee biology and habitat. It became clear to all during the workshops that Amazonian manatees breed quite slowly. This realisation, coupled with the admission from some fishers that the manatee population seemed to be in decline, promoted a community agreement to stop the hunting of manatees.

By 2005 some local fishers reported that there were more manatees in the area. But the work of the Manatee Conservation Program, highlighted by Airuwe as an ambassador for Amazonian manatees, had changed perceptions of manatees and the cultural tradition of using them for meat.

Using a variety of posters, booklets and other materials, the success of the Manatee Conservation Program has spread beyond the original town where it began (Puerto Nariño). Schools and communities (and agencies) now endorse the approach of the programme for manatee conservation.

This model contrasts with a pressing but unresolved dugong conservation issue in East Africa. Directed hunting and incidental taking in gill and other mesh nets and other gear have reduced dugong numbers to insignificant levels in most countries in that region (Chapter 8). Nonetheless, Bazaruto Archipelago National Park in Mozambique supports several hundred dugongs that are legally protected (Chapters 7 and 8). Local fishing, including use of illegal gill nets, takes several dugongs annually; exact numbers are uncertain but are unlikely to be sustainable. The primary targets of the gill netters are large sharks from which fins are taken for sale at an illegal Chinese market in the area. As explained in Chapter 8, Mozambique ranks among the poorest countries in the world, with an annual per capita income equivalent to US$196 in 2010 (World Bank 2010). Compounding the difficulty of dugong conservation is the fact that shark fins reportedly sell for up to the equivalent of US$200 per kilogram, representing an enormous and otherwise unattainable source of income (plus considerable meat from the remainder of the shark) for local people. The dugongs that drown in the nets represent a much-desired source of meat and are consumed locally. The benefits of netting sharks and dugongs (among other things) outweigh the risks of being punished for illegal netting activities. Until the value of a living shark or dugong surpasses the value of a dead one, conservation of dugongs in Mozambique cannot succeed.

In May 2009 a workshop was held in Maputo, Mozambique, to discuss how the government of Mozambique or involved non-governmental organisations could turn this situation around and prevent the extirpation of the local dugong population (Attwell in press). The workshop organisers had assumed that a high-priority action would be monitoring of the dugong population. Instead, participants indicated forcefully that the most urgent issue was *not* further monitoring of dugongs but rather getting the gill nets out of the water and protecting the remaining dugongs. Thus, the discussion ultimately focused on relevance of fishing practices, socioeconomic issues and education, rather than survey methods. Unfortunately, and as a lesson for other such workshops, the contributions of the majority of the workshop participants from outside Mozambique were limited by their background as natural scientists. Experts in sociology, economics or anthropology would have provided important insights.

The primary challenge is to replace a major source of (illegally based) income from shark-finning in a country where poverty and need for food dominate. The workshop participants discussed how several changes, working together, could provide a cluster of solutions, including but not limited to those in Box 9.6 (see also 'Economic tools' below). Implementing these solutions seems an essential first step to solving or reversing the decline of dugongs in Bazaruto Archipelago National Park and Mozambique more generally.

It is uncertain whether dugongs can be conserved in Mozambique. Success depends on the interest and substantive involvement of non-governmental organisations and the government of Mozambique (which could provide funds to support activities such as those described above), local business to develop jobs in which the value of living resources outweighs the value of dead ones, strong leadership and the subsistence communities of the Bazaruto Archipelago. This situation is but one of many examples

---

Box 9.6. Solutions to the problem of dugong mortality in Bazaruto Archipelago National Park as suggested by participants at a workshop in Maputo in May 2009

- Development and implementation of locally based enforcement (involving 'community rangers') who would be paid to police fishing practices;
- development and implementation of alternative fishing methods or livelihoods for local people;
- education/awareness programmes designed to communicate effectively with local communities;
- ensuring that relevant stakeholders have opportunities to meet and communicate.

---

where local activities in a poor, developing country have the potential to eradicate (or conserve) a sirenian population in the very near future.

Cross-species initiatives

Sirenians share habitats with other megafauna such as river dolphins (Amazonian and Antillean manatees), coastal dolphins (dugongs, West Indian and West African manatees) and sea turtles (dugongs, West Indian and West African manatees). Single-species conservation initiatives are too often developed at the expense of more cost-effective and potentially influential synergies. It is pleasing to note that the Memorandum of Understanding (MoU) to conserve the West African manatee under the Convention on Migratory Species (CMS) also covers small cetaceans.

Flagship species

Certain species of wildlife are designated as 'flagship species' when people perceive that they can represent an environmental cause – such as an ecosystem in need of conservation or a desire to encourage regional bio-diversity (Simberloff 1998). Home et al. (2009), among others, characterise such designations as 'marketing tools' designed to engender public support. The rationale is that concern for a particular species will bring about more assiduous and persistent efforts on its behalf, benefiting not only the flagship species itself, but improving prospects for the other species that also share its habitat or are subject to the same threatening processes.

However, it is unclear whether designation as a flagship species is simply arbitrarily assigned based on the interests of a particular environmental or interest group and, in fact, whether the designation is compelling to the public. In reality, both charismatic and uncharismatic species can have very positive influences on public perceptions and preferences for habitat factors that can promote biodiversity (Home *et al.* 2009). In addition, Bowen-Jones and Entwhistle (2002) underscore the importance of engaging local communities in conservation activities and the resultant wisdom of ensuring that choice of species to be designated as flagship includes input from local people regarding their perceptions and values of different organisms. As noted by Home *et al.* (2009, citing Lorimer 2006), charisma is culturally dependent. The point of these and other publications is that the marketing decision to create a flagship species to promote optimal conservation is one that should be done in a measured and strategic way, rather than simply selecting a species that certain (non-local) people consider charismatic or cute.

Like other marine mammals, sirenians have been dubbed as 'flagship species' due, in part at least, to their large size and widely perceived charismatic nature. Laws such as the US Marine Mammal Protection Act of 1972 and its subsequent amendments note that the primary goal of marine mammal conservation is to maintain the health and stability of marine ecosystems on which these animals depend, overtly supporting the idea that conservation of marine mammals is linked tightly to conservation of whole ecosystems, as is assumed to be the case for flagship species in general.

We began this book with a case study involving dugong conservation in Okinawa, Japan (Chapter 1). That issue involves both social and political components, but the employment of the dugong as a cultural icon for conservation of both the species and the coastal environment of Okinawa gained global attention.

The selection of sirenians as flagship species has already facilitated establishment of protected areas or other conservation actions in a number of countries (e.g. see Chapter 8; Marsh *et al.* 2002 for dugongs; UNEP 2010 for West Indian manatees). We believe that this process is occurring or is likely to occur in a number of additional locations as a result of the cultural or iconic importance and charismatic nature of sirenians, as well as the fact that effective conservation of manatees and dugongs simultaneously preserves habitat for species of importance for subsistence or commercial harvest, ecologically vital nursery grounds for many species, and resources of great aesthetic importance.

Reinforcement of cultural protocols

As explained in Chapter 8, the Torres Strait region supports a globally significant dugong population (Figure 9.1). Dugong hunting dates back

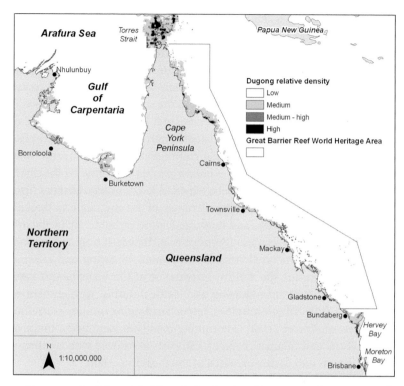

Figure 9.1. Spatial model of the relative density of dugongs on the east coast of Queensland and the Gulf of Carpentaria, Australia, based on the 20-year time series of aerial surveys commencing in the mid-1980s (Grech *et al.* 2011a). The model confirms the importance of Torres Strait as dugong habitat. Drawn by Alana Grech, reproduced with permission.

4000 years (Crouch *et al.* 2007). The right of traditional inhabitants to hunt (but not to sell the catch) has been established by decisions of Australia's highest court and the traditional fishery for dugongs is authorised by an international treaty, the 1984 Torres Strait Treaty, between Australia and Papua New Guinea (Chapters 7 and 8). In recent years, both Traditional Owners and scientists have expressed concern about the sustainability of the contemporary harvest (Heinsohn *et al.* 2004; Marsh *et al.* 2004). The Statutory Management Regulations associated with the Torres Strait Fisheries Act of 1984 place some controls on the dugong fishery in addition to the exclusivity of the hunting rights of Traditional Owners: (1) dugongs may only be taken by traditional inhabitants; (2) dugongs must be caught with a traditional harpoon with a detachable head or *wap*; (3) dugongs must only be caught from a vessel less than 6 m long; (4) dugongs must not be caught in the Dugong Sanctuary, a large area in western Torres Strait distant from the sea countries of most communities; and (5) the sale of dugong meat is prohibited. To date, enforcement of these restrictions has been limited.

The peoples of Torres Strait value the dugong for many reasons. The meat is an important source of protein both for ceremonies and *kai-kai* (everyday meat). The region is isolated and the cost of store-bought food is much higher than on the Australian mainland. In addition, the income of Islanders is much lower than the Australian average and the Islanders live in a welfare economy. The region's commercial fisheries are the only resources available for the people to build a market economy, yet the rights to most of these resources have been allocated to commercial fishers from other areas of Australia. The cultural values of the dugong and dugong hunting are extremely high, and dugongs feature prominently in the art-work of Torres Strait Islanders (Frontispiece). Delisle (2009 and unpub-lished) used semi-structured interviews followed by rating and ranking exercises to determine the relative importance of the benefits and costs associated with traditional dugong and turtle hunting to members of Mabuiag and St Paul's communities, important dugong hunting commun-ities in western Torres Strait. Community members identified a range of social, cultural and financial benefits and costs associated with hunting – cultural benefits and costs were rated as the most important.

Project officers employed by the Torres Strait Regional Authority (TSRA) have worked with 15 indigenous communities to develop community-based Turtle and Dugong Management Plans with funding from the Australian government. These plans are now being implemented with substantial funding totalling some A$20 million over nine years from the Australian government to the TSRA to support a community-based ranger programme.

These plans are not publicly available and thus cannot be formally cited, but copies were provided to research collaborators including H. Marsh. Document analysis of the plans confirms the value that the Islanders place on the cultural values of the dugong. The Islanders see community-based-management of their dugong and marine turtle fisheries as an important means of revitalising their culture. Each of the plans sets out objectives and management arrangements that aim to achieve sustainable use of dugong and turtle resources through implementing cultural practices and protocols.

The Islanders describe the tools contained in the plans as cultural tools, fisheries tools or closures (F. Loban, personal communication, 2009). Some of these categories overlap (Figure 9.2). Whereas each plan acknowledges the Statutory Management Regulations imposed by the Torres Strait Fisheries

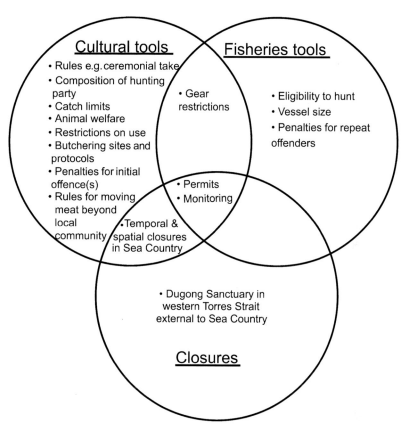

Cultural tools
- Rules e.g. ceremonial take
- Composition of hunting party
- Catch limits
- Animal welfare
- Restrictions on use
- Butchering sites and protocols
- Penalties for initial offence(s)
- Rules for moving meat beyond local community

- Gear restrictions

Fisheries tools
- Eligibility to hunt
- Vessel size
- Penalties for repeat offenders

- Permits
- Monitoring

- Temporal & spatial closures in Sea Country

- Dugong Sanctuary in western Torres Strait external to Sea Country

Closures

Figure 9.2. Conceptual model of the tools contained in the Turtle and Dugong Management Plans developed by 15 Torres Strait Islander communities, 2008–2010. Whereas each plan acknowledges the Statutory Management Regulations imposed by the Torres Strait Fisheries Act of 1984, cultural tools predominate in line with the Islanders' vision of the Plans as instruments of cultural revival.

Act 1984, the cultural tools predominate and the only new non-cultural tools are: (1) commitments to report breaches of the Statutory Management Regulations to the relevant government agencies; and (2) the referral of community members who persistently breach the traditional management rules outlined in the relevant plan to Commonwealth and State authorities for prosecution. Only some plans are explicit about both of these matters. However, all the plans reinforce the expectations that Islander rangers will be provided with opportunities for training and joint patrols with State and Commonwealth fisheries enforcement agencies with a view to eventually taking on fisheries enforcement roles and responsibilities. The plans also make explicit the aspirations of the Islanders to be involved in comprehensive ranger training programmes, community education, monitoring and research in partnership with relevant agencies and research institutions.

This approach includes some of the features identified by Nasi *et al.* (2008) as important for the sustainable harvest of bushmeat, especially the need to increase the capacity for local people to manage their own resources in association with exclusive use rights. Naturally, such an emphasis on reinforcing cultural values will only be relevant in places where the cultural values of sirenians remain strong.

The relationships between biological and cultural diversity are crystal-lising around the concept of biocultural diversity, the total variety exhibited by the world's natural and cultural systems. Loh and Harmon (2005) developed a global measure of biocultural diversity, using a country-level index. The index identifies three areas of exceptional biocultural diversity. All support sirenians, suggesting that programmes that reinforce tradi-tional values, cultural protocols and ethics might be worth exploring in the range states in these regions:

(1) *The Amazon Basin*: Peru and Ecuador (Amazonian manatee); French Guiana, Suriname and Guyana (Antillean manatee); Brazil and Colombia (Amazonian and Antillean manatees)
(2) *Central Africa*: Nigeria, Cameroon, the Democratic Republic of Congo and Gabon (West African manatee); Tanzania (dugong)
(3) *Indomalaysia/Melanesia*: Papua New Guinea, Indonesia, Malaysia, Brunei and Solomon Islands (dugong).

Nonetheless, there are two obvious impediments to this approach. At present population sizes, any hunting of sirenians will be unsustainable in most areas (Chapter 8). In addition, a programme such as the one implemented in Torres Strait is expensive. Thus this approach may be beyond the reach of govern-ments and non-governmental organisations in most sirenian range states.

Research

Orr (2002) stated that 'science on its own can give no reason for sustaining humankind (or anything else). It can . . . . create the knowledge that will cause our demise or that is necessary to live at peace with each another and nature.'

This statement reinforces the notion expressed by other authors (e.g. Marsh *et al.* 2003; Ragen *et al.* 2005; Lavigne *et al.* 2006; Reynolds *et al.* 2009; Hines *et al.* 2012c; among others) that science can and does play a crucial role in conservation by providing information to decision makers. However, the presence of scientific information of high quality does not ensure that it will have a positive effect.

Box 9.7. Suggested actions to promote the value of scientific research programmes for conservation (Ragen *et al.* 2005)

- Develop long-term, multidisciplinary research and management programmes suitably scaled to ecosystem complexity;
- ensure that population and ecosystem assessment programmes are sufficient to inform management decisions regarding current and future threats;
- develop and validate specific, measurable and robust management standards to achieve conservation goals;
- identify marine mammal conservation units essential to ecosystem health and function;
- increase international cooperation in studying and addressing human-related threats;
- properly assess and communicate the strengths and limitations of the scientific process, including measures of uncertainty that are an essential element of high-quality science;
- address ultimate as well as proximate causes of environmental problems.

Ensuring that science *is* of high quality and addresses the critical uncertainties to provide clear answers to conservation questions seems like an obvious approach. However, shortcomings exist, and Ragen *et al.* (2005) recommended that scientific research programmes take several actions to promote their value for conservation (Box 9.7).

There are several issues regarding scientific input to decision making. First, biological or ecological monitoring often takes place for years or decades, without the data being used to solve relevant conservation issues. The failure to resolve the imminent loss of some critical industrial warm-water refugia in Florida (Chapter 7 and below) is a good example in which manatee use of the areas is well documented, but solutions to mitigate the loss of needed warm-water resources have been largely ignored. In addition, although natural scientists have particular skills and insights, they are often not the best individuals to represent a conservation issue. Part of the problem is that scientists are neither trained for, nor typically very good at, communicating to general audiences. They generally receive their rewards and incentives for communicating to technical audiences of professional peers. However, the inability to communicate effectively to non-technical audiences sometimes limits scientists' ability to compel decision makers to prioritise conservation over other values.

As we have indicated above, conservation efforts for sirenians can and should proceed even when knowledge is incomplete, especially when data for related species may be applicable. However important scientific information may be, integration of the social, economic and other factors noted by Meffe *et al.* (1999, quoted above) is essential for success. This fact brings up the final limitation on natural scientists and their science: in our own experience, workshops and other fora designed to address pressing conservation problems often bring a number of biologists to the table, but neglect to involve professionals trained in other, and perhaps more essential, disciplines. When called upon to discuss social and cultural phenomena, natural scientists are not the most informed or appropriate choices. Among others, a recent paper by Moore *et al.* (2010) underscores the need to involve social scientists in projects and programmes at an early stage.

Natural scientists have an important role to play in sirenian conservation efforts. However, even in the face of inadequate scientific knowledge, conservation must proceed with an emphasis on including and empowering communities and regional organisations. Conservation should not be stalled by a perceived lack of scientific information.

Stock assessment

Since 1994, the US Marine Mammal Protection Act of 1972 has required relevant US government agencies to prepare draft assessment reports for each stock of marine mammal that occurs in waters under US jurisdiction in consultation with regional scientific review groups. Each stock assessment report has to contain various items of information, including: (1) a description of the stock, including its geographic range; (2) a minimum population estimate, a maximum net productivity rate and a description of the current population trend; (3) an estimate of the total annual human-related mortality (including fishing mortality) and serious injury to the stock and, for a strategic stock, other factors that may be causing a decline or impeding recovery of the stock, including effects on marine mammal habitat and prey; (4) a description of the commercial fisheries that interact with the stock, including the estimated level of incidental mortality and serious injury of the stock by each fishery on an annual basis; (5) a justification for categorising the stock as strategic or not; and (6) an estimate of the potential biological removal (PBR) level for the stock (Wade 1998) and the information used to calculate this statistic (Barlow *et al.* 1995). The Florida manatee and the Antillean manatee in Puerto Rico are the only sirenians for which such assessments are performed. We believe that this approach, although scientifically rigorous, is impractical for most other sirenian populations for which the budget priorities must be action, not research.

Managing for multiple threats

Given the multiple threats to sirenians, managers are faced with decisions about setting priorities for the utilisation of limited resources towards solving problems. Certainly, management must focus on all actions that can move conservation forward opportunistically, particularly actions that can be implemented without controversy. However, analytical tools are becoming available to help in deciding how management may lead to the greatest long-term improvement in conservation outcomes. On a population level, the ability to model future changes in abundance based on life history stages has been developed for the Florida manatee (Runge *et al.* 2007a,b), and application of geospatial tools has been employed in the study of dugongs in Australia (see Grech and Marsh 2008 and 'Spatial management of the risks to sirenians: a practical approach to conservation' below).

In the Florida case, demographers developed a 'Core Biological Model' (Runge *et al.* 2007a) built on the life history stage-based modelling of population growth rates described in Chapter 6. The model includes annual variability in parameter estimates, stochastic variation, changes in warm-water availability, and catastrophes, and can be applied to separate geographic regions using inputs for abundance and other scenarios to simulate predicted population levels given reductions or increases in various threat factors (Runge *et al.* 2007b). Core model results predict initially increasing populations in three areas with declines occurring over different time periods as warm-water capacity is lost; in the fourth area (south-western Florida) declines occur at the onset and are exacerbated by loss of warm-water capacity (Runge *et al.* 2007a). Statewide, the model predicted a 56% probability of a decline of at least 20% in population size over two manatee generations, given the input data available at the time. The model application for threats analysis, used to quantify risk in terms of probability of quasi-extinction (reduction to populations below a minimum effective population size), was based on five threat factors: mortality due to boat strikes; loss of warm-water refugia; incidental mortality caused by locks and gates; deaths due to entanglement; and deaths due to brevetoxin exposure (Runge *et al.* 2007b; see Chapter 7 for descriptions of these threats). The model allows complete or partial removal of various threats alone or in combination, and uses Bayesian modelling techniques to estimate fractions due to various mortality sources using data from the carcass salvage and necropsy programme (e.g. Bonde *et al.* 1983; Lightsey *et al.* 2006). Application of the model was based on a then-current estimate of the Florida manatee population of 3300 and must be revised based on 2010 estimates (see Chapter 8). Nevertheless, the analysis showed that removal of mortality due to boats, first and foremost, would dramatically lower the probability of extinction, followed by eliminating or reducing the threat due to loss of warm-water refugia. If both threats were substantially reduced then the risk of quasi-extinction would change from 8.6% over 100 years to 0.1% (Runge *et al.* 2007b). Both threats affect population dynamics in different ways over different time periods: reducing the threat of loss of warm-water refugia increases population buffering capacity, whereas reductions in mortality due to boats act to increase intrinsic growth rates (as noted above, boats kill mostly adults, and as mentioned in Chapter 6, adult survival is the most critical parameter contributing to population growth). Changing the variability in the threat factors can result in a variety of interesting and sometimes non-linear results applicable to management (reductions in quasi-extinction results can be achieved by combinations in reductions of lesser threats), detailed more comprehensively by Runge *et al.* (2007b). The strong, intuitive management concern described above regarding both mortality due to boat collisions and loss of warm-water refugia in Florida are well supported by this quantitative analytical tool.

Spatial management of the risks to sirenians: a practical approach to conservation

In all range states other than the United States, the impediments to sirenian conservation are much more immediate than the lack of scientifically robust data. Even data on the local distribution and relative abundance of sirenians and their habitats at the spatial scales required for effective conservation planning are unavailable for at least some locations. Collecting such information is expensive, logistically difficult (as outlined in Chapter 8) and beyond the capacity of most sirenian range states. In addition, constraints of time, expertise and cost often mean that monitoring programmes cannot be conducted at spatial scales that are large enough, or over time frames that are long enough to determine whether management interventions are working. These problems exist even in developed countries such as Australia, including high-profile regions such as the Great Barrier Reef World Heritage Area.

Alana Grech and her co-workers recently used a novel approach to address the challenges associated with informing the management of dugongs and their habitats in the Great Barrier Reef World Heritage Area using spatial models and risk assessments in geographical information systems (GIS). They developed spatial models using coastal seagrass mapping and dugong distribution and abundance data (Figure 9.1) at the scale of the coastal waters (approximately 22 600 km²) of the entire World Heritage Area and overlaid information on the spatial distributions of threats such as gill-netting, hunting, vessel strike and low-quality terrestrial run-off in the GIS (Grech and Marsh 2007, 2008; Grech et al. 2008, 2011a,b; Grech 2009; Grech and Coles 2010; A. Grech, personal communication, 2010).

Grech et al. (2008) then used expert knowledge to evaluate the relative risks of various threats to the dugongs and their seagrass habitats. They used expert opinion, spatial information on the distribution of threats, and the spatial model of seagrass and dugongs to identify areas where human impacts posed low, medium and high relative risks to dugongs and their habitats. This technique allowed Grech and her group to explore method-ically the ways in which the systematic removal of various threats would likely affect dugong status. The approach identified sites where gill-netting was still occurring in areas of high dugong density (e.g. Figure 9.3; see Grech et al. 2008), despite the 2003 re-zoning of the Great Barrier Reef Marine Park (McCook et al. 2010). Information concerning these sites has been provided to a review of the inshore gill net fishery. The risk assessment has also been used to identify areas in the community management areas of indigenous peoples where hunting does not occur at present because the areas are difficult to access in small boats. Several workshops have been held with local indigenous peoples, who have Native Title rights to hunt (Chapter 7), to discuss the possibility of declaring these regions 'no hunting areas' to pre-empt hunting expanding with improved technology.

The data used to develop the spatial models of threats, seagrass distribu-tions and dugong distribution and abundance (Grech and Marsh 2007; Grech et al. 2011a,b) were collected over many years using expensive and extensive vessel surveys (seagrasses), aerial surveys (dugongs; see Chapter 8) and government records of threats. The modelling was done using sophisti-cated techniques. However, similar but less robust information could be collected using less elaborate techniques, such as interviews with local fishers (Moore et al. 2010; Ortega-Argueta et al. 2012), provided the data were collected consistently and at ecologically appropriate spatial scales. Threats can be ranked using existing information, including information from other sirenians. We know, for example, from the information in Chapter 6, that the greatest risks to all sirenians are from anthropogenic activities that kill adult animals, such as hunting, capture in gill nets and vessel strikes (Chapter 7).

The prospect of conserving any sirenian species throughout its range is daunting. However, the relative severity of threats is not consistent across the ranges, and if one were able to identify the 'hotspots', conservation (including but not limited to science, capacity building and stakeholder involvement) could be focused in ways that could have a disproportionate effect on the wellbeing of the entire species. The locations of hotspots would depend on

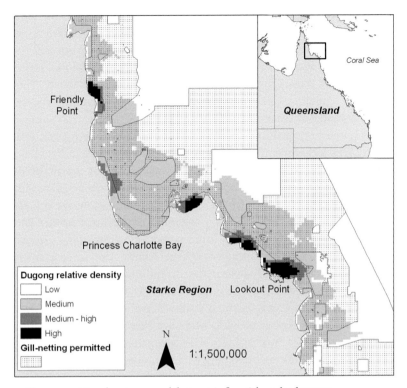

Figure 9.3. Map showing several 'hotspots' of spatial overlap between
gill-netting based on logbook data collected by commercial fishers and
high-density dugong habitat based on the spatial model of the relative density of
dugongs in the Cape York region on the east coast of Queensland, based on the
20-year time series of aerial surveys commencing in the mid-1980s (see Figure 9.1 for
the model for the entire east coast of Queensland and the Gulf of Carpentaria). The
information on the 'hotspots' has been provided to managers negotiating the revision
of regulations for this fishery. Drawn by Alana Grech, reproduced with permission.

factors such as number of animals present, type and severity of threats, and
extent to which those threats are being addressed and mitigated.

Moore *et al.* (2010) developed a rapid interview-based approach to assess
where marine mammals and sea turtles seem especially vulnerable to catch
in artisanal fisheries. This approach was applied in seven countries: three in
West Africa; two in East Africa; Malaysia; and Jamaica. Among other
interesting findings, the surveys found that, for the countries surveyed,
incidental catch of sirenians was most frequent in West Africa.

Naturally, an interview-based system will not provide all the information
one needs for effective conservation. However, it can allow managers and
scientists to focus their resources to either learn more about the situation or

to take rapid steps to mitigate threats. The dugong project being implemented under the CMS Dugong MoU (Chapter 8) has refined the survey instrument of Moore *et al.* (2010) and is using it across the range of the dugong to identify locations where species or population vulnerability is highest to enable the limited human and financial resources to be used most effectively (D. Kwan, personal communication, 2010).

The 'hotspot' concept may also be applied to people as well. As has been documented more generally by studies of the bushmeat crisis (Nasi *et al.* 2008), the deliberate hunting or incidental killing of sirenians is typically done by relatively few people within subsistence communities. Not only do such people have direct and measurable impacts on sirenian populations in particular areas, but they can be valuable to efforts to identify threats to sirenians and their habitat because of their expert knowledge, acquired over a lifetime of hunting or fishing (see Box 9.3). It seems prudent, therefore, to develop a community-based programme that includes all stakeholders (as described above) but to take special pains to involve the hunters/experts. This is the approach that Kendall used so successfully in her manatee conservation efforts in Colombia (described under 'Community partnerships' above).

Adaptive management

Adaptive management is learning by doing. A leader in applying adaptive management to sirenian conservation is Michael Runge of the US Geological Survey. For example, Runge has approached the issue of managing warm-water refugia for Florida manatees as an exercise in adaptive management in which scientists, managers and power companies will collectively develop predictive models, monitor manatee behaviour in relation to management actions, evaluate effectiveness of the management actions, and make appropriate changes in the models and future actions based on the results. Adaptive management provides an explicit structure that optimises the chances that conservation actions will be effective by engaging key stakeholders in critical processes such as goal setting, planning, management, enforcement and evaluation (Buck *et al.* 2001). Adaptive management requires effective monitoring, which can be challenging for small populations of sirenians, as discussed in Chapter 8 and under 'Monitoring' below.

Monitoring

Adaptive management requires monitoring to determine whether the approach adopted is successful. As outlined in Chapter 8, determining trends for small sirenian populations in time frames useful to management is usually impossible, so proxy indicators have to be developed to inform adaptive management. A technical workshop in Townsville, Australia, in 2010 critically evaluated a series of tools for monitoring the status of dugong populations in the Great Barrier Reef World Heritage Area, including: (1) large-scale aerial surveys; (2) spatial models and risk assessments; (3) broad-scale seagrass surveys and Seagrass-Watch (see below); (4) catch per unit effort data; and (5) the Queensland Marine Strandings Database. The major outcome of the technical workshop (Grech and Marsh 2010) was the recognition of the applicability and validity of all of these monitoring tools. It was acknowledged that there is not a 'one size fits all' monitoring solution; a combination of monitoring tools is required to inform management. The workshop agreed that a report card would provide an integrated assessment of the performance of management actions by taking into consideration the outputs of multiple monitoring tools. Such a dugong report card would need to be linked to report cards associated with water quality and seagrass to: (1) inform the status of the dugong's habitats; and (2) provide for assessments on the status of the relevant ecosystems. We consider that this situation is likely to be typical and that different suites of monitoring tools will need to be tailored for different populations of sirenians and the resources available.

Seagrass-Watch (2010) is a global, scientific, non-destructive, community-based seagrass assessment and monitoring programme that is important to the conservation of dugong and manatee habitats. Since its genesis in 1998 in Australia, Seagrass-Watch has expanded internationally, with participants in other sirenian range states including Papua New Guinea, Solomon Islands, New Caledonia, Palau, Japan, China, Vietnam, the Philippines, Thailand, Malaysia, Indonesia, Myanmar, India, Eritrea, the Comoros Islands, the United States and Singapore. Monitoring is now occurring at approximately 260 sites across 17 countries; an additional nine countries participate but are currently at the resource-identification stage.

Seagrass-Watch aims to raise awareness of the condition and trend of nearshore seagrass ecosystems and provide an early warning of major coastal environment changes (Seagrass-Watch 2010). The Seagrass-Watch programme involves collaboration/partnerships among the community, qualified scientists and the data users (environmental management agencies). People involved in the programme develop a deep sense of custodianship and understanding of their local marine environments, which reaches throughout the wider community. Coastal communities work in partnership with government agencies to play a primary information-gathering role. Participants are from a wide variety of backgrounds and all share a common interest in marine conservation. Most participants are associated with universities and research institutions, government (local and state) or non-governmental organisations, established local community groups and schools.

The level of involvement depends on local resources, local coordination, local support, available capital and scientific expertise (Seagrass-Watch 2010). Seagrass-Watch also integrates with existing education, government, non-government and scientific programmes to raise awareness and conserve seagrass ecosystems for the benefit of all. Participants collect quantitative data on seagrasses and their associated fauna using simple yet scientifically rigorous monitoring techniques.

The programme has a strong scientific underpinning with an emphasis on consistent data collection, recording and reporting. Scientific, statistical, data management, data interpretation and logistic support underpin all monitoring efforts. Seagrass-Watch identifies areas important for seagrass species diversity and conservation and the information collected is used to assist the management of coastal environments and to prevent significant areas and species being lost.

Seagrass-Watch is a model for habitat monitoring programmes that could be developed for sirenians and illustrates the need for sirenian conservation scientists and managers to partner with overlapping initiatives in habitat and wildlife conservation.

Economic tools

Perhaps the greatest challenge in sirenian conservation is to provide incentives to reduce the likelihood of: (1) low-income hunters killing manatees or dugongs deliberately; (2) low-income fishers killing the manatees or dugongs that they catch incidentally rather than releasing them alive; and (3) destruction of the habitats of these animals in developing countries through destructive fishing practices (Chapter 8). As explained in Chapter 7, the distinction between deliberate and incidental captures has blurred in many countries as gill nets are increasingly used to target sirenians.

In many developing countries, sirenian conservation is inextricably linked to poverty alleviation and the provision of alternative livelihoods for fishers and hunters. Efforts to alleviate poverty in low-income nations may produce incentives to degrade the local environment as discussed above (see 'Problem recognition and definition'). For example, increasing local incomes adjacent to ecologically valuable areas often increases land clearing for agriculture (Wünder 2001). Conversely, efforts to protect biodiversity through ecotourism ventures may not result in improvement in local livelihoods (Kiss 2004). Although integrated conservation and development projects were touted as a solution to biodiversity conservation in developing countries, they have had mixed success and alternatives and modifications to the original concept are being investigated (Wells *et al.* 2004).

Economic approaches to environmental protection can be positive or negative, direct or indirect, and can be designed as incentives or deterrents. Indirect, positive incentives include support for alternative livelihoods that value environmental assets, such as ecotourism. For example, both dugongs and manatees have proved to be tourism drawcards at clear-water sites in several developing countries. Antillean manatee tourism takes place in Belize (UNEP 2010) and dugongs have been promoted as tourist attractions in locations such as Vanuatu and the Philippines (Marsh *et al.* 2002). Nonetheless, sirenian tourism is likely to have limited appeal in most developing countries, except as part of the overall wildlife attractions of an area. Dugongs and manatees often occur in areas of high water turbidity and generally only the nose or back of the animal is visible very briefly as it surfaces to breathe (Figure 8.1). Especially in areas where they are hunted, both manatees and dugongs are generally cautious and tend to avoid humans. In addition, indirect approaches such as ecotourism often fail to protect biodiversity and ecosystems to the extent needed (Mandel *et al.* 2009) or to provide alternative livelihoods for local people (Kiss 2004), as noted above.

Direct incentive payment approaches to conserve biodiversity have been advocated and explored. These include payment for ecosystem services, restricted land easements and direct performance-based payments (Ferraro and Kiss 2002; Mandel *et al.* 2009). Lindsey West has successfully used performance-based payments to engage local people in sea turtle and dugong conservation in Tanzania (Box 9.8). However, incentive payments do not necessarily result in improved livelihoods (Mandel *et al.* 2009). Rather, they tend to be short-term payments that rely on a long-term funding stream and can result in ephemeral incentives. A lump-sum or one-time payment does not guarantee a lasting incentive for protection of an environmental asset. Direct payment schemes can also be complicated by: limited or no enforceable property rights such as fishing permits and contractual laws; restrictions on or regulations controlling foreign ownership; and ethical issues resulting from the difference in purchasing power between the 'buyer' and 'seller' (Mandel *et al.* 2009).

Microfinance seeks to eliminate poverty by providing fair, safe and ethical financial services for people who, because of their circumstances, are not able to access mainstream financial services. Microfinance institutions have had considerable success in alleviating poverty over the past two decades, particularly in Bangladesh (Davis and Kosla 2007). This approach is being extended to use debt as a finance mechanism for conserving biodiversity by combining microfinance lending approaches with a performance-based

Box 9.8.   Example of a cash payment incentive scheme for conserving turtles and dugongs operated in Tanzania by the non-government organisation, Sea Sense

Tanzania in East Africa has a very small and apparently depleted dugong population and a low Human Development Index (see Chapter 8, especially Figures 8.2–8.3 and Table 8.9). The non-government organisation Sea Sense has conducted a very successful direct payment community incentive scheme for sea turtles since 2001. Sea Sense pays US\$12–15 per nest, the amount depending on hatching success. The scheme has reduced nest poaching from 80% to less than 3%, with more than 2500 sea turtle nests monitored and protected, enabling an additional 170 000 sea turtle hatchlings to successfully reach the sea.

Sea Sense has also introduced a direct payment incentive scheme designed to reduce the capture of dugongs in gill nets by offering a significant payment (US\$140) for the release of a live dugong from a gill net and a similar amount for surrendering a dead dugong. The scheme has: (1) enabled specimens for genetic analysis to be collected from several dead dugongs; (2) reduced the sale and consumption of dugong meat; (3) increased public awareness of the conservation concern for the status of the dugong in the region; and (4) led to the reporting of consistent sightings of a live dugong in the Mafia Island region by an artisanal fisher, despite opposition from his peers who feared that the region would be converted to a 'no-take' marine protected area. (L. West, personal communication, 2010).

incentive structure for environmental stewardship (Mandel *et al.* 2009). Although this approach – termed 'environmental mortgages' – has not yet been applied to sirenian conservation, it is being investigated in the dugong project being developed under the CMS (D. Kwan, personal communication, 2010). Environmental mortgages are a promising approach for conserving sirenian habitats and reducing mortality from incidental capture in fishing gear.

Conditional cash transfers are another alternative worth investigating. Cash payments could be made (e.g. for school fees and the opportunity cost of sending the fisher's children to school; the income from sirenian hunting is sometimes used for this purpose in Africa; see Chapter 7) if, and only if, the fisher stopped the fishing practice that reduced habitat quality or killed manatees or dugongs. The programme could combine conditional cash transfers (immediate incentive) with environmental mortgages (longer-term incentive). The loans could be used for a range of ventures that reflected community needs, aspirations or economic possibilities.

For example, a community could agree to close an environmental asset such as a seagrass bed that was locally important dugong habitat to fishing practices that reduce the habitat quality of the area such as push-netting, in exchange for a reduced-interest loan. A pool of capital would be raised based on a combination of the international conservation community's willingness to pay for seagrass and dugong conservation and the amount the local community would need to receive to forgo benefits associated with degradation of the habitat such as the income earned from the push net fishery. This pool of capital would then be placed in a financial trust, under the partial control of the community, with the express purpose of making loans to the stakeholders in the community-held area (Mandel *et al.* 2009).

The status of the seagrass bed closed to fishing would be monitored at agreed intervals and the terms of the loan revised accordingly, including loan termination if necessary. The community could be involved in Seagrass-Watch (see 'Monitoring' above) and provided with the capacity to monitor changes in the extent of: (1) fishing damage to the seagrass bed; and (2) dugong feeding trails. The credibility of the programme would be reliant on the robustness of the monitoring scheme and its capacity to distinguish the effects of long-term change from the noise of environmental perturbations. From that perspective, monitoring damage to the seagrass bed from fishing would be the more reliable indicator. The density of dugong feeding trails could change for natural reasons unrelated to the closure (see Chapters 4, 5 and 6).

A similar approach could be used to reduce the impact of hunting or the incidental capture of dugongs or manatees in gill nets. In such cases, it would probably be necessary to couple economic incentives with increased enforcement to combat illegal practices. Again it would be important to consider what could realistically be monitored. Any sirenian population will almost certainly be too small for visual or acoustic monitoring to have the statistical power to detect change in abundance at a local level (Chapter 8), and attempts to monitor sales of dugong or manatee meat would be likely to drive that activity underground. However, it should be possible to monitor changes in fishing practices, such as the use of gill nets.

Mandel *et al.* (2009) identified challenges associated with these economic approaches and noted that careful biological and sociopolitical assessments of potential scenarios will be required to determine when they are appropriate. For example, to repay a loan, the fisher or their family would need to earn more from the alternative livelihood than the livelihood that threatens sirenians or their habitats. The approach would also be likely to be more successful in communities with robust social networks where social pressures to repay loans are strong. Nonetheless, given the success of microfinance institutions on poverty alleviation over the past two decades, environmental mortgages seem a promising approach for linking sustainable development and sirenian conservation. If challenges in design and implementation can be overcome, environmental mortgages could provide monetary values for the conservation of sirenians and provide not only the incentive but also the means for low-impact livelihoods and economic development.

OPPORTUNISM AND CREATIVITY

This chapter has underscored that effective conservation does not happen by accident. It happens because dedicated people approach conservation in a measured fashion that includes and respects a variety of stakeholders. It happens when managers and decision makers have the will to take a precautionary approach that places a high value on conserving species or ecosystems, even in the face of insufficient scientific information. Conservation can also happen when informed people are willing to act boldly and to take advantage of rare opportunities. We provide two examples, below.

Possible reintroduction of Antillean manatees to Guadeloupe (French Antilles)

The Parc National de la Guadeloupe (National Park of Guadeloupe) and other entities in Guadeloupe are interested in re-establishing manatees in the waters of the natural reserve called 'Grand Cul-de-Sac Marin'. The species has been extinct in the waters of Guadeloupe for several decades (see Chapter 8), having been wiped out locally by hunting; some elders apparently recall seeing manatee meat available for food. Manatees remain a part of local folklore and the history of Guadeloupe (e.g. the town of Lamantin (French for manatee), is located near Grand Cul-de-Sac Marin). In addition, a former 'manatee processing area' exists near Vieux Bourg (Grand Terre). The goal of reintroducing manatees has existed since a feasibility study was conducted in 2002.

The national park and the government of France are strongly behind the proposed initiative and are working to develop funding and stakeholder support. The scientists, park administrators and government officials are interested, knowledgeable and committed. The Grand Cul-de-sac Marin is a lagoon of 15 000 ha (150 km^2). It contains lush seagrass meadows (mostly *Thalassia*, but some *Syringodium*), mangroves, tidal creeks and freshwater sources. The human population of the scattered villages rimming the lagoon is approximately 200 000 people, mostly fishers. By comparison, Sarasota Bay, Florida, is approximately two-thirds the size of the Grand Cul-

> **Box 9.9.** Actions that should be considered before attempting to reintroduce sirenians to areas from which they have been extirpated
>
> - Ensure legal status and transparency of all activities;
> - take advantage of lessons learned from reintroductions of other species;
> - use only animals from appropriate genetic stocks for reintroductions;
> - minimise impacts to wild populations by either using healthy captive animals or removing as few individuals as possible from wild populations;
> - do not remove animals from wild populations estimated to number 100 or fewer;
> - only remove and use animals for which health status has been assessed and found to be good;
> - seek as much genetic variation among the introduced animals as possible;
> - ensure that threats to sirenians (contaminants, fishing) are identified and mitigated;
> - involve all possible stakeholders to ensure wide buy-in for the reintroduction programme;
> - ensure that long-term funding exists to monitor introduced animals, create and maintain education and awareness programmes and facilitate regular communication and interaction among stakeholder groups.

de-sac Marin, seasonally provides habitat for 100 or so manatees, and is accessible to several hundred thousand people. Most of the Grand Cul-de-sac Marin is included in the national park, including some locations that are designated as no-entry zones for boats.

The goal of re-establishing a species that has been extirpated by human activities is a noble one. It is also extremely difficult to achieve. In a report based on site visits and interviews, J. Reynolds and D. Wetzel (unpublished) indicated that such a venture should not be undertaken unless there is reasonable assurance of success. Otherwise, animals selected for the reintroduction may simply be lost – a tragedy, especially when dealing with a threatened species. Categories of topics or actions that should be considered prior to seriously undertaking such a difficult and risky process are summarised in Box 9.9.

The Guadeloupe manatee reintroduction project needs to answer important questions and develop improved communication and ties with stakeholders if it is to succeed. However, given the multiple, poorly controlled threats to Antillean manatees throughout most of their range (Chapter 8), the successful establishment of a manatee population in Guadeloupe, where anthropogenic threats to manatees and habitat and other impediments to conservation are *relatively* minor, could be extremely important for conservation of the Antillean subspecies. Populations of Antillean manatees are likely to disappear from many countries within the next 50–100 years (see Chapter 8, especially Table 8.9). A well-funded, well-integrated programme in Guadeloupe could lead to one of the larger and more stable manatee populations in the wider Caribbean. The reintroduction has the potential to be an extremely important and proactive step – if done properly. The requirements for successful reintroduction of Antillean manatees to Guadeloupe apply to sirenians more generally.

Protecting and enhancing locations with warm water in south-western Florida

As discussed in Chapters 7 and 8, Florida experienced some of the most prolonged cold weather in recent history in January–February 2010. As a result, unprecedented numbers of Florida manatees died from cold stress, or required capture and rehabilitation. As of the end of April 2010, the unofficial count of documented deaths related to cold stress for 2010 stood at 236 manatees (Florida Fish and Wildlife Conservation Commission: www.MyFWC.com); this figure does not include carcasses too decomposed to assess cause of death, but is still an order of magnitude higher than the five-year average for this cause of death. Perusal of maps showing the distribution of the cold-stressed, dead manatees showed something very interesting: relatively few deaths or rescues were documented for manatees occupying waters of natural springs. This observation of low winter mortality at natural springs was concordant with findings during other cold winters beginning over 30 years ago (e.g. Campbell and Irvine 1981).

Laist and Reynolds (2005b) have been among the advocates for enhancing access to natural springs as warm-water refugia for manatees in winter. The intense winter of 2010 underscored the advisability of this management action. One of the primary regions of Florida in which manatee conservation is jeopardised by lack of access for manatees to natural springs (coupled with lack of proactive management actions to mitigate the loss of industrially generated warm water) is south-western Florida, where more than 900 manatees have been observed at a power plant discharge area during a single aerial survey (J. Reynolds, unpublished). As of 7 April 2010, in Lee County waters, scientists had recovered 20 manatees that died due to cold stress, and another ten carcasses that were too decomposed to determine cause of death.

There are three types of locations in which manatees find refuge from cold in south-western Florida: the power plant discharge along the Orange River; passive thermal refugia, such as deep canals or basins; and a natural spring (Warm Mineral Spring) near Salt Creek and the Myakka River. Given the impermanence of the power plant and other industrial sources of warm water, managers have emphasised the need to maintain or even enhance access to the other types of refugia (US Fish and Wildlife Service 2001).

The network of passive thermal refugia currently provides warm-water alternatives for hundreds of manatees in south-western Florida (Stith *et al.* 2010). Some of these areas are maintained by the presence near the bottom of a layer of dense, salty water, which is warmer than the overlying fresh water. The stratification can be lost without some freshwater flow over a persistent tidal wedge of saltier water. By knowing how the warm bottom temperatures can be maintained, managers can ensure the existence of critical passive thermal refugia and enable proactive management to develop additional such areas for manatees in the future (Stith *et al.* 2010).

However, an unusual opportunity exists to take a proactive step to assuring adequate access to warm water in perpetuity for large numbers of manatees in south-western Florida. An exceptionally warm natural spring called Warm Mineral Spring has a long, narrow and shallow spring run where a few tens of manatees find refuge from cold weather in winter. Much of the land on either side of the spring run is privately owned and undeveloped and, unlike a lot of property in Florida, the community around Warm Mineral Springs is not affluent and property values are relatively modest. The spring run is sufficiently shallow that boat traffic is non-existent for much of the length of the run.

J. Reynolds has suggested that county, state and federal agencies, working in concert with interested non-governmental organisations, should take the opportunity while prices are moderate to purchase both the land around the main spring (to safeguard it from further development that could affect water flow and quality) and the undeveloped land along the spring run, with emphasis on multiple adjoining properties. With the land in hand, permits could be acquired to enhance the run by deepening or widening it to accommodate far more manatees than can currently gain access. Since the run is not currently used much for boats (and what use there is involves very small watercraft), local people do not consider it their 'right' to use fast, powerful boats there, so it could be relatively easy to maintain the enhanced run as an area seasonally off-limits to boaters. A research and education facility could be developed on the acquired properties, and members of the local community could be hired to help staff it.

The Orange River power plant that currently provides warm water for manatees in winter was 'modernised' as a natural-gas-burning plant several years ago. The plant is likely to exist and produce warm water for a number of years. Thus, there is potentially a decades-long window in which manatees will have industrially generated warm water, while they also learn of the existence of an enhanced natural alternative. There is ample evidence that Florida manatees will take advantage of sanctuaries set aside on their behalf at natural springs: numbers of manatees wintering at Blue Spring, Crystal River and Homosassa River have increased by an order of magnitude since their designation as protected zones in the 1970s.

Eventually, the power plant will be retired and the warm water it generates will disappear. If managers wait until that time to develop and implement mitigation options, it will be too late for many of the manatees. The most proactive and effective management options to protect manatees in south-western Florida for the long-term are to act soon to acquire property and enhance the spring run of Warm Mineral Springs; to identify and safeguard existing passive thermal basins; and to create new passive refugia. In fact, staff of the Florida Fish and Wildlife Conservation Commission (Ron Mezich, Carol Knox and colleagues) are already starting to work (along with other stakeholders) to make the safeguarding of the spring run at Warm Mineral Spring a reality for manatees.

SOLUTIONS AND FINAL THOUGHTS

If conservation were easy, people would already be doing it effectively. The fact is that the presence of multiple goals, values and perceptions among stakeholder groups has led to more divisiveness than unity in terms of appropriate human behaviours and approaches consistent with sound species conservation (Reynolds and Wells 2003).

As we have noted, the sirenian populations in two developed countries (Australia and the United States) are relatively well studied, and funds exist for their conservation and management (although that does not mean that conservation and management are optimal!). In addition, both Australian dugongs and Florida manatees are relatively abundant, compared with sirenians in most other parts of their range (Table 8.9). The critical and pressing conservation issues for sirenians globally exist for the 'other' species or populations for which information and funds are sparse, numbers are small and threats are severe and unmitigated.

Other components that may be useful to employ with communities include: using local people as 'rangers' to promote awareness and enforcement (see 'Reinforcement of cultural protocols' above); using local teachers to develop culturally relevant education and awareness programmes; and using economic instruments (see 'Economic tools' above) to protect key habitats and provide fishing gear alternatives to gill nets. In addition, authorities can enforce bans on the sale of sirenian products and valuable products associated with shark-finning or other fishing that is driving the demise of species, including dugongs and manatees.

The specifics of a successful approach to conservation will vary from place to place, and they will depend in part on developing an open dialogue among communities, managers and scientists. An explicit goal must be the creation of alternative livelihoods for members of the affected communities. That takes time and dedication. It is important to note that involvement of researchers and other stakeholders in developing countries is vital to that process, although international teams can be developed that succeed admirably in regional capacity development and the creation of balanced solutions for conservation. The key is to ensure that the researchers and others from the developing countries are neither marginalised nor taken advantage of, as has unfortunately happened at times in the past, and that the carefully blended team works towards goals that include the welfare of the local communities, not simply of sirenians and other wildlife.

Conservation crises did not develop overnight; nor will solutions to those crises. However, dedicated individuals and groups, armed with practical approaches that integrate the dual goals of conservation and the welfare of local communities, have a chance to succeed slowly but surely.

Holloway (2010) notes that ecosystems have tremendous resiliency and ability to recover from damage. Compared with many other species of large mammals, sirenians are able to exist near humans and are not wilderness-requiring animals (Chapter 5). If people can adjust their values and processes to include biodiversity conservation, there is a 'ray of hope – and a respite from apocalyptic headlines' (Holloway 2010, p. 28). But the time for such adjustments is now.

NOTE

1. Our focus on the CBD is not meant to indicate that it is the only significant international agreement focusing on conservation. There is a large number of such agreements, some regional and some global (e.g. see Chapter 8).

References

Ackerman, B. B. (1995). Aerial surveys of manatees: a summary and progress report. In *Population Biology of the Florida Manatee*, T. J. O'Shea, B. B. Ackerman and H. F. Percival (eds), Washington, DC: US Department of the Interior, National Biological Service, pp. 13–33.

Ackerman, B. B., Wright, S. D., Bonde, R. K., Odell, D. K. and Banowetz, D. J. (1995). Trends and patterns in mortality of manatees in Florida, 1974–1992. In *Population Biology of the Florida Manatee*, T. J. O'Shea, B. B. Ackerman and H. F. Percival (eds), Washington, DC: US Department of the Interior, National Biological Service, pp. 223–258.

Adulyanukosol, K. and Thongsukdee, S. (2005). Report of the results of the survey on dugongs, dolphins, sea turtles, and seagrass in Trang Province. Phuket Marine Biological Center, Department of Marine and Coastal Resources.

Adulyanukosol, K., Thongsukdee, S., Hara, T., Arai, N. and Tsuchiya, M. (2007). Observations of dugong reproductive behaviour in Trang Province, Thailand: further evidence of intraspecific variation in dugong behaviour. *Marine Biology*, **151**, 1887–1891.

Aipanjiguly, S., Jacobson, S. K. and Flamm, R. (2003). Conserving manatees: knowledge, attitudes, and intentions of boaters in Tampa Bay, Florida. *Conservation Biology*, **17**, 1098–1105.

Aketa, K., Asano, S., Wakai, Y. and Kawamura, A. (2003). Apparent digestibility of eelgrass *Zostera marina* by captive dugongs (*Dugong dugon*) in relation to the nutritional content of eelgrass and dugong feeding parameters. *Mammal Study*, **28**, 23–30.

Alber, M. (2002). A conceptual model of estuarine freshwater inflow management. *Estuaries and Coasts*, **25**, 1246–1261.

Albert, S., O'Neil, J. M., Udy, J. W., et al. (2005). Blooms of the cyanobacterium *Lyngbya majuscula* in coastal Queensland, Australia: disparate sites, common factors. *Marine Pollution Bulletin*, **51**, 428–437.

Alicea-Pou, J. A. (2001). Vocalizations and behavior of the Antillean and Florida manatee (*Trichechus manatus*): individual variability and geographical comparison. MS Thesis, Moss Landing Marine Laboratories and San Francisco State University.

Allen, S., Marsh, H. and Hodgson, A. (2004). Occurrence and conservation of the dugong (Sirenia: Dugongidae) in New South Wales. *Proceedings of the Linnean Society of New South Wales*, **125**, 211–216.

Allison, E. H., Perry, A. L., Badjeck, M.-C. *et al.* (2009). Vulnerability of national economies to the impacts of climate change on fisheries. *Fish and Fisheries*, **10**, 173–196.

Allmon, W. D., Emslie, S. D., Jones, D. S. and Morgan, G. S. (1996). Late Neogene oceanographic change along Florida's west coast: evidence and mechanisms. *Journal of Geology*, **104**, 143–162.

Allsopp, W. H. L. (1960). The manatee: ecology and use for weed control. *Nature*, **188**, 762.

Allsopp, W. H. L. (1969). Aquatic weed control by manatees: its prospects and problems. In *Man-Made Lakes*, L. E. Obeng (ed.), Accra: Ghana University Press, pp. 344–351.

Alvarez-Alemán, A., Beck, C. A. and Powell, J. A. (2010). First report of a Florida manatee (*Trichechus manatus latirostris*) in Cuba. *Aquatic Mammals*, **36**, 148–153.

Alves-Stanley, C. D. and Worthy, G. A. J. (2009). Carbon and nitrogen stable isotope turnover rates and diet-tissue discrimination in Florida manatees (*Trichechus manatus latirostris*). *Journal of Experimental Biology*, **212**, 2349–2355.

Alves-Stanley, C. D., Worthy, G. A. J. and Bonde, R. K. (2010). Feeding preferences of West Indian manatees in Florida, Belize, and Puerto Rico as indicated by stable isotope analysis. *Marine Ecology Progress Series*, **402**, 255–267.

Amaral, R. S., Rosas, F. C. W., Viau, P., *et al.* (2009). Noninvasive monitoring of androgens in male Amazonian manatee (*Trichechus inunguis*): biologic validation. *Journal of Zoo and Wildlife Medicine*, **40**, 458–465.

Ambati, B. K., Nozaki, M., Singh, N., *et al.* (2006). Corneal avascularity is due to soluble VEGF receptor-1. *Nature*, **443**, 993–997.

Ames, A. L. (2002). Lipid chemistry of the Florida manatee, *Trichechus manatus latirostris*. PhD Thesis, University of South Florida, Tampa.

Ames, A. L. and Van Vleet, E. S. (1996). Organochlorine residues in the Florida manatee, *Trichechus manatus latirostris*. *Marine Pollution Bulletin*, **32**, 374–377.

Ames, A. L., Van Vleet, E. S. and Sackett, W. M. (1996). The use of stable carbon isotope analysis for determining the dietary habits of the Florida manatee *Trichechus manatus latirostris*. *Marine Mammal Science*, **12**, 555–563.

Ames, A. L., Van Vleet, E. S. and Reynolds III, J. E. (2002). Comparison of lipids in selected tissues of the Florida manatee (Order Sirenia) and bottlenose dolphin (Order Cetacea; Suborder Odontoceti). *Comparative Biochemistry and Physiology Part B: Biochemistry and Molecular Biology*, **132**, 625–634.

Amrine, H. M. and Springer, M. S. (1999). Maximum-likelihood analysis of the tethythere hypothesis based on a multigene data set and comparison of different models of sequence evolution. *Journal of Mammalian Evolution*, **6**, 161–176.

Anderson, D. R., Wywialowski, A. P. and Burnham, K. P. (1981). Tests of the assumptions underlying life table methods for estimating parameters from cohort data. *Ecology*, **62**, 1121–1124.

Anderson, P. K. (1979). Dugong behaviour: on being a marine mammalian grazer. *The Biologist*, **61**, 113–144.

Anderson, P. K. (1981a). The behavior of the dugong (*Dugong dugon*) in relation to conservation and management. *Bulletin of Marine Science*, **31**, 640–647.

Anderson, P. K. (1981b). Dugong behaviour: observations, extrapolations and speculations. In *The Dugong: Proceedings of a Seminar/Workshop held at James Cook University 8–13 May 1979, Townsville*, H. Marsh (ed.), Townsville: James Cook University, pp. 91–111.

Anderson, P. K. (1982). Studies of dugongs at Shark Bay, Western Australia: II. Surface and subsurface observations. *Australian Wildlife Research*, **9**, 85–99.

Anderson, P. K. (1984). Suckling in *Dugong dugon*. *Journal of Mammalogy*, **65**, 510–511.

Anderson, P. K. (1986). Dugongs of Shark Bay Australia: seasonal migration, water temperature and forage. *National Geographic Research*, **2**, 473–490.

Anderson, P. K. (1989). Deliberate foraging on macroinvertebrates by dugongs. *National Geographic Research*, **5**, 4–6.

Anderson, P. K. (1994). Dugong distribution, the seagrass *Halophila spinulosa*, and thermal environment in winter in deeper waters of Eastern Shark Bay, Western Australia. *Wildlife Research*, **21**, 381–388.

Anderson, P. K. (1995a). Competition, predation, and the evolution and extinction of Steller's sea cow, *Hydrodamalis gigas*. *Marine Mammal Science*, **11**, 391–394.

Anderson, P. K. (1995b). Scarring and photo identification of dugongs in Shark Bay, Western Australia. *Aquatic Mammals*, **21**, 205–211.

Anderson, P. K. (1997). Shark Bay dugongs in summer: I. Lek mating. *Behaviour*, **134**, 433–462.

Anderson, P. K. (1998). Shark Bay dugongs (*Dugong dugon*) in summer: II. Foragers in a *Halodule*-dominated community. *Mammalia*, **62**, 409–425.

Anderson, P. K. (2001). Marine mammals in the next one hundred years: twilight for a Pleistocene megafauna? *Journal of Mammalogy*, **83**, 623–629.

Anderson, P. K. (2002). Habitat, niche, and evolution of sirenian mating systems. *Journal of Mammalian Evolution*, **9**, 55–98.

Anderson, P. K. and Barclay, R. M. R. (1995). Acoustic signals of solitary dugongs: physical characteristics and behavioral correlates. *Journal of Mammalogy*, **76**, 1226–1237.

Anderson, P. K. and Birtles, A. (1978). Behaviour and ecology of the dugong, *Dugong dugon* (Sirenia): observations in Shoalwater and Cleveland Bays, Queensland. *Australian Wildlife Research*, **5**, 1–23.

Anderson, P. K. and Domning, D. P. (2002). Steller's sea cow *Hydrodamalis gigas*. In *Encyclopedia of Marine Mammals*, W. F. Perrin, B. Wursig and J. G. M. Thewissen (eds), San Diego: Academic Press, pp. 1178–1181.

Anderson, P. K. and Prince, R. I. T. (1985). Predation on dugongs: attacks by killer whales. *Journal of Mammalogy*, **66**, 554–556.

André, J. and Lawler, I. R. (2003). Near infrared spectroscopy as a rapid and inexpensive means of dietary analysis for a marine herbivore, dugong *Dugong dugon*. *Marine Ecology Progress Series*, **257**, 259–266.

André, J., Gyuris, E. and Lawler, I. R. (2005). Comparison of the diets of sympatric dugongs and green turtles on the Orman Reefs, Torres Strait, Australia. *Wildlife Research*, **32**, 53–62.

Andrews, C. W. (1906). *A Descriptive Catalogue of the Tertiary Vertebrata of the Fayûm, Egypt*, London: British Museum (Natural History).

Annandale, N. (1905). Notes on the species and external characters of the dugong (*Halicore dugong*). *Journal of the Asiatic Society of Bengal*, **1**, 238–243.

Aragones, L. V. (1994). Observations on dugongs at Caluit Island, Busuanga, Palawan, Philippines. *Wildlife Research*, **21**, 709–717.

Aragones, L. V. (1996). Dugongs and green turtles: grazers in the tropical seagrass ecosystem. PhD Thesis, James Cook University, Townsville.

Aragones, L. V. and Marsh, H. (2000). Impact of dugong grazing and turtle cropping on tropical seagrass communities. *Pacific Conservation Biology*, **5**, 278–288.

Aragones, L. V., Lawler, I. R., Foley, W. J. and Marsh, H. (2006). Dugong grazing
and turtle cropping: grazing optimization in tropical seagrass systems?
Oecologia, **149**, 635–647.

Aragones, L. V., Lawler, I. R., Marsh, H., Domning, D. P. and Hodgson, A. (2012a).
The role of sirenians in aquatic ecosystems. In *Sirenian Conservation: Issues and
Strategies in Developing Countries*, E. Hines, J. Reynolds, A. Mignucci-Giannoni,
L. Aragones and M. Marmontel (eds), Gainesville: University Press of Florida.

Aragones, L. V., Marmontel, M. and Kendall, S. (2012b). Working with communities
for sirenian conservation. In *Sirenian Conservation: Issues and Strategies in
Developing Countries*, E. Hines, J. Reynolds, A. Mignucci-Giannoni, L. Aragones
and M. Marmontel (eds), Gainesville: University Press of Florida.

Aranda-Manteca, F. J., Domning, D. P. and Barnes, L. G. (1994). A new middle
Miocene sirenian of the genus *Metaxytherium* from Baja California and California:
relationships and paleobiogeographic implications. In *Contributions in Marine
Mammal Paleontology Honoring Frank C. Whitmore Jr*, A. Berta and T. A. Deméré
(eds), San Diego: San Diego Society of Natural History, pp. 191–204.

Arcese, P., Keller, L. F. and Cary, J. R. (1997). Why hire a behaviorist into a con-
servation or management team? In *Behavioral Approaches to Conservation in the
Wild*, J. R. Clemmons and R. Buchholz (eds), Cambridge: Cambridge University
Press, pp. 48–71.

Arraut, E. M., Marmontel, M., Mantovani, J. E., *et al.* (2010). The lesser of two evils:
seasonal migrations of Amazonian manatees in the western Amazon. *Journal of
Zoology*, **280**, 247–256.

Arrow, K., Bolin, B., Costanza, R. et al. (1996). Economic growth, carrying capacity,
and the environment. *Ecological Applications*, **6**, 13–15.

Asher, R. J. (2007). A web-database of mammalian morphology and a reanalysis of
placental phylogeny. *BMC Evolutionary Biology*, **7**, 108.

Asher, R. J. and Lehman, T. (2008). Dental eruption in afrotherian mammals. *BMC
Biology*, **6**, 14–25.

Asher, R. J., Novacek, M. J. and Geisler, J. G. (2003). Relationships of endemic
African mammals and their fossil relatives based on morphological and molecular evidence. *Journal of Mammalian Evolution*, **10**, 131–194.

Asher, R. J., Bennett, N. and Lehmann, T. (2009). The new framework for understanding placental mammal evolution. *BioEssays*, **31**, 853–864.

Attenborough, D. (1956). *Zoo Quest to Guiana*, London: Lutterworth.

Attwell, C. A. M. (ed.) (in press). Annex 2: report on a dugong workshop.
Conservation ecology of dugongs in Mozambique: precursor to a management
plan. Unpublished report to Mozambique DNAC.

Augustine, D. J., McNaughton, S. J. and Frank, D. A. (2003). Feedbacks between
large herbivores and soil nutrients in a managed savanna ecosystem. *Ecological
Applications*, **13**, 1325–1337.

Auil, N. (1998). Belize Manatee Recovery Plan. UNDP/GEF Coastal Management
Project, Belize (BZE/92/G31), 1–67.

Awobamise, A. (2008). Nigeria. In *Conservation Strategy for the West African
Manatee*, T. Dodman, N. M. D. Diop and S. Khady (eds), Nairobi: UNEP; and
Dakar: Wetlands International Africa, pp. 74–77.

Bacchus, M.-L. C., Dunbar, S. G. and Self-Sullivan, C. (2009). Characterization of resting holes and their use by the Antillean Manatee (*Trichechus manatus manatus*) in the Drowned Cayes, Belize. *Aquatic Mammals*, **35**, 62–71.

Bachteler, D. and Dehnhardt, G. (1999). Active touch performance in the Antillean manatee: evidence for a functional differentiation of facial tactile hairs. *Zoology*, **102**, 61–69.

Bajpai, S. and Domning, D. P. (1997). A new dugongine sirenian from the early Miocene of India. *Journal of Vertebrate Paleontology*, **17**, 219–228.

Bajpai, S., Thewissen, J. G. M., Kapur, V. V., Tiwari, B. N. and Sahni, A. (2006). Eocene and Oligocene sirenians (Mammalia) from Kachchh, India. *Journal of Vertebrate Paleontology*, **26**, 400–410.

Bajpai, S., Domning, D. P., Das, D. P. and Mishra, V. P. (2009). A new middle Eocene sirenian (Mammalia, Protosirenidae) from India. *Neues Jahrbuch für Geologie und Paläontologie, Abhandlungen*, 252–253, 257–267.

Bakker, E. S., Ritchie, M. E., Olff, H., Milchunas, D. G. and Knops, J. M. H. (2006). Herbivore impact on grassland plant diversity depends on habitat productivity and herbivore size. *Ecology Letters*, **9**, 780–788.

Ball, I. and Possingham, H. P. (2000). *Marxan (V1.8.2) Marine Reserve Design Software using Spatially Explicit Annealing*. A manual prepared for the Great Barrier Reef Marine Park Authority. Brisbane: University of Queensland.

Bandeira, S. O. and Gell, F. (2003). The seagrasses of Mozambique and Southeastern Africa. In *Seagrass Atlas of the World*, E. P. Green and F. T. Short (eds), Berkeley: University of California Press, pp. 105–112.

Bangs, O. (1895). The present standing of the Florida manatee, *Trichechus latirostris* (Harlan), in the Indian River waters. *American Naturalist*, **29**, 783–787.

Barlow, J. and Reeves, R. R. (2001). Population status and trends. In *Encyclopedia of Marine Mammals*, W. F. Perrin, B. Wursig and J. G. M. Thewissen (eds), San Diego: Academic Press, pp. 979–982.

Barlow, J., Swartz, S. L., Eagle, T. C. and Wade, P. R. (1995). U.S. marine mammal stock assessments: guidelines for preparation, background, and a summary of the 1995 assessments. NOAA Technical Memorandum NMFS-OPR-95-6.

Barrett, O. W. (1935). Notes concerning manatees and dugongs. *Journal of Mammalogy*, **16**, 216–220.

Bass, D. (2009). Dugong surveys of Manus and Bougainville Islands, Papua New Guinea. *Conservation International*. Available at: http://www.sprep.org/att/irc/ecopies/countries/papua_new_guinea/53.pdf.

Bass, D. K. (2010). Status of Dugong *Dugong dugon* and Australian Snubfin Dolphin *Orcaella heinsohni*, in the Solomon Islands. *Pacific Conservation Biology*, **16**, 133–143.

Batrawi, A. (1953). The external features of the dugong kidney. *Bulletin Zoological Society of Egypt*, **11**, 12–13.

Batrawi, A. (1957). The structure of the dugong kidney. *Publication Marine Biological Station Al-Ghardaqa Red Sea*, **9**, 51–68.

Bauer, G. B., Colbert, D. E., Gaspard III, J. C., Littlefield, B. and Fellner, W. (2003). Underwater visual acuity of Florida manatees (*Trichechus manatus latirostris*). *International Journal of Comparative Psychology*, **16**, 130–142.

Bauer, G. B., Colbert, D. E. and Gaspard, J. C. I. (2010). Learning about manatees: a collaborative program between New College of Florida and Mote Marine

Laboratory to conduct laboratory research for manatee conservation. *International Journal of Comparative Psychology*, **23**, 811–825.

Baugh, T. M., Valade, J. A. and Zoodsma, B. J. (1989). Manatee use of *Spartina alterniflora* in Cumberland Sound. *Marine Mammal Science*, **5**, 88–90.

Bayliss, P. (1986). Factors affecting aerial surveys of marine fauna, and their relationship to a census of dugongs in the coastal waters of the Northern Territory. *Australian Wildlife Research*, **13**, 27–37.

Beal, W. P. B. (1939). The manatee as a food animal. *Nigerian Field*, **8**, 124–126.

Beck, C. A. and Barros, N. B. (1991). The impact of debris on the Florida manatee. *Marine Pollution Bulletin*, **22**, 508–510.

Beck, C. A. and Clark, A. (2012). Individual identification of sirenians. In *Sirenian Conservation: Issues and Strategies in Developing Countries*, E. Hines, J. Reynolds, A. Mignucci-Giannoni, L. Aragones and M. Marmontel (eds), Gainesville: University Press of Florida.

Beck, C. A. and Clementz, M. T. (2012). Techniques for the study of the food habits of Sirenians. In *Sirenian Conservation: Issues and Strategies in Developing Countries*, E. Hines, J. Reynolds, A. Mignucci-Giannoni, L. Aragones and M. Marmontel (eds), Gainesville: University Press of Florida.

Beck, C. A. and Forrester, D. J. (1988). Helminths of the Florida manatee, *Trichechus manatus latirostris*, with a discussion and summary of the parasites of sirenians. *Journal of Parasitology*, **74**, 628–637.

Beck, C. A. and Reid, J. P. (1995). An automated photo-identification catalog for studies of the life history of the Florida manatee. In *Population Biology of the Florida Manatee*, T. J. O'Shea, B. B. Ackerman and H. F. Percival (eds), Washington, DC: US Department of the Interior, National Biological Service, pp. 120–134.

Beck, C. A., Bonde, R. K. and Rathbun, G. B. (1982). Analyses of propeller wounds on manatees in Florida. *Journal of Wildlife Management*, **46**, 531–535.

Beissinger, S. R. and Westphal, M. I. (1998). On the use of demographic models of population viability in endangered species management. *Journal of Wildlife Management*, **62**, 821–841.

Bender, M. A., Knutson, T. R., Tuleya, R. E., *et al.* (2010). Modeled impact of anthropogenic warming on the frequency of intense Atlantic hurricanes. *Science*, **327**, 454–458.

Bengtson, J. L. (1981). Ecology of manatees (*Trichechus manatus*) in the St Johns River, Florida. PhD Thesis, University of Minnesota.

Bengtson, J. L. (1983). Estimating food consumption of free-ranging manatees in Florida. *Journal of Wildlife Management*, **47**, 1186–1192.

Bengtson, J. L. and Fitzgerald, S. M. (1985). Potential role of vocalizations in West Indian manatees. *Journal of Mammalogy*, **66**, 816–819.

Bengtson Nash, S. M., McMahon, K., Eaglesham, G. and Müller, J. F. (2005). Application of a novel phytotoxicity assay for the detection of herbicides in Hervey Bay and the Great Sandy Straits. *Marine Pollution Bulletin*, **51**, 351–360.

Benson, J. M., Hahn, F. F., March, T. H., *et al.* (2005). Inhalation toxicity of brevetoxin 3 in rats exposed for twenty-two days. *Environmental Health Perspectives*, **113**, 626–631.

Bentley, B. L. and Johnson, N. D. (1991). Plants as food for herbivores: the roles of nitrogen fixation and carbon dioxide enrichment. In *Plant–Animal Interactions: Evolutionary Ecology in Tropical and Temperate Regions*, P. W. Price,

T. M. Lewinsohn, G. W. Fernandes and W. W. Benson (eds), New York: John Wiley and Sons, pp. 257–272.

Benzecry, A. and Brack-Hanes, S. D. (2008). A new hydrocharitacean seagrass from the Eocene of Florida. *Botanical Journal of the Linnean Society*, **157**, 19–30.

Bertram, G. C. L. and Bertram, C. K. R. (1968). Bionomics of dugongs and manatees. *Nature*, **218**, 423–426.

Bertram, G. C. L. and Bertram, C. K. R. (1973). The modern Sirenia: their distribution and status. *Biological Journal of the Linnean Society*, **5**, 297–338.

Best, R. C. (1981). Foods and feeding habits of wild and captive Sirenia. *Mammal Review*, **11**, 3–29.

Best, R. C. (1982). Seasonal breeding in the Amazonian manatee, *Trichechus inunguis* (Mammalia: Sirenia). *Biotropica*, **14**, 76–78.

Best, R. C. (1983). Apparent dry-season fasting in Amazonian manatees (Mammalia, Sirenia). *Biotropica*, **15**, 61–64.

Best, R. C. (1984). The aquatic mammals and reptiles of the Amazon. In *The Amazon: Limnology and Landscape Ecology of a Mighty Tropical River and its Basin*, H. Sioli (ed.), Dordrecht: Junk, pp. 371–412.

Best, R. C. and Teixeira, D. M. (1982). Notas sobre a distribuição e status aparente dos peixes-bois (Mammalia: Sirenia) nas costas amapaenses brasileiras. *Boletim Fundação Brasileira para a Conservação da Natureza, Rio de Janeiro*, **17**, 41–47.

Betts, R. A., Malhi, Y. and Roberts, J. T. (2008). The future of the Amazon: new perspectives from climate, ecosystem and social sciences. *Philosophical Transactions of the Royal Society B: Biological Sciences*, **363**, 1729–1735.

Beuchat, C. A. (2002). Kidney, structure and function. In *Encyclopedia of Marine Mammals*, W. F. Perrin, B. Würsig and J. G. M. Thewissen (eds), San Diego: Academic Press, pp. 666–669.

Beusse, D. O. Jr, Asper, E. D. and Searles, S. W. (1981a). Some causes of manatee mortality. In *The West Indian Manatee in Florida: Proceedings of a Workshop held in Orlando, Florida 27–29 March 1978*, R. L. Brownell Jr, and K. Ralls (eds), Tallahassee: Florida Department of Natural Resources, pp. 98–101.

Beusse, D. O. Jr, Asper, E. D. and Searles, S. W. (1981b). Diagnosis and treatment of manatees at Sea World of Florida. In *The West Indian Manatee in Florida: Proceedings of a Workshop held in Orlando, Florida 27–29 March 1978*, R. L. Brownell Jr and K. Ralls (eds), Tallahassee: Florida Department of Natural Resources, pp. 111–120.

Bianucci, G., Carone, G., Domning, D. P., *et al.* (2008). Peri-Messinian dwarfing in Mediterranean *Metaxytherium* (Mammalia: Sirenia): evidence of habitat degradation related to the Messinian salinity crisis. *Garyounis Scientific Bulletin*, **5**, 145–157.

Birch, W. R. (1975). Some chemical and calorific properties of tropical marine angiosperms compared with those of other plants. *Journal of Applied Ecology*, **12**, 201–212.

Bishop, F. (2010). *Lyngbya* outbreak. *Seagrass-Watch*, **40**, 10–11.

Bjorndal, K. A. (1980). Nutrition and grazing of the Green turtle, *Chelonia mydas*. *Marine Biology*, **56**, 147–156.

Blackman, A. J. and Jones, D. C. (2010). Queensland's recreational marine industry. Report by Marine Queensland in collaboration with the Centre for Independent Business Research. Available at: http://businessgc.com.au/uploads/file/

Marine/Queensland%20Marine%20Industry_Annual%20Report_2009.pdf (accessed 10 October 2010).

Blainville, H. M. D. (1836). Classification presented in 1834. In *Dictionnaire Pittoreque d'Histoire Naturelle*, Vol. 4., F.-E. Guérin (ed.), Paris: Bureau de Souscription, p. 619.

Blair, D. (1981). The monostome flukes (Digenea: Families Opisthotrematidae Poche and Rhabdiopoeidae Poche) parasitic in sirenians (Mammalia: Sirenia). *Australian Journal of Zoology*, **81**, 1–54.

Blair, D. and Hudson, B. E. T. (1992). Population structure of *Lankatrematoides gardneri* (Digenea: Opisthotrematidae) in the pancreas of the dugong (*Dugong dugon*) (Mammalia: Sirenia). *The Journal of Parasitology*, **78**, 1077–1079.

Bledsoe, E. L., Harr, K. E., Cichra, M. F., *et al.* (2006). A comparison of biofouling communities associated with free-ranging and captive Florida manatees (*Trichechus manatus latirostris*). *Marine Mammal Science*, **22**, 997–1003.

Boever, W. J., Thoen, C. O. and Wallach, J. D. (1976). *Mycobacterium chelonei* infection in a natterer manatee. *Journal of the American Veterinary Medical Association*, **169**, 927–929.

Bonde, R. K. (2009). Population genetics and conservation of the Florida manatee: past, present and future. PhD Dissertation, University of Florida, Gainesville, Florida.

Bonde, R. K. and Potter, C. W. (1995). Manatee butchering sites in Port Honduras. *Sirenews: Newsletter of the International Union for Conservation of Nature and Natural Resources/Species Survival Commission*, **24**, 7.

Bonde, R. K., O'Shea, T. J. and Beck, C. A. (1983). *Manual of Procedures for the Salvage and Necropsy of Carcasses of the West Indian Manatee (*Trichechus manatus*)*, Springfield, Virginia: National Technical Information Service.

Bonde, R. K., Aguirre, A. A. and Powell, J. (2004). Manatees as sentinels of marine ecosystem health: are they the 2000-pound canaries? *EcoHealth*, **1**, 255–262.

Borges, J. C. G., Alves, L. C., Vergara-Parente, J. E., Faustino, M. A. G. and Machado, E. C. L. (2009). Ocorrência de infecção *Cryptosporidium* spp. em peixe-boi marinho (*Trichechus manatus*). [Occurrence of *Cryptosporidium* spp. infection in Antillean manatee (*Trichechus manatus*)]. *Revista Brasileira de Parasitologia Veterinaria*, **18**, 60–61.

Bossart, G. D. (2001). Manatees. In *Marine Mammal Medicine*, 2nd edn, L. A. Dierauf, and F. M. D. Gulland (eds), Boca Raton: CRC Press, pp. 939–960.

Bossart, G. D. (2006). Case study: marine mammals as sentinel species for oceans and human health. *Oceanography*, **19**, 134–137.

Bossart, G. D. (2011). Marine mammals as sentinel species for oceans and human health. *Veterinary Pathology*, **48**, 676–690.

Bossart, G. D., Baden, D. G., Ewing, R. Y., Roberts, B. and Wright, S. D. (1998). Brevetoxicosis in manatees (*Trichechus manatus latirostris*) from the 1996 epizootic: gross, histologic, and immunohistochemical features. *Toxicologic Pathology*, **26**, 276–282.

Bossart, G. D., Reidarson, T. H., Dierauf, L. A. and Duffield, D. A. (2001). Clinical pathology. In *Marine Mammal Medicine*, 2nd edn, L. A. Dierauf and F. M. D. Gulland (eds), Boca Raton: CRC Press, pp. 383–436.

Bossart, G. D., Meisner, R. A., Rommel, S. A., Ghim, S. and Jenson, A. B. (2002a). Pathological features of the Florida manatee cold stress syndrome. *Aquatic Mammals*, **29**, 9–17.

Bossart, G. D., Ewing, R. Y., Lowe, M., *et al.* (2002b). Viral papillomatosis in Florida manatees (*Trichechus manatus latirostris*). *Experimental and Molecular Pathology*, **72**, 37–48.

Bossart, G. D., Baden, D. G., Ewing, R. Y. and Wright, S. D. (2002c). Manatees and brevetoxicosis. In *Molecular and Cell Biology of Marine Mammals*, C. J. Pfeiffer (ed.), Melbourne, Florida: Krieger Publishing Company, pp. 205–212.

Bossart, G. D., Meisner, R. A., Rommel, S. A., *et al.* (2004). Pathologic findings in Florida manatees (*Trichechus manatus latirostris*). *Aquatic Mammals*, **30**, 434–440.

Bowen-Jones, E. and Entwhistle, A. (2002). Identifying appropriate flagship species: the importance of culture and local contexts. *Oryx*, **36**, 189–195.

Boyd, I. L., Lockyer, C. and Marsh, H. (1999). Reproduction in marine mammals. In *Biology of Marine Mammals*, J. E. Reynolds III and S. A. Rommel (eds), Washington, DC: Smithsonian Institution Press, pp. 218–286.

Bradbury, J. W. (1977). Lek mating behavior in the hammer-headed bat. *Zeitschrift für Tierpsychologie*, **45**, 225–255.

Bradbury, J. W. and Vehrencamp, S. L. (1977). Social organization and foraging in emballonurid bats: III. Mating systems. *Behavioral Ecology and Sociobiology*, **2**, 1–17.

Brand, L. E. and Compton, A. (2007). Long-term increase in *Karenia brevis* abundance along the Southwest Florida Coast. *Harmful Algae*, **6**, 232–252.

Brandt, J. F. (1846). *Symbolae Sirenologicae Quibus Praecipue Rhytinae Historia Naturalis Illustratur*. Memoirs Academie Imperiale Des Sciences, Series 6, Science Naturelle, **5**, 160 pp. + pls. I–V, St Petersburg, Russia. (Translated 1974 and cited as 1849 by Alison Barlow for the Smithsonian Institution and the National Science Foundation, Washington, District of Columbia, as *Contributions to Sirenology, Being Principally an Illustrated Natural History of Rhytina*, Springfield, Virginia: U.S. National Technical Information Service and Belgrade: Nolit Publishing House.

Brashares, J. S., Arcese, P., Sam, M. K., *et al.* (2004). Bushmeat hunting, wildlife declines, and fish supply in West Africa. *Science*, **306**, 1180–1183.

Braulik, G. T., Ranjbar, S., Owfi, F., *et al.* (2010). Marine mammal records from Iran. *Journal of Cetacean Research and Management*, **11**, 49–63.

Broderick, A. C., Frauenstein, R., George, T., *et al.* (2006). Are green turtles globally endangered? *Global Ecology and Biogeography*, **15**, 21–26.

Broderick, D., Ovenden, J., Slade, R. and Lanyon, J. M. (2007). Characterization of 26 new microsatellite loci in the dugong (*Dugong dugon*). *Molecular Ecology Notes*, **7**, 1275–1277.

Bronner, G. N. and Jenkins, P. D. (2005). Order Afrosoricida. In *Mammal Species of the World*, 3rd edn, D. E. Wilson, and D. M. Reeder (eds), Baltimore: Johns Hopkins University Press, pp. 70–81.

Brouns, J. J. (1994). *Seagrasses and Climate Change*. Gland, Switzerland: IUCN.

Brownell, R. L. Jr, Anderson, P. K., Owen, R. P. and Ralls, K. (1981). The status of dugongs at Palau, an isolated island group. In *The Dugong: Proceedings of a*

Seminar/workshop held at James Cook University 8–13 May 1979, Townsville, H. Marsh, (ed.), Townsville: James Cook University, pp. 19–42.

Buck, L. E., Geisler, C. G., Schelhas, J. and Wollenberg, E. (2001). *Biological Diversity: Balancing Interests through Adaptive Collaborative Management*, Boca Raton: CRC Press.

Buckingham, C. A., Lefebvre, L. W., Schaefer, J. M. and Kochman, H. I. (1999). Manatee response to boating activity in a thermal refuge. *Wildlife Society Bulletin*, **27**, 514–522.

Buckland, S. T., Anderson, D. R., Burnham, K. P., *et al.* (2001). *Introduction to Distance Sampling: Estimating Abundance of Biological Populations*, New York: Oxford University Press.

Budge, S. M., Iverson, S. J. and Koopman H. N. (2006). Studying trophic ecology in marine ecosystems using fatty acids: a primer on analysis and interpretation. *Marine Mammal Science*, **22**, 759–801.

Buergelt, C. D. and Bonde, R. K. (1983). Toxoplasmic meningoencephalitis in a West Indian manatee. *Journal of the American Veterinary Medical Association*, **183**, 1294–1296.

Buergelt, C. D., Bonde, R. K., Beck, C. A. and O'Shea, T. J. (1984). Pathologic findings in manatees in Florida. *Journal of the American Veterinary Medical Association*, **185**, 1331–1334.

Buergelt, C. D., Bonde, R. K., Beck, C. A. and O'Shea, T. J. (1990). Myxomatous transformation of heart valves in Florida manatees (*Trichechus manatus latirostris*). *Journal of Zoo and Wildlife Medicine*, **21**, 220–227.

de Buffrénil, V., Astibia, H., Suberbiola, X. P., Berreteaga, A. and Bardet, N. (2008). Variation in bone histology of middle Eocene sirenians from western Europe. *Geodiversitas*, **30**, 425–432.

Bujang, J. S. and Zakaria, M. H. (2003). The seagrasses of Malaysia. In *World Atlas of Seagrasses*, E. P. Green and F. F. Short (eds), Berkeley: University of California Press, pp. 153–157.

Bullock, T. H., Domning, D. P. and Best, R. C. (1980). Evoked brain potentials demonstrate hearing in a manatee (*Trichechus inunguis*). *Journal of Mammalogy*, **61**, 130–133.

Bullock, T. H., O'Shea, T. J. and McClune, M. C. (1982). Auditory evoked potentials in the West Indian manatee (Sirenia: *Trichechus manatus*). *Journal of Comparative Physiology A: Sensory, Neural, and Behavioral Physiology*, **148**, 547–554.

Burgess, E. A., Blanshard, W. H., Keeley, T. and Lanyon, J. M. (2009). Reproductive hormone monitoring of dugongs: 2009 ARAZPA conference proceedings. Australasian Regional Association of Zoological Parks and Aquaria (ARAZPA) Conference, Gold Coast, Queensland, 22–26 March 2009.

Burn, D. M. (1986). The digestive strategy and efficiency of the West Indian manatee, *Trichechus manatus*. *Comparative Biochemistry and Physiology Part A: Physiology*, **85**, 139–142.

Burnham, K. P. and Anderson, D. R. (2002). *Model Selection and Multimodel Inference: A Practical Information-Theoretic Approach*, 2nd edn, New York: Springer–Verlag.

Butchart, S. H. M., Walpole, M., Collen, B. *et al.* (2010). Global biodiversity: indicators of recent declines. *Science*, **328**, 1164–1168.

Cadenat, J. (1957). Observations de cetaces, sireniens, cheloniens et sauriens en 1955–1956. *Bulletin de Institut Français D'Afrique Noire*, **19**, 1358–1375.

Cahn, A. R. (1940). Manatees and the Florida freeze. *Journal of Mammalogy*, **21**, 222–223.

Calambokidis, J. and Barlow, J. (2004). Abundance of blue and humpback whales in the eastern North Pacific estimated by capture–recapture and line-transect methods. *Marine Mammal Science*, **20**, 63–85.

Calleson, C. S. and Frohlich, R. K. (2007). Slower boat speeds reduce risks to manatees. *Endangered Species Research*, **3**, 295–304.

Campbell, H. W. and Irvine, A. B. (1981). Manatee mortality during the unusually cold winter of 1976–1977. In *The West Indian Manatee in Florida: Proceedings of a Workshop held in Orlando, Florida 27–29 March 1978*, R. L. Brownell Jr and K. Ralls (eds), Tallahassee: Florida Department of Natural Resources, pp. 86–91.

Campbell, R. S. F. and Ladds, P. W. (1981). Diseases of the dugong in north-eastern Australia: a preliminary report. In *The Dugong: Proceedings of a Seminar/Workshop held at James Cook University 8–13 May 1979, Townsville*, H. Marsh (ed.), Townsville: James Cook University, pp. 100–102.

Campbell, S. J., McKenzie, L. J. and Kerville, S. P. (2006). Photosynthetic responses of seven tropical seagrasses to elevated seawater temperature. *Journal of Experimental Marine Biology and Ecology*, **330**, 455–468.

Cansdale, G. S. (1970). *All the Animals of the Bible Lands*, Grand Rapids: Zondervan Publishing House.

Cantanhede, A. M., Ferreira Da Silva, V. M., Farias, I. P., *et al.* (2005). Phylogeography and population genetics of the endangered Amazonian manatee, *Trichechus inunguis* Natterer, 1883 (Mammalia, Sirenia). *Molecular Ecology*, **14**, 401–413.

Cardeilhac, P. T., Walker, C. M., Jenkins, R. L., *et al.* (1981). Complications in the formula-rearing of infant manatees associated with bacterial infections. In *The West Indian Manatee in Florida: Proceedings of a Workshop held in Orlando, Florida 27–29 March 1978*, R. L. Brownell Jr and K. Ralls (eds), Tallahassee: Florida Department of Natural Resources, pp. 141–146.

Cardeilhac, P. T., White, J. R. and Francis-Floyd, R. (1984). Initial information on the reproductive biology of the Florida manatee. *Proceedings of the International Association for Aquatic Animal Medicine*, **1**, 35–42.

Carter, A. M., Enders, A. C., Künzle, H., Oduor-Okelo, D. and Vogel, P. (2004). Placentation in species of phylogenetic importance: the Afrotheria. *Animal Reproduction Science*, **82–83**, 35–48.

Castelblanco-Martínez, D. N., Bermudez-Romero, A. L., Gomez-Camelo, I. V., *et al.* (2009). Seasonality of habitat use, mortality and reproduction of the vulnerable Antillean manatee *Trichechus manatus manatus* in the Orinoco River, Colombia: implications for conservation. *Oryx*, **43**, 235–242.

Caughley, G. (1977). *Analysis of Vertebrate Populations*, Chichester: John Wiley and Sons.

Charnock-Wilson, J. (1968). The manatee in British Honduras. *Oryx*, **9**, 293–295.

Cheke, A. S. (1987). Ecological history of the Mascarenes with particular reference to extinction and introduction of vertebrates. In *Studies of Mascrene Island Birds*, A. W. Diamond, (ed.), Cambridge: Cambridge University Press, pp. 5–89.

Chilvers, L. B., Delean, S., Gales, N. J., *et al.* (2004). Diving behaviour of dugongs, *Dugong dugon. Journal of Experimental Marine Biology and Ecology*, **304**, 203–224.

CITES. (2010). CITES (Convention on International Trade in Endangered Species of Wild Fauna and Flora). Available at: http://www.cites.org.

Clarke, R. V. (1997). Introduction. In *Situational Crime Prevention: Successful Case Studies*, 2nd edn, R. V. Clarke (ed.), New York: Harrow and Heston, pp. 1–43.

Clausen, R. and York, R. (2008). Global biodiversity decline of marine and fresh-water fish: a cross-national analysis of economic, demographic, and ecological influences. *Social Science Research*, **37**, 1310–1320.

Clauss, M. and Hummel, J. (2005). The digestive performance of mammalian herbivores: why big may not be that much better. *Mammal Review*, **35**, 174–187.

Clauss, M., Schwarm, A., Ortmann, S., Streich, W. J. and Hummel, J. (2007). A case of non-scaling in mammalian physiology? Body size, digestive capacity, food intake, and ingesta passage in mammalian herbivores. *Comparative Biochemistry and Physiology Part A: Physiology*, **148**, 249–265.

Clementz, M. T., Holden, P. and Koch, P. L. (2003a). Are calcium isotopes a reliable monitor of trophic level in marine settings? *International Journal of Osteoarchaeology*, **13**, 29–36.

Clementz, M. T., Hoppe, K. A. and Koch, P. L. (2003b). A paleoecological paradox: the habitat and dietary preferences of the extinct tethythere *Desmostylus*, inferred from stable isotope analysis. *Paleobiology*, **29**, 506–519.

Clementz, M. T., Goswami, A., Gingerich, P. D. and Koch, P. L. (2006). Isotopic records from early whales and sea cows: contrasting patterns of ecological transition. *Journal of Vertebrate Paleontology*, **26**, 355–370.

Clementz, M. T., Koch, P. L. and Beck, C. A. (2007). Diet induced differences in carbon isotope fractionation between sirenians and terrestrial ungulates. *Marine Biology*, **151**, 1773–1784.

Clementz, M. T., Sorbi, S. and Domning, D. P. (2009). Evidence of Cenozoic environmental and ecological change from stable isotope analysis of sirenian remains from the Tethys-Mediterranean region. *Geology*, **37**, 307–310.

Clifton, K. B., Yan, J., Mecholsky, J. J. Jr and Reep, R. L. (2008a). Material properties of manatee rib bone. *Journal of Zoology*, **274**, 150–159.

Clifton, K. B., Reep, R. L. and Mecholsky, J. J. Jr (2008b). Quantitative fractography for estimating whole bone properties of manatee rib bones. *Journal of Materials Science*, **43**, 2026–2034.

Clutton-Brock, T. H. (1989). Mammalian mating systems. *Proceedings of the Royal Society of London, B: Biological Sciences*, **236**, 339–372.

CMS. (2010). Convention on Migratory Species. Available at: http://www.cms.int.

Cockcroft, V. G., Guissamulo, A. and Findlay, K. (2008). Dugongs in the Bazaruto Archipelago, Mozambique. Unpublished report, Centre for Dolphin Studies, Plettenberg, South Africa.

Coe, M. T., Costa, M. H. and Soares-Filho, B. S. (2009). The influence of historical and potential future deforestation on the stream flow of the Amazon river–land surface processes and atmospheric feedbacks. *Journal of Hydrology*, **369**, 165–174.

Cohen, J. L., Tucker, G. S. and Odell, D. K. (1982). The photoreceptors of the West Indian manatee. *Journal of Morphology*, **173**, 197–202.

Colares, I. G. and Colares, E. P. (2002). Food plants eaten by Amazonian manatees (*Trichechus inunguis*, Mammalia: Sirenia). *Brazilian Archives of Biology and Technology*, **45**, 67–72.

Colbert, D. E., Gaspard III, J. C., Reep, R. L., Mann, D. A. and Bauer, G. B. (2009). Four-choice sound localization abilities of two Florida manatees, *Trichechus manatus latirostris*. *Journal of Experimental Biology*, **212**, 2105–2112.

Coles, R., McKenzie, L. and Campbell, S. (2003a). The seagrasses of Eastern Australia. In *World Atlas of Seagrasses*, E. P. Green and F. F. Short (eds), Berkeley: University of California Press, pp. 119–133.

Coles, R., McKenzie, L., Campbell, S., Fortes, M. and Short, F. F. (2003b). The seagrasses of the Western Pacific Islands. In *World Atlas of Seagrasses*, E. P. Green and F. F. Short (eds), Berkeley: University of California Press, pp. 161–170.

Committee on Taxonomy. (2009). List of marine mammal species and subspecies. Society for Marine Mammalogy. Available at: http://www.marinemammal science.org (accessed 24 September 2010).

Conradt, L. and Roper, T. J. (2000). Activity synchrony and social cohesion: a fission–fusion model. *Proceedings of the Royal Society of London B: Biological Sciences*, **267**, 2213–2218.

Convention on International Trade in Endangered Species of Wild Fauna and Flora. (2010). See CITES.

Convention on Migratory Species. (2010). See CMS.

Conway, D., Persechino, A., Ardoin-Bardin, S., *et al.* (2009). Rainfall and water resources variability in sub-Saharan Africa during the twentieth century. *Journal of Hydrometeorology*, **10**, 41–59.

Cope, E. D. (1889). Synopsis of the families of Vertebrata. *The American Naturalist*, **23**, 849–877.

Coral Triangle Initiative. (2009). Coral triangle initiative on coral reefs, fisheries and food security. Available at: http://www.cti-secretariat.net/about-cti/about-cti.

Cormack, R. M. (1964). Estimates of survival from the sighting of marked animals. *Biometrika*, **51**, 429–438.

Correa-Viana, M. and O'Shea, T. J. (1992). El manati en la tradicion y folklore de Venezuela. *Revista Unellez de Ciencia y Tecnologia*, **10**, 7–13.

Correa-Viana, M., O'Shea, T. J., Ludlow, M. E. and Robinson, J. G. (1990). Distribucion y abundancia del manati, *Trichechus manatus*, en Venezuela. *Biollania*, **7**, 101–123.

Couzin, I. D. (2006). Behavioral ecology: social organization in fission–fusion societies. *Current Biology*, **16**, 169–171.

Cox, P. M., Harris, P. P., Huntingford, C., *et al.* (2008). Increasing risk of Amazonian drought due to decreasing aerosol pollution. *Nature*, **453**, 212–215.

Craig, B. A. and Reynolds III, J. E. (2004). Determination of manatee population trends along the Atlantic coast of Florida using a Bayesian approach with temperature-adjusted aerial survey data. *Marine Mammal Science*, **20**, 386–400.

Craig, B. A., Newton, M. A., Garrott, R. A., Reynolds III, J. E. and Wilcox, J. R. (1997). Analysis of aerial survey data on Florida manatee using Markov Chain Monte Carlo. *Biometrics*, **53**, 524–541.

Crouch, J., McNiven, I. J., David, B., Rowe, C. and Weisler, M. I. (2007). Berberass: marine resource specialisation and environmental change in Torres Strait during the past 4000 years. *Archaeology in Oceania*, **42**, 49–64.

Crownhart-Vaughan, E. A. P. (1972). Introduction. In *Explorations of Kamchatka 1735–1741*, S. P. Krasheninnikov, Portland: Oregon Historical Society, pp. v–xxxiii.

Cumbaa, S. L. (1980). Aboriginal use of marine mammals in the southeastern United States. *Southeastern Archaeological Conference Bulletin*, **17**, 6–10.

Cumming, G. S., Cumming, D. H. M. and Redman, C. L. (2006). Scale mismatches in social–ecological systems: causes, consequences, and solutions. *Ecology and Society*, **11**, 14.

Czech, B., Krausman, P. R. and Devers, P. K. (2000). Economic associations among causes of species endangerment in the United States. *Bioscience*, **50**, 593–601.

Dai, A., Qian, T., Trenberth, K. E. and Milliman, J. D. (2009). Changes in continental freshwater discharge from 1948 to 2004. *Journal of Climate*, **22**, 2773–2792.

Daley, B., Griggs, P. and Marsh, H. (2008). Exploiting marine wildlife in Queensland: the commercial dugong and marine turtle fisheries, 1847–1969. *Australian Economic History Review*, **48**, 227–265.

Dampier, W. (2005). *The Buccaneer Explorer: William Dampier's Voyages*, edited and with an introduction by Gerald Norris, Woodbridge: Boydell Press.

Das, H. S. and Dey, S. C. (1999). Observations on the dugong, *Dugong dugon* (Müller), in the Andaman and Nicobar Islands, India. *Journal of the Bombay Natural History Society*, **96**, 195–198.

Davis, S. and Khosla, V. (2007). The architecture of audacity: assessing the impact of the Microcredit Summit Campaign. *Innovations*, **2**, 159–180.

de Iongh, H. H. (1996). Plant–herbivore interactions between seagrasses and dugongs in a tropical small island ecosystem. PhD Thesis, Katholieke Universiteit Nijmegen.

de Iongh, H. H., Wenno, B. J. and Meelis, E. (1995). Seagrass distribution and seasonal biomass changes in relation to dugong grazing in the Moluccas, East Indonesia. *Aquatic Botany*, **50**, 1–19.

de Iongh, H. H., Kiswara, W., Kustiawan, W. and Loth, P. E. (2007). A review of research on the interactions between dugongs (*Dugong dugon* Müller 1776) and intertidal seagrass beds in Indonesia. *Hydrobiologia*, **591**, 73–83.

de Iongh, H., Hutosomo, M., Moraal, M. and Kiswara, W. (2009a). National dugong conservation strategy and action plan for Indonesia. Part 1: scientific report. Jakarta: Institute of Environmental Sciences Leiden and Research Centre for Oceanographic Research Jakarta.

de Iongh, H., Hutosomo, M., Moraal, M. and Kiswara, W. (2009b). National dugong conservation strategy and action plan for Indonesia. Part 2: strategy report, Jakarta: Institute of Environmental Sciences Leiden and Research Centre for Oceanographic Research Jakarta.

de Jong, W. W., Zweers, A. and Goodman, M. (1981). Relationship of aardvaark to elephants, hyraxes, and sea cows from α-crystallin sequences. *Nature*, **292**, 538–540.

de la Mare, W. K. (1986). On the estimation of mortality rates from whale age data, with particular reference to minke whales (*Balaenoptera acutorostrata*) in the Southern Hemisphere. *Report of the International Whaling Commission*, **36**, 239–250.

De Souza Amaral, R., Rosas, F. C. W., Viau, P., *et al.* (2009). Noninvasive monitoring of androgens in male Amazonian manatee (*Trichechus inunguis*): biologic validation. *Journal of Zoo and Wildlife Medicine*, **40**, 458–465.

Debusk, T. A. and Ryther, J. H. (1987). Biomass production and yields of aquatic plants. In *Aquatic Plants for Water Treatment and Resource Recovery*, K. R. Reddy and W. H. Smith (eds), Orlando: Magnolia Publishing, pp. 579–598.

Deevey, E. S. Jr. (1947). Life tables for natural populations of animals. *Quarterly Review of Biology*, **22**, 283–314.

DeGabriel, J. L., Wallis, I. R., Moore, B. D. and Foley, W. J. (2008). A simple, integrative assay to quantify nutritional quality of browses for herbivores. *Oecologia*, **156**, 107–116.

DeGabriel, J. L., Moore, B. D., Foley, W. J. and Johnson, C. N. (2009). The effects of plant defensive chemistry on nutrient availability predict reproductive success in a mammal. *Ecology*, **90**, 711–719.

Delisle, A. (2009). Perceived costs and benefits of Indigenous hunting of dugongs and marine turtles: Mabuiag Island as a case study. Available at: http://www.ecoeco.org/anzsee09/cd_view_detail.php?id=773.

DeMaster, D. P. (1981). Estimating the average age of first birth in marine mammals. *Canadian Journal of Fisheries and Aquatic Sciences*, **38**, 237–239.

Denkinger, J. (2010). Status of the Amazonian manatee (*Trichechus inunguis*) in the Cuyabeno Reserve, Ecuador. *Avances*, **2**, B29–B34.

Denton, G. R. W., Marsh, H., Heinsohn, G. E. and Burdon-Jones, C. (1980). The unusual metal status of the dugong, *Dugon dugon*. *Marine Biology*, **57**, 201–219.

Department of Environment and Conservation. (2007). Dugong (Dugong dugon) management program for Western Australia 2007–2016. Western Australian Government.

Deutsch, C. J. and Reynolds III, J. E. (2012). Florida manatee status and conservation issues. In *Sirenian Conservation: Issues and Strategies in Developing Countries*, E. Hines, J. Reynolds, A. Mignucci-Giannoni, L. Aragones and M. Marmontel (eds), Gainesville: University Press of Florida.

Deutsch, C. J., Reid, J. P., Bonde, R. K., *et al.* (2003). Seasonal movements, migratory behavior, and site fidelity of West Indian manatees along the Atlantic Coast of the United States. *Wildlife Monographs*, **151**, 1–77.

Deutsch, C. J., Self-Sullivan, C. and Mignucci-Giannoni, A. (2008). *Trichechus manatus*. In *IUCN 2010: IUCN Red List of Threatened Species*. Version 2010.3. Available at: http://www.iucnredlist.org (accessed 19 September 2010).

DiBattista, J. D. (2008). Patterns of genetic variation in anthropogenically impacted populations. *Conservation Genetics*, **9**, 141–156.

Diedrich, C. G. (2008). The food of the miosiren *Anomotherium langenwieschei* (Siegfried): indirect proof of seaweed or seagrass by xenomorphic oyster fixation structures in the Upper Oligocene (Neogene) of the Doberg, Bünde (NW Germany) and comparisons to modern *Dugong dugon* (Müller) feeding strategies. *Senckenbergiana Maritima*, **38**, 59–73.

Dietz, T. (1992). *The Call of the Siren: Manatees and Dugongs*. Golden, CO: Fulcrum Publishing.

Dixon, P. M. and Pechmann, J. H. K. (2005). A statistical test to show negligible trend. *Ecology*, **86**, 1751–1756.

Dobbs, K., Fernandes, L., Slegers, S., *et al.* (2008). Incorporating dugong habitats into the marine protected area design for the Great Barrier Reef Marine Park, Queensland, Australia. *Ocean and Coastal Management*, **51**, 368–375.

Dodman, T., Ndiaye, M. D. D. and Sarr, K. (eds). (2008). *Conservation Strategy for the West African Manatee*. Nairobi: UNEP and Dakar: Wetlands International Africa.

Domning, D. (1972). Steller's sea cow and the origin of North Pacific aboriginal whaling. *Syesis*, **5**, 187–189.

Domning, D. P. (1976). An ecological model for late tertiary sirenian evolution in the north Pacific Ocean. *Systematic Zoology*, **25**, 352–362.

Domning, D. P. (1978a). Sirenian evolution in the north Pacific Ocean. *University of California Publications in Geological Sciences*, **118**, 1–176.

Domning, D. P. (1978b). Sirenia. In *Evolution of African Mammals*, V. J. Maglio and H. B. S. Cooke (eds), Cambridge, MA: Harvard University Press, pp. 573–581.

Domning, D. P. (1978c). The myology of the Amazonian manatee, *Trichechus inunguis* (Natterer) (Mammalia: Sirenia). *Acta Amazonica*, **8**, Supl 1, 1–80.

Domning, D. P. (1982a). Evolution of manatees: a speculative history. *Journal of Paleontology*, **56**, 599–619.

Domning, D. P. (1982b). Commercial exploitation of manatees *Trichechus* in Brazil c.1785–1973. *Biological Conservation*, **22**, 101–126.

Domning, D. P. (1988). Fossil Sirenia of the West Atlantic and Caribbean region: I. *Metaxytherium floridanum* Hay, 1922. *Journal of Vertebrate Paleontology*, **8**, 395–426.

Domning, D. P. (1989a). Fossil Sirenia of the West Atlantic and Caribbean region: II. *Dioplotherium manigaulti* Cope, 1883. *Journal of Vertebrate Paleontology*, **9**, 415–428.

Domning, D. P. (1989b). Fossil Sirenia of the West Atlantic and Caribbean region: III. *Xenosiren yucateca*, gen. et sp. nov. *Journal of Vertebrate Paleontology*, **9**, 429–437.

Domning, D. P. (1989c). Kelp evolution: a comment. *Paleobiology*, **15**, 53–56.

Domning, D. P. (1990). Fossil Sirenia of the West Atlantic and Caribbean region: IV. *Corystosiren varguezi*, gen. et sp. nov. *Journal of Vertebrate Paleontology*, **10**, 361–371.

Domning, D. P. (1994). A phylogenetic analysis of the Sirenia. In *Contributions in Marine Mammal Paleontology Honoring Frank C. Whitmore, Jr*, A. Berta and T. A. Démére (eds), San Diego: San Diego Society of Natural History, pp. 177–189.

Domning, D. P. (1996). Bibliography and index of the Sirenia and Desmostylia. *Smithsonian Contributions to Paleobiology*, **80**, 1–611.

Domning, D. P. (1997a). Fossil Sirenia of the West Atlantic and Caribbean region: VI. *Crenatosiren olseni* (Reinhart, 1976). *Journal of Vertebrate Paleontology*, **17**, 397–412.

Domning, D. P. (1997b). Sirenia. In *Vertebrate Paleontology in the Neotropics: the Miocene Fauna of La Venta, Colombia*, R. F. Kay, R. H. Madden, R. L. Difelli and J. J. Flynn (eds), Washington, DC: Smithsonian Institution Press, pp. 383–391.

Domning, D. P. (2000). The readaptation of Eocene sirenians to life in water. *Historical Biology*, **14**, 115–119.

Domning, D. P. (2001a). Evolution of the Sirenia and Desmostylia. In *Secondary Adaptation of Tetrapods to Life in Water*, J.-M. Mazin and V. de Buffrénil (eds), Munich: Verlag, pp. 151–168.

Domning, D. P. (2001b). Sirenians, seagrasses, and Cenozoic ecological change in the Caribbean. *Palaeogeography, Palaeoclimatology, Palaeoecology*, **166**, 27–50.

Domning, D. P. (2001c). The earliest known fully quadrupedal sirenian. *Nature*, **413**, 625–627.

Domning, D. P. (2002). The terrestrial posture of desmostylians. In *Cenozoic Mammals of Land and Sea: Tributes to the Career of Clayton E. Ray*, R. J. Emry (ed.), Washington, DC: Smithsonian Institution Press, pp. 99–111.

Domning, D. P. (2005). Fossil Sirenia of the West Atlantic and Caribbean region: VII. Pleistocene *Trichechus manatus* Linneaus, 1758. *Journal of Vertebrate Paleontology*, **25**, 685–701.

Domning, D. P. (2010). Bibliography and index of the Sirenia and Desmostylia. Available at: http://www.sirenian.org/biblio (accessed 19 September 2010).

Domning, D. P. and Aguilera, O. A. (2008). Fossil Sirenia of the West Atlantic and Caribbean region: VIII. *Nanosiren garciae*, gen. et sp. nov. and *Nanosiren sanchezi*, sp. nov. *Journal of Vertebrate Paleontology*, **28**, 479–500.

Domning, D. P. and Beatty, B. L. (2007). Use of tusks in feeding by dugongid sirenians: observations and tests of hypotheses. *Anatomical Record*, **290**, 523–538.

Domning, D. P. and Buffrénil, V. de. (1991). Hydrostasis in the Sirenia: quantitative data and functional interpretations. *Marine Mammal Science*, **7**, 331–368.

Domning, D. P. and Furusawa, H. (1994). Summary of taxa and distribution of Sirenia in the north Pacific Ocean. *The Island Arc*, **3**, 506–512.

Domning, D. P. and Gingerich, P. D. (1994). *Protosiren smithae*, new species (Mammalia, Sirenia), from the late Middle Eocene of Wadi Hitan, Egypt. *Contributions from the Museum of Paleontology, University of Michigan*, **29**, 69–87.

Domning, D. P. and Hayek, L.-A. C. (1984). Horizontal tooth replacement in the Amazonian manatee (*Trichechus inunguis*). *Mammalia*, **48**, 105–127.

Domning, D. P. and Hayek, L.-A. C. (1986). Interspecific and intraspecific morphological variation in manatees (Sirenia: *Trichechus*). *Marine Mammal Science*, **2**, 87–144.

Domning, D. P. and Myrick, A. C. Jr. (1980). Tetracycline marking and the possible layering rate of bone in an Amazonian manatee, *Trichechus inunguis*. In *Age Determination of Toothed Whales and Sirenians*, W. F. Perrin and A. C. Myrick Jr (eds), *Reports of the International Whaling Commission*, special issue, **3**, 203–207.

Domning, D. P. and Pervesler, P. (2001). The osteology and relationships of *Metaxytherium krahuletzi* Depéret, 1895 (Mammalia: Sirenia). *Abhandlungen der Senckenbergischen Naturforschenden Gesellschaft*, **553**, 1–89.

Domning, D. P. and Thomas, H. (1987). *Metaxytherium serresii* (Mammalia: Sirenia) from the Early Pliocene of Libya and France: a reevaluation of its morphology, phyletic position, and biostratigraphic and paleoecological significance. In *Neogene Paleontology and Geology of Sahabi*, N. T. Boaz, A. El-Arnauti, W. de Heinzelin and D. D. Boaz (eds), New York: Alan R. Liss Publishers, pp. 205–232.

Domning, D. P., Morgan, G. S. and Ray, C. E. (1982). North American Eocene sea cows (Mammalia: Sirenia). *Smithsonian Contributions to Paleobiology*, **52**, 1–69.

Domning, D. P., Ray, C. E. and McKenna, M. C. (1986). Two new Oligocene desmostylians and a discussion of tethytherian systematics. *Smithsonian Contributions to Paleobiology*, **59**, 1–56.

Domning, D. P., Gingerich, P. D., Simons, E. L. and Ankel-Simons, F. A. (1994). A new early Oligocene dugongid (Mammalia: Sirenia) from Fayum Province,

Egypt. *Contributions from the Museum of Paleontology, University of Michigan*, **29**, 89–108.

Domning, D. P., Thomason, J. and Corbett, D. G. (2007). Steller's sea cow in the Aleutian Islands. *Marine Mammal Science*, **23**, 976–983.

Duarte, C. M. (1992). Nutrient concentration of aquatic plants: patterns across species. *Limnology and Oceanography*, **37**, 882–889.

Dubey, J. P., Zarnke, R., Thomas, N. J., *et al.* (2003). *Toxoplasma gondii, Neospora caninum, Sarcocystis neurona,* and *Sarcocystis canis*-like infections in marine mammals. *Veterinary Parasitology*, **116**, 275–296.

Dudgeon, D., Arthington, A. H., Gessner, M. O., *et al.* (2006). Freshwater biodiversity: importance, threats, status and conservation challenges. *Biological Reviews*, **81**, 163–182.

Duignan, P. J., House, C., Walsh, M. T., *et al.* (1995). Morbillivirus infection in manatees. *Marine Mammal Science*, **11**, 441–451.

Dunshea, G. (2004). Development of telomere based aging method for the dugong. BSc(Hons), James Cook University, Townsville.

Easterling, D. R., Meehl, G. A., Parmesan, C., *et al.* (2000). Climate extremes: observations, modeling, and impacts. *Science*, **289**, 2068–2074.

Eberhardt, L. L. (1977). Optimal policies for the conservation of large mammals, with special reference to marine ecosystems. *Environmental Conservation*, **4**, 205–212.

Eberhardt, L. L. (1982). *Censusing Manatees: Manatee Population Research Report 1.* Gainesville: Florida Cooperative Fish and Wildlife Research Unit.

Eberhardt, L. L. (2002). A paradigm for population analysis of long-lived vertebrates. *Ecology*, **83**, 2841–2854.

Eberhardt, L. L. and O'Shea, T. J. (1995). Integration of manatee life-history data and population modeling. In *Population Biology of the Florida Manatee*, T. J. O'Shea, B. B. Ackerman and H. F. Percival (eds), Washington, DC: US Department of the Interior, National Biological Service, pp. 269–279.

Eberhardt, L. L. and Siniff, D. B. (1977). Population dynamics and marine mammal management policies. *Journal of the Fisheries Research Board of Canada*, **34**, 183–190.

Edwards, H. H., Pollock, K. H., Ackerman, B. B., Reynolds III, J. E. and Powell, J. A. (2007). Estimation of detection probability in manatee aerial surveys at a winter aggregation site. *Journal of Wildlife Management*, **71**, 2052–2060.

Egerton, F. N. (2008). A history of the ecological sciences, part 27: naturalists explore Russia and the North Pacific during the 1700s. *Bulletin of the Ecological Society of America*, **89**, 39–60.

Elliot, H. A., Thomas, A., Ladds, P. W. and Heinsohn, G. E. (1981). A fatal case of salmonellosis in a dugong. *Journal of Wildlife Diseases*, **17**, 203–208.

Emlen, J. M. (1966). The role of time and energy in food preference. *American Naturalist*, **100**, 611–617.

Erftemeijer, P. L. A., Moka, D. and Moka, W. (1993). Stomach content analysis of a dugong (*Dugong dugon*) from South Sulawesi, Indonesia. *Australian Journal of Marine and Freshwater Research*, **44**, 229–233.

Eros, C., Marsh, H., Bonde, R., *et al.* (2007). *Procedures for the Salvage and Necropsy of the Dugong (Dugong dugon)*, 2nd edn, Townsville: Great Barrier Reef Marine Park Authority.

Estes, J. A. and Duggins, D. O. (1995). Sea otters and kelp forests in Alaska: generality and variation in a community ecological paradigm. *Ecological Monographs*, **65**, 75–100.

Estes, J. A. and Palmisano, J. F. (1974). Sea otters: their role in structuring nearshore communities. *Science*, **185**, 1058–1060.

Estes, J. A., Duggins, D. O. and Rathbun, G. B. (1989). The ecology of extinctions in kelp forest communities. *Conservation Biology*, **3**, 252–264.

Etheridge, K., Rathbun, G. B., Powell, J. A. and Kochman, H. I. (1985). Consumption of aquatic plants by the West Indian manatee. *Journal of Aquatic Plant Management*, **23**, 21–25.

Evans, W. E. and Herald, E. S. (1970). Underwater calls of a captive Amazon manatee, *Trichechus inunguis*. *Journal of Mammalogy*, **51**, 820–823.

Falkenmark, M. (1997a). Meeting water requirements of an expanding world population. *Philosophical Transactions of the Royal Society B: Biological Sciences*, **352**, 929–936.

Falkenmark, M. (1997b). Society's interaction with the water cycle: a conceptual framework for a more holistic approach. *Hydrological Sciences Journal*, **42**, 451–466.

Falkenmark, M. (1998). Dilemma when entering 21st century: rapid change but lack of sense of urgency. *Water Policy*, **1**, 421–436.

Federal Register. (2009). Marine Mammal Protection Act: stock assessment report. Department of the Interior, Fish and Wildlife Service.

Fernandes, L., Day, J., Lewis, A., *et al.* (2005). Establishing representative no-take areas in the Great Barrier Reef: large-scale implementation of theory on marine protected areas. *Conservation Biology*, **19**, 1733–1744.

Ferraro, P. J. and Kiss, A. (2002). Direct payments to conserve biodiversity. *Science*, **298**, 1718–1719.

Fischer, M. S. (1986). Die stellung der schliefer (Hyracoidea) im phylogenetischen system der Eutheria. *Courier Forschunginstitut Senckenberg*, **84**, 1–132.

Flamm, R. O., Weigle, B. L., Wright, I. E., Ross, M. and Aglietti, S. (2005). Estimation of manatee (*Trichechus manatus latirostris*) places and movement corridors using telemetry data. *Ecological Applications*, **15**, 1415–1426.

Fleming, L. E., Kirkpatrick, B., Backer, L. C., *et al.* (2005). Initial evaluation of the effects of aerosolized Florida red tide toxins (brevetoxins) in persons with asthma. *Environmental Health Perspectives*, **113**, 650–657.

Flewelling, L. J., Naar, J. P., Abbott, J. P., *et al.* (2005). Brevetoxicosis: red tides and marine mammal mortalities. *Nature*, **435**, 755–756.

Florida Fish and Wildlife Conservation Commission. (2002). Final biological status review of the Florida manatee (*Trichechus manatus latirostris*). Florida Fish and Wildlife Conservation Commission, Florida Marine Research Institute.

Florida Fish and Wildlife Conservation Commission. (2007). Florida Manatee Management Plan *Trichechus manatus latirostris*. Available at: http://www.myfwc.com/docs/WildlifeHabitats/Manatee_Mgmt_Plan.pdf.

Florida Fish and Wildlife Conservation Commission. (2010a). Manatee synoptic surveys. Available at: http://research.myfwc.com/features/view_article.asp?id=15246 (accessed 9 October 2010).

Florida Fish and Wildlife Conservation Commission. (2010b). 2010 manatee mortality. Available at: http://research.myfwc.com/features/category_sub.asp?id=2241 (accessed 9 October 2010).

Florida Fish and Wildlife Conservation Commission. (2010c). YTD preliminary manatee mortality table by county, Florida from 01/01/2009 to 12/31/2009. Florida Fish and Wildlife Conservation Commission, Marine Mammal Pathobiology Laboratory. Available at: http://research.myfwc.com/engine/download_redirection_process.asp?file=2009_Cumulative_Category_Summary23Mar2010.pdfandobjid=11693anddltype=article.

Florida Fish and Wildlife Conservation Commission. (2010d). YTD preliminary manatee mortality table by county, Florida from 01/01/2010 to 07/31/2010. Florida Fish and Wildlife Commission, Marine Mammal Pathobiology Laboratory. Available at: http://research.myfwc.com/engine/download_redirection_process.asp?file=2010_Cumulative_Category_Summary17Aug.pdf&objid=14855&dl type=article.

Florida Oceans and Coastal Council. (2009). The effects of climate change on Florida's ocean & coastal resources: a special report to the Florida Energy and Climate Commission and the people of Florida. Available at: http://www.floridaoceanscouncil.org/reports/Climate_Change_Report_v2.pdf.

Flowerdew, J. R. (1987). *Mammals: Their Reproductive Biology and Population Ecology*, London: Edward Arnold.

Foley, W. J. and Moore, B. D. (2005). Plant secondary metabolites and vertebrate herbivores: from physiological regulation to ecosystem function. *Current Opinion in Plant Biology*, **8**, 430–435.

Forrester, D. J. (1992). *Parasites and Diseases of Wild Mammals in Florida*, Gainesville: University Press of Florida.

Forrester, D. J., White, F. H., Woodard, J. C. and Thompson, N. P. (1975). Intussusception in a Florida manatee. *Journal of Wildlife Diseases*, **11**, 566–568.

Fowler, C. W. (1984). Density dependence in cetacean populations. *Reports of the International Whaling Commission*, special issue, **6**, 373–379.

Frape, D. (2004). *Equine Nutrition and Feeding*, 3rd edn, Oxford: Blackwell Publishing Ltd.

Frappart, F., Seyler, F., Martinez, J. M., León, J. G. and Cazenave, A. (2005). Floodplain water storage in the Negro River basin estimated from microwave remote sensing of inundation area and water levels. *Remote Sensing of the Environment*, **99**, 387–399.

Frazier, J. G. and Mundkur, T. (1991). Dugong, *Dugong dugon* (Müller) in the Gulf of Kutch, Gujarat. *Journal of the Bombay Natural History Society*, **87**, 368–379.

Frye, F. L. and Herald, E. S. (1969). Osteomyelitis in a manatee. *Journal of the American Veterinary Medical Association*, **155**, 1073–1076.

Gales, N., McCauley, R. D., Lanyon, J. and Holley, D. (2004). Change in the abundance of dugongs in Shark Bay, Ningaloo and Exmouth Gulf, Western Australia: evidence for large-scale migration. *Wildlife Research*, **31**, 283–290.

Galliard, J.-M., Festa-Bianchet, M. and Yoccoz, N. G. (1998). Population dynamics of large herbivores: variable recruitment with constant adult survival. *Trends in Ecology and Evolution*, **13**, 58–63.

Galliard, J.-M., Festa-Bianchet, M., Yoccoz, N. G., Loison, A. and Toïgo, C. (2000). Temporal variation in fitness components and population dynamics of large herbivores. *Annual Review of Ecology and Systematics*, **31**, 367–393.

Gallivan, G. J. and Best, R. C. (1980). Metabolism and respiration of the Amazonian manatee (*Trichechus inunguis*). *Physiological Zoology*, **53**, 245–253.

Gannon, J. G., Scolardi, K. M., Reynolds III, J. E., Koelsch, J. K. and Kessenich, T. J. (2007). Habitat selection by manatees in Sarasota Bay, Florida. *Marine Mammal Science*, **23**, 133–143.

Garcia-Rodriguez, A. I., Bowen, B. W., Domning, D., *et al.* (1998). Phylogeography of the West Indian manatee (*Trichechus manatus*): how many populations and how many taxa? *Molecular Ecology*, **7**, 1137–1149.

Garcia-Rodriguez, A. I., Moraga-Amador, D., Farmerie, W., McGuire, P. and King, T. L. (2000). Isolation and characterization of microsatellite DNA markers in the Florida manatee (*Trichechus manatus latirostris*) and their application in selected sirenian species. *Molecular Ecology*, **9**, 2161–2163.

Gärdenfors, U. (2001). Classifying threatened species at national versus global levels. *Trends in Ecology and Evolution*, **16**, 511–516.

Garner, A., Rachlow, J. L. and Hicks, J. F. (2005). Patterns of genetic diversity and its loss in mammalian populations. *Conservation Biology*, **19**, 1215–1221.

Garrigue, C., Patenaude, N. and Marsh, H. (2008). Distribution and abundance of the dugong in New Caledonia, southwest Pacific. *Marine Mammal Science*, **24**, 81–90.

Garrott, R. A., Ackerman, B. B., Cary, J. R., *et al.* (1995). Assessment of trends in sizes of manatee populations at several Florida aggregation sites. In *Population Biology of the Florida Manatee*, T. J. O'Shea, B. B. Ackerman and H. F. Percival (eds), Washington, DC: US Department of the Interior, National Biological Service, pp. 34–55.

GBRMPA. (1981). Nomination of the Great Barrier Reef by the Commonwealth of Australia for inclusion in the World Heritage List. Great Barrier Reef Marine Park Authority, Townsville.

Geist, V. (1974). On the relationship of ecology and behavior in the evolution of ungulates: theoretical considerations. In *The Behavior of Ungulates and its Relation to Management*, V. Geist and F. Walther (eds), Morges: IUCN, pp. 235–246.

Geraci, J. R. and Lounsbury, V. J. (2005). *Marine Mammals Ashore: A Field Guide for Strandings*, 2nd edn, Baltimore: National Aquarium in Baltimore.

Geraci, J. R., Anderson, D. M., Timperi, R. J., *et al.* (1989). Humpback whales (*Megaptera novaeangliae*) fatally poisoned by dinoflagellate toxin. *Canadian Journal of Fisheries and Aquatic Sciences*, **46**, 1895–1898.

Gerges, M. A. (2002). The Red Sea and Gulf of Aden action plan: facing the challenges of an ocean gateway. *Ocean and Coastal Management*, **45**, 885–903.

Gerrodette, T. (1987). A power analysis for detecting trends. *Ecology*, **68**, 1364–1372.

Gerrodette, T., Taylor, B. L., Swift, R., *et al.* (2011). A combined visual and acoustic estimate of 2008 abundance, and change in abundance since 1997, for the vaquita, *Phocaena sinus*. *Marine Mammal Science*, **27**, E79–E100.

Gerstein, E. (2002). Manatees, bioacoustics and boats. *American Scientist*, **90**, 154–163.

Gerstein, E. R., Gerstein, L., Forsythe, S. E. and Blue, J. E. (1999). The underwater audiogram of the West Indian manatee (*Trichechus manatus*). *Journal of the Acoustical Society of America*, **105**, 3575–3583.

Gheerbrant, E. (2009). Paleocene emergence of elephant relatives and the rapid radiation of African ungulates. *Proceedings of the National Academy of Sciences USA*, **106**, 10717–10721.

Gheerbrant, E., Domning, D. P. and Tassy, P. (2005). Paenungulata (Sirenia, Proboscidea, Hyracoidea, and relatives). In *The Rise of Placental Mammals*, K. D. Rose and J. D. Archibald (eds), Baltimore: Johns Hopkins University Press, pp. 84–105.

Gill, T. N. (1872). Arrangement of the families of mammals. *Smithsonian Miscellaneous Collections*, **11**, 1–98.

Gillespie, A. (2005). The dugong action plan for the South Pacific: an evaluation based on the need for international and regional conservation of sirenians. *Ocean Development and International Law*, **36**, 135–158.

Gingerich, P. D., Arif, M., Bhatti, M. A., Raza, H. A. and Raza, S. M. (1995). *Protosiren* and *Babiacetus* (Mammalia, Sirenia and Cetacea) from the Middle Eocene Drazinda formation, Sulaiman Range, Punjab (Pakistan). *Contributions of the Museum of Paleontology, University of Michigan*, **29**, 331–357.

Gladstone, W. (1996). *Trace Metals in Sediments, Indicator Organisms and Traditional Seafoods of the Torres Strait*, Townsville: Great Barrier Reef Marine Park Authority.

Gladstone, W., Tawfiq, N., Nasr, D., *et al.* (1999). Sustainable use of renewable resources and conservation in the Red Sea and Gulf of Aden: issues, needs and strategic actions. *Ocean and Coastal Management*, **42**, 671–697.

Glaser, K. S. (photographs) and Reynolds III, J. E. (text). (2003). *Mysterious Manatees*, Gainesville: University Press of Florida.

Godfrey, M. H. and Godley, B. J. (2008). Seeing past the red: flawed IUCN global listings for sea turtles. *Endangered Species Research*, **6**, 155–159.

Gohar, H. A. F. (1957). The Red Sea dugong. *Publication of the Marine Biological Station, Al-Ghardaqa (Red Sea)*, **9**, 3–49.

Golder, F. A. (1968). *Bering's Voyages: An Account of the Efforts of the Russians to Determine the Relations of Asia and America*, New York: Octagon Books. [Reprint of Golder's (1922, 1925) two-volume original publication by the American Geographical Society].

Goodwin, G. C. (1946). The end of the great northern sea cow. *Natural History*, **55**, 56–61.

Goodwin, M. B., Domning, D. P., Lipps, J. H. and Benjamini, C. (1998). The first record of an Eocene (Lutetian) marine mammal from Israel. *Journal of Vertebrate Paleontology*, **18**, 813–815.

Gordon, I. J. and Lindsay, W. K. (1990). Could mammalian herbivores 'manage' their resources?. *Oikos*, **59**, 270–280.

Gorzelany, J. F. (2004). Evaluation of boater compliance with manatee speed zones along the Gulf Coast of Florida. *Coastal Management*, **32**, 215–226.

Goto, M., Ito, C., Yahaya, M. S., *et al.* (2004a). Characteristics of microbial fermentation and potential digestibility of fiber in the hindgut of dugongs (*Dugong dugon*). *Marine and Freshwater Behaviour and Physiology*, **37**, 99–107.

Goto, M., Ito, C., Yahaya, M. S., *et al.* (2004b). Effects of age, body size and season on food consumption and digestion of captive dugongs (*Dugong dugon*). *Marine and Freshwater Behaviour and Physiology*, **37**, 89–97.

Goulden, M., Conway, D. and Persechino, A. (2009). Adaptation to climate change in international river basins in Africa: a review. *Hydrological Sciences Journal*, **54**, 805–828.

Grayson, J., Marsh, H., Delean, S. and Hagihara, R. (2008). Improving knowledge of dugong life history using surrogate data. Great Barrier Reef Marine Park Authority. Available at: http://dugong.id.au/publications/Unpublished/Tusk%20and%20Fecundity%20project%20final%2023%20May.pdf.

Grech, A. (2009). Spatial models and risk assessment to inform marine planning at ecosystem-scales: seagrasses and dugongs as a case study. PhD Thesis, James Cook University, Townsville.

Grech, A. and Coles, R. G. (2010). An ecosystem-scale predictive model of coastal seagrass distribution. *Aquatic Conservation: Marine and Freshwater Ecosystems*, **20**, 437–444.

Grech, A. and Marsh, H. (2007). Prioritising areas for dugong conservation in a marine protected area using a spatially explicit population model. *Applied GIS*, **3**, 1–14.

Grech, A. and Marsh, H. (2008). Rapid assessment of risks to a mobile marine mammal in an ecosystem-scale marine protected area. *Conservation Biology*, **22**, 711–720.

Grech, A. and Marsh, H. (2010). Condition, trends and predicted futures of dugong populations in the Great Barrier Reef World Heritage Area: including an evaluation of the potential and cost-effectiveness of indicators of the status of these populations. Final report to the Marine and Tropical Research Facility.

Grech, A., Marsh, H. and Coles, R. (2008). A spatial assessment of the risk to a mobile marine mammal from bycatch. *Aquatic Conservation: Marine and Freshwater Ecosystems*, **18**, 1127–1139.

Grech, A., Sheppard, J. and Marsh, H. (2011a). Informing species conservation at multiple scales using data collected for marine mammal stock assessments. *PLoS ONE*, **6**, e17993.

Grech, A., Coles, R. and Marsh, H. (2011b). A broad-scale assessment of the risk to coastal seagrasses from cumulative threats. *Marine Policy*, **35**, 560–567.

Green, D. M. (2005). Designatable units for status assessment of endangered species. *Conservation Biology*, **19**, 1813–1820.

Green, E. P. and Short, F. T. (2003). *World Atlas of Seagrasses*, Berkeley: University of California Press.

Gregory, W. K. (1910). The orders of mammals. *Bulletin of the American Museum of Natural History*, **27**, 1–524.

Griebel, U. and Schmid, A. (1996). Color vision in the manatee (*Trichechus manatus*). *Vision Research*, **36**, 2747–2757.

Griebel, U. and Schmid, A. (1997). Brightness discrimination ability in the West Indian manatee (*Trichechus manatus*). *Journal of Experimental Biology*, **200**, 1587–1592.

Grigione, M. M. (1996). Observations on the status and distribution of the West African manatee in Cameroon. *African Journal of Ecology*, **34**, 189–195.

Gunter, G. (1942). Further miscellaneous notes on American manatees. *Journal of Mammalogy*, **23**, 89–90.

Gur, M. B. and Niezrecki, C. (2009). A source separation approach to enhancing marine mammal vocalizations. *Journal of the Acoustical Society of America*, **126**, 3062–3070.

Guterres, M. G., Marmontel, M., Ayub, D. M., Singer, R. F. and Singer, R. B. (2008). *Anatomia e Morfhologia de Plantas Aquáticas da Amazonia*. Utilizadas Como Potencial Alimento por Peixe-Boi Amazônico, Belem/PA.

Haley, D. (1980). The great northern sea cow: Steller's gentle siren. *Oceans*, **13**, 7–11.

Hall, L. S., Krausman, P. R. and Morrison, M. L. (1997). The habitat concept and a plea for standard terminology. *Wildlife Society Bulletin*, **25**, 173–182.

Hall, M. O., Durako, M. J., Fourqurean, J. W. and Zieman, J. C. (1999). Decadal changes in seagrass distribution and abundance in Florida Bay. *Estuaries*, **22**, 445–459.

Halpern, B. S., Walbridge, S., Selkoe, K. A., *et al.* (2008). A global map of human impact on marine ecosystems. *Science*, **319**, 948–952.

Hammer, A. S., Klausen, B., Knold, S., Dietz, H. H. and Dutoit, S. J. H. (2005). Malignant lymphoma in a West Indian manatee (*Trichechus manatus*). *Journal of Wildlife Diseases*, **41**, 834–838.

Hanks, J. and McIntosh, J. E. A. (1973). Population dynamics of the African elephant (*Loxodonta africana*). *Journal of Zoology*, **169**, 29–38.

Harino, H., Ohji, M., Wattayakorn, G., *et al.* (2007). Concentrations of organotin compounds in tissues and organs of dugongs from Thai coastal waters. *Archives of Environmental Contamination and Toxicology*, **53**, 495–502.

Harr, K., Harvey, J., Bonde, R., *et al.* (2006). Comparison of methods used to diagnose generalized inflammatory disease in manatees (*Trichechus manatus latirostris*). *Journal of Zoo and Wildlife Medicine*, **37**, 151–159.

Harr, K. E., Szabo, N. J., Cichra, M. and Philips, E. J. (2008a). Debromoaplysiatoxin in *Lyngbya*-dominated mats on manatees (*Trichechus manatus latirostris*) in the Florida King's Bay ecosystem. *Toxicon*, **52**, 385–388.

Harr, K., Allison, K., Bonde, R., Murphy, D. and Harvey, J. (2008b). Comparison of blood aminotransferase methods for assessment of myopathy and hepatopathy in Florida manatees (*Trichechus manatus latirostris*). *Journal of Zoo and Wildlife Medicine*, **39**, 180–187.

Harris, P. P., Huntingford, C. and Cox, P. M. (2008). Amazon Basin climate under global warming: the role of the sea surface temperature. *Philosophical Transactions of the Royal Society B: Biological Sciences*, **363**, 1753–1759.

Hartman, D. S. (1979). Ecology and behavior of the manatee (*Trichechus manatus*) in Florida. *American Society of Mammalogists Special Publication*, **5**, 1–153.

Harvell, C. D., Mitchell, C. E., Ward, J. R., *et al.* (2002). Climate warming and disease risks for terrestrial and marine biota. *Science*, **296**, 2158–2162.

Harvell, D., Altizer, S., Cattadori, I. M., Harrington, L. and Weil, E. (2009). Climate change and wildlife diseases: when does the host matter the most? *Ecology*, **90**, 912–920.

Harvey, J. W., Harr, K. E., Murphy, D., *et al.* (2007). Clinical biochemistry in healthy manatees (*Trichechus manatus latirostris*). *Journal of Zoo and Wildlife Medicine*, **38**, 269–279.

Harvey, J. W., Harr, K. E., Murphy, D., *et al.* (2009). Hematology of healthy Florida manatees (*Trichechus manatus*). *Veterinary Clinical Pathology*, **38**, 183–193.

Harvey, P. H. and Zammuto, R. M. (1985). Patterns of mortality and age of first reproduction in natural populations of mammals. *Nature*, **315**, 319–320.

Haubold, E. M., Deutsch, C. J. and Fonnesbeck W. C. (2006). Final biological status review of the Florida manatee (*Trichechus manatus latirostris*). St Petersburg: Florida Fish and Wildlife Conservation Commission.

Hauxwell, J., Osenberg, C. W. and Frazer, T. K. (2004). Conflicting management goals: manatees and invasive competitors inhibit restoration of a native macrophyte. *Ecological Applications*, **14**, 571–586.

Havemann, P., Thiriet, D., Marsh, H. and Jones, C. (2005). Traditional use of marine resources agreements and dugong hunting in the Great Barrier Reef World Heritage Area. *Environmental and Planning Law Journal*, **22**, 258–280.

Haynes, D. and Johnson, J. E. (2000). Organochlorine, heavy metal and polyaromatic hydrocarbon pollutant concentrations in the Great Barrier Reef (Australia) environment: a review. *Marine Pollution Bulletin*, **41**, 267–278.

Haynes, D., Müller, J. F. and McLachlan, M. S. (1999). Polychlorinated dibenzo-*p*-dioxins and dibenzofurans in Great Barrier Reef (Australia) dugongs (*Dugong dugon*). *Chemosphere*, **38**, 255–262.

Haynes, D., Müller, J. and Carter, S. (2000a). Pesticide and herbicide residues in sediments and seagrasses from the Great Barrier Reef World Heritage Area and Queensland coast. *Marine Pollution Bulletin*, **41**, 279–287.

Haynes, D., Ralph, P., Prange, J. and Dennison, B. (2000b). The impact of the herbicide diuron on photosynthesis in three species of tropical seagrass. *Marine Pollution Bulletin*, **41**, 288–293.

Haynes, D., Carter, S., Gaus, C., Müller, J. and Dennison, W. (2005). Organochlorine and heavy metal concentrations in blubber and liver tissue collected from Queensland (Australia) dugong (*Dugong dugon*). *Marine Pollution Bulletin*, **51**, 361–369.

Hays, G. C., Akesson, S., Broderick, A. C., *et al.* (2001). The diving behaviour of green turtles undertaking oceanic migration to and from Ascension Island: dive durations, dive profiles and depth distribution. *Journal of Experimental Biology*, **204**, 4093–4098.

Heinsohn, G. E. (1972). A study of dugongs (*Dugong dugon*) in northern Queensland, Australia. *Biological Conservation*, **4**, 205–213.

Heinsohn, G. E. and Birch, W. R. (1972). Foods and feeding habits of the dugong, *Dugong dugon* (Erxleben), in northern Queensland Australia. *Mammalia*, **36**, 414–422.

Heinsohn, G. E. and Marsh, H. (1984). Sirens of northern Australia. In *Vertebrate Zoogeography and Evolution in Australia: Animals in Space and Time*, M. Archer and G. Clayton (eds), Carlisle, WA: Hesperion Press, pp. 1003–1010.

Heinsohn, G. E. and Spain, A. V. (1974). Effects of a tropical cyclone on littoral and sub-littoral biotic communities and on a population of dugongs (*Dugong dugon* [Müller]). *Biological Conservation*, **6**, 143–152.

Heinsohn, G. E., Spain, A. V. and Anderson, P. K. (1976). Populations of dugong (Mammalia: Sirenia): aerial survey of the inshore waters of tropical Australia. *Biological Conservation*, **9**, 21–23.

Heinsohn, G. E., Wake, J., Marsh, H. and Spain, A. V. (1977). The dugong (*Dugong dugon* [Müller]) in the seagrass system. *Aquaculture*, **12**, 235–248.

Heinsohn, G. E., Lear, R. J., Bryden, M. M., Marsh, H. and Gardner, B. R. (1978). Discovery of a large population of dugongs off Brisbane, Australia. *Environmental Conservation*, **5**, 91–92.

Heinsohn, R., Lacy, R. C., Lindenmayer, D. B., *et al.* (2004). Unsustainable harvest of dugongs in Torres Strait and Cape York (Australia) waters: two case studies using population viability analysis. *Animal Conservation*, **7**, 417–425.

Heisler, J., Glibert, P. M., Burkholder, J. M., *et al.* (2008). Eutrophication and harmful algal blooms: a scientific consensus. *Harmful Algae*, **8**, 3–13.

Heithaus, M. R. (2001). The biology of tiger sharks, *Galeocerdo cuvier*, in Shark Bay, Western Australia: sex ratio, size distribution, diet, and seasonal changes in catch rates. *Environmental Biology of Fishes*, **61**, 25–36.

Heithaus, M. R., Wirsing, A. J., Dill, L. M. and Heithaus, L. I. (2007). Long-term movements of tiger sharks satellite-tagged in Shark Bay, Western Australia. *Marine Biology*, **151**, 1455–1461.

Heithaus, M. R., Wirsing, A. J., Burkholder, D., Thomson, J. and Dill, L. M. (2009). Towards a predictive framework for predator risk effects: the interaction of landscape features and prey escape tactics. *Journal of Animal Ecology*, **78**, 556–562.

Henzi, S. P., Lycett, J. E. and Piper, S. E. (1997). Fission and troop size in a mountain baboon population. *Animal Behaviour*, **53**, 525–535.

Hermanussen, S., Matthews, V., Päpke, O., Limpus, C. J. and Gaus, C. (2008). Flame retardants (PBDEs) contamination of marine turtles, dugongs and seafood from Queensland, Australia. *Marine Pollution Bulletin*, **57**, 409–418.

Hernandez, M., Robinson, I., Aguilar, A., *et al.* (1998). Did algal toxins cause monk seal mortality?. *Nature*, **393**, 28–29.

Hernandez, P., Reynolds III, J. E., Marsh, H. and Marmontel, M. (1995). Age and seasonality in spermatogenesis of Florida manatees. In *Population Biology of the Florida Manatee*, T. J. O'Shea, B. B. Ackerman and H. F. Percival (eds), Washington, DC: US Department of the Interior, National Biological Service, pp. 84–97.

Hess, L. L., Melack, J. M., Novo, E. M. L. M., Barbosa, C. C. F. and Gastil, M. (2003). Dual-season mapping of wetland inundation and vegetation for the central Amazon basin. *Remote Sensing of Environment*, **87**, 404–428.

Hildebrand, J. (2005). Impacts of anthropogenic sound. In *Marine Mammal Research: Conservation Beyond Crisis*, J. E. Reynolds III, W. F. Perrin, R. R. Reeves, S. Montgomery and T. J. Ragen (eds), Baltimore: Johns Hopkins University Press, pp. 101–123.

Hill, B. D., Fraser, I. R. and Prior, H. C. (1997). *Cryptosporidium* infection in a dugong (*Dugong dugon*). *Australian Veterinary Journal*, **75**, 670–671.

Hill, D. A. and Reynolds III, J. E. (1989). Gross and microscopic anatomy of the kidney of the West Indian manatee, *Trichechus manatus* (Mammalian: Sirenia). *Acta Anatomica*, **135**, 53–56.

Hines, E., Adulyanukosol, K. and Duffus, D. A. (2005a). Dugong abundance along the Andaman coast of Thailand. *Marine Mammal Science*, **21**, 536–549.

Hines, E., Adulyanukosol, K., Duffus, D. and Dearden, P. (2005b). Community perspectives and conservation needs for dugongs (*Dugong dugon*) along the Andaman Coast of Thailand. *Environmental Management*, **36**, 654–664.

Hines, E., Adulyanukosol, K., Somany, P., Sam Ath, L. and Cox, N. (2008). Conservation needs of the dugong *Dugong dugon* in Cambodia and Phu Quoc Island, Vietnam. *Oryx*, **42**, 113–121.

Hines, E., Adulyanukosol, K., Poochaviranon, S., *et al.* (2012a). Dugongs in Asia. In *Sirenian Conservation: Issues and Strategies in Developing Countries*, E. Hines, J. Reynolds, A. Mignucci-Giannoni, L. Aragones and M. Marmontel (eds), Gainesville: University Press of Florida.

Hines, E., Reynolds, J., Mignucci-Giannoni, A., Aragones, L. and Marmontel, M. (eds). (2012b). *Sirenian Conservation: Issues and Strategies in Developing Countries*, Gainesville: University Press of Florida.

Hines, E., Domning, D., Aragones, L., *et al.* (2012c). The role of scientists in sirenian conservation in developing countries. In *Sirenian Conservation: Issues and Strategies in Developing Countries*, E. Hines, J. Reynolds, A. Mignucci-Giannoni, L. Aragones and M. Marmontel (eds), Gainesville: University Press of Florida.

Hirasaka, K. (1932). The occurrence of the dugong in Formosa. *Memoirs of the Faculty of Science and Agriculture, Taihoku Imperial University, Formosa Japan*, **8**, 1–3.

Hobbs, J.-P. A., Frisch, A. J., Hender, J. and Gilligan, J. J. (2007). Long-distance oceanic movement of a solitary dugong (*Dugong dugon*) to the Cocos (Keeling) Islands. *Aquatic Mammals*, **33**, 175–178.

Hodgson, A. J. (2004). Dugong behaviour and responses to human influences. PhD Thesis, James Cook University, Townsville.

Hodgson, A. J. and Marsh, H. (2007). Response of dugongs to boat traffic: the risk of disturbance and displacement. *Journal of Experimental Marine Biology and Ecology*, **340**, 50–61.

Hodgson, A. J., Marsh, H., Delean, S. and Marcus, L. (2007). Is attempting to change marine mammal behaviour a generic solution to the bycatch problem? A dugong case study. *Animal Conservation*, **10**, 263–273.

Hodgson, A. J., Marsh, H., Gales, N., Holley, D. K. and Lawler, I. (2008). Dugong population trends across two decades in Shark Bay, Ningaloo Reef and Exmouth Gulf. Denham, Western Australia: WA Department of Environment and Conservation.

Hoegh-Guldberg, O. and Bruno, J. F. (2010). The impact of climate change on the world's marine ecosystems. *Science*, **328**, 1523–1528.

Hoffmann, J. P. (2004). Social and environmental influences on endangered species: a cross-national study. *Sociological Perspectives*, **47**, 79–107.

Holley, D. K. (2006). Movement patterns and habitat usage of Shark Bay dugongs. MS Thesis, Edith Cowan University, Joondalup, Western Australia.

Holley, D. K., Lawler, I. R. and Gales, N. J. (2006). Summer survey of dugong distribution and abundance in Shark Bay reveals additional key habitat area. *Wildlife Research*, **33**, 243–250.

Holloway, M. (2010). Wounds that can heal. *Conservation Magazine*, **11**, 28–32.

Home, R., Keller, C., Nagel, P., Bauer, N. and Hunziker, M. (2009). Selection criteria for flagship species by conservation organizations. *Environmental Conservation*, **36**, 139–148.

Hooker, S. K. and Gerber, L. R. (2004). Marine reserves as a tool for ecosystem-based management: the potential importance of megafauna. *Bioscience*, **54**, 27–39.

Hudson, B. E. T. (1977). *Dugong: Distribution, Hunting, Protective Legislation and Cultural Significance in Papua New Guinea*, Konedobu: Wildlife Division, Department of Lands and Environment.

Hudson, B. E. T. (1981). The dugong conservation, management and public education programme in Papua New Guinea: working with people to conserve their dugong resources. In *The Dugong: Proceedings of a Seminar/Workshop held at James Cook University 8–13 May 1979, Townsville*, H. Marsh (ed.), Townsville: James Cook University, pp. 70–79.

Hudson, B. E. T. (1986). The hunting of dugong in Daru, Papua New Guinea, during 1978–82: community management and education initiatives. In *Torres Strait Fisheries Seminar, Port Moresby, 11–14 February 1985*, A. K. Haines, G. C. Williams and D. Coates (eds), Canberra: Australian Government Printing Service, pp. 77–94.

Huggett, A. St G. and Widdas, W. F. (1951). The relationship between mammalian foetal weight and conception age. *Journal of Physiology*, **114**, 306–317.

Human Development Report. (2009). Human Poverty Index (HPI-1) value (%). Available at: http://hdrstats.undp.org/en/indicators/97.html (accessed 10 October 2010).

Hunter, M., Broderick, D., Ovenden, J. R., *et al.* (2010a). Characterization of highly informative cross-species microsatellite panels for the Australian dugong (*Dugong dugon*) and Florida manatee (*Trichechus manatus latirostris*) including five novel primers. *Molecular Ecology Resources*, **10**, 368–377.

Hunter, M., Auil-Gomez, N. E., Tucker, K. P., *et al.* (2010b). Low genetic variation and evidence of limited dispersal in the regionally important Belize manatee. *Animal Conservation*, **13**, 592–602.

Hunter, M., King, T. L., Tucker, K. P., *et al.* (in press). Comprehensive genetic investigation recognizes evolutionary divergence in the Florida (*Trichechus manatus latirostris*) and Puerto Rico (*T. m. manatus*) manatee populations and subtle substructure in Puerto Rico. *Molecular Ecology*.

Husar, S. L. (1977). *Trichechus inunguis*: mammalian species. *The American Society of Mammalogists*, **72**, 1–4.

Husar, S. L. (1978). *Dugong dugon*: mammalian species. *The American Society of Mammalogists*, **88**, 1–7.

Ichikawa, K., Tsutsumi, C., Arai, N., *et al.* (2006). Dugong (*Dugong dugon*) vocalization patterns recorded by automatic underwater sound monitoring systems. *Journal of the Acoustical Society of America*, **119**, 3726–3733.

Ichikawa, K., Akamatsu, T., Shinke, T., *et al.* (2009). Detection probability of vocalizing dugongs during playback of conspecific calls. *Journal of the Acoustical Society of America*, **126**, 1954–1959.

Ikeda, K. and Mukai, H. (2012). Dugongs in Japan. In *Sirenian Conservation: Issues and Strategies in Developing Countries*, E. Hines, J. Reynolds, A. Mignucci-Giannoni, L. Aragones and M. Marmontel (eds), Gainesville: University Press of Florida.

Ilangakoon, A. and Tun, T. (2007). Rediscovering the dugong (*Dugong dugon*) in Myanmar and capacity building for research and conservation. *Raffles Bulletin of Zoology*, **55**, 195–199.

Ilangakoon, A. D., Sutaria, D., Hines, E. and Raghavan, R. (2008). Community interviews on the status of the dugong (*Dugong dugon*) in the Gulf of Mannar (India and Sri Lanka). *Marine Mammal Science*, **24**, 704–710.

Illiger, C. (1811). *Prodromus systematis mammalium et avium additis terminis zoographicus utriusque classis, erumque versione Germanica*, Berlin: C. Salfeld.

Irvine, A. B. (1983). Manatee metabolism and its influence on distribution in Florida. *Biological Conservation*, **25**, 315–334.

Irvine, A. B. and Campbell, H. W. (1978). Aerial census of the West Indian manatee, *Trichechus manatus*, in the southeastern United States. *Journal of Mammalogy*, **59**, 613–617.

Irvine, A. B., Caffin, J. E. and Kochman, H. I. (1982). Aerial surveys for manatees and dolphins in western peninsular Florida. *Fishery Bulletin*, **80**, 621–630.

Issa, A. M. (2008). Niger. In *Conservation Strategy for the West African Manatee*, T. Dodman, M. D. D. Ndiaye and K. Sarr (eds), Nairobi: UNEP and Dakar: Wetlands International Africa, pp. 82–86.

Iturralde-Vinent, M. A. (2003). A brief account of the evolution of the Caribbean Seaway: Jurassic to present. In *From Greenhouse to Icehouse: The Marine Eocene–Oligocene Transition*, D. R. Prothero, L. C. Ivany and E. A. Nesbitt (eds), New York: Columbia University Press, pp. 386–396.

IUCN. (1986). *1986 IUCN Red List of Threatened Animals*, Gland, Switzerland and Cambridge: IUCN Conservation Monitoring Centre.

IUCN. (1994). *Guidelines for Protected Area Management Categories*, Gland, Switzerland and Cambridge: IUCN.

IUCN. (2001). *IUCN Red List Categories and Criteria: Version 3.1*, Gland, Switzerland and Cambridge: IUCN.

IUCN. (2003). *Guidelines for Application of IUCN Red List Criteria at Regional Levels: Version 3.0*, Gland, Switzerland and Cambridge: IUCN.

IUCN. (2005). Conservation of dugong (Dugong dugon), Okinawa woodpecker (Sapheopipo noguchii), and Okinawa rail (Gallirallus okinawae) in Japan. RECo32 in resolutions and recommendations, 3rd IUCN World Conservation Congress 17–25 November 2004, Bangkok, Thailand.

IUCN. (2009). 2009 IUCN Red List of threatened species: version 2009.2. Available at: http://www.iucnredlist.org (accessed 26 November 2009).

Ivany, L. C., Nesbitt, E. A. and Prothero, D. R. (2003). The marine Eocene–Oligocene transition: a synthesis. In *From Greenhouse to Icehouse: The Marine Eocene–Oligocene Transition*, D. R. Prothero, L. C. Ivany and E. A. Nesbitt (eds), New York: Columbia University Press, pp. 522–534.

Jaaman, S. A., Lah-Anyi, Y. U. and Pierce, G. J. (2009). The magnitude and sustainability of marine mammal by-catch in fisheries in East Malaysia. *Journal of the Marine Biological Association of the UK*, **89**, 907–920.

Jackson, J. B. C., Budd, A. F. and Coates, A. G. (eds). (1996). *Evolution and Environment in Tropical America*, Chicago: University of Chicago Press.

Jackson, J. B. C., Kirby, M. X., Berger, W. H., *et al*. (2001). Historical overfishing and the recent collapse of coastal ecosystems. *Science*, **293**, 629–638.

Jacobs, S. W. L., Les, D. H. and Moody, M. L. (2006). New combinations in Australasian *Zostera* (Zosteraceae). *Telopea*, **11**, 127–128.

Jagtap, T. G., Komarpant, D. S. and Rodrigues, R. (2003). The seagrasses of India. In *World Atlas of Seagrasses*, E. P. Green and F. F. Short (eds), Berkeley: University of California Press, pp. 101–108.

Janis, C. (2008). An evolutionary history of browsing and grazing ungulates. In *The Ecology of Browsing and Grazing*, Vol. 195, I. J. Gordon and H. H. T. Prins (eds), Berlin: Springer, pp. 21–45.

Jarman, P. J. (1966). The status of the dugong (*Dugong dugon* Müller) Kenya, 1961. *East African Wildlife Journal*, 4, 82–88.

Jefferson, T., Webber, M. and Pitman, R. (2008). *Marine Mammals of the World: A Comprehensive Guide to their Identification*, London: Academic Press.

Jickells, T. D., An, Z. S., Andersen, K. K., *et al.* (2005). Global iron connections between desert dust, ocean biogeochemistry, and climate. *Science*, 308, 67–71.

Jiménez, I. (2005). Development of predictive models to explain the distribution of the West Indian manatee *Trichechus manatus* in tropical watercourses. *Biological Conservation*, 125, 491–503.

Johannes, R. E. and MacFarlane, W. (1991). *Traditional Fishing in the Torres Strait*, Hobart: CSIRO Division of Fisheries.

Johnstone, I. M. and Hudson, B. E. T. (1981). The dugong diet: mouth sample analysis. *Bulletin of Marine Science*, 31, 681–690.

Jolly, G. M. (1965). Explicit estimates from capture–recapture data with both death and immigration: stochastic model. *Biometrika*, 52, 225–247.

Jones, A. (2005). Family Paramphistomidae Fischoeder, 1901. In *Keys to the Trematoda*, Vol. 2, A. Jones, R. A. Bray and D. I. Gibson (eds), London: CABI and the Natural History Museum, pp. 229–246.

Jones, C. (1978). *Dendrohyrax dorsalis. Mammalian Species*, 113, 1–4.

Jones, M.-A., Stauber, J., Apte, S., *et al.* (2005). A risk assessment approach to contaminants in Port Curtis, Queensland, Australia. *Marine Pollution Bulletin*, 51, 448–458.

Jones, S. (1981). Distribution and status of the dugong, *Dugong dugon* (Müller), in the Indian Region. In *The Dugong: Proceedings of a Seminar/Workshop held at James Cook University 8–13 May 1979, Townsville*, H. Marsh (ed.), Townsville: James Cook University, pp. 24–30.

Jonklaas, R. (1961). Some observations on dugongs (*Dugong dugon* – Erxleben). *Loris*, 9, 1–8.

Jordan, D. S., Stejneger, L., Lucas, F. A., *et al.* (1899). *The Fur Seals and Fur-Seal Islands of the North Pacific Ocean, Part 3*, Washington, DC: US Government Printing Office.

Junk, W. J. (1997). Structure and function of the large Central Amazonian river floodplains: synthesis and discussion. In *The Central Amazon Floodplain: Ecology of a Pulsing System*, W. J. Junk (ed.), Berlin: Springer-Verlag, pp. 455–472.

Junk, W. J. and Piedade, M. T. F. (1997). Plant life in the floodplain with special references to herbaceous plants. In *The Central Amazon Floodplain: Ecology of a Pulsing System*, W. J. Junk (ed.), Berlin: Springer-Verlag, pp. 147–185.

Kaiser, H. (1974). *Morphology of the Sirenia: A Macroscopic and X-ray Atlas of the Osteology of Recent Species*, Basel: S. Karger.

Kamiya, T. and Yamasaki, F. (1981). A morphological note on the sinus hair of the dugong. In *The Dugong: Proceedings of a Seminar/Workshop held at James Cook University 8–13 May 1979, Townsville*, H. Marsh (ed.), Townsville: James Cook University.

Kan-Atireklap, S., Tanabe, S. and Sanguansin, J. (1997). Contamination by butyltin compounds in sediments from Thailand. *Marine Pollution Bulletin*, 34, 894–899.

Keith, L. C. and Collins, T. (2007). West African manatee (*Trichechus senegalensis*) 2006 survey activities in Gabon. *SireNews*, 47, 10.

Keller, M., Moliner, J. L., Vasquez, G., *et al.* (2008). Nephrolithiasis and pyelo-nephritis in two West Indian manatees (*Trichechus manatus* spp.). *Journal of Wildlife Diseases*, **44**, 707–711.

Kellogg, E. (2001). The evolutionary history of grasses. *Plant Physiology*, **125**, 1198–1205.

Kellogg, M. E., Burkett, S., Dennis, T. R., *et al.* (2007). Chromosome painting in the manatee supports Afrotheria and Paenungulata. *BMC Evolutionary Biology*, **7**, 6.

Kendall, S. (2001). Distribution and conservation of the Amazonian manatee (*Trichechus inunguis*) in the area of Puerto Narino, Colombia. Final Report to Fauna and Flora International, Project Reference 99/54/4.

Kendall, S. and Orozco, D. L. (2003). El árbol de los manatíes: concertación y conservación en la Amazonia colombiana. In *Fauna Socializada- tendencias en el manejo participativo de la fauna en America Latina*, C. Campos-Rozo and A. Ulloa (eds), Columbia: Fundación Natura, pp. 215–237.

Kendall, W. L., Hines, J. E. and Nichols, J. D. (2003). Adjusting multistate capture–recapture models for misclassification bias: manatee breeding proportions. *Ecology*, **84**, 1058–1066.

Kendall, W. L., Langtimm, C. A., Beck, C. A. and Runge, M. C. (2004). Capture–recapture analysis for estimating manatee reproductive rates. *Marine Mammal Science*, **20**, 424–437.

Kennedy, S. (1998). Morbillivirus infections in marine mammals. *Journal of Comparative Pathology*, **119**, 201–225.

Kenworthy, W. J., Durako, M. J., Fatemy, S. M. R., Valavi, H. and Thayer, G. W. (1993). Ecology of seagrasses in northeastern Saudi Arabia one year after the Gulf War oil spill. *Marine Pollution Bulletin*, **27**, 213–222.

Kenyon, K. (1969). The sea otter in the eastern Pacific Ocean. *North American Fauna*, **68**, 1–352.

Kenyon, K. W. (1977). Caribbean monk seal extinct. *Journal of Mammalogy*, **58**, 97–98.

Ketten, D. R., Odell, D. K. and Domning, D. P. (1992). Structure, function, and adaptation of the manatee ear. In *Marine Mammal Sensory Systems*, J. A. Thomas, R. A. Kastelein and A. Y. Supin (eds), New York: Plenum Press, pp. 77–95.

Keys, J. E. Jr, Van Soest, P. J. and Young, E. P. (1969). Comparative study of the digestibility of forage cellulose and hemicellulose in ruminants and nonrumi-nants. *Journal of Animal Science*, **29**, 11–15.

Kimura, S. (1957). The twinning in southern fin whales. *Scientific Report of the Whales Research Institute, Tokyo*, **12**, 103–125.

Kinch, J. (2008). An assessment of dugong (Dugong dugon) resources in the autonomous region of Bouganville, Papua New Guinea. Unpublished report prepared for Conservation International, Melanesian Centre for Biodiversity Conservation, Atherton, Queensland Australia.

King, J. M. and Heinen, J. T. (2004). An assessment of the behaviors of overwin-tering manatees as influenced by interactions with tourists at two sites in central Florida. *Biological Conservation*, **117**, 227–234.

King, T. F. (2006). Creatures and culture: some implications of dugong v. Rumsfeld. *International Journal of Cultural Property*, **13**, 235–240.

Kingdon, J. (1971). Dugong. In *East African Mammals: An Atlas of Evolution in Africa*, London: Academic Press.

Kinnaird, M. F. (1985). Aerial census of manatees in northeastern Florida. *Biological Conservation*, **32**, 59–79.

Kirkpatrick, B., Fleming, L. E., Squicciarini, D., *et al.* (2004). Literature review of Florida red tide: implications for human health effects. *Harmful Algae*, **3**, 99–115.

Kiss, A. (2004). Making biodiversity a land-use priority. In *Getting Biodiversity Projects to Work*, T. O. McShane and M. P. Wells (eds), New York: Columbia University Press, pp. 98–123.

Klas, M. (2010a). After months of controversy, PSC to decide FPL's rate request. *St Petersburg Times*.

Klas, M. (2010b). FPL faces key vote on rate increase of up to 30 percent. *The Palm Beach Post*.

Kleinschmidt, T., Czelusniak, J., Goodman, M. and Braunitzer, G. (1986). Paenungulata: a comparison of the hemoglobin sequences from elephant, hyrax, and manatee. *Molecular Biology and Evolution*, **3**, 427–435.

Klevezal, G. A. (1980). Layers in the hard tissues of mammals as a record of growth rhythms of individuals. *Reports of the International Whaling Commission*, special issue, **3**, pp. 89–94.

Klishin, V. O., Diaz, R. P., Popov, V. V. and Supin, A. Y. (1990). Some characteristics of hearing of the Brazilian manatee, *Trichechus inunguis*. *Aquatic Mammals*, **16**, 139–144.

Kobayashi, S., Horikawa, H. and Miyazaki, S. (1995). A new species of Sirenia (Mammalia: Hydrodamalinae) from the Shiotsubo Formation in Takasato, Aizu, Fukushima Prefecture, Japan. *Journal of Vertebrate Paleontology*, **15**, 815–829.

Kochman, H. I., Rathbun, G. B. and Powell, J. A. (1985). Temporal and spatial distribution of manatees in Kings Bay, Crystal River, Florida. *Journal of Wildlife Management*, **49**, 921–924.

Koelsch, J. K. (2001). Reproduction in female manatees observed in Sarasota Bay, Florida. *Marine Mammal Science*, **17**, 331–342.

Kojeszewski, T. and Fish, F. E. (2007). Swimming kinematics of the Florida manatee (*Trichechus manatus latirostris*): hydrodynamic analysis of an undulatory mammalian swimmer. *Journal of Experimental Biology*, **210**, 2411–2418.

Kouadio, A. (2012). The West African manatee. In *Sirenian Conservation: Issues and Strategies in Developing Countries*, E. Hines, J. Reynolds, A. Mignucci-Giannoni, L. Aragones and M. Marmontel (eds), Gainesville: University Press of Florida.

Kovacs, K. (2008). *Monachus tropicalis*. IUCN 2010 Red List of Threatened Species. Available at: http://www.iucnredlist.org (accessed 25 September 2010).

Kuriandewa, T. E., Kiswara, W., Hutomo, M. and Soemodihardjo, S. (2003). The seagrasses of Indonesia. In *World Atlas of Seagrasses*, E. P. Green and F. F. Short (eds), Berkeley: University of California Press, pp. 171–182.

Kwan, D. (2002). Towards a sustainable Indigenous fishery for dugongs in Torres Strait: a contribution of empirical data and process. PhD Thesis, James Cook University, Townsville.

Lack, M. and Sant, G. (2008). *Illegal, Unreported, and Unregulated Shark Catch: A Review of Current Knowledge and Action*, Department of the Environment, Water, Heritage and the Arts, Australia and TRAFFIC.

Lainson, R., Naiff, R. D., Best, R. C. and Shaw, J. J. (1983). *Eimeria trichechi* n. sp. from the Amazonian manatee, *Trichechus inunguis* (Mammalia: Sirenia). *Systematic Parasitology*, **5**, 287–289.

Laist, D. W. and Reynolds III, J. E. (2005a). Influence of power plants and other warm-water refuges on Florida manatees. *Marine Mammal Science*, **21**, 739–764.

Laist, D. W. and Reynolds III, J. E. (2005b). Florida manatees, warm-water refuges, and an uncertain future. *Coastal Management*, **33**, 279–295.

Laist, D. W. and Shaw, C. (2006). Preliminary evidence that boat speed restrictions reduce deaths of Florida manatees. *Marine Mammal Science*, **22**, 472–479.

Laist, D. W., Knowlton, A. R., Mead, J. G., Collet, A. S. and Podesta, M. (2001). Collisions between ships and whales. *Marine Mammal Science*, **17**, 35–75.

Landsberg, J. H., Flewelling, L. J. and Naar, J. (2009). *Karenia brevis* red tides, brevetoxins in the food web, and impacts on natural resources: decadal advancements. *Harmful Algae*, **8**, 598–607.

Langeland, K. A. (1996). *Hydrilla verticillata* (L.F.) Royle (Hydrocharitaceae), 'the perfect aquatic weed'. *Castanea*, **61**, 293–304.

Langer, P. (1984). Comparative anatomy of the stomach in mammalian herbivores. *Quarterly Journal of Experimental Physiology*, **69**, 615–625.

Langer, P. (2003). Lactation, weaning period, food quality, and digestive tract differentiations in eutheria. *Evolution*, **57**, 1196–1215.

Langtimm, C. A. (2009). Non-random temporary emigration and the robust design: conditions for bias at the end of a time series. In *Modeling Demographic Processes in Marked Populations*, Vol. 3, D. L. Thomson, E. G. Cooch and M. J. Conroy (eds), New York: Springer, pp. 745–761.

Langtimm, C. A. and Beck, C. A. (2003). Lower survival probabilities for adult Florida manatees in years with intense coastal storms. *Ecological Applications*, **13**, 257–268.

Langtimm, C. A., O'Shea, T. J., Pradel, R. and Beck, C. A. (1998). Estimates of annual survival probabilities for adult Florida manatees (*Trichechus manatus latirostris*). *Ecology*, **79**, 981–997.

Langtimm, C. A., Beck, C. A., Edwards, H. H., *et al.* (2004). Survival estimates for Florida manatees from the photo-identification of individuals. *Marine Mammal Science*, **20**, 438–463.

Langtimm, C. A., Krohn, M. D., Reid, J. P., Stith, B. M. and Beck, C. A. (2006). Possible effects of the 2004 and 2005 hurricanes on manatee survival rates and movement. *Estuaries and Coasts*, **29**, 1026–1032.

Lanyon, J. (1991). The nutritional ecology of the dugong (Dugong dugon) in tropical North Queensland. PhD Thesis, Monash University, Melbourne.

Lanyon, J. M. (2003). Distribution and abundance of dugongs in Moreton Bay, Queensland, Australia. *Wildlife Research*, **30**, 397–409.

Lanyon, J. M. and Marsh, H. (1995a). Temporal changes in the abundance of some tropical intertidal seagrasses in North Queensland. *Aquatic Botany*, **49**, 217–237.

Lanyon, J. M. and Marsh, H. (1995b). Digesta passage times in the dugong. *Australian Journal of Zoology*, **43**, 119–127.

Lanyon, J. M. and Sanson, G. D. (2006a). Mechanical disruption of seagrass in the digestive tract of the dugong. *Journal of Zoology*, **270**, 277–289.

Lanyon, J. M. and Sanson, G. D. (2006b). Degenerate dentition of the dugong (*Dugong dugon*), or why a grazer does not need teeth: morphology, occlusion and wear of mouthparts. *Journal of Zoology*, **268**, 133–152.

Lanyon, J. M., Sneath, H. L., Kirkwood, J. M. and Slade, R. W. (2002). Establishing a mark–recapture program for dugongs in Moreton Bay, south-east Queensland. *Australian Mammalogy*, **24**, 51–56.

Lanyon, J. M., Johns, T. and Sneath, H. L. (2005a). Year-round presence of dugongs in Pumicestone Passage, southeast Queensland, examined in relation to water temperature and seagrass distribution. *Wildlife Research*, **32**, 361–368.

Lanyon, J. M., Smith, K. M. and Carrick, F. N. (2005b). Reproductive steroids are detectable in the faeces of dugongs. *Australian Zoologist*, **33**, 247–250.

Lanyon, J.M., Newgrain, K. and Sahir Syah Alli, T. (2006a). Estimation of water turnover rate in captive dugongs (*Dugong dugon*). *Aquatic Mammals*, **32**, 103–108.

Lanyon, J. M., Slade, R. W., Sneath, H. L. *et al.* (2006b). A method for capturing dugongs (*Dugong dugon*) in open water. *Aquatic Mammals*, **32**, 196–201.

Lanyon, J. M., Sneath, H. L., Ovenden, J. R., Broderick, D. and Bonde, R. K. (2009). Sexing sirenians: validation of visual and molecular sex determination in both wild dugongs (*Dugong dugon*) and Florida manatees (*Trichechus manatus latirostris*). *Aquatic Mammals*, **35**, 187–192.

Lanyon, J. M., Sneath, H. L., Long, T. and Bonde, R. K. (2010). Physiological response of wild dugongs (*Dugong dugon*) to out-of-water sampling for health assessment. *Aquatic Mammals*, **36**, 46–58.

Larkin, I. L. V. (2000). Reproductive endocrinology of the Florida manatee (*Trichechus manatus latirostris*): estrous cycles, seasonal patterns and behavior. PhD Thesis, University of Florida, Gainesville.

Larkin, I. L. V., Gross, T. S. and Reep, R. L. (2005). Use of faecal testosterone concentrations to monitor male Florida manatee (*Trichechus manatus latirostris*) reproductive status. *Aquatic Mammals*, **31**, 52–61.

Larkin, I. L. V., Fowler, V. F. and Reep, R. L. (2007). Digesta passage rates in the Florida manatee (*Trichechus manatus latirostris*). *Zoo Biology*, **26**, 503–515.

Larkum, A. W. D. and den Hartog, C. (1989). Evolution and biogeography of seagrasses. In *Biology of Seagrasses: A Treatise on the Biology of Seagrasses with Special Reference to the Australian Region*, A. W. D. Larkum, A. J. McComb and S. A. Sheppard (eds), Amsterdam and New York: Elsevier, pp. 112–156.

Lau, K.-M. and Wu, H.-T. (2007). Detecting trends in tropical rainfall characteristics, 1979–2003. *International Journal of Climatology*, **27**, 979–988.

Laurance, W. F. (2004). The perils of payoff: corruption as a threat to global biodiversity. *Trends in Ecology and Evolution*, **19**, 399–401.

Lavigne, D. M., Cox, R. K., Menon, V. and Wamithi, M. (2006). Reinventing wildlife conservation for the 21st century. In *Gaining Ground: In Pursuit of Ecological Sustainability*, D. M. Lavigne (ed.), Guelph, Canada and Limerick: International Fund for Animal Welfare and University of Limerick, pp. 379–406.

Lawler, I., Aragones, L., Berding, N., Marsh, H. and Foley, W. (2006). Near-infrared reflectance spectroscopy as a rapid, cost-effective predictor of seagrass nutrients. *Journal of Chemical Ecology*, **32**, 1353–1365.

Lawler, I. R., Parra, G. and Noad, M. (2007). Vulnerability of marine mammals in the Great Barrier Reef to climate change. In *Climate Change and the Great Barrier Reef: A Vulnerability Assessment*, J. E. Johnson and P. A. Marshall (eds), Australia: Great Barrier Reef Marine Park Authority and Australian Greenhouse Office, pp. 498–513.

Laws, R. M. (1959). Foetal growth rates of whales with special reference to the fin whale, *Balaenoptera physalus* (L.). *Discovery Reports*, **29**, 281–308.

Lebreton, J.-D., Burnham, K. P., Clobert, J. and Anderson, D. R. (1992). Modeling survival and testing biological hypotheses using marked animals: a unified approach with case studies. *Ecological Monographs*, **62**, 67–118.

Lécuyer, C., Grandjean, P., Paris, F., Robardet, M. and Robineau, D. (1996). Deciphering 'temperature' and ' salinity' from biogenic phosphates: the $\delta^{18}O$ of coexisting fishes and mammals of the Middle Miocene sea of western France. *Palaeogeography, Palaeoclimatology, Palaeoecology*, **126**, 61–74.

Ledder, D. A. (1986). Food habits of the West Indian manatee, *Trichechus manatus latirostris*, in South Florida. MS Thesis, University of Miami, Miami.

Lee Long, W. J., Coles, R. G. and McKenzie, L. J. (1996). Deepwater seagrasses in northeastern Australia: how deep, how meaningful?. In *Seagrass Biology: Proceedings of an International Workshop, Rottnest Island, Western Australia, 25–29 January 1996*, J. Kuo, R. C. Phillips, D. I. Walker and H. Kirkman (eds), Perth: University of Western Australia, pp. 41–50.

Lefebvre, L. W. and Powell, J. A. (1990). Manatee grazing impacts on seagrasses in Hobe Sound and Jupiter Sound in southeast Florida during the winter of 1988–89, Gainesville, Florida.

Lefebvre, L. W., Domning, D. P., Kenworthy, W. J., McIvor, D. E. and Ludlow, M. E. (1988). Appendix B: manatee grazing impacts on seagrasses in Hobe Sound and Jupiter Sound January–February 1988. In *The Submarine Light Regime and Ecological Studies of Seagrasses in Hobe Sound, Florida. 1988*, W. J. Kenworthy, M. S. Fonesca, D. E. McIvor and G. W. Thayer (eds), pp. 1–18.

Lefebvre, L. W., O'Shea, T. J., Rathbun, G. B. and Best, R. C. (1989). Distribution, status, and biogeography of the West Indian manatee. In *Biogeography of the West Indies: Past, Present, and Future*, C. A. Woods (ed.), Gainesville: Sandhill Crane Press, pp. 567–610.

Lefebvre, L. W., Ackerman, B. B., Portier, K. M. and Pollock, K. H. (1995). Aerial survey as a technique for estimating trends in manatee population size: problems and prospects. In *Population Biology of the Florida Manatee*, T. J. O'Shea, B. B. Ackerman and H. F. Percival (eds), Washington, DC: US Department of the Interior, National Biological Service, pp. 63–74.

Lefebvre, L. W., Reid, J. P., Kenworthy, W. J. and Powell, J. A. (2000). Characterizing manatee habitat use and seagrass grazing in Florida and Puerto Rico: implications for conservation and management. *Pacific Conservation Biology*, **5**, 289–298.

Lefebvre, L. W., Marmontel, M., Reid, J. P., Rathbun, G. B. and Domning, D. P. (2001). Status and biogeography of the West Indian manatee. In *Biogeography of the West Indies: Patterns and Perspectives*, C. A. Woods and F. E. Sergile (eds), Boca Raton: CRC Press, pp. 425–474.

Leistra, W. H. G., Hoyer, M. J., Kik, M. and Sinke, J. D. (2003). Recidiverende huidproblemen bij een zeekoe: de diagnostische aanpak. *Tijdschrift Voor Diergeneeskunde*, **128**, 140–144.

Lenes, J. M., Darrow, B. A., Walsh, J. J., *et al.* (2008). Saharan dust and phosphatic fidelity: a three-dimensional biogeochemical model of *Trichodesmium* as a nutrient source for red tides on the West Florida Shelf. *Continental Shelf Research*, **28**, 1091–1115.

Leopold, A. (1949). *A Sand County Almanac and Sketches Here and There*. New York: Oxford University Press.

Leverington, F., Hockings, M., Pavese, H., Costa, K. L. and Courrau, J. (2008). *Management Effectiveness Evaluation in Protected Areas: A Global Study. Overview of Approaches and Methodologies*. Supplementary report No. 1, Australia: University of Queensland, Gatton, TNC, WWF, IUCN-WCPA.

Levin, M. J. and Pfeiffer, C. J. (2002). Gross and microscopic observations on the lingual structure of the Florida manatee *Trichechus manatus latirostris*. *Anatomia, Histologia, Embryologia*, **31**, 278–285.

Lewis III, R. R., Carlton, J. M. and Lombardo, R. (1984). Algal consumption by the manatee *Trichechus manatus latirostris* in Tampa Bay, Florida, USA. *Florida Scientist*, **47**, 189–191.

Lightsey, J. D., Rommel, S. A., Costidis, A. M. and Pitchford, T. D. (2006). Methods used during gross necropsy to determine watercraft-related mortality in the Florida manatee (*Trichechus manatus latirostris*). *Journal of Zoo and Wildlife Medicine*, **37**, 262–275.

Ligon, S. H. (1976). A survey of dugongs (*Dugong dugon*) in Queensland. *Journal of Mammalogy*, **57**, 580–582.

Lima, R. P., Luna, F. O., Castro, D. F. and Vianna, J. A. (2001). Levantamento da Distribuição, *Status* de Conservação e Campanhas Conservacionistas do Peixe-Boi Amazônico (*Trichechus inunguis*). Relatório Final do Convênio IBAMA/CMA, FNMA e FMA – IBAMA/CMA.

Limpus, C. J. and Nichols, N. (1988). The southern oscillation regulates the annual numbers of green turtles (*Chelonia mydas*) breeding around northern Australia. *Australian Wildlife Research*, **15**, 157–161.

Limpus, C. J., Currie, K. J., Haines, J. (2003). *Marine Wildlife Stranding and Mortality Database Annual Report 2002: I. Dugong*, Brisbane: Environmental Protection Agency and Queensland Parks and Wildlife Service.

Lindsay, D. S., Collins, M. V., Mitchell, S. M., *et al.* (2003). Sporulation and survival of *Toxoplasma gondii* oocysts in seawater. *Journal of Eukaryotic Microbiology*, **50**, 687–688.

Linnaeus, C. (1758). *Systema naturae per regna tria naturae, secundum classes, ordines, genera, species, cum characteribus, differentiis, synonymis, locis*, Stockholm: Laurentii Salvii.

Lipkin, Y. (1975). Food of Red-Sea dugong (Mammalia-Sirenia) from Sinai. *Israel Journal of Zoology*, **24**, 81–98.

Logan, M. and Sanson, G. D. (2002). The effect of tooth wear on the feeding behaviour of free-ranging koalas (*Phascolarctos cinereus*, Goldfuss). *Journal of Zoological Society of London*, **256**, 63–69.

Loh, J. and Harmon, D. (2005). A global index of biocultural diversity. *Ecological Indicators*, **5**, 231–241.

Lomolino, M. V. and Ewel, K. C. (1984). Digestive efficiencies of the West Indian manatee, *Trichechus manatus*. *Florida Scientist*, **47**, 176–179.

Long, B. G. and Poiner, I. R. (1997). Sea grass communities of Torres Strait, northern Australia. Marine Research CSIRO Report, MR-GIS 97/6.

Longstaff, B. J. and Dennison, W. C. (1999). Seagrass survival during pulsed turbidity events: the effects of light deprivation on the seagrasses *Halodule pinifolia* and *Halophila ovalis*. *Aquatic Botany*, **65**, 105–121.

Lorimer, J. (2006). Nonhuman charisma: which species trigger our emotions and why?. *ECOS* **27**, 20–27.

Lowenstein, J. M. and Scheuenstuhl, G. (1991). Immunological methods in molecular palaeontology. *Philosophical Transactions of the Royal Society B: Biological Sciences*, **333**, 375–380.

MacArthur, R. H. and Pianka, E. R. (1966). On the optimal use of a patchy habitat. *American Naturalist*, **100**, 603–609.

Mace, G. M. and Lande, R. (1991). Assessing extinction threats: toward a re-evaluation of IUCN threatened species categories. *Conservation Biology*, **5**, 148–157.

Mace, G. M., Collar, N., Cooke, J., *et al.* (1992). The development of new criteria for listing species on the IUCN Red List. *Species*, **19**, 16–22.

MacFadden, B. J., Higgins, P., Clementz, M. T. and Jones, D. S. (2004). Diets, habitat preferences, and niche differentiation of Cenozoic sirenians from Florida: evidence from stable isotopes. *Paleobiology*, **30**, 297–324.

Mackay-Sim, A., Duvall, D. and Graves, B. M. (1985). The West Indian manatee (*Trichechus manatus*) lacks a vomeronasal organ. *Brain, Behavior and Evolution*, **27**, 186–194.

MacMillan, L. (1955). The dugong. *Walkabout*, **17**, 17–20.

Magaña, H. A., Contreras, C. and Villareal, T. A. (2003). A historical assessment of *Karenia brevis* in the western Gulf of Mexico. *Harmful Algae*, **2**, 163–171.

Mahmud, S. (2010). The status of dugongs in Eritrea. Unpublished report to Eritrean Ministry of Marine Resources.

Malhi, Y., Roberts, J. T., Betts, R. A., *et al.* (2008). Climate change, deforestation, and the fate of the Amazon. *Science*, **319**, 169–172.

Malmqvist, B. and Rundle, S. (2002). Threats to the running water ecosystems of the world. *Environmental Conservation*, **29**, 134–153.

Maluf, N. S. R. (1989). Renal anatomy of the manatee, *Trichechus manatus*, Linneaus. *American Journal of Anatomy*, **184**, 269–286.

Mandel, J. T., Donlan, C. J., Wilcox, C., *et al.* (2009). Debt investment as a tool for value transfer in biodiversity conservation. *Conservation Letters*, **2**, 233–239.

Mani, S. B. (1960). Occurrence of the sea cow, *Halicore dugong* (Erxl.) off the Saurashtra Coast. *Journal of the Bombay Natural History Society*, **57**, 216–217.

Mann, D., Colbert, D. E., Gaspard, J. C., *et al.* (2005). Temporal resolution of the Florida manatee (*Trichechus manatus latirostris*) auditory system. *Journal of Comparative Physiology A: Neuroethology, Sensory, Neural, and Behavioral Physiology*, **191**, 903–908.

Mann, D., O'Shea, T. J. and Nowacek, D. P. (2006). Nonlinear dynamics in manatee vocalizations. *Marine Mammal Science*, **22**, 548–555.

Mann, D., Bauer, G., Colbert, D., Gaspard, J. and Reep, R. L. (2007). *Sound Localization Abilities of the West Indian Manatee*, St Petersburg: Fish and Wildlife Conservation Commission.

Marengo, J. A., Nobre, C. A., Tomasella, J., *et al.* (2008). The drought of Amazonia in 2005. *Journal of Climate*, **21**, 495–516.

Marine Mammal Commission. (2007). *The Biological Viability of the most Endangered Marine Mammals and the Cost-effectiveness of Protection Programs: A Report to Congress by the Marine Mammal Commission*, Maryland: US Marine Mammal Commission. Available at: www.mmc.gov.

Marine Mammal Commission. (2009). *Annual Report to Congress 2008*, Bethesda: US Marine Mammal Commission. Available at: www.mmc.gov.

Marmontel, M. (1993). Age determination and population biology of the Florida manatee *Trichechus manatus latirostris*. PhD Dissertation, University of Florida, Gainesville.

Marmontel, M. (1995). Age and reproduction in female Florida manatees. In *Population Biology of the Florida Manatee*, T. J. O'Shea, B. B. Ackerman and H. F. Percival (eds), Washington, DC: US Department of the Interior, National Biological Service, pp. 98–119.

Marmontel, M. (2008). *Trichechus inunguis*. In IUCN Red List of Threatened Species. Version 2010.3. Available at: http://www.iucnredlist.org (accessed 19 September 2010).

Marmontel, M., Odell, D. K. and Reynolds III, J. E. (1992). Reproductive biology of South American manatees. In *Reproductive Biology of South American Veterbrates*, W. C. Hamlett (ed.), New York: Springer-Verlag, pp. 295–312.

Marmontel, M., O'Shea, T. J., Kochman, H. I. and Humphrey, S. R. (1996). Age determination in manatees using growth-layer group counts in bone. *Marine Mammal Science*, **12**, 54–88.

Marmontel, M., Humphrey, S. R. and O'Shea, T. J. (1997). Population viability analysis of the Florida manatee (*Trichechus manatus latirostris*), 1976–1991. *Conservation Biology*, **11**, 467–481.

Marmontel, M., Rosas, F. C. W. and Kendall, S. (2012). The Amazonian manatee. In *Sirenian Conservation: Issues and Strategies in Developing Countries*, E. Hines, J. Reynolds, A. Mignucci-Giannoni, L. Aragones and M. Marmontel (eds), Gainesville: University Press of Florida.

Marsh, H. (1980). Age determination of the dugong (*Dugong dugon* [Müller]) in northern Australia and its biological implications. *Report of the International Whaling Commission*, special issue, **3**, 181–201.

Marsh, H. (1986). The status of the dugong in Torres Strait. In *Torres Strait Fisheries Seminar, Port Moresby, 11–14 February 1985*, A. K. Haines, G. C. Williams and D. Coates (eds), Canberra: Australian Government Printing Service, pp. 53–76.

Marsh, H. (1989). Mass stranding of dugongs by a tropical cyclone in northern Australia. *Marine Mammal Science*, **5**, 78–84.

Marsh, H. (1995). The life history, pattern of breeding, and population dynamics of the dugong. In *Population Biology of the Florida Manatee*, T. J. O'Shea, B. B. Ackerman and H. F. Percival (eds), Washington, DC: US Department of the Interior, National Biological Service, pp. 75–83.

Marsh, H. (2000). Evaluating management initiatives aimed at reducing the mortality of dugongs in gill and mesh nets in the Great Barrier Reef World Heritage Area. *Marine Mammal Science*, **16**, 684–694.

Marsh, H. (2008). *Dugong dugon*. In IUCN Red List of Threatened Species. Version 2010.3. Available at: http://www.iucnredlist.org (accessed 19 September 2010).

Marsh, H. and Corkeron, P. (1997). The status of the dugong in the Great Barrier Reef Marine Park. In *State of the Great Barrier Reef World Heritage Area: Proceedings of a Technical Workshop held in Townsville, Queensland, Australia, 27–29 November 1995. Workshop Series 23*, D. Wachenfeld, J. Oliver and J. Davis (eds), Towsville: Great Barrier Marine Park Authority, pp. 231–247. Available at: http://www.gbrmpa.gov.au/__data/assets/pdf_file/0005/4289/ws023_paper_17.pdf.

Marsh, H. and Eisentraut, M. (1984). Die Gaumenfaltern des Dugong. *Z f Saugetierkunde*, **49**, 314–315.

Marsh, H. and Kasuya, T. (1986). Evidence for reproductive senescence in female cetaceans. *Reports of the International Whaling Commission*, special issue, **8**, 57–74.

Marsh, H. and Kwan, D. (2008). Temporal variability in the life history and reproductive biology of female dugongs in Torres Strait: the likely role of sea grass dieback. *Continental Shelf Research*, **28**, 2152–2159.

Marsh, H. and Lawler, I. (2001). *Dugong Distribution and Abundance in the Southern Great Barrier Reef and Hervey Bay: Results of an Aerial Survey in October–December 1999*. Townsville: Great Barrier Marine Park Authority.

Marsh, H. and Morales-Vela, B. (2012). Guidelines for developing protected areas for sirenians. In *Sirenian Conservation: Issues and Strategies in Developing Countries*, E. Hines, J. Reynolds, A. Mignucci-Giannoni, L. Aragones and M. Marmontel (eds), Gainesville: University Press of Florida.

Marsh, H. and Rathbun, G. B. (1990). Development and application of conventional and satellite radio tracking techniques for studying dugong movements and habitat use. *Australian Wildlife Research*, **17**, 83–100.

Marsh, H. and Saalfeld, W. K. (1989). Distribution and abundance of dugongs in the northern Great Barrier Reef Marine Park. *Australian Wildlife Research*, **16**, 429–440.

Marsh, H. and Saalfeld, W. K. (1990). The distribution and abundance of dugongs in the Great Barrier Reef Marine Park south of Cape Bedford. *Australian Wildlife Research*, **17**, 511–524.

Marsh, H. and Sinclair, D. F. (1989a). An experimental evaluation of dugong and sea turtle aerial survey techniques. *Australian Wildlife Research*, **16**, 639–650.

Marsh, H. and Sinclair, D. F. (1989b). Correcting for visibility bias in strip transect aerial surveys of aquatic fauna. *Journal of Wildlife Management*, **53**, 1017–1024.

Marsh, H., Heinsohn, G. and Spain, A. V. (1977). The stomach and duodenal diverticula of the dugong. In *Functional Analysis of Marine Mammals*, Vol. 3, R. J. Harrison (ed.), London: Academic Press, pp. 271–295.

Marsh, H., Spain, A. V. and Heinsohn, G. E. (1978). Minireview: physiology of the dugong. *Comparative Biochemistry and Physiology*, **61**, 159–168.

Marsh, H., Gardner, B. R. and Heinsohn, G. E. (1981). Present-day hunting and distribution of dugongs in the Wellesley Islands (Queensland): implications for conservation. *Biological Conservation*, **19**, 255–267.

Marsh, H., Channells, P. W., Heinsohn, G. E. and Morrissey, J. (1982). Analysis of stomach contents of dugongs from Queensland. *Australian Wildlife Research*, **9**, 55–68.

Marsh, H., Heinsohn, G. E. and Channells, P. W. (1984a). Changes in the ovaries and uterus of the dugong, *Dugong dugon* (Sirenia: Dugongidae), with age and reproductive activity. *Australian Journal of Zoology*, **32**, 743–766.

Marsh, H., Heinsohn, G. E. and Glover, T. D. (1984b). Changes in the male reproductive organs of the dugong, *Dugong dugon* (Sirenia: Dugongidae) with age and reproductive activity. *Australian Journal of Zoology*, **32**, 721–742.

Marsh, H., Heinsohn, G. E. and Marsh, L. M. (1984c). Breeding cycle, life history and population dynamics of the dugong *Dugong dugon* (Sirenia: Dugongidae). *Australian Journal of Zoology*, **32**, 767–788.

Marsh, H., Freeland, W. J., Limpus, C. J. and Reed, P. C. (1986). *The Stranding of Dugongs and Sea Turtles Resulting from Cyclone Kathy, March 1984: A Report on the Rescue Effort and the Biological Data Obtained*, Darwin: Conservation Commission of the Northern Territory.

Marsh, H., Prince, R. I. T., Saalfeld, W. K. and Shepherd, R. (1994). The distribution and abundance of the dugong in Shark Bay, Western Australia. *Wildlife Research*, **21**, 149–161.

Marsh, H., Rathbun, G. B., O'Shea, T. J. and Preen, A. R. (1995). Can dugongs survive in Palau? *Biological Conservation*, **72**, 85–89.

Marsh, H., Harris, A. N. M. and Lawler, I. R. (1997). The sustainability of the indigenous dugong fishery in Torres Strait, Australia/Papua New Guinea. *Conservation Biology*, **11**, 1375–1386.

Marsh, H., Beck, C. A. and Vargo, T. (1999). Comparison of the capabilities of dugongs and West Indian manatees to masticate seagrasses. *Marine Mammal Science*, **15**, 250–255.

Marsh, H., Penrose, H., Eros, C. and Hugues, J. (2002). The dugong (*Dugong dugon*) status reports and action plans for countries and territories in its range. *Early Warning and Assessment Reports*. Nairobi: United Nations Environment Programme.

Marsh, H., Arnold, P., Freeman, M., *et al.* (2003). Strategies for conserving marine mammals. In *Marine Mammals: Fisheries, Tourism and Management Issues*, N. Gales, M. Hindell and R. Kirkwood (eds), Collingwood: CSIRO Publishing, pp. 1–19.

Marsh, H., Lawler, I. R., Kwan, D., *et al.* (2004). Aerial surveys and the potential biological removal technique indicate that the Torres Strait dugong fishery is unsustainable. *Animal Conservation*, **7**, 435–443.

Marsh, H., De'ath, G., Gribble, N. and Lane, B. (2005). Historical marine population estimates: triggers or targets for conservation? The dugong case study. *Ecological Applications*, **15**, 481–492.

Marsh, H., Lawler, I., Hodgson, A. and Grech, A. (2006). Is dugong management in the coastal waters of urban Queensland effective species conservation? Final report to the Marine and Tropical Research Facility.

Marsh, H., Hodgson, A., Lawler, I., Grech, A. and Delean, S. (2007a). Condition, status and trends and projected futures of the dugong in the Northern Great Barrier Reef and Torres Strait: including identification and evaluation of the key threats and evaluation of available management option to improve its status. Final Report to Marine and Tropical Science Research Facility. Available at: http://www.rrrc.org.au/publications/downloads/141-JCU-Marsh-2007-NGBR–Torres-Strait-Final-Report.pdf (accessed 10 October 2010).

Marsh, H., Dennis, A., Hines, H., *et al.* (2007b). Optimizing the allocation of management resources for wildlife. *Conservation Biology*, **21**, 387–399.

Marsh, H., Grech, A., Hodgson, A. and Delean, S. (2008). Distribution and abundance of the dugong in Gulf of Carpentaria waters: a basis for cross-jurisdictional conservation planning and management. Report to the Department of the Environment, Water, Heritage and the Arts.

Marshall, C. D., Huth, G. D., Edmonds, V. M., Halin, D. L. and Reep, R. L. (1998). Prehensile use of perioral bristles during feeding and associated behaviors of the Florida manatee (*Trichechus manatus latirostris*). *Marine Mammal Science*, **14**, 274–289.

Marshall, C. D., Kubilis, P. S., Huth, G. D., *et al.* (2000). Food-handling ability and feeding-cycle length of manatees feeding on several species of aquatic plants. *Journal of Mammalogy*, **81**, 649–658.

Marshall, C. D., Maeda, H., Iwata, M., *et al.* (2003). Orofacial morphology and feeding behaviour of the dugong, Amazonian, West African and Antillean manatees (Mammalia: Sirenia): functional morphology of the muscular-vibrissal complex. *Journal of Zoology*, **259**, 245–260.

Marshall, C. D., Vaughn, S. D., Sarko, D. K. and Reep, R. L. (2007). Topographical organization of the facial motor nucleus in Florida manatees (*Trichechus manatus latirostris*). *Brain, Behavior and Evolution*, **70**, 164–173.

Masini, R. J., Anderson, P. K. and McComb, A. J. (2001). A *Halodule*-dominated community in a subtropical embayment: physical environment, productivity, biomass, and impact of dugong grazing. *Aquatic Botany*, **71**, 179–197.

Mattioli, S. and Domning, D. P. (2006). An annotated list of extant skeletal material of Steller's sea cow (*Hydrodamalis gigas*) (Sirenia: Dugongidae) from the Commander Islands. *Aquatic Mammals*, **32**, 273–288.

Mayor, S. J., Schneider, D. C., Schaefer, J. A. and Mahoney, S. P. (2009). Habitat selection at multiple scales. *Ecoscience*, **16**, 238–247.

McClenaghan, L. R. Jr. and O'Shea, T. J. (1988). Genetic variability in the Florida manatee (*Trichechus manatus*). *Journal of Mammalogy*, **69**, 481–488.

McCook, L. J., Ayling, T., Cappo, M., *et al.* (2010). Adaptive management of the Great Barrier Reef: a globally significant demonstration of the benefits of networks of marine reserves. *Proceedings of the National Academy of Sciences*, **107**, 18278.

McDonald, B. J. (2005). Population genetics of dugongs around Australia: implications of gene flow and migration. PhD Thesis, James Cook University, Townsville.

McKenna, M. C. (1975). Toward a phylogenetic classification of the Mammalia. In *Phylogeny of the Primates: A Multidisciplinary Approach*, W. P. Luckett and F. S. Szalay (eds), New York: Plenum Press, pp. 21–46.

McKenna, M. C. and Bell, S. K. (1997). *Classification of Mammals above the Species Level*, New York: Columbia University Press.

McKenzie, L., Lee Long, W., Roelofs, A., Roder, C. and Coles, R. (1998). *Port of Mourilyan Seagrass Monitoring: First 4 Years*, Brisbane: Ports Corporation of Queensland.

McKillop, H. (1984). Prehistoric Maya reliance on marine resources: analysis of a midden from Moho Cay, Belize. *Journal of Field Archaeology*, **11**, 25–35.

McKillop, H. I. (1985). Prehistoric exploitation of the manatee in the Maya and circum-Caribbean areas. *World Archaeology*, **16**, 337–353.

McLachlan, M. S., Haynes, D. and Müller, J. F. (2001). PCDDs in the water–sediment–seagrass–dugong (*Dugong dugon*) food chain on the Great Barrier Reef (Australia). *Environmental Pollution*, **113**, 129–134.

McMahon, K. (2005). Recovery of subtropical seagrasses from natural disturbance. PhD Thesis, University of Queensland, Brisbane.

McMahon, K., Nash, S. B., Eaglesham, G., *et al.* (2005). Herbicide contamination and the potential impact to seagrass meadows in Hervey Bay, Queensland, Australia. *Marine Pollution Bulletin*, **51**, 325–334.

McNaughton, S. J. (1984). Grazing lawns: animals in herds, plant form, and coevolution. *American Naturalist*, **124**, 863–886.

McNaughton, S. J., Tarrants, J. L., McNaughton, M. M. and Davis, R. H. (1985). Silica as a defense against herbivory and a growth promotor in African grasses. *Ecology*, **66**, 528–535.

McNiven, I. J. (2010). Navigating the human–animal divide: marine mammal hunters and rituals of sensory allurement. *World Archaeology*, **42**, 215–230.

McNiven, I. J. and Bedingfield, A. C. (2008). Past and present marine mammal hunting rates and abundances: dugong (*Dugong dugon*) evidence from Dabangai Bone Mound, Torres Strait. *Journal of Archaeological Science*, **35**, 505–515.

McNiven, I. J. and Feldman, R. (2003). Ritually orchestrated seascapes: hunting magic and dugong bone mounds in Torres Strait, NE Australia. *Cambridge Archaeological Journal*, **13**, 169–194.

Mech, L. D. (1966). *The Wolves of Isle Royale*, Washington, DC: US Government Printing Office.

Meffe, G. K., Perrin, W. F. and Dayton, P. K. (1999). Marine mammal conservation: guiding principles and their implementation. In *Conservation and Management of Marine Mammals*, J. R. Twiss Jr, and R. R. Reeves (eds), Washington, DC: Smithsonian Institution Press, pp. 437–454.

Melville, H. (1851). *Moby-Dick or The Whale*. Available at: http://ebooks.gutenberg.us/DjVu_Collection/DJEDS/.../MOBY/Download.pdf (accessed 15 August 2010).

Méry, S., Charpentier, V., Auxiette, G. and Pelle, E. (2009). A dugong bone mound: the Neolithic ritual site on Akab in Umm al-Quwain, United Arab Emirates. *Antiquity*, **83**, 696–708.

Meybeck, M. (2003). Global analysis of river systems: from Earth system controls to Anthropocene syndromes. *Philosophical Transactions of the Royal Society B: Biological Sciences*, **358**, 1935–1955.

Miall, L. C. and Greenwood, F. (1878). The anatomy of the Indian elephant. *Journal of Anatomy and Physiology*, **12**, 261–287.

Michael, M. A. (1962). Introduction. In *The Russian Expedition to America by Sven Waxell*, New York: Collier Books, pp. 9–26.

Midorikawa, S., Arai, T., Harino, H., *et al.* (2004). Concentrations of organotin compounds in sediment and clams collected from coastal areas in Vietnam. *Environmental Pollution*, **131**, 401–408.

Mignucci-Giannoni, A. A. (1999). Assessment and rehabilitation of wildlife affected by an oil spill in Puerto Rico. *Environmental Pollution*, **104**, 323–333.

Mignucci-Giannoni, A. A., Montoya-Ospina, R. A., Jiménez-Marrero, N. M., *et al.* (2000). Manatee mortality in Puerto Rico. *Environmental Management*, **25**, 189–198.

Miksis-Olds, J. L. and Tyack, P. L. (2009). Manatee (*Trichechus manatus*) vocalization usage in relation to environmental noise levels. *Journal of the Acoustical Society of America*, **125**, 1806–1815.

Miksis-Olds, J. L., Donaghay, P. L., Miller, J. H., Tyack, P. L. and Nystuen, J. A. (2007a). Noise level correlates with manatee use of foraging habitats. *Journal of the Acoustical Society of America*, **121**, 3011–3020.

Miksis-Olds, J. L., Donaghay, P. L., Miller, J. H., Tyack, P. L. and Reynolds III, J. E. (2007b). Simulated vessel approaches elicit differential responses from manatees. *Marine Mammal Science*, **23**, 629–649.

Miller, D. L., Ewing, R. Y. and Bossart, G. D. (2001). Emerging and resurging diseases. In *Marine Mammal Medicine*, 2nd edn, L. A. Dierauf and F. M. D. Gulland (eds), Boca Raton: CRC Press, pp. 15–30.

Miller, K. E., Ackerman, B. B., Lefebvre, L. W. and Clifton, K. B. (1998). An evaluation of strip-transect aerial survey methods for monitoring manatee populations in Florida. *Wildlife Society Bulletin*, 26, 561–570.

Milly, P. C. D., Dunne, K. A. and Vecchia, A. V. (2005). Global patterns of trends in streamflow and water availability in a changing climate. *Nature*, 438, 347–350.

Miraglia, N., Bergero, D., Bassano, B., Tarantola, M. and Ladetto, G. (1999). Studies of apparent digestibility in horses and the use of internal markers. *Livestock Production Science*, 60, 21–25.

Miyamoto, M. M. and Goodman, M. (1986). Biomolecular systematics of eutherian mammals: phylogenetic patterns and classification. *Systematic Zoology*, 35, 230–240.

Mok, W. Y. and Best, R. C. (1979). Saprophytic colonization of a hyphomycete on the Amazonian manatee, *Trichechus inunguis* (Mammalia: Sirenia). *Aquatic Mammals*, 7, 79–82.

Montoya-Ospina, R. A., Caicedo-Herrera, D., Millán-Sánchez, S. L., Mignucci-Giannoni, A. A. and Lefebvre, L. W. (2001). Status and distribution of the West Indian manatee, *Trichechus manatus manatus*, in Colombia. *Biological Conservation*, 102, 117–129.

Moore, D. P., Williams, E. H., Mignucci-Giannoni, A. A., *et al.* (2008). Acute myocarditis in a West Indian Manatee, *Trichechus manatus* (Sirenia: Trichechidae), from Puerto Rico. *Revista De Biologia Tropical*, 56, 277–283.

Moore, J. C. (1951a). The range of the Florida manatee. *Quarterly Journal of the Florida Academy of Sciences*, 14, 1–19.

Moore, J. C. (1951b). The status of the manatee in the Everglades National Park, with notes on its natural history. *Journal of Mammalogy*, 32, 22–35.

Moore, J. C. (1953). Distribution of marine mammals to Florida waters. *American Midland Naturalist*, 49, 117–158.

Moore, J. C. (1956). Observations of manatees in aggregations. *American Museum Novitates*, 1811, 1–24.

Moore, J. E., Cox, T. M., Lewison, R. L., *et al.* (2010). An interview-based approach to assess marine mammal and sea turtle captures in artisanal fisheries. *Biological Conservation*, 143, 795–805.

Morales, P., Madin, S. H. and Hunter, A. (1985). Systemic *Mycobacterium marinum* infection in an Amazon manatee. *Journal of the American Veterinary Medical Association*, 187, 1230–1231.

Morales-Vela, B. and Olivera-Gomez, L. D. (1992). De sirenas a manatíes. Cuaderna de divulgación 4, Centro de Investigaciones de Quintana Roo, Chetumal, 1–30.

Morales-Vela, B., Olivera-Gomez, D., Reynolds III, J. E. and Rathbun, G. B. (2000). Distribution and habitat use by manatees (*Trichechus manatus manatus*) in Belize and Chetumal Bay, Mexico. *Biological Conservation*, 95, 67–75.

Morales-Vela, B., Suárez-Morales, E., Padilla-Saldìvar, J. and Heard, R. W. (2008). The tanaid *Hexapleomera robusta* (Crustacea: Peracarida) from the Caribbean manatee, with comments on other crustacean epibionts. *Journal of the Marine Biological Association of the United Kingdom*, 88, 591–596.

Morgan, U. M., Xiao, L., Hill, B. D., *et al.* (2000). Detection of the *Cryptosporidium parvum* 'human' genotype in a dugong (*Dugong dugon*). *Journal of Parasitology*, **86**, 1352–1354.

Morrison, M. L., Marcot, B. G. and Mannan, R. W. (2006). *Wildlife–Habitat Relationships: Concepts and Applications*, 2nd edn, Madison: University of Wisconsin Press.

Mou Sue, L. L., Chen, D. H., Bonde, R. K. and O'Shea, T. J. (1990). Distribution and status of manatees (*Trichechus manatus*) in Panama. *Marine Mammal Science*, **6**, 234–241.

Mrosovsky, N. (1997). IUCN's credibility critically endangered. *Nature*, **389**, 436.

Muanke, P. B. and Niezrecki, C. (2007). Manatee position estimation by passive acoustic localization. *Journal of the Acoustical Society of America*, **121**, 2049–2059.

Muir, C. and Kiszka, J. (2012). East African dugongs. In *Sirenian Conservation: Issues and Strategies in Developing Countries*, E. Hines, J. Reynolds, A. Mignucci-Giannoni, L. Aragones and M. Marmontel (eds), Gainesville: University Press of Florida.

Muizon, C. de and Domning, D. P. (1985). The first records of fossil sirenians in the southeastern Pacific Ocean. *Bulletin du Museum Nationale d'Historie Naturelle (Paris) C*, **3**, 189–213.

Mukhametov, L. M., Lyamin, O. I., Chetybrok, I. S., Vassilyev, A. A. and Diaz, R. P. (1992). Sleep in an Amazonian manatee, *Trichechus inunguis*. *Experientia*, **48**, 417–419.

Murie, J. (1872). On the form and structure of the manatee (*Manatus americanus*). *Transactions of the Zoological Society of London*, **8**, 127–202.

Murphy, W. J., Pevzner, P. A. and O'Brien, S. J. (2004). Mammalian phylogenomics comes of ages. *Trends in Genetics*, **20**, 631–639.

Murray, D. L. and Patterson, B. R. (2006). Wildlife survival estimation: recent advances and future directions. *Journal of Wildlife Management*, **70**, 1499–1503.

Murray, R. M. (1981). The importance of VFA in dugong nutrition. In *The Dugong: Proceedings of a Seminar/Workshop held at James Cook University 8–13 May 1979, Townsville*, H. Marsh (ed.), Townsville: James Cook University, pp. 96–98.

Murray, R. M., Marsh, H., Heinsohn, G. E. and Spain, A. V. (1977). Role of midgut caecum and large intestine in digestion of sea grasses by the dugong (Mammalia-Sirenia). *Comparative Biochemistry and Physiology*, **56**, 7–10.

Naidoo, R. and Adamowicz, W. L. (2001). Effects of economic prosperity on numbers of threatened species. *Conservation Biology*, **15**, 1021–1029.

NAILSMA. (2010). Northern Australian Indigenous Land and Sea Management Alliance Dugong and Marine Turtle Project. National Heritage Trust regional competitive component project final report. Available at: http://www.nailsma.org.au/nailsma/publications/downloads/Final-Report-web.pdf (accessed 4 September 2010).

Nair, R. V. and Lal Mohan, R. S. (1975). Studies on the vocalisation of the sea cow *Dugong dugon* in captivity. *Indian Journal of Fisheries*, **22**, 277–278.

Nakagawa, S., Gemmel, N. J. and Burke, T. (2004). Measuring vertebrate telomeres: applications and limitations. *Molecular Ecology*, **13**, 2523–2533.

Nakanishi, Y., Adulyanukosol, K., Arai, N., *et al.* (2008). Dugong grazing scars confirmed in *Enhalus acoroides* meadows. *Journal of Advanced Marine Science and Society*, **14**, 1–8.

Nakaoka, M. and Aioi, K. (1999). Growth of seagrass *Halophila ovalis* at dugong trails compared to existing within-patch variation in a Thailand intertidal flat. *Marine Ecology Progress Series*, **184**, 97–103.

Nakaoka, M., Mukai, H. and Chunhabundit, S. (2002). Impacts of dugong foraging on benthic animal communities in a Thailand seagrass bed. *Ecological Research*, **17**, 625–638.

Nasi, R., Brown, D., Wilkie, D., *et al.* (2008). *Conservation and Use of Wildlife-based Resources: The Bushmeat Crisis*, Montreal and Bogor: Secretariat of the Convention on Biological Diversity and Center for International Forestry Research (CIFOR).

Natiello, M. and Samuelson, D. (2005). Three-dimensional reconstruction of the angioarchitecture of the ciliary body of the West Indian manatee (*Trichechus manatus*). *Veterinary Ophthalmology*, **8**, 367–373.

Newman, L. A. and Robinson, P. R. (2006). The visual pigments of the West Indian manatee (*Trichechus manatus*). *Vision Research*, **46**, 3326–3330.

Nico, L. G., Loftus, W. F. and Reid, J. P. (2009). Interactions between non-native armored suckermouth catfish (Loricariidae: *Pterygoplichthys*) and native Florida manatee (*Trichechus manatus latirostris*) in artesian springs. *Aquatic Invasions*, **4**, 511–519.

Nietschmann, B. (1984). Hunting and ecology of dugongs and green turtles, Torres Strait, Australia. *National Geographic Society Research Reports*, **17**, 625–651.

Nietschmann, B. and Nietschmann, J. (1981). Good dugong, bad dugong; bad turtle, good turtle. *Natural History*, **90**, 54–63.

Nishihara, H., Satta, Y., Nikaido, M., *et al.* (2005). A retroposon analysis of afrotherian phylogeny. *Molecular Biology and Evolution*, **22**, 1823–1833.

Noreña-Barroso, E., Simá-Álvarez. R., Gold-Bouchot, G., and Zapata-Pérez, O. (2004). Persistent organic pollutants and histological lesions in Mayan catfish *Ariopsis assimilis* from the Bay of Chetumal, Mexico. *Marine Pollution Bulletin*, **48**, 263–269.

Norris, C. E. (1960). The distribution of the dugong in Ceylon. *Loris*, **8**, 296–300.

Norton, T., Beer, T. and Dovers, S. (eds). (1996). *Risk and Uncertainty in Environmental Management. Proceedings of the 1995 Australian Academy of Science Fenner Conference on the Environment*, Canberra: Australian National University.

Novacek, M. J. (2001). Mammalian phylogeny: genes and supertrees. *Current Biology*, **11**, R573–R575.

Novacek, M. J., Wyss, A. R. and McKenna, M. C. (1988). The major groups of eutherian mammals. In *The Phylogeny and Classification of the Tetrapods*, Vol. 2, M. J. Benton (ed.), Oxford: Clarendon Press, pp. 31–71.

Nowacek, D. P., Casper, B. M., Wells, R. S., Nowacek, S. M. and Mann, D. A. (2003). Intraspecific and geographic variation of West Indian manatee (*Trichechus manatus* spp.) vocalizations. *Journal of the Acoustical Society of America*, **114**, 66–69.

Nowacek, S. M., Wells, R. S., Owen, E. C. G., *et al.* (2004). Florida manatees, *Trichechus manatus latirostris*, respond to approaching vessels. *Biological Conservation*, **119**, 517–523.

Noyes, P. D., McElwee, M. E., Miller, H. D., *et al.* (2009). The toxicology of climate change: environmental contaminants in a warming world. *Environmental International*, **35**, 971–986.

Nursey-Bray, M., Marsh, H. and Ross, H. (2010). Exploring discourses in environmental decision making: an indigenous hunting case study. *Society and Natural Resources*, **23**, 366–382.

Ochieng, C. A. and Erftemeijer, P. L. A. (2003). The seagrasses of Kenya and Tanzania. In *World Atlas of Seagrasses*, E. P. Green and F. F. Short (eds), Berkeley: University of California Press, pp. 82–92.

Odell, D. K. (1982). West Indian manatee, *Trichechus manatus*. In *Wild Mammals of North Ameria: Biology, Management and Economics*, J. A. Chapman and G. A. Feldhammer (eds), Baltimore: Johns Hopkins University Press, pp. 828–837.

Odell, D. K. and Reynolds III, J. E. (1979). Observations of manatee mortality in south Florida. *Journal of Wildlife Management*, **43**, 572–577.

Odell, D. K., Forrester, D. J. and Apser, E. D. (1981). A preliminary analysis of organ weights and sexual maturity in the West Indian manatee (*Trichechus manatus*). In *The West Indian Manatee in Florida: Proceedings of a Workshop held in Orlando, Florida 27–29 March 1978*, R. L. Brownell Jr and K. Ralls (eds), Tallahassee: Florida Department of Natural Resources, pp. 52–65.

Odell, D. K., Bossart, G. D., Lowe, M. T. and Hopkins, T. D. (1995). Reproduction of the West Indian manatee in captivity. In *Population Biology of the Florida Manatee*, T. J. O'Shea, B. B. Ackerman and H. F. Percival (eds), Washington, DC: US Department of the Interior, National Biological Service, pp. 192–193.

Ogden, A. (1941). *The California Sea Otter Trade 1784–1848*, Berkeley: University of California Press.

O'Hara, T. M. and O'Shea, T. J. (2005). Assessing impacts of environmental contaminants. In *Marine Mammal Research: Conservation Beyond Crisis*, J. E. Reynolds III, W. F. Perrin, R. R. Reeves, S. Montgomery and T. J. Ragen (eds), Baltimore: Johns Hopkins University Press, pp. 63–83.

Okumura, N., Ichikawa, K., Akamatsu, T., *et al.* (2006). Stability of call sequence in dugongs' vocalization. *Oceans: ASIA Pacific*, **1–2**, 726–729.

Olds, N. and Shoshani, J. (1982). *Procavia capensis*. *Mammalian Species*, **171**, 1–7.

Olff, H. and Ritchie, M. E. (1998). Effects of herbivores on grassland plant diversity. *Trends in Ecology and Evolution*, **13**, 261–265.

Olivera-Gómez, L. D. and Mellink, E. (2005). Distribution of the Antillean manatee (*Trichechus manatus manatus*) as a function of habitat characteristics, in Bahía de Chetumal, Mexico. *Biological Conservation*, **121**, 127–133.

O'Malley, L. S. S. (1908). Eastern Bengal District Gazetteers, Chittagong. Chittagong: Bengal Secretariat Book Depot.

O'Riordan, T. and Stoll-Kleemann, S. (eds). (2002). *Biodiversity, Sustainability and Human Communities: Protecting Beyond the Protected*. Cambridge: Cambridge University Press.

Orozco, D. L. (2001). Manatí *Trichechus inunguis*: Caza, percepción y conocimiento de las comunidades *del* municipio de Puerto Nariño, Amazonas. Thesis in Ecology, Pontificia Universidad Javeriana, Bogotá.

Orr, D. W. (2002). Four challenges of sustainability. *Conservation Biology*, **16**, 1457–1460.

Ortega-Argueta, A., Hines, E. and Calvimontes, J. (2012). Using interviews in sirenian research. In *Sirenian Conservation: Issues and Strategies in Developing Countries*, E. Hines, J. Reynolds, A. Mignucci-Giannoni, L. Aragones and M. Marmontel (eds), Gainesville: University Press of Florida.

Orth, R. J., Carruthers, T. J. B., Dennison, W. C., *et al.* (2006). A global crisis for seagrass ecosystems. *Bioscience*, **56**, 987–996.

Ortiz, R. M., Worthy, G. A. J. and MacKenzie, D. S. (1998). Osmoregulation in wild and captive West Indian manatees (*Trichechus manatus*). *Physiological Zoology*, **71**, 449–457.

Ortiz, R. M., Worthy, G. A. J. and Byers, F. M. (1999). Estimation of water turnover rates of captive West Indian manatees (*Trichechus manatus*) held in fresh and salt water. *Journal of Experimental Biology*, **202**, 33–38.

O'Shea, T. J. (1986). Mast foraging by West Indian manatees (*Trichechus manatus*). *Journal of Mammalogy*, **67**, 183–185.

O'Shea, T. J. (1988). The past, present, and future of manatees in the southeastern United States: realities, misunderstandings, and enigmas. In *Proceedings of the Third Southeastern Nongame and Endangered Wildlife Symposium*, R. R. Odom, K. A. Riddleberger and J. C. Ozier (eds), Social Circle, GA: Georgia Department of Natural Resources Game and Fish Division, pp. 184–204.

O'Shea, T. J. (1995). Waterborne recreation and the Florida manatee. In *Wildlife and Recreationists: Coexistence through Management and Research*, R. L. Knight and K. J. Gutzwiller (eds), Washington, DC: Island Press, pp. 297–311.

O'Shea, T. J. (1999). Environmental contaminants and marine mammals. In *Biology of Marine Mammals*, J. E. Reynolds III and S. A. Rommel (eds), Washington, DC: Smithsonian Institution Press, pp. 485–564.

O'Shea, T. J. (2003). Toxicology of sirenians. In *Toxicology of Marine Mammals*, J. G. Vos, G. D. Bossart, M. Fournier and T. J. O'Shea (eds), London: Taylor and Francis, pp. 270–287.

O'Shea, T. J. and Ackerman, B. B. (1995). Population biology of the Florida manatee: an overview. In *Population Biology of the Florida Manatee*, T. J. O'Shea, B. B. Ackerman and H. F. Percival (eds), Washington, DC: US Department of the Interior, National Biological Service, pp. 280–287.

O'Shea, T. J. and Hartley, W. C. (1995). Reproduction and early-age survival of manatees at Blue Spring, Upper St John's River, Florida. In *Population Biology of the Florida Manatee*, T. J. O'Shea, B. B. Ackerman and H. F. Percival (eds), Washington, DC: US Department of the Interior, National Biological Service, pp. 157–170.

O'Shea, T. J. and Langtimm, C. A. (1995). Estimation of survival of adult Florida manatees in the Crystal River, at Blue Spring and on the Atlantic Coast. In *Population Biology of the Florida Manatee*, T. J. O'Shea, B. B. Ackerman and H. F. Percival (eds), Washington, DC: US Department of the Interior, National Biological Service, pp. 194–222.

O'Shea, T. J. and Poche, L. B. (2006). Aspects of underwater sound communication in Florida manatees (*Trichechus manatus latirostris*). *Journal of Mammalogy*, **87**, 1061–1071.

O'Shea, T. J. and Reep, R. L. (1990). Encephalization quotients and life-history traits in the Sirenia. *Journal of Mammalogy*, **71**, 534–543.

O'Shea, T. J. and Salisbury, C. A. (1991). Belize: a last stronghold for manatees in the Caribbean. *Oryx*, **25**, 156–164.

O'Shea, T. J. and Tanabe, S. (2003). Persistent ocean contaminants and marine mammals: a retrospective overview. In *Toxicology of Marine Mammals*, J. G. Vos,

G. D. Bossart, M. Fournier and T. J. O'Shea (eds), London: Taylor and Francis, pp. 99–134.

O'Shea, T. J., Moore, J. F. and Kochman, H. I. (1984). Contaminant concentrations in manatees in Florida. *Journal of Wildlife Management*, **48**, 741–748.

O'Shea, T. J., Beck, C. A., Bonde, R. K., Kochman, H. I. and Odell, D. K. (1985). An analysis of manatee mortality patterns in Florida, 1976–81. *Journal of Wildlife Management*, **49**, 1–11.

O'Shea, T. J., Correa-Viana, M., Ludlow, M. E. and Robinson, J. G. (1988). Distribution, status, and traditional significance of the West Indian manatee *Trichechus manatus* in Venezuela. *Biological Conservation*, **46**, 281–301.

O'Shea, T. J., Rathbun, G. B., Bonde, R. K., Buergelt, C. D. and Odell, D. K. (1991). An epizootic of Florida manatees associated with a dinoflagellate bloom. *Marine Mammal Science*, **7**, 165–179.

O'Shea, T. J., Lefebvre, L. W. and Beck, C. A. (2001). Florida manatees: perspectives on populations, pain, and protection. In *Marine Mammal Medicine*, 2nd edn, L. A. Dierauf and F. M. D. Gulland (eds), Boca Raton: CRC Press, pp. 31–43.

Owen, R. (1855). On the fossil skull of a mammal (*Prorastomus sirenoides*, Owen), from the Island of Jamaica. *Quarterly Journal of the Geological Society of London*, **11**, 541–543.

Owen-Smith, N. (1990). Demography of a large herbivore, the greater kudu *Tragelaphus strepsiceros*, in relation to rainfall. *Journal of Animal Ecology*, **59**, 893–913.

Owen-Smith, N. and Novellie, P. (1982). What should a clever ungulate eat? *American Naturalist*, **119**, 151–178.

Owings, D. H. and Morton, E. S. (1998). *Animal Vocal Communication: A New Approach*, Cambridge: Cambridge University Press.

Ozawa, T., Hayashi, S. and Mikhelson, V. M. (1997). Phylogenetic position of mammoth and Steller's sea cow within Tethytheria demonstrated by mitochondrial DNA sequences. *Journal of Molecular Evolution*, **44**, 406–413.

Pabst, D. A., Rommel, S. A. and Mclellan, W. A. (1999). The functional morphology of marine mammals. In *Biology of Marine Mammals*, J. E. Reynolds III and S. A. Rommel (eds), Washington, DC: Smithsonian Institution Press, pp. 15–72.

Packard, J. M. (1984). Impact of manatees (*Trichechus manatus*) on seagrass communities in eastern Florida. *Acta Zoologica Fennica*, **172**, 21–22.

Packard, J. M., Summers, R. C. and Barnes, L. B. (1985). Variation of visibility bias during aerial surveys of manatees. *Journal of Wildlife Management*, **49**, 347–351.

Paerl, H. W. and Huisman, J. (2008). Blooms like it hot. *Science*, **320**, 57–58.

Paeth, H., Born, K., Girmes, R., Podzun, R. and Jacob, D. (2009). Regional climate change in tropical and northern Africa due to greenhouse forcing and land use changes. *Journal of Climate*, **22**, 114–132.

Pallas, P. S. (1811). *Zoographia Rosso-Asiatica, sistens omnium animalium in extenso Imperis Rossico et adjacentibus maribus observatorum recensionem, domicilia, mores et descriptiones, anatomen atque icones plurimorum*. Vol. 1. St Petersburg: Academiae Scientiarum (Academy of Sciences).

Palmer, M. A., Reidy Liermann, C. A., Nilsson, C., *et al.* (2008). Climate change and the world's river basins: anticipating management options. *Frontiers in Ecology and the Environment*, **6**, 81–89.

Palmer, M. A., Lettenmaier, D. P., Poff, N. L., *et al.* (2009). Climate change and river ecosystems: protection and adaptation options. *Environmental Management*, **44**, 1053–1068.

Pardini, A. T., O'Brien, P. C. M., Fu, B., *et al.* (2007). Chromosome painting among Proboscidea, Hyracoidea and Sirenia: support for Paenungulata (Afrotheria, Mammalia) but not Tethytheria. *Proceedings of the Royal Society B: Biological Sciences*, **274**, 1333–1340.

Parent, S. and Poonian, C. (2009). Dugongs without borders. *SeagrassWatch Magazine*, **39**, 15.

Parente, C. L., Vergara-Parente, J. E. and Lima, R. P. (2004). Strandings of Antillean manatees. *Latin American Journal of Aquatic Mammals*, **3**, 69–75.

Parra, A. (1978). Comparison of foregut and hindgut fermentation in herbivores. In *The Ecology of Arboreal Folivores*, G. G. Montgomery (ed.), Washington, DC: Smithsonian Institution Press, pp. 205–229.

Patel, J. (2010). Regulators slash FPL's rate-hike request: in response to the decision, FPL announced it would halt $10 billion in projects. *Sun Sentinel*. Available at: http://www.sun-sentinel.com/business/fl-psc-increase-vote-20100113,0,1053253. story.

Patterson, E. K. (1939). The dugong hunters. *Walkabout*, **5**(12), 43–44.

Pauly, D. (2006). Major trends in small-scale marine fisheries, with emphasis on developing countries, and some implications for the social sciences. *Maritime Studies* (MAST), **4**(2), 7–22.

Pause, K. C. (2007). Conservation genetics of the Florida manatee, *Trichechus manatus latirostris*. PhD Dissertation, University of Florida, Gainesville.

Pause, K. C., Nourisson, C., Clark, A., *et al.* (2007). Polymorphic microsatellite DNA markers for the Florida manatee (*Trichechus manatus latirostris*). *Molecular Ecology Notes*, **7**, 1073–1076.

Perrin, W. F. and Myrick, A. (eds). (1980). Growth of odontocetes and sirenians: problems in age determination. *Reports of the International Whaling Commission*, special issue, **3**, 144.

Perrin, W. F. and Reilly, S. B. (1984). Reproductive parameters of dolphins and small whales of the family Delphinidae. *Reports of the International Whaling Commission*, special issue, **6**, 97–125.

Perrin, W. F., Reeves, R. R., Dolar, M. L. L., *et al.* (2005) *Report of the Second Workshop on the Biology and Conservation of Small Cetaceans and Dugongs of South East Asia*, CMS Technical Series Publication No. 9.

Perry, C. J. and Dennison, W. C. (1996). Effects of dugong grazing on microbial processes in seagrass sediments. In *Seagrass Biology: Proceedings of an International Workshop, Rottnest Island, Western Australia, 25–29 January 1996*, J. Kuo, R. C. Phillips, D. I. Walker and H. Kirkman (eds), Perth: University of Western Australia, p. 371.

Perry, C. J. and Dennison, W. C. (1999). Microbial nutrient cycling in seagrass sediments. *AGSO Journal of Australian Geology and Geophysics*, **17**, 227–231.

Peterken, C. J. and Conacher, C. A. (1997). Seed germination and recolonisation of *Zostera capricorni* after grazing by dugongs. *Aquatic Botany*, **59**, 333–340.

Pew Oceans Commission. (2003). America's living oceans: charting a course for sea change. Available at: http://www.pewtrusts.org/uploadedFiles/wwwpewtrustsorg/Reports/Protecting_ocean_life/env_pew_oceans_final_report.pdf.

Phillips, O. L., Lewis, S. L., Baker, T. R., Chao, K.-J. and Higuchi, N. (2008). The changing Amazon forest. *Philosophical Transactions of the Royal Society B: Biological Sciences*, **363**, 1819–1827.

Phillips, O. L., Aragão, L. E. O. C., Lewis, S. L., *et al.* (2009). Drought sensitivity of the Amazon rainforest. *Science*, **323**, 1344–1347.

Phillips, R. C. (2003). The seagrasses of the Arabian Gulf and Arabian region. In *World Atlas of Seagrasses*, E. P. Green and F. F. Short (eds), Berkeley: University of California Press, pp. 74–81.

Pierce, G. J. and Ollason, J. G. (1987). Eight reasons why optimal foraging theory is a complete waste of time. *Oikos*, **49**, 111–117.

Piggins, D., Muntz, W. R. A. and Best, R. C. (1983). Physical and morphological aspects of the eye of the manatee, *Trichechus inunguis* Natterer 1883 (Sirenia: Mammalia). *Marine Behaviour and Physiology*, **9**, 111–130.

Poiner, I. R. and Peterken, C. (1996). Seagrasses. In *The State of the Marine Environment Report for Australia. Technical Annex 1*, L. P. Zann and P. Kailola (eds), Townsville: Great Barrier Marine Park Authority, pp. 40–45.

Pollock, K. H. (1982). A capture–recapture design robust to unequal probability of capture. *Journal of Wildlife Management*, **46**, 752–757.

Pollock, K. H. and Kendall, W. L. (1987). Visibility bias in aerial surveys: a review of estimation procedures. *Journal of Wildlife Management*, **51**, 502–510.

Pollock, K. H., Marsh, H. D., Lawler, I. R. and Alldredge, M. W. (2006). Estimating animal abundance in heterogeneous environments: an application to aerial surveys for dugongs. *Journal of Wildlife Management*, **70**, 255–262.

Pomeroy, R. S., Parks, J. E. and Watson, L. M. (2004). *How is Your MPA Doing? A Guidebook of Natural and Social Indicators for Evaluating Marine Protected Area Management Effectiveness*. Gland, Switzerland and Cambridge: IUCN.

Popov, V. and Supin, A. (1990). Electrophysiological studies of hearing in some cetaceans and a manatee. In *Sensory Ability of Cetaceans: Laboratory and Field Evidence*, J. Thomas and R. Kastelstein (eds), New York: Plenum Press, pp. 405–415.

Possingham, H. P., Ball, I. and Adelman, S. (2000). Mathematical methods for identifying representative reserve networks. In *Quantitative Methods for Conservation Biology*, S. Ferson and M. A. Burgman (eds), New York: Springer-Verlag, pp. 291–307.

Possingham, H. P., Andelman, S. J., Burgman, M. A., *et al.* (2002). Limits to the use of threatened species lists. *Trends in Ecology and Evolution*, **17**, 503–507.

Powell, J. A. (1978). Evidence of carnivory in manatees (*Trichechus manatus*). *Journal of Mammalogy*, **50**, 442.

Powell, J. A. (1981). The manatee population in Crystal River, Citrus County, Florida. In *The West Indian Manatee in Florida: Proceedings of a Workshop held in Orlando, Florida 27–29 March 1978*, R. L. Brownell Jr and K. Ralls (eds), Tallahassee: Florida Department of Natural Resources, pp. 33–40.

Powell, J. A. (1996). The distribution and biology of the West African manatee (*Trichechus senegalensis* Link, 1795). In *Regional Seas Program, Oceans and Coastal Areas*, Nairobi: United Nations Environmental Program.

Powell, J. and Kouadio, A. (2008). *Trichechus senegalensis*. In IUCN Red List of Threatened Species. Version 2010.3. Available at: http://www.iucnredlist.org (accessed 19 September 2010).

Powell, J. A. and Rathbun, G. B. (1984). Distribution and abundance of manatees along the northern coast of the Gulf of Mexico. *Northeast Gulf Science*, **7**, 1–28.

Powell, J. A. and Waldron, J. C. (1981) The manatee population in Blue Spring, Volusia County, Florida. In *The West Indian Manatee in Florida: Proceedings of a Workshop held in Orlando, Florida 27–29 March 1978*, R. L. Brownell Jr and K. Ralls (eds), Tallahassee: Florida Department of Natural Resources, pp. 41–51.

Powell, J. A., Belitsky, D. W. and Rathbun, G. B. (1981). Status of the West Indian manatee (*Trichechus manatus*) in Puerto Rico. *Journal of Mammalogy*, **62**, 642–646.

Preen, A. (1989a). Observations of mating behavior in dugongs (*Dugong dugon*). *Marine Mammal Science*, **5**, 382–387.

Preen, A. (1989b). The status and conservation of dugongs in the Arabian region. MEPA coastal and marine management series report, No. 10.

Preen, A. (1992). Interactions between dugongs and seagrasses in a subtropical environment. PhD Thesis, James Cook University, Townsville.

Preen, A. (1995a). Diet of dugongs: are they omnivores? *Journal of Mammalogy*, **76**, 163–171.

Preen, A. (1995b). Impacts of dugong foraging on seagrass habitats: observational and experimental evidence for cultivation grazing. *Marine Ecology Progress Series*, **124**, 201–213.

Preen, A. (2004). Distribution, abundance and conservation status of dugongs and dolphins in the southern and western Arabian Gulf. *Biological Conservation*, **118**, 205–218.

Preen, A. and Marsh, H. (1995). Response of dugongs to large-scale loss of seagrass from Hervey Bay, Queensland, Australia. *Wildlife Research*, **22**, 507–519.

Preen, A. R., Lee Long, W. J. and Coles, R. G. (1995). Flood and cyclone related loss, and partial recovery, of more than 1000 km² of seagrass in Hervey Bay, Queensland, Australia. *Aquatic Botany*, **52**, 3–17.

Preen, A. R., Marsh, H., Lawler, I. R., Prince, R. I. T. and Shepherd, R. (1997). Distribution and abundance of dugongs, turtles, dolphins and other megafauna in Shark Bay, Ningaloo Reef and Exmouth Gulf, Western Australia. *Wildlife Research*, **24**, 185–208.

Preen, A., Das, H., Al-Rumaidh, M. and Hodgson, A. (2012). Dugongs in Arabia. In *Sirenian Conservation: Issues and Strategies in Developing Countries*, E. Hines, J. Reynolds, A. Mignucci-Giannoni, L. Aragones and M. Marmontel (eds), Gainesville: University Press of Florida.

Prince, R. I. T., Lawler, I. R. and Marsh, H. D. (2001). The distribution and abundance of dugongs and other megavertebrates in Western Australian coastal waters extending seaward to the 20 m isobath between North West Cape and the DeGrey river mouth, Western Australia, April 2000. Report for Environment Australia.

Prospero, J. M. and Lamb, P. J. (2003). African droughts and dust transport to the Caribbean: climate change implications. *Science*, **302**, 1024–1027.

Prothero, D. R., Manning, E. M. and Fischer, M. (1988). The phylogeny of the ungulates. In *The Phylogeny and Classification of the Tetrapods*, Vol. 2, M. J. Benton (ed.), Oxford: Clarendon Press, pp. 201–234.

Provancha, J. A. and Hall, C. R. (1991). Observations of associations between seagrass beds and manatees in east central Florida. *Florida Scientist*, **54**, 87–98.

Provancha, J. A. and Provancha, M. J. (1988). Long-term trends in abundance and distribution of manatees (*Trichechus manatus*) in the northern Banana River, Brevard County, Florida. *Marine Mammal Science*, **4**, 323–338.

Provancha, J. A. and Stolen, E. D. (2008) Dugong aerial survey report May 25–29, 2008 Bazaruto Archipelago National Park Inhambane Province, Mozambique. World Wide Fund for Nature.

Purvis, A., Agapow, P.-M., Gittleman, J. L. and Mace, G. M. (2000). Nonrandom extinction and the loss of evolutionary history. *Science*, **288**, 328–330.

Pyenson, N. D. (2009). Requiem for *Lipotes*: an evolutionary perspective on marine mammal extinction. *Marine Mammal Science*, **25**, 714–724.

Pyke, G. H. (1984). Optimal foraging theory: a critical review. *Annual Review of Ecology and Systematics*, **15**, 523–575.

Qi Jingfen. (1984). Breeding in the West Indian manatee (*Trichechus manatus* Linn.). *Acta Theriologica Sinica*, **4**, 27–33.

Ragen, T. J., Reeves, R. R., Reynolds III, J. E. and Perrin, W. F. (2005). Future directions in marine mammal research. In *Marine Mammal Research: Conservation beyond Crisis*, J. E. Reynolds III, W. F. Perrin, R. R. Reeves, S. Montgomery and T. J. Ragen (eds), Baltimore: Johns Hopkins University Press, pp. 179–183.

Rainey, W. E., Lowenstein, J. M., Sarich, V. M. and Magor, D. M. (1984). Sirenian molecular systematics: including the extinct Steller's sea cow (*Hydrodamalis gigas*). *Naturwissenschaften*, **71**, 586–588.

Rajamani, L. (2008). The conservation biology of the dugong (*Dugong dugon*) and its seagrass habitat in Sabah, Malaysia: a basis for conservation planning. Unpublished PhD Thesis Universiti Malaysia, Sabah, Kota Kinabalu, Sabah.

Rajamani, L., Cabanban, A. S. and Rahman, R. A. (2006). Indigenous use and trade of dugong (*Dugong dugon*) in Sabah, Malaysia. *Ambio*, **35**, 266–268.

Ralph, C. L., Young, S., Gettinger, R. and O'Shea, T. J. (1985). Does the manatee have a pineal body? *Acta Zoologica*, **66**, 55–60.

Ramsar. (2010). The Ramsar convention on wetlands. Available at: http://www.ramsar.org/cda/en/ramsar-home/main/ramsar/1_4000_0_ (accessed 10 October 2010).

Rasheed, M. (2000). Recovery and succession in north Queensland Tropical seagrass communities. PhD Thesis, James Cook University, Townsville.

Rathbun, G. B. and O'Shea, T. J. (1984). The manatee's simple social life. In *The Encyclopedia of Mammals*, D. Macdonald (ed.), New York: Facts on File, pp. 300–301.

Rathbun, G. B., Powell, J. A. and Cruz, G. (1983). Status of the West Indian manatee in Honduras. *Biological Conservation*, **26**, 301–308.

Rathbun, G. B., Reid, J. P. and Bourassa, J. B. (1987). *Design and Construction of a Tethered, Floating Radio-tag Assembly for Manatees*, Springfield, VA: National Technical Information Service.

Rathbun, G. B., Reid, J. P. and Carowan, G. (1990). Distribution and movement patterns of manatees (*Trichechus manatus*) in Northwestern Peninsular Florida. *Florida Marine Research Publications*, **48**, 1–33.

Rathbun, G. B., Reid, J. P., Bonde, R. K. and Powell, J. A. (1995). Reproduction in free-ranging Florida manatees. In *Population Biology of the Florida Manatee*,

T. J. O'Shea, B. B. Ackerman and H. F. Percival (eds), Washington, DC: US Department of the Interior, National Biological Service, pp. 135–156.

Raven, M. (1990). The point of no diminishing returns: hunting and resource decline on Boigu Island, Torres Strait. Unpublished PhD Dissertation, University of California, Davis.

Ray, C. E. (1960). The manatee in the Lesser Antilles. *Journal of Mammalogy*, **41**, 412–413.

Read, A. J. (2008). The looming crisis: interactions between marine mammals and fisheries. *Journal of Mammalogy*, **89**, 541–548.

Rector, A., Bossart, G. D., Ghim, S. J., *et al.* (2004). Characterization of a novel close-to-root papillomavirus from a Florida manatee by using multiply primed rolling-circle amplification: *Trichechus manatus latirostris* Papillomavirus Type 1. *The Journal of Virology*, **78**, 12698–12702.

Reef Water Quality Protection Plan. (2009). For the Great Barrier Reef World Heritage Area and adjacent catchments. Available at: http://www.reefplan.qld.gov.au/library/pdf/reef-plan-2009.pdf (accessed 25 September 2010).

Reep, R. L. and Bonde, R. K. (2006). *The Florida Manatee: Biology and Conservation*, Gainesville: University Press of Florida.

Reep, R. L., Marshall, C. D., Stoll, M. L. and Whitaker, D. M. (1998). Distribution and innervation of facial bristles and hairs in the Florida manatee (*Trichechus manatus latirostris*). *Marine Mammal Science*, **14**, 257–273.

Reep, R. L., Stoll, M. L., Marshall, C. D., Homer, B. L. and Samuelson, D. A. (2001). Microanatomy of facial vibrissae in the Florida manatee: the basis for specialized sensory function and oripulation. *Brain, Behavior and Evolution*, **58**, 1–14.

Reep, R. L., Marshall, C. D. and Stoll, M. L. (2002). Tactile hairs on the postcranial body in Florida manatees: a mammalian lateral line? *Brain, Behavior and Evolution*, **59**, 141–154.

Reep, R. L., Finlay, B. L. and Darlington, R. B. (2007). The limbic system in mammalian brain evolution. *Brain, Behavior and Evolution*, **70**, 57–70.

Reeves, R. R., Tuboku-Metzger, D. and Kapindi, R. A. (1988). Distribution and exploitation of manatees in Sierra Leone. *Oryx*, **22**, 75–84.

Reeves, R. R., Leatherwood, S., Jefferson, T. A., Curry, B. E. and Henningsen, T. (1996). Amazonian manatees, *Trichechus inunguis*, in Peru: distribution, exploitation, and conservation status. *Interciencia*, **21**, 246–254.

Reich, K. J. and Worthy, G. A. J. (2006). An isotopic assessment of the feeding habits of free-ranging manatees. *Marine Ecology Progress Series*, **322**, 303–309.

Reid, J. P. (1995). Chessie's most excellent adventure: the 1995 East Coast tour. *Sirenews*, **24**, 9–11.

Reid, J. P. (2006). Cooperative manatee reserach in Puerto Rico. *Endangered Species Bulletin*, **31**, 18–19.

Reid, J. P. and O'Shea, T. J. (1989). Three years operational use of satellite telemetry on Florida manatees: tag improvements based on challenges from the field. In *Proceedings of the 1989 North America Argos Users Conference*, Landover, MD: Service Argos, pp. 217–232.

Reid, J. P., Rathbun, G. B. and Wilcox, J. R. (1991). Distribution patterns of individually identifiable West Indian manatees (*Trichechus manatus*) in Florida. *Marine Mammal Science*, **7**, 180–190.

Reid, J. P., Bonde, R. K. and O'Shea, T. J. (1995). Reproduction and mortality of radio-tagged and recognizable manatees on the Atlantic coast of Florida. In *Population Biology of the Florida Manatee*, T. J. O'Shea, B. B. Ackerman and H. F. Percival (eds), Washington, DC: US Department of the Interior, National Biological Service, pp. 171–191.

Reid, W. V. (2002). Epilogue. In *Biodiversity, Sustainability and Human Communities: Protecting Beyond the Protected*, T. O'Riordan and S. Stoll-Kleemann (eds), Cambridge: Cambridge University Press, pp. 311–314.

Reinhart, R. (1953). Diagnosis of the new mammalian order Desmostylia. *Journal of Geology*, **61**, 187.

Reinhart, R. H. (1959). A review of the Sirenia and Desmostylia. *University of California Publications in Geological Sciences*, **36**, 1–146.

Reyero, M., Cacho, E., Martinez, A., *et al.* (1999). Evidence of saxitoxin derivatives as causative agents in the 1997 mass mortality of monk seals in the Cape Blanc Peninsula. *Natural Toxins*, **7**, 311–315.

Reynolds III, J. E. (1977). Aspects of the social behavior and ecology of a semi-isolated colony of Florida manatees, *Trichechus manatus*. MS Thesis, University of Miami.

Reynolds III, J. E. (1980). Aspects of the structural and functional anatomy of the gastrointestinal tract of the West Indian manatee, *Trichechus manatus*. PhD Thesis, University of Miami.

Reynolds III, J. E. (1981a). Behavior patterns in the West Indian manatee, with emphasis on feeding and diving. *Florida Scientist*, **44**, 233–242.

Reynolds III, J. E. (1981b). Aspects of the social behaviour and herd structure of a semi-isolated colony of West Indian manatees, *Trichechus manatus*. *Mammalia*, **45**, 431–451.

Reynolds III, J. E. (1995). Florida manatee population biology: research progress, infrastructure, and applications for conservation and management. In *Population Biology of the Florida Manatee*, T. J. O'Shea, B. B. Ackerman and H. F. Percival (eds), Washington, DC: US Department of the Interior, National Biological Service, pp. 6–12.

Reynolds III, J. E. (1999). Efforts to conserve the manatees. In *Conservation and Management of Marine Mammals*, J. Twiss Jr and R. R. Reeves (eds), Washington, DC: Smithsonian Institution Press, pp. 267–295.

Reynolds III, J. E. (2000). *Possible Locations for Long-term, Warmwater Refugia for Manatees in Florida: Alternatives to Power Plants*, Juno Beach: Florida Power & Light Company.

Reynolds III, J. E. (2010). Distribution and abundance of Florida manatees (*Trichechus manatus latirostris*) around selected power plants following winter cold fronts: 2009–2010. Final report to Florida Power & Light Company, June Beach, Florida.

Reynolds III, J. E. and Ferguson, J. C. (1984). Implications of the presence of manatees (*Trichechus manatus*) near the Dry Tortugas Islands. *Florida Scientist*, **44**, 187–189.

Reynolds III, J. E. and Krause, W. J. (1982). A brief anatomical note on the duodenum of the West Indian manatee, *Trichechus manatus*, with emphasis on the duodenal glands. *Acta anatomica*, **114**, 33–40.

Reynolds III, J. E. and Odell, D. K. (1991). *Manatees and Dugongs*, New York: Facts on File, Inc.

Reynolds III, J. E. and Rommel, S. A. (1996). Structure and function of the gastro-intestinal tract of the Florida manatee, *Trichechus manatus latirostris*. *Anatomical Record*, **245**, 539–558.

Reynolds III, J. E. and Wells, R. S. (2003). *Dolphins, Whales and Manatees of Florida: A Guide to Sharing Their World*, Gainesville: University Press of Florida.

Reynolds III, J. E. and Wilcox, J. R. (1994). Observations of Florida manatees (*Trichechus manatus latirostris*) around selected power plants in winter. *Marine Mammal Science*, **10**, 163–177.

Reynolds III, J. E., Szelistowski, W. A. and Leon, M. A. (1995). Status and conservation of manatees *Trichechus manatus manatus* in Costa Rica. *Biological Conservation*, **71**, 193–196.

Reynolds III, J. E., Odell, D. K. and Rommel, S. A. (1999). Marine mammals of the world. In *Biology of Marine Mammals*, J. E. Reynolds III and S. A. Rommel (eds), Washington, DC: Smithsonian Institution Press, pp. 1–14.

Reynolds III, J. E., Rommel, S. A. and Pitchford, M. E. (2004). The likelihood of sperm competition in manatees: explaining an apparent paradox. *Marine Mammal Science*, **20**, 464–476.

Reynolds III, J. E., Perrin, W. F., Reeves, R. R., Montgomery, S. and Ragen, T. J. (eds). (2005). *Marine Mammal Research: Conservation Beyond Crisis*, Baltimore: Johns Hopkins University Press.

Reynolds III, J. E., Marsh, H. and Ragen, T. J. (2009). Marine mammal conservation. *Endangered Species Research*, **7**, 23–28.

Reynolds III, J. E., Morales-Vela, B., Lawler, I. R. and Edwards, H. (2012). Utility and design of aerial surveys for sirenians. In *Sirenian Conservation: Issues and Strategies in Developing Countries*, E. Hines, J. Reynolds, A. Mignucci-Giannoni, L. Aragones and M. Marmontel (eds), Gainesville: University Press of Florida.

Richman, L. K., Montali, R. J., Garber, R. L., *et al.* (1999). Novel endotheliotropic herpesviruses fatal for Asian and African elephants. *Science*, **283**, 1171–1176.

Richman, L. K., Montali, R. J., Cambre, R. C., *et al.* (2000). Clinical and pathological findings of a newly recognized disease of elephants caused by endotheliotropic herpesviruses. *Journal of Wildlife Diseases*, **36**, 1–12.

Ricqlès A. de and Buffrénil, V. de. (1995). Sur la présence de pachyostéosclérose chez la rhytine de Steller (*Rhytina [Hydrodamalis] gigas*), sirénien récent éteint. *Annales des Sciences Naturelles-Zoologie et Biologie Animale*, **16**, 47–53.

Riedman, M. L. and Estes, J. A. (1990). The sea otter (*Enhydra lutris*): behavior, ecology, and natural history. *US Fish and Wildlife Service Biological Report*, **90**, 1–126.

Riegel, R. J. and Hakola, S. E. (1996). *Illustrated Atlas of Clinical Equine Anatomy and Common Disorders of the Horse*, Marysville: Equistar Publications.

River Zoo Farm. (2010). River View Farm wildlife breeding farm & fauna sanctuary. Available at: http://www.riverzoofarm.com.

Robertson, J. B. and Van Soest, P. J. (1981). The detergent system of analysis and its application to human food. In *The Analysis of Dietary Fiber in Food*, W. P. T. James and O. Theander (eds), New York and Basel: Marcel Dekker, pp. 123–158.

Robineau, D. (1969). Morphologie externe du complexe osseux temporal chez les sirenens. *Mémoire du Muséum National d'Histoire Naturelle (Paris)*. Series A Zoologie, **60**, 1–32.

Rodrigues, A. S. L., Pilgrim, J. D., Lamoreux, J. F., Hoffmann, M. and Brooks, T. M. (2006). The value of the IUCN Red List for conservation. *Trends in Ecology and Evolution*, **21**, 71–76.

Roelofs, A., Coles, R. and Smit, N. (2005). A survey of intertidal seagrass from Van Diemen Gulf to Castlereagh Bay, Northern Territory, and from Gove to Horn Island, Queensland. Australian Government, Department of the Environment and Heritage, National Oceans Office.

Rommel, S. and Reynolds III, J. E. (2000). Diaphragm structure and function in the Florida manatee (*Trichechus manatus latirostris*). *Anatomical Record*, **259**, 41–51.

Rommel, S. A., Reynolds III, J. E. and Lynch, H. A. (2003). Adaptations of the herbivorous marine mammals. In *Matching Herbivore Nutrition to Ecosystems Biodiversity*, L.-t. Mannetje, L. Ramirez-Aviles, C. Sandoval-Castro and J. C. Ku-Vera (eds), Mexico: Universidad Autonoma de Yucatan, pp. 287–308.

Rommel, S. A., Costidis, A. M., Pitchford, T. D., *et al.* (2007). Forensic methods for characterizing watercraft from watercraft-induced wounds on the Florida manatee (*Trichechus manatus latirostris*). *Marine Mammal Science*, **23**, 110–132.

Ronald, K., Selley, L. J. and Amoroso, E. C. (1978). *Biological Synopsis of the Manatee*, Ottawa: IDRC.

Rosas, F. C. W. (1994). Biology, conservation and status of the Amazonian manatee *Trichechus inunguis*. *Mammal Review*, **24**, 49–59.

Rose, P. M. (1997). Manatees and the future of electric utilities deregulation in Florida. *Sirenews*, **28**, 1–3.

Rovero, F., Rathbun, G. B., Perkin, A., *et al.* (2008). A new species of giant sengi or elephant-shrew (genus: *Rhynchocyon*) highlights the exceptional biodiversity of the Udzungwa Mountains of Tanzania. *Journal of Zoology (London)*, **274**, 126–133.

Rowcliffe, J. M., Milner-Gulland, E. J. and Cowlishaw, G. (2005). Do bushmeat consumers have other fish to fry? *Trends in Ecology and Evolution*, **20**, 274–276.

Runge, M. C., Langtimm, C. A. and Kendall, W. L. (2004). A stage-based model of manatee population dynamics. *Marine Mammal Science*, **20**, 361–385.

Runge, M. C., Sanders-Reed, C. A. and Fonnesbeck, C. J. (2007a). A core stochastic population projection model for Florida manatees (*Trichechus manatus latirostris*). US Geological Survey Open-File Report 2007-1082. Available at: http://www.pwrc.usgs.gov/resshow/manatee/documents/OFR2007–1082.pdf.

Runge, M. C., Sanders-Reed, C. A., Langtimm, C. A. and Fonnesbeck, C. J. (2007b). A quantitative threats analysis for the Florida manatee (*Trichechus manatus latirostris*). US Geological Survey Open-File Report 2007-1086. Available at: http://www.pwrc.usgs.gov/resshow/manatee/documents/OFR2007–1086.pdf.

Ryther, J. H., Williams, L. O., Hanisak, M. D., Stenberg, R. W. and Debusk, T. A. (1978). Biomass production by some marine and freshwater plants. In *Proceedings of Fuels from Biomass Symposium*, Vol. 2, Troy: Rensselaer Polytechnic Institute, pp. 947–989.

Saalfeld, K. and Marsh, H. (2004). Dugong. *Key Species: A Description of Key Species Groups in the Northern Planning area*, Hobart: National Oceans Office, pp. 93–112. Available at: http://www.environment.gov.au/coasts/mbp/publications/north/pubs/n-key-species.pdf.

Saether, B. E. (1997). Environmental stochasticity and population dynamics of large herbivores: a search for mechanisms. *Trends in Ecology and Evolution*, **12**, 143–149.

Sagne, C. (2001a). *Halitherium taulannense*, nouveau sirénien (Sirenia, Mammalia) de l'Éocène supérieur provenant du domaine Nord-Téthysien (Alpes-de-Haute-Provence, France). *Comptes Rendus Sciences de la Terre et des Planètes Paléontologie*, **333**, 471–476.

Sagne, C. (2001b). La diversification des siréniens à l'Eocène (Sirenia, Mammalia): Etude morphologique et analyse phylogénétique du sirénien de Taulanne, *Halitherium taulannense*. PhD Thesis, Muséum National d'Histoire Naturelle, Paris.

Salkind, J. H. (1998). Etude sur les lamantins au Tchad. *Revue Scientifique du Tchad*, **5**(1), 41–49.

Salm, R. V., Clark, J. and Siirilla, E. (2000). *Marine and Coastal Protected Areas: A Guide for Planners and Managers*, Washington, DC: IUCN.

Samonds, K. E., Zalmout, I. S., Irwin, M. T., *et al.* (2009). *Eotheroides lambondrano*, new Middle Eocene seacow (Mammalia, Sirenia) from the Mahajanga Basin, northwestern Madagascar. *Journal of Vertebrate Paleontology*, **29**, 1233–1243.

Sarko, D. K. and Reep, R. L. (2007). Somatosensory areas of manatee cerebral cortex: histochemical characterization and functional implications. *Brain, Behavior and Evolution*, **69**, 20–36.

Sarko, D. K., Johnson, J. I., Switzer III, R. C., Welker, W. I. and Reep, R. L. (2007a). Somatosensory nuclei of the manatee brainstem and thalamus. *Anatomical Record*, **290**, 1138–1165.

Sarko, D. K., Reep, R. L., Mazurkiewicz, J. E. and Rice, F. L. (2007b). Adaptations in the structure and innervation of follicle-sinus complexes to an aquatic environment as seen in the Florida manatee (*Trichechus manatus latirostris*). *Journal of Comparative Neurology*, **504**, 217–237.

Sato, T., Shibuya, H., Ohba, S., Nojiri, T. and Shirai, W. (2003). Mycobacteriosis in two captive Florida manatees (*Trichechus manatus latirostris*). *Journal of Zoo and Wildlife Medicine*, **34**, 184–188.

Savage, R. J. G. (1976). Review of early Sirenia. *Systematic Zoology*, **25**, 344–351.

Savage, R. J. G., Domning, D. P. and Thewissen, J. G. M. (1994). Fossil Sirenia of the West Atlantic and Caribbean region: V. The most primitive known sirenian, *Prorastomus sirenoides* Owen, 1855. *Journal of Vertebrate Paleontology*, **14**, 427–449.

Scheffer, V. B. (1972) The weight of the Steller sea cow. *Journal of Mammalogy*, **53**, 912–914.

Scheffer, V. B. (1973) The last days of the sea cow. *Smithsonian*, **3**, 64–67.

Schevill, W. E. and Watkins, W. A. (1965). Underwater calls of *Trichechus* (manatee). *Nature*, **205**, 373–374.

Schiedek, D., Sundelin, B., Readman, J. W. and Macdonald, R. W. (2007). Interactions between climate change and contaminants. *Marine Pollution Bulletin*, **54**, 1845–1856.

Scholander, P. F. and Irving, L. (1941). Experimental investigations on the respiration and diving of the Florida manatee. *Journal of Cellular and Comparative Physiology*, **17**, 169–191.

Scholin, C. A., Gulland, F., Doucette, G. J., *et al.* (2000). Mortality of sea lions along the central California coast linked to a toxic diatom bloom. *Nature*, **403**, 80–84.

Schultz, J. K., Baker, J. D., Toonen, R. J. and Bowen, B. W. (2009). Extremely low genetic diversity in the endangered Hawaiian monk seal (*Monachus schauinslandi*). *Journal of Heredity*, **100**, 25–33.

Schwarz, L. K. (2008). Methods and models to determine perinatal status of Florida manatee carcasses. *Marine Mammal Science*, **24**, 881–898.

Schwarz, L. K. and Runge, M. C. (2009). Hierarchical Bayesian analysis to incorporate age uncertainty in growth curve analysis and estimates of age from length: Florida manatee (*Trichechus manatus*) carcasses. *Canadian Journal of Fisheries and Aquatic Sciences*, **66**, 1775–1789.

Scott, M. D. and Powell, J. A. (1982). Commensal feeding of little blue herons with manatees. *Wilson Bulletin*, **94**, 215–216.

Scott, P. (1965). Section XIII: preliminary list of rare mammals and birds. In *The Launching of a New Ark*, London: Collins, pp. 15–207.

Seagrass-Watch. (2010). What is Seagrass-Watch? Available at: http://www.seagrasswatch.org/about.html.

Searle, K. R., Hobbs, N. T. and Shipley, L. A. (2005). Should I stay or should I go? Patch departure decisions by herbivores at multiple scales. *Oikos*, **111**, 417–424.

Seber, G. A. F. (1965). A note on the multiple recapture census. *Biometrika*, **52**, 249–259.

Seiffert, E. R. (2007). A new estimate of afrotherian phylogeny based on simultaneous analysis of genomic, morphological, and fossil evidence. *BMC Evolutionary Biology*, **7**, 224–237.

Self-Sullivan, C. and Mignucci-Giannoni, A. (2008). *Trichechus manatus* ssp. *manatus*. In IUCN Red List of Threatened Species. Version 2010.3. Available at: http://www.iucnredlist.org (accessed 10 October 2010).

Sengco, M. R. (2009). Prevention and control of *Karenia brevis* blooms. *Harmful Algae*, **8**, 623–628.

Sguros, P. (1966). Use of the Florida manatee as an agent for the suppression of aquatic and bankweed growth in essential inland waterways. Research and extension proposal submitted to the Central and Southern Florida Flood Control Board.

Shane, S. H. (1984). Manatee use of power plant effluents in Brevard County, Florida. *Florida Scientist*, **47**, 180–187.

Sheppard, J. K., Preen, A. R., Marsh, H., *et al.* (2006). Movement heterogeneity of dugongs, *Dugong dugon* (Müller), over large spatial scales. *Journal of Experimental Marine Biology and Ecology*, **344**, 64–83.

Sheppard, J. K., Lawler, I. R. and Marsh, H. (2007). Seagrass as pasture for seacows: landscape-level dugong habitat evaluation. *Estuarine, Coastal and Shelf Science*, **71**, 117–132.

Sheppard, J. K., Carter, A. B., McKenzie, L. J., Pitcher, C. R. and Coles, R. G. (2008). Spatial patterns of sub-tidal seagrasses and their tissue nutrients in the Torres Strait, northern Australia: implications for management. *Continental Shelf Research*, **28**, 2282–2291.

Sheppard, J. K., Jones, R. E., Marsh, H. and Lawler, I. R. (2009). Effects of tidal and diel cycles on dugong habitat use. *Journal of Wildlife Management*, **73**, 45–59.

Sheppard, J. K., Marsh, H., Jones, R. E. and Lawler, I. R. (2010). Dugong habitat use in relation to seagrass nutrients, tides, and diel cycles. *Marine Mammal Science*, **26**, 855–879.

Shirakihara, M., Yoshida, H., Yokochi, H., *et al.* (2007). Current status and conservation needs of dugongs in southern Japan. *Marine Mammal Science*, **23**, 694–706.

Short, F. T. and Neckles, H. A. (1999). The effects of global climate change on seagrasses. *Aquatic Botany*, **63**, 169–196.

Shoshani, J. (1986). Mammalian phylogeny: comparison of morphological and molecular results. *Molecular Biology and Evolution*, **3**, 222–242.

Sickenberg, O. (1934). Beiträge zur Kenntnis tertiärer Sirenen. *Mémoires du Musée Royal d'Histoire Naturelle de Belgique*, **63**, 1–352.

Siegal-Willott, J., Estrada, A., Bonde, R., *et al.* (2006). Electrocardiography in two subspecies of manatee (*Trichechus manatus latirostris* and *T. m. manatus*). *Journal of Zoo and Wildlife Medicine*, **37**, 447–453.

Silva, M. A. and Araújo, A. (2001). Distribution and current status of the West African manatee (*Trichechus senegalensis*) in Guinea-Bissau. *Marine Mammal Science*, **17**, 418–424.

Silverberg, D. J. (1988). The role of nutrients and energy in the diet selection of the West Indian manatee (*Trichechus manatus*) in the winter refugium at Homosassa Springs, Citrus County, Florida. MS Thesis, Florida Institute of Technology.

Simberloff, D. (1998). Flagships, umbrellas, and keystones: is single-species management passé in the landscape era? *Biological Conservation*, **83**, 247–257.

Simenstad, C. A., Estes, J. A. and Kenyon, K. W. (1978). Aleuts, sea otters, and alternate stable-state communities. *Science*, **200**, 403–411.

Simpfendorfer, C. A., Goodreid, A. B. and McAuley, R. B. (2001). Size, sex and geographic variation in the diet of the tiger shark, *Galeocerdo cuvier*, from Western Australian waters. *Environmental Biology of Fishes*, **61**, 37–46.

Simpson, G. G. (1932). Fossil Sirenia of Florida and the evolution of the Sirenia. *Bulletin of the American Museum of Natural History*, **59**, 419–503.

Simpson, G. G. (1945). The principles of classification and a classification of mammals. *Bulletin of the American Museum of Natural History*, **85**, 1–350.

Sippel, S. J., Hamilton, S. K., Melack, J. M. and Novo, E. M. M. (1998). Passive microwave observations of inundation area and the area/stage relation in the Amazon River floodplain. *International Journal of Remote Sensing*, **19**, 3055–3074.

Siung-Chang, A. (1997). A review of marine pollution issues in the Caribbean. *Environmental Geochemistry and Health*, **19**, 45–55.

Skarpe, C. and Hester, A. J. (2008). Plant traits, browsing and grazing herbivores and vegetation dynamics. In *The Ecology of Browsing and Grazing*, Vol. 195, I. J. Gordon and H. H. T. Prins (eds), Berlin: Springer, pp. 217–261.

Skilleter, Greg A., Wegscheidl, C. and Lanyon, J. M. (2007). Effects of grazing by a marine mega-herbivore on benthic assemblages in a subtropical seagrass bed. *Marine Ecology-Progress Series*, **351**, 287–300.

Smethurst, D. and Nietschmann, B. (1999). The distribution of manatees (*Trichechus manatus*) in the coastal waterways of Tortuguero, Costa Rica. *Biological Conservation*, **89**, 267–274.

Smith, A. G., Smith, D. G. and Funnell, B. M. (1994) *Atlas of Mesozoic and Cenozoic Coastlines*. Cambridge: Cambridge University Press.

Smith, K. N. (1993). *Manatee Habitat and Human-related Threats to Seagrass in FLORIDA: A Review*, Tallahassee: Florida Department of Environmental Protection.

Snipes, R. L. (1984). Anatomy of the cecum of the West Indian manatee, *Trichechus manatus* (Mammalia, Sirenia). *Zoomorphology*, **104**, 67–78.

Sonoda, S. and Takemura, A. (1973). Underwater sounds of the manatees, *Trichechus manatus manatus* and *T. inunguis* (Trichechidae). *Report of the Institute for Breeding Research, Tokyo University of Agriculture*, **4**, 19–24.

Sosulski, F. W. and Imafidon, G. I. (1990). Amino acid composition and nitrogen-to-protein conversion factors for animal and plant foods. *Journal of Agricultural and Food Chemistry*, **38**, 1351–1356.

Sousa-Lima, R. S., Paglia, A. P. and Da Fonseca, G. A. B. (2002). Signature information and individual recognition in the isolation calls of Amazonian manatees, *Trichechus inunguis* (Mammalia: Sirenia). *Animal Behaviour*, **63**, 301–310.

South East Queensland Healthy Waterways Strategy. (2007–2012). Available at: http://www.healthywaterways.org/TheStrategy.aspx.

Spain, A. V. and Heinsohn, G. E. (1973). Cyclone associated feeding changes in the dugong (Mammalia: Sirenia). *Mammalia*, **37**, 678–680.

Spain, A. V. and Heinsohn, G. E. (1975). Size and weight allometry in a North Queensland Population of *Dugong dugon* (Müller) (Mammalia – Sirenia). *Australian Journal of Zoology*, **23**, 159–168.

Spinage, C. A. (1972). African ungulate life tables. *Ecology*, **53**, 645–652.

Sprent, J. F. A. (1983). Ascaridoid nematodes of sirenians: a new species in the Senegal manatee. *Journal of Helminthology*, **57**, 69–76.

SPREP. (2003). Regional marine species programme framework 2003–2007. Attachment 3. Dugong (RMMCP) strategic action plan (2003–2007), pp. 17–21.

Springer, M. S. and Murphy, W. J. (2007). Mammalian evolution and biomedicine: new views from phylogeny. *Biological Reviews*, **82**, 375–392.

Springer, M. S., Murphy, W. J., Eizirik, E. and O'Brien, S. J. (2003). Placental mammal diversification and the Cretaceous–Tertiary boundary. *Proceedings of the National Academy of Sciences USA*, **100**, 1056–1061.

Springer, M. S., Murphy, W. J., Eizirik, E. and O'Brien, S. J. (2005). Molecular evidence for major placental clades. In *The Rise of Placental Mammals*, K. D. Rose and J. D. Archibald (eds), Baltimore: Johns Hopkins University Press, pp. 37–49.

Springer, M. S., Burk-Herrick, A., Meredith, R., *et al.* (2007). The adequacy of morphology for reconstructing the early history of placental mammals. *Systematic Biology*, **56**, 673–684.

Stachowicz, J. J., Terwin, J. R., Whitlatch, R. B. and Osman, R. W. (2002). Linking climate change and biological invasions: ocean warming facilitates nonindigenous species invasions. *Proceedings of the National Academy of Sciences USA*, **99**, 15497–15500.

Stanhope, M. J., Waddell, V. G., Madsen, O., *et al.* (1998a). Molecular evidence for multiple origins of Insectivora and for a new order of endemic African insectivore mammals. *Proceedings of the National Academy of Sciences USA*, **95**, 9967–9972.

Stanhope, M. J., Madsen, O., Waddell, V. G., *et al.* (1998b). Highly congruent molecular support for a diverse superordinal clade of endemic African mammals. *Molecular Phylogenetics and Evolution*, **9**, 501–508.

Stavros, H. C. W., Bonde, R. K. and Fair, P. A. (2008). Concentrations of trace elements in blood and skin of Florida manatees (*Trichechus manatus latirostris*). *Marine Pollution Bulletin*, **56**, 1221–1225.

Steel, C. (1982). Vocalization patterns and corresponding behavior of the West Indian manatee (*Trichechus manatus*). PhD Thesis, Florida Institute of Technology, Melbourne.

Steidinger, K. A. (2009). Historical perspective on *Karenia brevis* red tide research in the Gulf of Mexico. *Harmful Algae*, **8**, 549–561.

Steinberg, P. D. (1985). Feeding preferences of *Tegula funebralis* and chemical defenses of marine brown algae. *Ecological Monographs*, **55**, 333–349.

Steinberg, P. D., Estes, J. A. and Winter, F. C. (1995). Evolutionary consequences of food chain length in kelp forest communities. *Proceedings of the National Academy of Sciences USA*, **92**, 8145–8148.

Stejneger, L. (1884). Contributions to the history of the Commander Islands: no. 2. Investigations relating to the date of the extermination of Steller's sea cow. *Proceedings of the United States National Museum*, **7**, 181–189.

Stejneger, L. (1886). On the extermination of the Great Northern Sea-Cow (*Rytina*): a reply to Professor AE Nordenskiöld. *Journal of the American Geographical Society of New York*, **18**, 317–328.

Stejneger, L. (1887). How the great northern sea-cow (Rytina) became exterminated. *The American Naturalist*, **21**, 1047–1054.

Stejneger, L. (1893). Skeletons of Steller's sea-cow preserved in the various museums. *Science*, **21**, 81.

Stejneger, L. (1936). *Georg Wilhelm Steller, the Pioneer of Alaskan Natural History*, Cambridge, MA: Harvard University Press.

Steller, G. W. (1751). De bestiis marinis. *Novi Commentarii Academiae Scientiarum Imperialis Petropolitanae*, **2**, 289–398. (Translated by W. Miller and J. E. Miller. (1899)). Part VIII: the early history of the northern fur seal. In *The Fur Seals and Fur-Seal Islands of the North Pacific Ocean*, D. S. Jordan et al. (eds), Washington, DC: US Government Printing Office, pp. 179–218. Available at: http://digitalcommons.unl.edu/cgi/viewcontent.cgi?article=1019andcontext=libraryscience (transcribed and edited by P. Royster).

Steller, G. W. (1925). Steller's journal of the sea voyage from Kamchatka to America and return on the second expedition 1741–1742. In *Bering's Voyages: An Account of the Efforts of the Russians to Determine the Relation of Asia and America*, Vol. 2, F. A. Golder (ed.), American Geographical Society Research Series, pp. 9–248. (Reprinted 1968, New York: Octagon Books).

Steller, G. W. (1988). *Journal of a Voyage with Bering 1741–1742*, Stanford, CA: Stanford University Press.

Stern, D. I., Common, M. S. and Barbier, E. B. (1996). Economic growth and environmental degradation: the environmental Kuznets curve and sustainable development. *World Development*, **24**, 1151–1160.

Stith, B. M., Slone, D. H. and Reid, J. P. (2006). Review and synthesis of manatee data in Everglades National Park. Gainesville: United States Geological Survey Florida Integrated Science Center. Available at: http://fl.biology.usgs.gov/Center_Publications/Manatee_Publications_p1/Stith_et_al_ENP_Manatee_Administrative_Report.pdf.

Stith, B. M., Reid, J. P., Langtimm, C. A., et al. (2010). Temperature inverted haloclines provide winter warm-water refugia for manatees in southwest Florida. *Estuaries and Coasts*, **34**, 106–119.

Strandnet. (2010). Queensland Department of Environment and Resource Management marine wildlife stranding and mortality database: operational policy. Available at: http://www.derm.qld.gov.au/register/p02153aa.pdf.

Strayer, D. L. and Dudgeon, D. (2010). Freshwater biodiversity conservation: recent progress and furture challenges. *Journal of the North American Benthological Society*, **29**, 344–355.

Strong, W. D. (1935). Archaeological investigations in the Bay Islands, Spanish Honduras. *Smithsonian Miscellaneous Collections*, **92**, 1–176.

Suárez-Morales, E., Morales-Vela, B., Padilla-Saldivar, J. and Silva-Briano, M. (2010). The copepod *Balaenophilus manatorum* (Ortíz, Lalana and Torres, 1992) (Harpacticoida), an epibiont of the Caribbean manatee. *Journal of Natural History*, **44**, 847–859.

Supanwanid, C. (1996). Recovery of the seagrass *Haliphola ovalis* after grazing by dugong. In *Seagrass Biology: Proceedings of an International Workshop, Rottnest Island*, J. Kuo, R. C. Phillips, D. I. Walker and H. Kirkman (eds), Perth: University of Western Australia, pp. 315–318.

Supanwanid, C. and Lewmanomont, K. (2003). The seagrasses of Thailand. In *World Atlas of Seagrasses*, E. P. Green and F. F. Short (eds), Berkeley: University of California Press, pp. 144–151.

Supanwanid, C., Albertsen, J. O. and Mukai, H. (2001). Methods of assessing the grazing effects of large herbivores on seagrasses. In *Global Seagrass Research Methods*, F. T. Short and R. G. Coles (eds), Amsterdam: Elsevier, pp. 293–312.

Sutaria, D. (2009). Species conservation in a complex socio-ecological system: Irrawaddy dolphins, *Orcaella brevirostris* in Chilika Lagoon, India. PhD Thesis, James Cook University, Townsville.

Swaisgood, R.R. and Sheppard, J. (2011). Reconnecting people to nature is a prerequisite for the future conservation agenda: response from Swaisgood and Sheppard. *BioScience*, **61**, 94–95.

Tabuce, R., Asher, R. J. and Lehmann, T. (2008). Afrotherian mammals: a review of current data. *Mammalia*, **72**, 2–14.

Tabuchi, K., Muku, T. and Satomichi, T., *et al.* (1974). A dermatosis in manatee (*Trichechus manatus*): mycological report of a case. *Bulletin of the Azabu Veterinary College*, **28**, 127–134.

Takahashi, E. M., Arthur, K. E. and Shaw, G. R. (2008). Occurrence of okadaic acid in the feeding grounds of dugongs (*Dugong dugon*) and green turtles (*Chelonia mydas*) in Moreton Bay, Australia. *Harmful Algae*, **7**, 430–437.

Takahashi, S., Domning, D. P. and Saito, T. (1986). *Dusisiren dewana*, n. sp. (Mammalia: Sirenia), a new ancestor of Steller's sea cow from the upper Miocene of Yamagata Prefecture, northeastern Japan. *Transactions and Proceedings of the Palaeontological Society of Japan, N.S.*, **141**, 296–321.

Tanabe, S. (1999). Butyltin contamination in marine mammals: a review. *Marine Pollution Bulletin*, **39**, 62–72.

Tanji, M. (2008). U.S. court rules in the 'Okinawa dugong' case. Implications for the U.S. military bases overseas. *Critical Asian Studies*, **40**, 475–487.

Taranto, T., Jacobs, D., Skewes, T. and Long, B. (1997). Torres Strait reef resources: maps of percentage cover of algae and percentage cover of seagrasses. *Torres Strait Conservation and Planning Report, Annex 8*. Canberra: CSIRO.

Tassy, P. and Shoshani, J. (1988). The Tethytheria: elephants and their relatives. In *The Phylogeny and Classification of the Tetrapods*, Vol. 2, M. J. Benton (ed.), Oxford: Clarendon Press, pp. 283–315.

Taylor, B. L., Martinez, M., Gerrodette, T., Barlow, J. and Hrovat, Y. N. (2007). Lessons from monitoring trends in abundance of marine mammals. *Marine Mammal Science*, **23**, 157–175.

Taylor, C. R. (2006). A survey of Florida Springs to determine accessibility to Florida manatees (*Trichechus manatus latirostris*): developing a sustainable thermal network. Final Report to the US Marine Mammal Commission.

Thayer, G. W., Bjorndal, K. A., Ogden, J. C., Williams, S. L. and Zieman, J. C. (1984). Role of larger herbivores in seagrass communities. *Estuaries*, **7**, 351–376.

Thomas, D. (1966). Natural history of the dugong in Rameswarum waters. *Madras Journal of Fish*, **2**, 80–82.

Thornback, J. and Jenkins, M. (1982). *The IUCN Mammal Red Data Book: Part 1. Threatened Mammalian Taxa of the Americas and the Australasian Zoogeographic Region (Excluding Cetacea)*. Gland, Switzerland: IUCN.

Timm, R. M., Albuja V, L. and Clauson, B. L. (1986). Ecology, distribution, harvest, and conservation of the Amazonian manatee *Trichechus inunguis* in Ecuador. *Biotropica*, **18**, 150–156.

Toledo, P. M. de and Domning, D. P. (1991). Fossil Sirenia (Mammalia: Dugongidae) from the Pirabas Formation (Early Miocene), northern Brazil. *Boletim do Museu Paraense Emílio Goeldi, Série Ciências da Terra*, **1**, 119–146.

TRAFFIC. (2009). Journal entry by TRAFFIC on 23 March 2009. *TRAFFIC Bulletin Seizures and Prosecutions*: **16**(3) (March 1997) to **22**(2) (June 2009), p. 160.

Tringali, M. D., Seyoum, S., Carney, S. L., *et al.* (2008). Eighteen new polymorphic microsatellite markers for the endangered Florida manatee, *Trichechus manatus latirostris*. *Molecular Ecology Resources*, **8**, 328–331.

Tripp, K. (2006). Manatees: ancient marine mammals in a modern coastal environment. *Ocean Challenge*, **16**, 14–22.

Tripp, K. M., Verstegen, J. P., Deutsch, C. J., *et al.* (2008). Validation of a serum immunoassay to measure progesterone and diagnose pregnancy in the West Indian manatee (*Trichechus manatus*). *Theriogenology*, **70**, 1030–1040.

Tripp, K. M., Dubois, M., Delahaut, P. and Verstegen, J. P. (2009). Detection and identification of plasma progesterone metabolites in the female Florida manatee (*Trichechus manatus latirostris*) using GC/MS/MS. *Theriogenology*, **72**, 365–371.

Tripp, K. M., Verstegen, J. P., Deutsch, C. J., *et al.* (2010). Evaluation of adrenocortical function in Florida manatees (*Trichechus manatus latirostris*). *Zoo Biology*, **29**, 1–15.

Tsukinowa, E., Karita, S., Asano, S., *et al.* (2008). Fecal microbiota of a dugong (*Dugong dugon*) in captivity at Toba Aquarium. *Journal of General and Applied Microbiology*, **54**, 25–38.

Tsutsumi, C., Ichikawa, K., Arai, N., *et al.* (2006). Feeding behavior of wild dugongs monitored by a passive acoustical method. *Journal of the Acoustical Society of America*, **120**, 1356–1360.

Tucker, C. S. and Debusk, T. A. (1981). Seasonal growth of *Eichhornia crassipes* (Mart.) solms: relationship to protein, fiber, and available carbohydrate content. *Aquatic Botany*, **11**, 137–141.

Tun, T. and Ilangakoon, A. (2007). Assessment of dugong (*Dugong dugon*) occurrence and distribution in an extended area off the Rakhine coast of western Myanmar. Report to the Society of Marine Mammalogy.

Tun, T. and Ilangakoon, A. (2008). Assessment of dugong (*Dugong dugon*) occurrence, distribution and threats in the north-western coastal waters of Rakhine State, Myanmar. Report to the Society of Marine Mammalogy.

Turvey, S. T. and Risley, C. L. (2006). Modelling the extinction of Steller's sea cow. *Biology Letters*, **2**, 94–97.

Turvey, S. T., Pitman, R. L., Taylor, B. L., *et al.* (2007). First human-caused extinction of a cetacean species? *Biology Letters*, **3**, 537–540.

Tyack, P. L. (2008). Implications for marine mammals of large-scale changes in the marine acoustic environment. *Journal of Mammalogy*, **89**, 549–558.

Udevitz, M. S. and Ballachey, B. E. (1998). Estimating survival rates with age-structure data. *Journal of Wildlife Management*, **62**, 779–792.

UNDP. (2009). *Human Development Report 2009: Overcoming Barriers. Human Mobility and Development.* Houndmills and Hampshire, NY: Palgrave Macmillan.

UNDP. (2010). *Human Development Report 2010. The Real Wealth of Nations: Pathways to Human Development.* United Nations Development Program. Available at http://hdr.undp.org/en/reports/global/hdr2010/.

UNEP. (2010). Regional management plan for the West Indian manatee (*Trichechus manatus*). Compiled by E. Quintana-Rizzo and J. Reynolds III. CEP Technical Report No. 48, UNEP Caribbean Environment Programme.

UNEP and CMS. (2008). Action plan for the conservation of the West African manatee: annex I to the memorandum of understanding concerning the conservation of the manatee and small cetaceans of Western Africa and Macronesia.

United Nations. (2003). Convention on biological diversity with annexes.

United Nations. (2004). *World Population to 2300*, New York: United Nations.

United States Census Bureau. (2009). International data base: demographic indicators. Available at http://www.census.gov/ipc/www/idb/country.php.

United States Geological Survey Geologic Names Committee. (2007). Divisions of geologic time: major chronostratigraphic and geochronologic units. US Geological Survey Fact Sheet 2007–3015, 1–2.

Upton, S. J., Odell, D. K., Bossart, G. D. and Walsh, M. T. (1989). Description of the oocysts of two new species of *Eimeria* (Apicomplexa: Eimeriidae) from the Florida manatee, *Trichechus manatus* (Sirenia: Trichechidae). *The Journal of Protozoology*, **36**, 87–90.

US Fish and Wildlife Service. (2001). *Florida Manatee Recovery Plan*, Trichechus manatus latirostris, *Third Revision*, Atlanta: US Fish and Wildlife Service.

Valentine, J. F. and Duffy, J. E. (2006). The central role of grazing in seagrass ecology. In *Seagrasses: Biology, Ecology and Conservation*, A. W. D. Larkum, R. J. Orth and C. M. Duarte (eds), Dordrecht: Springer, pp. 463–501.

Valentine, J. F. and Heck, K. L. J. (1999). Seagrass herbivory: evidence for the continued grazing of marine grasses. *Marine Ecology Progress Series*, **176**, 291–302.

Van Dolah, F. M. (2000). Marine algal toxins: origins, health effects, and their increased occurrence. *Environmental Health Perspectives*, **108**, 133–141.

Van Dolah, F. M., Doucette, G. J., Gulland, F. M. D., Rowles, T. L. and Bossart, G. D. (2003). Impacts of algal toxins on marine mammals. In *Toxicology of Marine Mammals*, J. G. Vos, G. D. Bossart, M. Fournier and T. J. O'Shea (eds), London: Taylor and Francis, pp. 247–269.

Van Soest, P. J. (1982). *Nutritional Ecology of the Ruminant*, Ithaca, NY: Cornell University Press.

Van Soest, P. J. (1994). *Nutritional Ecology of the Ruminant*, Ithaca, NY: Comstock Publishing Associates.

van Tussenbroek, B. I., Vonk, J. A., Stapel, J., *et al.* (2006). The biology of *Thalassia*: paradigms and recent advances in research. In *Seagrasses: Biology, Ecology and Conservation*, A. W. D. Larkum, R. J. Orth and C. M. Duarte (eds), Dordrecht: Springer, pp. 409–439.

Varela, R. A., Schmidt, K., Goldstein, J. D. and Bossart, G. D. (2007). Evaluation of cetacean and sirenian cytologic samples. *Veterinary Clinics of North America: Exotic Animal Practice*, **10**, 79–130.

Vergara-Parente, J. E., Sidrim, J. J. C., Pessa, A. B. G. d. P., *et al.* (2003a). Bacterial flora of upper respiratory tract of captive Antillean manatees. *Aquatic Mammals*, **29**, 124–130.

Vergara-Parente, J. E., Sidrim, J. J. C., Teixeira, M. F. d. S., Marcondes, M. C. C. and Rocha, M. F. G. (2003b). Salmonellosis in an Antillean manatee (*Trichechus manatus manatus*) calf: a fatal case. *Aquatic Mammals*, **29**, 131–136.

Vetter, W., Scholz, E., Gaus, C., Müller, J. F. and Haynes, D. (2001). Anthropogenic and natural organohalogen compounds in blubber of dolphins and dugongs (*Dugong dugon*) from northeastern Australia. *Archives of Environmental Contamination and Toxicology*, **41**, 221–231.

Vianna, J. A., Bonde, R. K., Caballero, S., *et al.* (2006). Phylogeography, phylogeny and hybridization in trichechid sirenians: implications for manatee conservation. *Molecular Ecology*, **15**, 433–447.

von Bodungen, B., John, H.-Ch., Lutjeharms, J. R. E., Mohrholz, V. and Veitch, J. (2008). Hydrographic and biological patterns across the Angola–Benguela Frontal Zone under undisturbed conditions. *Journal of Marine Systems*, **74**, 189–215.

Vörösmarty, C. J., McIntyre, P. B., Gessner, M. O., *et al.* (2010). Global threats to human water security and river biodiversity. *Nature* **467**, 555–561.

Vos, J. G., Bossart, G. D., Fournier, M. and O'Shea, T. J. (eds). (2003). *Toxicology of Marine Mammals*, London: Taylor and Francis.

Wade, P. R. (1998). Calculating limits to the allowable human-caused mortality of cetaceans and pinnipeds. *Marine Mammal Science*, **14**, 1–37.

Wake, J. A. (1975). A study of the habitat requirements and feeding biology of the dugong *Dugong dugon* (Müller). Honours thesis, James Cook University, Townsville.

Walker, D. I. and McComb, A. J. (1988). Seasonal variation in the production, biomass and nutrient status of *Amphibolis antarctica* (Labill.) Sonder ex Aschers. and *Posidonia australis* hook. f. in Shark Bay, Western Australia. *Aquatic Botany*, **31**, 259–275.

Walsh, C. J., Luer, C. A. and Noyes, D. R. (2005). Effects of environmental stressors on lymphocyte proliferation in Florida manatees, *Trichechus manatus latirostris*. *Veterinary Immunology and Immunopathology*, **103**, 247–256.

Walsh, C. J., Stuckey, J. E., Cox, H., *et al.* (2007). Production of nitric oxide by peripheral blood mononuclear cells from the Florida manatee, *Trichechus manatus latirostris*. *Veterinary Immunology and Immunopathology*, **118**, 199–209.

Walsh, J. J., Jolliff, J. K., Darrow, B. P., *et al.* (2006). Red tides in the Gulf of Mexico: where, when, and why? *Journal of Geophysical Research: Oceans*, **111**, 1–46.

Walsh, M. T., Bossart, G. D., Young, W. G. Jr and Rose, P. M. (1987). Omphalitis and peritonitis in a young West Indian manatee (*Trichechus manatus*). *Journal of Wildlife Diseases*, **23**, 702–704.

Waser, P. M. and Jones, W. T. (1983). Natal philopatry among solitary animals. *Quarterly Review of Biology*, **58**, 355–390.

Watson, A. G. and Bonde, R. K. (1986). Congenital malformations of the flipper in three West Indian manatees, *Trichechus manatus*, and a proposed mechanism for development of ectrodactyly and cleft hand in mammals. *Clinical Orthopaedics and Related Research*, **202**, 294–301.

Waxell, S. (1962). *The Russian Expedition to America* (trans. M. A. Michael), New York: Collier Books.

Waycott, M., McMahon, K., Mellors, J., Calladine, A. and Kleine, D. (2004). *A Guide to Tropical Seagrasses of the Indo-West Pacific*, Townsville: James Cook University.

Waycott, M., Collier, C., McMahon, K., *et al.* (2007). Vulnerability of seagrasses in the Great Barrier Reef to climate change. In *Climate Change and the Great Barrier Reef: A Vulnerability Assessment*, J. Johnson and P. Marshall (eds), Townsville: Great Barrier Marine Park Authority, pp. 193–235.

Waycott, M., Duarte, C. M., Carruthers, T. J. B., *et al.* (2009). Accelerating loss of seagrasses across the globe threatens coastal ecosystems. *Proceedings of the National Academy of Sciences USA*, **106**, 12377–12381.

Webster, P. J., Holland, G. J., Curry, J. A. and Chang, H. R. (2005). Changes in tropical cyclone number, duration, and intensity in a warming environment. *Science*, **309**, 1844–1846.

Wellehan, J. F. X, Johnson, A. J., Childress, A. L., Harr, K. E. and Isaza, R. (2008). Six novel gammaherpesviruses of Afrotheria provide insight into the early divergence of the Gammaherpesvirinae. *Veterinary Microbiology*, **127**, 249–257.

Wells, M. P., McShane, T. O., Dublin, H. T., O'Connor, S. and Redford, K. H. (2004). The future of integrated conservation and development projects: building on what works. In *Getting Biodiversity Projects to Work: Towards More Effective Conservation and Development*, T. O. McShane and M. P. Wells (eds), New York: Columbia University Press, pp. 397–421.

Wells, R. S., Boness, D. J. and Rathbun, G. B. (1999). Behavior. In *The Biology of Marine Mammals*, J. E. Reynolds III and S. A. Rommel (eds), Washington, DC: Smithsonian Institution Press, pp. 324–422.

West, J. A., Sivak, J. G., Murphy, C. J. and Kovacs, K. M. (1991). A comparative study of the anatomy of the iris and ciliary body in aquatic mammals. *Canadian Journal of Zoology*, **69**, 2594–2607.

Wetmore, A. (1946). Biographical memoir of Leonhard Hess Stejneger, 1851–1943. *Biographical Memoirs of the National Academy of Sciences*, **24**, 145–195.

Wetzel, D. L., Reynolds III, J. E., Sprinkel, J. M., *et al.* (2010). Fatty acid signature analysis as a potential forensic tool for Florida manatees (*Trichechus manatus latirostris*). *Science of the Total Environment*, **408**, 6124–6133.

Wetzel, D. L., Pulster, E. and Reynolds III, J. E. (2012). Organic contaminants and sirenians. In *Sirenian Conservation: Issues and Strategies in Developing Countries*, E. Hines, J. Reynolds, A. Mignucci-Giannoni, L. Aragones and M. Marmontel (eds), Gainesville: University Press of Florida.

White, G. C. and Burnham, K. P. (1999). Program MARK: survival estimation from populations of marked animals. *Bird Study*, **46** (Supplement), 120–139.

White, G. C., Burnham, K. P. and Anderson, D. R. (2001). Advanced features of Program MARK. In *Wildlife, Land, and People: Priorities for the 21st Century. Proceedings of the Second International Wildlife Management Congress*, R. Field, R. J. Warren, H. Okarma and P. R. Sievert (eds), Bethesda: The Wildlife Society, pp. 368–377.

Whitehead, H., Waters, S. and Lyrholm, T. (1991). Social organization in female sperm whales and their offspring: constant companions and casual acquaintances. *Behavioral Ecology and Sociobiology*, **29**, 385–389.

Whitehead, P. J. P. (1977). The former southern distribution of New World manatees (*Trichechus* spp.). *Biological Journal of the Linnean Society*, **9**, 165–189.

Whitham, T. G., Maschinski, J., Larson, K. C. and Paige, K. N. (1991). Plant responses to herbivory: the continuum from negative to positive and underlying physiological mechanisms. In *Plant–Animal Interactions: Evolutionary Ecology in Tropical and Temperate Regions*, P. W. Price, T. M. Lewinsohn, G. W. Fernandes and W. W. Benson (eds), New York: John Wiley and Sons, pp. 227–256.

Whiting, S. D. (1999). Use of the remote Sahul Banks, Northwestern Australia, by dugongs, including breeding females. *Marine Mammal Science*, **15**, 609–615.

Whiting, S. D. (2002a). Dive times for foraging dugongs in the Northern Territory. *Australian Mammalogy*, **23**, 167–168.

Whiting, S. D. (2002b). Rocky reefs provide foraging habitat for dugongs in the Darwin region of Northern Australia. *Australian Mammalogy*, **24**, 147–150.

Whiting, S. D. (2008). Movements and distribution of dugongs (*Dugong dugon*) in a macro-tidal environment in northern Australia. *Australian Journal of Zoology*, **56**, 215–222.

Whiting, S. D. and Guinea, M. L. (2003). Dugongs of Ashmore Reef and the Sahul Banks: a review of current knowledge and a distribution of sightings. *The Beagle: Records of the Museums and Art Galleries of the Northern Territory*, **1**, 207–210.

Wible, J. R., Rougier, G. W., Novacek, M. J. and Asher, R. J. (2007). Cretaceous eutherians and Laurasian origin for placental mammals near the K/T boundary. *Nature*, **447**, 1003–1006.

Wilkinson, C. and Souter, D. (eds). (2008). *Status of Caribbean Coral Reefs after Bleaching and Hurricanes in 2005*. Townsville: Global Coral Reef Monitoring Network and Reef and Rainforest Research Centre. Available at: http://www.coris.noaa.gov/activities/caribbean_rpt/.

Williams, B. K., Nichols, J. D. and Conroy, M. J. (2002). *Analysis and Management of Animal Populations*, San Diego: Academic Press..

Williams, E. H. Jr, Mignucci-Giannoni, A. A., Bunkley-Williams, L., *et al.* (2003). Echeneid-sirenian associations, with information on sharksucker diet. *Journal of Fish Biology*, **63**, 1176–1183.

Williams, L. E. (2005). Individual distinctiveness, short- and long-term comparisons, and context specific rates of Florida manatee vocalizations. MS Thesis, University of North Carolina-Wilmington.

Williams, M. E. and Domning, D. P. (2004). Pleistocene or post-Pleistocene manatees in the Mississippi and Ohio River valleys. *Marine Mammal Science*, 20, 167–176.

Willis, C. K. R. and Brigham, R. M. (2004). Roost switching, roost sharing and social cohesion: forest-dwelling big brown bats, *Eptesicus fuscus*, conform to the fission–fusion model. *Animal Behaviour*, 68, 495–505.

Wilson, D. E. and Culik, B. M. (1991). The cost of a hot meal: facultative specific dynamic action may ensure temperature homeostasis in post-ingestive endotherms. *Comparative Biochemistry and Physiology A*, 100, 151–154.

Wing, E. S. and Reitz, E. J. (1982). Prehistoric fishing communities of the Caribbean. *Journal of New World Archaeology*, 5, 13–32.

Wirsing, A. J., Heithaus, M. R. and Dill, L. M. (2007a). Can you dig it? Use of excavation, a risky foraging tactic, by dugongs is sensitive to predation danger. *Animal Behaviour*, 74, 1085–1091.

Wirsing, A. J., Heithaus, M. R. and Dill, L. M. (2007b). Fear factor: do dugongs (*Dugong dugon*) trade food for safety from tiger sharks (*Galeocerdo cuvier*)? *Oecologia*, 153, 1031–1040.

Wirsing, A. J., Heithaus, M. R. and Dill, L. M. (2007c). Can measures of prey availability improve our ability to predict the abundance of large marine predators? *Oecologia*, 153, 563–568.

Wirsing, A. J., Heithaus, M. R. and Dill, L. M. (2007d). Living on the edge: dugongs prefer to forage in microhabitats that allow escape from rather than avoidance of predators. *Animal Behaviour*, 74, 93–101.

Woodruff, R. A., Bonde, R. K., Bonilla, J. A. and Romero, C. H. (2005). Molecular identification of a papilloma virus from cutaneous lesions of captive and free-ranging Florida manatees. *Journal of Wildlife Diseases*, 41, 437–441.

World Bank. (2010). Gross national income per capita 2008. Available at: http://siteresources.worldbank.org/DATASTATISTICS/Resources/GNIPC.pdf.

World Wildlife Fund Eastern African Marine Ecoregion. (2004). Towards a Western Indian Ocean dugong conservation strategy. In *The Status of Dugongs in the Western Indian Ocean and Priority Conservation Actions*, Dar es Salaam: World Wildlife Fund.

Wright, I. E., Reynolds III, J. E., Ackerman, B. B., *et al.* (2002). Trends in manatee (*Trichechus manatus latirostris*) counts and habitat use in Tampa Bay, 1987–1994: implications for conservation. *Marine Mammal Science*, 18, 259–274.

Wright, S. D., Ackerman, B. B., Bonde, R. K., Beck, C. A. and Banowetz, D. J. (1995). Analysis of watercraft-related mortality of manatees in Florida, 1979–1991. In *Population Biology of the Florida Manatee*, T. J. O'Shea, B. B. Ackerman and H. F. Percival (eds), Washington, DC: US Department of the Interior, National Biological Service, pp. 259–268.

Wűnder, S. (2001). *The Economics of Deforestations: The Example of Ecuador*, Basingstoke: Palgrave MacMillan.

Wyss, A. R., Novacek, M. J. and McKenna, M. C. (1987). Amino acid sequence versus morphological data and the interordinal relationships of mammals. *Molecular Biology and Evolution*, 4, 99–116.

Yamasaki, F., Komatsu, S. and Kamiya, T. (eds). (1980). A comparative morphological study on the tongues of manatee and dugong (Sirenia). *Scientific Reports of the Whales Research Institute*, 32, 127–144.

Yan, J., Clifton, K. B., Mecholsky, J. J. Jr. and Reep, R. L. (2006a). Fracture toughness of manatee rib and bovine femur using a chevron-notched beam test. *Journal of Biomechanics*, **39**, 1066–1074.

Yan, J., Clifton, K. B., Reep, R. L. and Mecholsky, J. J. Jr. (2006b). Application of fracture mechanics to failure in manatee rib bone. *Journal of Biomechanical Engineering*, **128**, 281–289.

Yen, R. (2006). Preliminary assessment of the status of the dugong population of Samarai, Milne Bay Province. The Milne Bay community-based coastal and marine conservation project.

Zalmout, I. S., Ul-Haq, M. and Gingerich, P. D. (2003). New species of *Protosiren* (Mammalia, Sirenia) from the early middle Eocene of Balochistan (Pakistan). *Contributions of the Museum of Paleontology, University of Michigan*, **31**, 79–87.

Zieman, J. C., Iverson, R. L. and Ogden, J. C. (1984). Herbivory effects on *Thalassia testudinum* leaf growth and nitrogen content. *Marine Ecology Progress Series*, **15**, 151–158.

Zoodsma, J. (1991). Distribution and behavioral ecology of manatees in southeastern Georgia. MS Thesis, University of Florida.

Online supplementary material available at
www.cambridge.org/9780521888288

CHAPTER 3

Appendix 3.1 A summary of genomic investigations of the Afrotheria pertinent to the Sirenia.
Appendix 3.2 Summaries of representative fossil species of sirenians, their temporal presence in the fossil record, geographical distributions and descriptive aspects of their morphology, ecology and relationships.
Appendix Table 3.2.1 Representative fossil taxa in the sirenian families Prorastomidae and Protosirenidae.
Appendix Table 3.2.2 Representative fossil taxa of the subfamily Halitheriinae in the family Dugongidae.
Appendix Table 3.2.3 Representative fossil and Recent taxa in the subfamilies Dugonginae and Hydrodamalinae in the family Dugongidae.
Appendix Table 3.2.4 Representative fossil and Recent taxa in the sirenian family Trichechidae.

CHAPTER 4

Appendix 4.1 A series of tables listing the foods eaten by the dugong, the Amazonian manatee, the West Indian manatee and the West African manatee.
Appendix Table 4.1 Plants and animals eaten by the dugong.
Appendix Table 4.2 Genera of freshwater food plants eaten by the Amazonian manatee.
Appendix Table 4.3 Plants and animals eaten by the West Indian manatee.
Appendix Table 4.4 Plants and animals eaten by the West African manatee.

CHAPTER 6

Appendix 6.1 Reproductive anatomy
Appendix 6.2 History of manatee population modelling

CHAPTER 7

Appendix 7.1 Helminth parasites of sirenians
Appendix Table 7.1 Helminth parasites of sirenians, after Beck and Forrester (1988), with updates.

Index

Printed in the United States
By Bookmasters